The oryx, or gemsbok, is a large antelope of the arid and semiarid sections of Africa, including the borders of the Sahara Desert. It has an amazing ability to survive intense desert heat. Under experimental conditions, the oryx can withstand exposure to an ambient temperature of 45 degrees C (113 degrees F!) for 12 hours.

Perhaps more remarkable is the animal's lack of dependence on drinking water. Its water needs are probably satisfied by its food. By feeding at night on leaves and shrubs that have absorbed moisture from humid night air, the oryx can manage an intake of some five liters. This is at best a minimal amount of water for an animal weighing roughly 200 kilograms (450 pounds), living in shelterless desert.

SECOND EDITION

MAMMALOGY

TERRY A. VAUGHAN
Northern Arizona University
Flagstaff, Arizona

W. B. SAUNDERS COMPANY
Philadelphia London Toronto

W. B. Saunders Company: West Washington Square
Philadelphia, PA 19105

1 St. Anne's Road
Eastbourne, East Sussex BN21 3UN, England

1 Goldthorne Avenue
Toronto, Ontario M8Z 5T9, Canada

Library of Congress Cataloging in Publication Data

Vaughan, Terry A

Mammalogy.

Bibliography: p.

Includes index.

1. Mammals.   I. Title.

QL703.V38 1978   599   77–2111

ISBN 0–7216–9009–2

Mammalogy   ISBN 0-7216-9009-2

Last digit is the print number: 9   8   7   6   5   4   3   2

PREFACE
to the Second Edition

In this second edition the original organization has been retained, but most of the chapters have been altered by the addition of recently published material, new photographs, and new drawings. A number of chapters have been rather drastically changed, and some have been essentially rewritten and expanded to present the most up-to-date material possible. The following subjects have been considerably expanded or changed: mammalian origins (Chapter 3), noneutherian mammals (Chapter 5), rodents (Chapter 10), ecology (Chapter 15), zoogeography (Chapter 16), behavior (Chapter 17), and metabolism and temperature regulation (Chapter 19).

After the completion of the first edition, I had the privilege of spending a 14-month sabbatical leave with my family in Kenya, East Africa. Many of the new photographs and skull drawings result from work done during this period, and many descriptions of specific African mammals or discussions of their ecology or behavior derive from talks I had with scientists working in East Africa, from the use of their published or, occasionally, unpublished research, and from personal observations.

Most of the photographs used in the first edition appear again, as well as much of the material reviewed by friends and colleagues. Because of this, the preface to the first edition, which thanks the people who assisted in a variety of ways, is reprinted here. The intent of the book, as outlined in the original preface, has not changed.

The help of the staff of the W. B. Saunders Company is gratefully acknowledged. The advice and friendship of Richard H. Lampert, former Biology Editor, were especially important.

Many friends and colleagues have contributed importantly to the second edition. For critical reviews of chapters or sections of chapters, I would like to thank Russell P. Balda, Gilbert L. Dryden, Jason A. Lillegraven, Larry G. Marshall, Thomas J. O'Shea, C. N. Slobodchikoff, William R. Willis, and Robert W. Wilson. Photographs and, in some cases, valuable discussions and opinions on specific points involving mammalogy were contributed by Margaret H. Bowker, Richard G. Bowker, Patricia Brown, Jack W. Bradbury, H. N. Hoeck, Donna J. Howell, Thomas R. Huels, Steven R. Humphrey, Jennifer U. M. Jarvis, George J. Kenagy, Hans Kruuk, R. L. Peterson, Galen Rathbun, and Timothy Strickler. A number of others, in a great variety of ways, helped to steer the course of revision of this second edition. They include E. H. Colbert, M. Brock Fenton, Jack W. Hudson, G. M. O. Maloiy, Gilbert C. Pogany, O. J. Reichman, James A. Simmons, and Philip L. Wright.

The following students variously helped with typing, darkroom and bibliographic work, and the figures: Elizabeth Aitchison, Karen Averitt, Phyllis Czarnecki, Elouise Elliott, Marcey Olajos, and Margaret Prenzno.

During our stay in Kenya, the late Mr. Hugh Stanton and Mrs. Jane Stanton provided lodging at their Bushwackers Safari Camp, essential advice of many sorts, and fascinating information on African mammals gained from their many years of trapping rhinos and other large African game for zoos around the world. Of greatest importance was their warm friendship, without which the difficult times would have been more bleak and the pleasant times could not have been so happily shared.

My wife, Hazel A. Vaughan, has helped in many material ways, such as by reading proof and working on the bibliography, but her most important contribution has been in retaining her balanced and optimistic outlook toward this project on the many occasions when I lost mine.

TERRY A. VAUGHAN

PREFACE
to the First Edition

This text is intended for use in upper division college or university courses on mammals to be taken by students who have a basic background in zoology. The approach has been to discuss structure and function together whenever possible, for students are far more interested in the functioning mammal than in lists of diagnostic morphological characters. Probably no two instructors cover the same material in presenting a course in mammalogy, and with this in mind I have tried to treat enough aspects of the biology of mammals to make this book useful to different instructors with contrasting approaches to the subject.

I have made liberal use of anatomical drawings because many schools lacking adequate teaching collections of mammals offer courses in mammalogy, and students at these schools can become familiar with the structures of certain groups of mammals only by referring to drawings. Some of the drawings may also be useful as reference material during laboratory work.

In mammalogy, as in many other fields of science, new knowledge is being gained at such a rate that a textbook becomes to some extent a progress report. At best, a book on any aspect of science is somewhat out of date by the time it is published. The major diagnostic structural features of most families and genera of mammals have been well known for some time, but new studies yielding information bearing on the relationships between mammals and new fossil material are constantly forcing revisions of taxonomic schemes and of our ideas as to patterns of evolution. Within the broad fields of mammalian behavior, ecology, physiology, paleontology, and functional morphology, much remains to be learned. Even basic life history information is lacking for many species of mammals. But the job of assembling information on mammals has been eased within the last few years by the publication of books with collections of articles on such subjects as primate behavior, hibernation, the biology of bats, and aquatic mammals.

I have had considerable help on this book from many people. The help of the staff of W. B. Saunders Company has been invaluable. Carl W. May, former Biology Editor, shepherded this book through all but its final stages of preparation and had a hand in planning the organization and coverage. Richard H. Lampert, present Biology Editor, gave encouragement, advice, and editorial assistance in the final stages of writing and during production of the book.

For critically reviewing chapters and offering important comments and advice I would like to thank Sydney Anderson, Russell P. Balda, Gary C. Bateman, Tyler Buchenau, William H. Burt, William A. Clemens, Mary R. Dawson, James S. Findley, Edwin Gould, E. Raymond Hall, Milton Hildebrand, C. D. Johnson, Karl F. Koopman, William Z. Lidicker, Richard E. MacMillen, Robert T. Orr, James L. Patton, Oliver P. Pearson, Gilbert C. Pogany, Frank Richardson, Alfred S. Romer, Constantine Slobodchikoff, Hobart M. Van Deusen, B. J. Verts, Warren F. Walker, Jr., Olwen Williams, and William O. Wirtz II.

Many students assisted with various aspects of this project. I would especially like to thank Celeste Haren, Thomas R. Huels, Jo Ann Mendolia, Larry G. Marshall,

Thomas J. O'Shea, O. J. Reichman, Samuel Semoff, Jan Smith, Roger B. Smith, Kent M. Van De Graaff, Kenneth Weber, and Gary J. Weisenberger.

For allowing me to use their photographs, I am indebted to many friends and colleagues whose names appear in the legends for their photographs. Special thanks are due to Hobart M. Van Deusen, who supplied photographs of New Guinea mammals; Fritz C. Eloff, who sent photographs of the Kalahari Desert and of gemsbok; Diana Harrison and Larry G. Marshall, who arranged for my use of photographs of marsupials and of Australian terrestrial communities; and W. Leslie Robinette, who kindly heeded my pleas for photographs of African mammals. For expert help in selecting appropriate photographs from the files of the San Diego Zoo, I extend my appreciation to Edalee Harwell.

Throughout much of the course of this project Mrs. Elouise Weisenberger did the typing, assisted with figures, and worked on a variety of laborious jobs that one finds it expedient to give to a capable assistant. In the latter stages of the work Miss Susan M. Beeston very ably took over this work, and Mrs. Karen Van De Graaff assisted with typing.

Finally and most important, I am indebted to my wife, Hazel A. Vaughan, who contributed optimism, tolerance, and a critical and artistic eye.

TERRY A. VAUGHAN

CONTENTS

INTRODUCTION

1

THE DOMAIN OF MAMMALOGY

Mammalogy—the division of zoology dealing with mammals—has occupied the efforts of scientists of many kinds. Vertebrate zoologists have studied such aspects as the structure, taxonomy, distribution, and life histories of mammals; physiologists have considered mammalian hibernation and water metabolism; physicists and engineers have studied mammalian echolocation and locomotion; geologists and vertebrate paleontologists have outlined the patterns of mammalian evolution; and anthropologists and psychologists have considered mammalian behavior. Indeed, many significant observations of mammals under natural conditions have been made by keen observers who lacked formal training in zoology. In mammalogy, as in many other fields of zoology, there is room for the nonprofessional. An individual willing to devote time to careful observations of the activities of mammals, and to devote effort to the accurate recording of these observations may make worthwhile contributions. Wide use of the observations of trappers and outdoorsmen has added importantly to the published works of well-known zoologists. As an example, many observations of carnivores by the late W. H. Parkinson, a trapper who worked on the western slope of the Sierra Nevadas of California, were included in Grinnell, Dixon, and Linsdale's *Fur-bearing Mammals of California* (1937). Our present knowledge of mammals, then, has been contributed by scientists trained in zoology and in diverse nonzoological fields, and by perceptive but untrained observers.

Mammals have been regarded as worthy of study by this wide variety of workers for many reasons. The practical aspects have attracted some. Much has been learned of mammalian histology and of the effects of diseases and drugs by the use of various kinds of laboratory mammals; work on domestic breeds of mammals has improved meat production; and research on game species has shown how sustained yields of these animals can be achieved through appropriate management techniques. But to most students and researchers, practicality is not foremost. To this group, mammals are simply fascinating creatures with physiological, structural, and behavioral adaptations to many different modes of life; living mammals in their natural settings are the focal point of interest. The adaptations themselves, how they enable mammals to efficiently exploit demanding environmental conditions, and how the adaptations evolved are all fascinating lines of inquiry. Studies of interactions between different species of mammals in their natural habitats, of population cycles and migrations, and of predator-prey relationships are ecological investigations frequently begun primarily because of a researcher's intense interest in a biological relationship rather than his preoccupation with solving a practical problem. The impressive literature on mammals has resulted largely from such basic research. In this book I deal primarily with the basic information about mammals; "practical mammalogy" is a secondary concern.

During the last 30 years our knowledge of mammalian biology has expanded tremendously: echolocation (natural "sonar") has been studied intensively in both bats and marine mammals; the remarkable ability of some mammals to live under conditions of extreme aridity with no drinking water has been explained; insulation and circulatory and metabolic adaptations in relation to temperature regulation and metabolic economy have been studied; adaptations to deep diving in marine mammals have been investigated; hibernation and migration and the mechanisms influencing them have received attention; important contributions have been made to our knowledge of population cycles of mammals and the factors that may control them; and studies of functional morphology have enlarged our understanding of mammalian terrestrial, aquatic, and aerial locomotion. Probably no field has been more tardy in developing than that of animal behavior, but within the last few years this has been a productive discipline. Considerable attention has been concentrated on the social behavior of anthropoid primates, and these and other mammals have been found to have remarkably complex patterns of social behavior. Explanations for the shapes of the horns and antlers of various deer and antelope and their kin have resulted from careful studies of the breeding behavior of these animals (see p. 390), and predation by mammals has been put in reasonable perspective partially by recent behavioral research. There are still great and vital gaps in our knowledge, but these gaps have been narrowed markedly in roughly the last three decades, and work continues.

CLASSIFICATION

In any careful study, one of the vital early steps is the organization and naming of objects. As stated by Simpson (1945:1) "It is impossible to examine their relationships to each other and their places among the vast, incredibly complex phenomena of the universe, in short to treat them scientifically, without putting them into some sort of formal arrangement." The arrangement of organisms is the substance of taxonomy. But the modern taxonomist, per-

haps better termed a systematist, is less interested in identifying and classifying animals than in studying their evolution. He brings information from such fields as genetics, ecology, behavior, and paleontology to bear on the subjects of his research. He attempts to include in each taxonomic category only animals that evolved from a common ancestor. Excellent discussions of the importance of systematics to our knowledge of animal evolution are given by Simpson (1945) and Mayr (1963).

Because of the difficulties arising from a single kind of animal or plant being recognized by different common names by people in different areas, or by many common names by people in one area, scientists more than 200 years ago adopted a system of naming organisms that would be recognized by biologists throughout the world. Each known kind of organism has been given a binomial (two part) scientific name. The first, the *generic name*, may be applied to a number of kinds; but the second name refers to a specific kind, a *species*. As an example, the blacktailed jackrabbit of the western United States is *Lepus californicus*. To the genus *Lepus* belong a number of similar, but distinct, long-legged species of hares, such as *L. othos* of Alaska, *L. europaeus* of Europe, and *L. capensis* of Africa. Because considerable geographic variation frequently occurs within a species, a third name is often added; this designates a particular *subspecies*. Thus, the large-eared and pale-colored subspecies of *L. californicus* that occurs in the deserts of the western United States is *L. c. deserticola;* the smaller-eared and dark-colored subspecies from coastal California is *L. c. californicus.*

The species is the basic unit of classification. A modern and widely accepted definition of a species is given by Mayr (1942:120): "Species are groups of actually or potentially interbreeding natural populations, which are reproductively isolated from other such groups." Each species is generally separated from all other species by a "reproductive gap," but within each species there is the possibility for gene exchange; as put by Dobzhansky

(1950), all members of a species "share a common gene pool."

Clearly, however, not all species resemble each other to the same degree or are equally closely related. The hierarchy of classification, based on the starting point of the species, has been developed to express degrees of structural similarity and, ideally, phylogenetic relationships between species and groups of species. The taxonomic scheme includes a series of categories, each higher category more inclusive than the one below. Using our example of the hares, a number of long-legged species are included in the genus *Lepus*; this genus, and other genera containing "rabbit-like" mammals, form the family Leporidae; this family, and the family Ochotonidae (the pikas), share certain structural features not possessed by the other mammals, and belong to the *order* Lagomorpha; this order, and all other mammalian orders, form the *class* Mammalia, members of which differ from all other animals in the possession of hair, mammary glands, and many other features. Mammals, birds, reptiles, amphibians, and fish all possess an endoskeleton, and these groups (in addition to some others) form the *phylum* Chordata. All of the phyla of animals (Protozoa, Porifera, Coelenterata, etc.) are united in the Animal Kingdom. The classification of our jackrabbit can be outlined as follows:

Kingdom Animal
  Phylum Chordata
    Class Mammalia
      Order Lagomorpha
        Family Leporidae
          Genus *Lepus*
            Species *Lepus californicus*

Further subdivision of this scheme of classification may result from the recognition of subgroups such as subclass, superorder, or subfamily.

Most ordinal names end in -*a*, as in Carnivora; all family names end in -*idae*, and all subfamily names end in -*inae*. In the following discussions, contractions of the names of orders, families, or subfamilies will often be used as adjectives for the sake of convenience: leporid will refer to Leporidae, leporine to Leporinae, and lagomorph to Lagomorpha.

Some similarities between different kinds of animals are due to parallelism or to convergence. Parallelism occurs when two closely related kinds of animals pursue similar modes of life for which similar structural adaptations have evolved. The similar specializations of the skull and dentition (elongate snouts and reduced teeth) that occur in a number of genera of nectar-feeding Neotropical bats are examples of parallelism. *Convergence* involves the development of similar adaptations to similar (or occasionally nearly identical) styles of life by distantly related species. The golden moles of Africa (see p. 84) and the marsupial "moles" of Australia (see p. 60) are examples of convergence. These animals belong to different mammalian infraclasses (Eutheria and Metatheria, respectively; see p. 42), and their lineages have been separate for over 70 million years; but their habits are much the same and structurally they resemble each other in many ways.

An outline of the classification of mammals used in this book is given in Chapter 4. It is not based on any single published classification, but in some ways it reflects current taxonomic thought. Although it departs from his system in many minor ways, the classification used here is based on that of Simpson (1945). It should be stressed that no universal agreement has been reached on the classification of mammals. Our knowledge of many groups of mammals is incomplete, and future study may demonstrate that some of the families listed here can be discarded because they contain animals best included in another family; perhaps a family or two are yet to be described. The present classification, then, is not used by all mammalogists, and it is by no means immutable.

DESCRIBING SPECIALIZATION

The terms "primitive," "specialized," and "advanced" are used repeatedly in the chapters on the orders and families. A primitive mammal is one that has not departed far from the ancestral type, or at least has retained many structural characters typical of the ancestral type. A monotreme (duck-billed platypus or spiny anteater) is more primitive than a house cat because it lays eggs, has a small brain, and has bones in the pectoral girdle that are found in reptiles but not in other mammals. In these ways the monotreme resembles reptiles, the ancestral group of mammals, whereas the cat does not.

A specialized mammal is one that, in becoming adapted to a particular mode of life, has departed strongly from the ancestral structural plan. A horse is specialized, because in becoming adapted to a life in which speed afoot and the ability to feed on grasses are important, the limbs became elongate, all digits but the third were essentially lost, and the cheek teeth developed complicated occlusal surfaces. In these ways (and many others) the specialized horse has departed strongly from the structure of the primitive mammal.

The term *advanced* is frequently used in comparisons of different members of an evolutionary line or taxonomic group. An advanced species is one in which the particular structural features that characterize the group are highly developed. As an example, the present-day big brown bat is more advanced than the early Eocene bat *Icaronycteris index*. Both are specialized for flight, but because of many refinements in the flight apparatus of the big brown bat it is the far more perfectly adapted flier.

PLAN OF THE BOOK

The first part of this book covers preliminary material: Chapter 2 deals with the characteristics of mammals, the structural features that characterize the group; Chapter 3 briefly covers the evolution of mammals from reptilian ancestry; and Chapter 4 is an introduction to the classification of mammals. In roughly the first half of the remainder of the book (Chapters 5 through 14) the orders and families of mammals are discussed; the second half (Chapters 15 through 21) treats selected aspects of the biology of mammals. My main hope is that from this coverage of the subject the student can gain a general understanding of, and an appreciation for, the form and function of mammals.

The anatomical drawings, which I regard as essential to the ordinal chapters, should help students understand the discussions of structure and function. I have made liberal use of these drawings, most of which illustrate skulls, teeth, or feet. The profiles of skulls are from the left side, and all occlusal views of teeth show the right upper or the left lower tooth row.

The bibliography in the appendix includes not only papers that are cited in the text, but also some additional publications that have material useful to students by authors whose papers are cited.

MAMMALIAN CHARACTERISTICS

2

Mammals owe their spectacular success to many features. Many of the most important and most diagnostic mammalian characteristics serve to further intelligence and sensory ability, to promote endothermy, or to increase the efficiency of reproduction or of securing and processing food. The senses of sight and smell are highly developed, and the sense of hearing has undergone greater specialization in mammals than in any other vertebrates. Efficient gathering and utilization of a tremendous variety of foods is aided by specializations of the dentition and the digestive system. The perfection of endothermy has allowed mammals to remain active under a wide array of environmental conditions, and specializations of the postcranial anatomy, particularly the limbs and feet, have enabled them to make effective use of this activity. Extended periods of association between the parents and young of some species have facilitated the use of demanding types of foraging and the development of complex social behavior.

The basic structural plan of mammals was inherited from a relatively unspectacular reptilian group, the mammal-like reptiles of the synapsid order Therapsida (see p. 25). Members of this ancient order followed an evolutionary path that diverged strongly from that of the reptiles from which arose the vastly more spectacular and more successful dinosaurs and other ruling reptiles (subclass Archosauria). The key to the marginal persistence of the therapsids through the Triassic (see Table 3–1) was perhaps their ability to move and to think more quickly than their dull-witted archosaurian contemporaries. These same abilities probably enabled the descendants of the therapsids,

the mammals, to survive through the Jurassic and Cretaceous, periods when the dinosaurs completely dominated the terrestrial scene. Also of major importance to early mammals was the highly specialized dentition, which probably allowed mammals to utilize certain foods more efficiently than could reptiles.

An important morphological trend in the therapsid-mammalian line was toward skeletal simplification. An engineer may redesign a machine and increase its efficiency by reducing the number of parts to the minimum consistent with the effective performance of a particular function; a similar type of simplification occurred in the therapsid skeleton. In the skull and lower jaw, which in primitive reptiles consisted of many bones, a number of bones were lost or reduced in size. The limbs and limb girdles were also simplified to some extent, and their massiveness was reduced. As a result, the advanced therapsid skeleton roughly resembles that of the egg-laying mammals (order Monotremata), but the limbs of some therapsids were less laterally splayed than are those of today's specialized monotremes.

When mammals first appeared in the Triassic, then, they represented no radical structural departure from the therapsid plan, but had simply attained a level of development (involving a dentary/squamosal jaw articulation; see p. 29) that is interpreted by most vertebrate paleontologists as indicating that the animals had crossed the mammalian-reptilian boundary. Many of the mammalian characters discussed in this chapter resulted from evolutionary trends clearly characteristic of therapsid reptiles. Unfortunately, the fossil record cannot indicate when various im-

5

portant features of the soft anatomy became established, and one can only speculate whether or not advanced therapsid reptiles had such features as mammae, hair, or a four-chambered heart.

The following sources have discussions of the characteristics of mammals: Romer and Parsons, 1977; Smith, 1960; Weichert, 1965; Young, 1957. Especially useful is the treatment by Romer in *Vertebrate Paleontology* (1966:187–196).

SOFT ANATOMY

SKIN GLANDS.   The skin of mammals contains several kinds of glands not found among other vertebrates; the most important of these are the *mammary glands.* The nutritious milk secreted from these glands provides the sole nourishment for young during their initial period of rapid postnatal growth. In most mammals the openings of the glands are in projecting nipples (mammae), from which milk is sucked by the young. In monotremes, nipples are lacking, and young suck milk from tufts of hair on the mammary areas (Burrell, 1927; Ewer, 1968:236). Cetaceans have muscles that force milk into the mouth of the young, a seemingly necessary adaptation in ani-

mals that have no lips and are therefore unable to suck. The number of nipples varies from 2 in a number of mammals to about 19 in the mouse-opossum (*Marmosa;* Tate, 1933:36).

Other types of skin glands are of importance in mammals. The watery secretion of the *sweat glands* functions primarily to promote evaporative cooling, but also eliminates some waste materials. In man and some ungulates, sweat glands are broadly distributed over the body surface, but in most mammals they are more restricted. In some insectivores, rodents, and carnivores, sweat glands occur only on the feet or on the venter, and they are completely lacking in the Cetacea and in some bats and rodents. Hair follicles are supplied with *sebaceous glands* (Fig. 2–1); their oil secretion lubricates the hair and the skin. A variety of *scent glands* and *musk glands* occur among mammals. These glands are variously used as attractants, for marking territories, for communication during social interactions, or for protection. "Skunk smell" is familiar to all but the most city-bound, and has caused the temporary banishment of many a farm dog. A musk gland marked by a patch of dark hairs occurs on the top of the tail of wolves and coyotes, as well as on

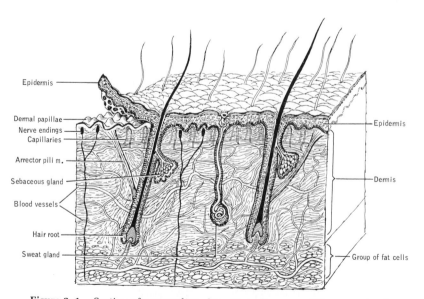

Figure 2–1.   Section of mammalian skin. (From Romer and Parsons, 1977.)

the tail of many domestic dogs. The functions of some of the mammalian scent glands in connection with social behavior are discussed in Chapter 17 (p. 363).

HAIR. The bodies of mammals are typically covered with hair, a unique mammalian feature that has no structural homologue among other vertebrates. Hair was perhaps developed by therapsid reptiles before a scaly covering was lost. Hairs arise from between scales on the tails of many small mammals and from between the bony plates of armadillos; they retain the same distribution on the skin of some scaleless mammals that they would have if scales were present.

A hair consists of dead epidermal cells that are strengthened by keratin, a tough horny tissue made of proteins. A hair grows from living cells in the hair root (Fig. 2–1). Each hair consists of an outer layer of cells arranged in a scalelike pattern, the *cuticle*, a deeper layer of highly packed cells, the *cortical layer*, and in some cases a central core of cuboidal cells, the *medulla* (Fig. 2–2). The color of hair depends on pigment in either the medulla or the cortical cells; the cuticle is usually transparent.

The coat of hair, termed the *pelage*, functions primarily as insulation. The dissipation of heat from the skin surface to the environment, and the absorption of heat from the environment, are retarded by pelage. Pinnipeds, many of which live in extremely cold water, are insulated by both hair and subcutaneous blubber. Some mammals are hairless, or nearly so. These either live in warm areas or have specialized means of insulation other than hair. The essentially hairless whales and porpoises have thick layers of blubber that provide insulation. Hair is sparse on elephants, rhinos, and hippopotami; these animals occupy warm areas, have thick skins that offer some insulation, and have such favorable mass/surface ratios because of their large size (see p. 418) that retention of body heat is no problem.

Hair, being nonliving material, is subject to considerable wear and bleaching of pigments. During periodic molts, usually

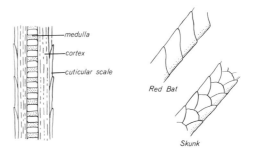

Figure 2–2. Structure of a hair and two types of reticular scale patterns. (After Storer, T. I., and Usinger, R. L.: *General Zoology*, 4th ed. Copyright 1965 by McGraw-Hill Book Company. Used with permission of McGraw-Hill Book Company.)

once or twice a year, old hairs are lost and new ones replace them. This often occurs in a regular pattern of replacement (Fig. 2–3). In many north-temperate species the molts are in the spring and fall, and the summer pelage is generally shorter and has less insulating ability than the winter pelage. In some species that occupy areas with continuous snow-cover in the winter, the summer pelage is brown and the winter coat is white. The arctic fox, several species of hares, and some weasels follow this pattern.

The color of most small, terrestrial mammals closely resembles the color of the soil on which they live. In his careful study of concealing coloration in desert rodents of the Tularosa Basin of New Mexico, Benson (1933) found that white sands were inhabited by nearly white rodents, whereas on adjoining stretches of black lava lived black rodents. Broadly speaking, mammals that occur in forests or beneath a thick canopy of vegetation are dark colored, whereas those that occur in more open situations are relatively pale.

Countershading is a color pattern common to mammals and many other vertebrates. Under most lighting conditions the back of an animal is more brightly illuminated than is the shaded underside. If a mammal were all of a single color, the underside would appear very dark relative to the back, and the form of the animal would be obvious. But when the back and sides are darkly colored and the underside and insides of the legs are white—an al-

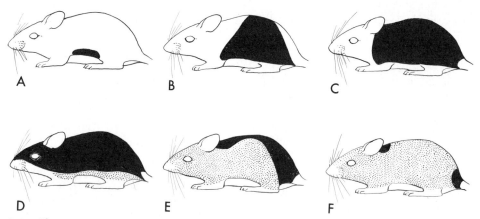

Figure 2–3. The pattern of post-juvenile molt in the golden mouse (*Ochrotomys nuttalli*). Black areas indicate portions where replacement of juvenile hair by adult hair is occurring. Stippled areas indicate new adult pelage. (After Linzey and Linzey, 1967.)

most universal color pattern among terrestrial mammals — the well-lighted back reflects little light and the shaded white venter tends to strongly reflect light. The result is that the form of the animal becomes obscured to some extent, and the animal becomes less conspicuous.

The color patterns of mammals serve a variety of purposes. The pelages of some ungulates and some rodents are marked by white stripes that tend to obliterate the shapes of the animals when they are against broken patterns of light and shade. The eye is one of the most conspicuous and unmistakable vertebrate features; some facial markings in mammals may serve importantly to obviate the bold pattern of the eye by superimposing a more dominant and disruptive pattern (Fig. 2–4). If these

markings only occasionally allow an animal to go unnoticed by a predator, or if they cause a predator to be indecisive in his attack for but a fraction of a second, they have adaptive value.

The stripes of zebras under conditions of dim light cause the animals to fade into their backgrounds, but another adaptive importance has been considered by Cott (1966). The stripes of zebras are so patterned to create an optical illusion: the animals' apparent size is increased (Fig. 2–5). In the dim light under which most predators hunt, this illusion may cause a slight miscalculation of range and an occasional inaccurate leap.

But what about the glaringly white rump patch of the pronghorn, and other conspicuous white markings belonging to cur-

A   B

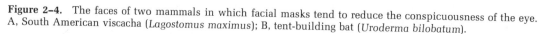

Figure 2–4. The faces of two mammals in which facial masks tend to reduce the conspicuousness of the eye. A, South American viscacha (*Lagostomus maximus*); B, tent-building bat (*Uroderma bilobatum*).

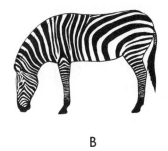

Figure 2–5. Markings of a zebra (B) in contrast to horizontal markings (A), showing the optical illusion of increased size in the zebra. (Partly after Cott, 1966.)

A   B

sorial grassland or open country species? These markings probably function as warning signals when an animal begins to run from danger, and the gaits that some ungulates use at these times serve to show off the markings. The black and white coloration of skunks, on the other hand, makes these defensively well-endowed animals conspicuous and unmistakable to their enemies.

CIRCULATORY SYSTEM. In keeping with their active lives and their endothermic ability, mammals have highly efficient circulatory systems. A complete separation of the systemic circulation and the pulmonary circulation has been achieved in mammals. The four-chambered heart functions as a double pump: the right side of the heart receives venous blood from the body and pumps it to the lungs; the left side receives oxygenated blood from the lungs and pumps it to the body. The fascinating evolution of the mammalian heart and circulatory pattern is described in detail by Hildebrand (1974:277–299).

As might be expected because of the great size differential between the smallest and the largest mammal (the weights of the 2-g shrew and the 160,000-kg whale differ by a factor of 80,000,000), the heart rate is highly variable between species. The rate in nonhibernating mammals varies from under 20 beats per minute in whales to over 1300 in a shrew (Table 2–1). Especially remarkable is the ability of some mammals to alter their heart rates rapidly. As an extreme example, a resting big brown bat (*Eptesicus*) has a rate of about 400 beats per minute; this rate increases almost instantly to about 1000 when the bat takes flight, and generally returns to the resting rate within one second after the flight (Fig. 2–6).

The erythrocytes (red blood cells) of mammals are biconcave discs. They extrude their nuclei when they mature, apparently as a means of increasing oxygen-carrying capacity.

RESPIRATORY SYSTEM. In mammals, the lungs are large and, together with the heart, virtually fill the thoracic cavity. Air passes down the trachea, into the bron-

Table 2–1. HEART RATES OF SELECTED MAMMALS

Species	Common Name	Weight	Heart Rate Beats/Min
Erinaceus europaeus	European hedgehog	500–900 g	246 (234–264)
Sorex cinereus	Gray shrew	3–4 g	782 (588–1320)
Eutamias minimus	Least chipmunk	40 g	684 (660–702)
Sciurus carolinensis	Gray squirrel	500–600 g	390
Phocaena phocaena	Harbor porpoise	170 kg	40–110
Mustela vison	Mink	0.7–1.4 kg	272–414
Phoca vitulina	Harbor seal	20–25 kg	18–25
Elephas maximus	Asiatic elephant	2,000–3,000 kg	25–50
Equus caballus	Horse	380–450 kg	34–55
Sus scrofa	Swine	100 kg	60–80
Ovis aries	Sheep	50 kg	70–80

(Data from Altman and Dittmer, 1964:235.)

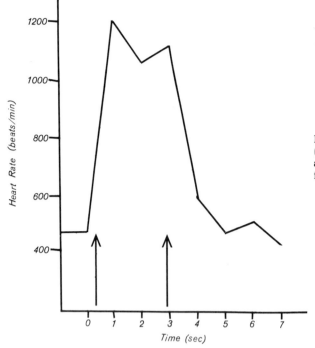

Figure 2–6. Heart rate of the big brown bat (*Eptesicus fuscus*) at rest and during flight. The arrows indicate the beginning and end of flight. (After Studier and Howell, 1969.)

chi, and through a series of branches of diminishing size into the bronchioles, from which branch alveolar ducts. Clustered around each alveolar duct is a series of tiny terminal chambers, the alveoli. Exchange of gases between inspired air and the blood stream occurs in the alveoli; the thin alveolar membranes are surrounded by dense capillary beds. In man, the lungs contain about 300 million alveoli, which provide a total respiratory surface of about 70 square meters (well over 600 square feet —some 40 times the surface area of the body).

Air is forced into the lungs by muscular action that increases the volume of the thoracic cavity and decreases the pressure within the cavity. Some increase is gained by the forward and outward movement of the ribs under the control of intercostal muscles, but of greater importance is the muscular diaphragm (a structure unique to mammals). When relaxed, the diaphragm is bowed forward, but when contracted its central part moves backward toward the celomic cavity, thus increasing the volume of the thoracic cavity.

REPRODUCTIVE SYSTEM. In mam-

mals, both ovaries are functional and the ova are fertilized in the oviducts. The embryo develops in the uterus and lies within a fluid-filled amniotic sac. Nourishment for the embryo comes from the maternal blood stream by way of the placenta. (The female reproductive cycle and the establishment of the placenta are discussed in Chapter 18.) The structure of the uterus is variable (Fig. 2–7).

The male copulatory organ, the penis, contains erectile tissue and is surrounded by a sheath of skin, the prepuce. In many species the penis contains a bone, the *os penis* or baculum, which may differ markedly even between closely related species (Fig. 2–8) and may therefore be of considerable use in taxonomic studies. The tip of the penis has an extremely complicated form in some species (Fig. 2–9). The testes of mammals, instead of lying in the celomic cavity as in other vertebrates, are typically contained in the scrotum, a sac-like structure that lies outside the body cavity but is an extension of the celomic cavity. The testes either descend permanently from the celomic cavity into the scrotum when the male reaches reproduc-

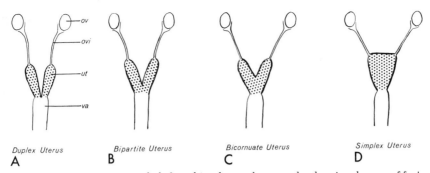

Duplex Uterus
A

Bipartite Uterus
B

Bicornuate Uterus
C

Simplex Uterus
D

Figure 2–7. Several types of uteri (stippled) found in placental mammals, showing degrees of fusion of the two "horns" of the uterus. A duplex uterus (A) occurs in the orders Lagomorpha, Rodentia, Tubulidentata, and Hyracoidea; a bipartite uterus (B) is known in the order Cetacea; a bicornuate uterus (C) is found in the order Insectivora, in some members of the orders Chiroptera and Primates, and in the orders Pholidota, Carnivora, Proboscidea, Sirenia, Perissodactyla, and Artiodactyla; a simplex uterus (D) is typical of some members of the orders Chiroptera and Primates, and of the order Edentata. (From Smith, H. M.: *Evolution of Chordate Structure: An Introduction to Comparative Anatomy.* © 1960 by Holt, Rinehart and Winston, Inc. Reproduced by permission of Holt, Rinehart and Winston, Inc.)

Figure 2–8. The bacula of several species of New Guinean murid rodents, showing differences in the structure of this bone in closely related mammals. (From Lidicker, 1968.)

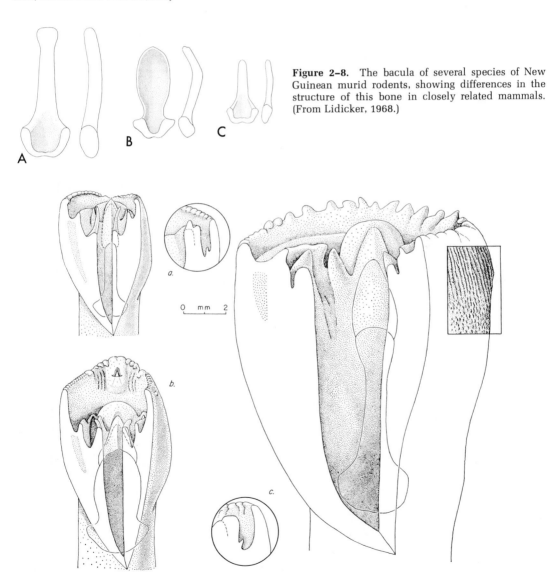

Figure 2–9. Ventral views of the penises of several species of New Guinean murid rodents, showing the complex structure of the organ in these mammals. (From Lidicker, 1968.)

tive maturity or are withdrawn into the body cavity between breeding seasons and descend when the animal again becomes fertile. In most mammals the maturation of sperm cannot proceed normally at the usual deep body temperatures, and the scrotum functions as a "cooler" for the testes and developing sperm.

BRAIN. Compared to the brains of other vertebrates, that of the mammal is unusually large. This greater size is due largely to a tremendous increase in the size of the cerebral hemispheres. These structures were ultimately derived from a part of the brain important in "lower" vertebrates in receiving and relaying olfactory stimuli. Most characteristic of the brain of higher mammals is the great development of the *neopallium*, a mantle of gray matter that first appeared as a small area in the front part of the cerebral hemispheres in some reptiles, which in mammals has expanded over the surface of the deeper, "primitive" vertebrate brain. The surface area of the neopallium is vastly increased in many mammals by a complex pattern of folding (Fig. 2–10). A new development

in placental mammals is the *corpus callosum*, a large concentration of nerve fibers that passes between the two halves of the neopallium and provides communication between them.

The unique behavior of mammals is largely a result of the development of the neopallium, which functions as a control center that has come to dominate the original brain centers. Sensory stimuli are relayed to the neopallium, where much motor activity originates. Present actions are influenced by past experience; learning and "intelligence" are important. The size of the brain relative to total body size seems not always to be a reliable guide to intelligence, for brain size apparently need not increase in proportion to increases in body size to maintain intelligence. The degree of development of convolutions on the surface of the neopallium is perhaps a better indication of intelligence.

SENSE ORGANS. The sense of smell is acute in many mammals, probably in part as a result of the development of turbinal bones in the nasal cavities (Fig. 2–11). The olfactory bulbs and olfactory

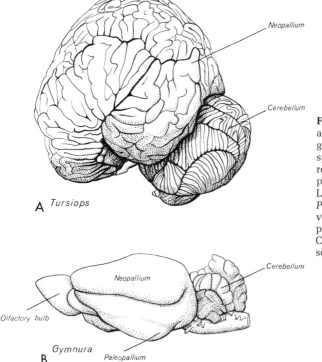

Figure 2–10. The brains of a porpoise (A) and a hedgehog (B). The neopallium is greatly enlarged and highly convoluted in the specialized and intelligent porpoise, but is relatively small and smooth-surfaced in the primitive hedgehog. (*Tursiops* after Kruger, L., in Norris, K. S.: *Whales, Dolphins and Porpoises*, originally published by the University of California Press, 1966, redrawn by permission of the Regents of the University of California. *Gymnura* after Romer and Parsons, 1977.)

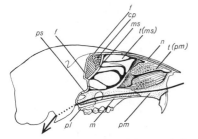

Figure 2–11. Cutaway view of the nasal chamber of the Abert's squirrel (*Sciurus aberti*), showing the complicated arrangement of turbinal bones. The entire right half of the nasal part of the skull is removed, exposing the left side of the nasal chamber. The arrow shows the main path air follows from the external to the internal nares, but some air circulates through the upper part of the chamber and over the turbinal bones. Abbreviations: *cp*, cribriform plate of the mesethmoid bone (through which the branches of the olfactory nerve pass out of the braincase); *f*, frontal; *m*, maxillary; *ms*, mesethmoid; *n*, nasal; *pl*, palatine; *pm*, premaxillary; *ps*, presphenoid; *t(ms)*, turbinals connected to the mesethmoid; *t(pm)*, turbinals connected to the premaxillary.

lobes form a great part of the brain in some insectivores and are reasonably large in carnivores and rodents. The sense of smell is poorly developed and the olfactory part of the brain is strongly reduced in whales and the higher primates; the olfactory system is absent in porpoises and dolphins (Kruger, 1966:247).

The sense of hearing is highly developed in mammals, and no other vertebrates seem to depend so heavily on this sense.

Mammals alone have an external structure (the *pinna*) that serves to intercept sound waves; the pinnae may be extremely large and elaborate in some mammals, particularly in bats (Fig. 6–33; p. 105). Pinnae are missing (presumably secondarily lost) in some insectivores, phocid seals, and cetaceans. The external auditory meatus, the tube leading from the pinna to the tympanic membrane, is typically long in mammals, and in cetaceans is extremely long. The middle ear is an air-filled chamber that houses the three *ossicles*, and is typically enclosed by a bony bulla (Fig. 2–12). The mammalian cochlea is more or less coiled. (Some variations in the structure of the mammalian ear are discussed on p. 467).

The eye of mammals resembles that of most amniote vertebrates. In most nocturnal mammals the *tapetum lucidum* is well developed. This is a reflective structure within the choroid that improves night vision by reflecting light back to the retina. This reflection accounts for the "eye shine" when a rabbit's eyes are picked up by the beams of headlights at night. Although in most mammals the eyes are well developed, in some insectivores and some cetaceans they are strongly reduced. In such species the eyes are only able to differentiate between light and dark and may serve primarily to aid the animals in main-

Figure 2–12. Lateral view of the right middle ear chamber (anterior is to the right) of the Abert's squirrel (*Sciurus aberti*), with the auditory bulla largely removed. The complex partitioning of the bulla, the positions of the ear bones and the inner ear, and the ligamentous bracing of the malleus and incus are shown. In life the manubrium of the malleus rests against the tympanic membrane. Abbreviations: *al*, anterior ligament (of the malleus); *as*, alisphenoid; *i*, incus; *m*, malleus; *mn*, manubrium; *p*, periotic; *pl*, posterior ligament; *pp*, paroccipital process; *pr*, partitions of bulla; *s*, stapes; *sq*, squamosal; *t*, tympanic.

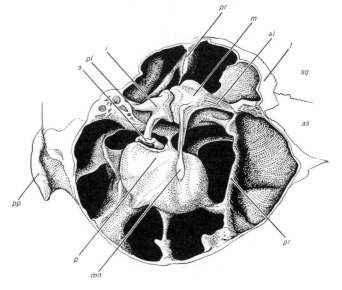

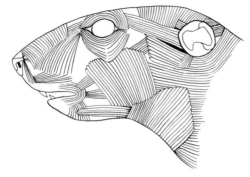

Figure 2–13. The superficial facial muscles of the cotton rat (*Sigmodon hispidus*); these muscles partly control facial expression. (After Rinker, 1954.)

taining the appropriate nocturnal or diurnal activity cycles (Herald, et al., 1969; Lund and Lund, 1965).

Most mammals have *vibrissae*. These are the whiskers on the muzzle and the long, stiff hairs that are present on the lower legs of some mammals. The vibrissae are tactile organs, and those on the face probably enable nocturnal species to detect obstacles near the face. The vibrissae on the muzzle generally arise from a structure termed the *mystacial pad*, and are controlled by a complex of muscles (Fig. 2–13).

DIGESTIVE SYSTEM. Salivary glands are present in mammals, and in some ant-eating species they are specialized for the production of a mucilaginous material that makes the tongue sticky. The stomach is a single saclike compartment in most species, but is complexly subdivided in ruminant artiodactyls, in cetaceans, and sirenians (Fig. 2–14). In herbivorous species, digestion is frequently accomplished

partly by microorganisms that inhabit the stomach or cecum (the cecum is a blind sac that opens into the posterior end of the small intestine).

MUSCULAR SYSTEM. The mammalian limb and trunk musculature has been highly plastic. Different evolutionary lines have developed muscular patterns beautifully adapted to diverse modes of locomotion. Cetaceans are the fastest marine animals, certain carnivores and ungulates are the most rapid runners, and bats as fliers maneuver better than birds. Some muscular specializations favoring specific types of locomotion are described in the chapters on the orders of mammals. Especially notable in mammals is the great development of dermal musculature. In many mammals these muscles form a sheath over most of the body and allow the skin to be moved. These dermal muscles have differentiated and have moved over much of the head (Fig. 2–13); these facial muscles control many essential actions. In mammals there are no more vital voluntary muscles than those that encircle the mouth; these function during suckling and are among the first voluntary muscles to be subjected to heavy use. Facial muscles move the ears, close the eyes, and control the subtle changes in expression that are so important in the social lives of many mammals.

THE SKELETON

GENERAL FEATURES. The mammalian skeleton differs from that of the reptile in several basic ways; all may well be

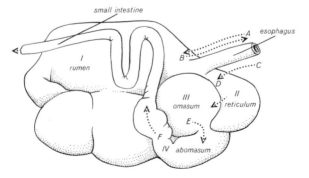

Figure 2–14. The four-chambered stomach of a ruminant artiodactyl. As the animal feeds, it swallows the vegetation, which is then stored in the rumen (I). While the animal rests, it regurgitates the food from the rumen and "chews its cud" (remasticates the food). The food then goes to the reticulum, omasum, and abomasum, where digestion is aided by a diverse microbiota. (After Storer, T. I., and Usinger, R. L.: *General Zoology*, 4th ed. Copyright 1965 by McGraw-Hill Book Company. Used with permission of McGraw-Hill Book Company.)

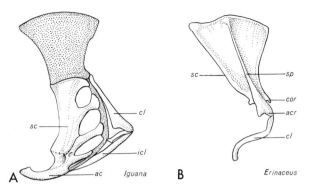

Figure 2–15. Lateral view of the right side of the pectoral girdles of a lizard (A) and a mammal (B), showing the greater ossification and simplification of the structure in the mammal. Abbreviations: *ac,* anterior coracoid; *acr,* acromion process; *cl,* clavicle; *cor,* coracoid process; *icl,* interclavicle; *sc,* scapula; *sp,* spine of scapula.

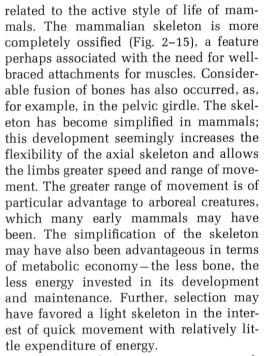

related to the active style of life of mammals. The mammalian skeleton is more completely ossified (Fig. 2–15), a feature perhaps associated with the need for well-braced attachments for muscles. Considerable fusion of bones has also occurred, as, for example, in the pelvic girdle. The skeleton has become simplified in mammals; this development seemingly increases the flexibility of the axial skeleton and allows the limbs greater speed and range of movement. The greater range of movement is of particular advantage to arboreal creatures, which many early mammals may have been. The simplification of the skeleton may have also been advantageous in terms of metabolic economy—the less bone, the less energy invested in its development and maintenance. Further, selection may have favored a light skeleton in the interest of quick movement with relatively little expenditure of energy.

To an animal as active as a mammal, well-formed articular surfaces on limb bones and solid points of attachment for muscles are highly advantageous during the period of growth of the skeleton as well as during adult life. Mammals have abandoned the pattern of bone growth typical of reptiles. In many reptiles, skeletal growth may continue throughout much of life. Growth in reptiles occurs at the ends of limb bones by ossification of the deep parts of a persistently growing cartilaginous cap; such a pattern clearly limits the establishment of a well-formed joint. In mammals, however, skeletal growth is generally restricted to the early part of life. The articular surfaces and some points of attachment of large muscles become well

formed and ossified early, while rapid growth is still under way. Growth continues at a cartilaginous zone where the end of the bone and its articular surface, the *epiphysis,* joins the shaft of the bone, the *diaphysis* (Fig. 2–16). When full growth is attained, this cartilaginous zone of growth becomes ossified, fusing the epiphysis and diaphysis. Because within a given species this fusion usually occurs at a certain age, the degree of closure of the "epiphyseal line" is useful in estimating the age of a mammal.

THE SKULL (Fig. 2–17). The braincase of the mammalian skull is large. In addition to its primary function of protecting the brain, it provides a surface from which the temporal muscles originate. (In many

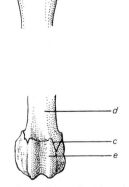

Figure 2–16. The proximal (above) and distal (below) ends of the right femur of a young hedgehog (*Erinaceus europaeus*), showing the epiphyses (*e*), the diaphysis (*d*), and the intervening cartilaginous zone (*c*).

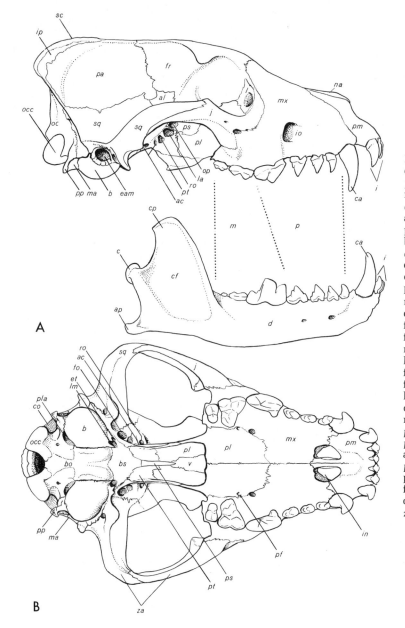

Figure 2–17. Skull of the African hunting dog *(Lycaon pictus)*, showing bones, foramina, and teeth. Abbreviations: *ac*, alisphenoid canal; *al*, alisphenoid; *ap*, angular process; *b*, auditory bulla; *bo*, basioccipital; *bs*, basisphenoid; *c*, condyle; *ca*, canines; *cf*, coronoid fossa; *co*, anterior condyloid foramen; *cp*, coronoid process; *d*, dentary; *eam*, external auditory meatus; *et*, eustachian tube; *fo*, foramen ovale; *fr*, frontal; *i*, incisors; *in*, incisive foramen; *io*, infraorbital foramen; *ip*, interparietal; *j*, jugal; *l*, lacrimal; *la*, anterior lacerate foramen; *lm*, medial lacerate foramen; *m*, molars; *mx*, maxillary; *na*, nasal; *oc*, occipital; *occ*, occipital condyle; *op*, optic foramen; *p*, premolars; *pa*, parietal; *pf*, posterior palatine foramen; *pl*, palatine; *pla*, posterior lacerate foramen; *pm*, premaxillary; *pp*, paroccipital process; *ps*, presphenoid; *pt*, pterygoid; *ro*, foramen rotundum; *sc*, sagittal crest; *sq*, squamosal; *v*, vomer; *za*, zygomatic arch.

mammals these are the most powerful muscles that close the jaws.) A *sagittal crest* increases the area of origin for the temporal muscles in many mammals; the *lambdoidal crest* gives origin to the temporal muscles and some cervical muscles. The *zygomatic arch* is usually present as a structure that flares outward from the skull. It serves to protect the eye and to provide origin for the masseter muscles; it forms the surface with which the condyle of the dentary (lower jaw) articulates. The zygomatic arch may be reduced or lost, as in some insectivores and cetaceans, or may be enlarged in those groups in which the masseter muscles largely supplant the temporals as the major jaw muscles (Fig. 10–3; p. 169). The mammalian skull has a secondary palate (see p. 25 and Fig. 3–4), and there are *turbinal bones* within the long nasal cavities (Fig. 2–11).

A number of *foramina* (openings) perforate the braincase and allow passage of the cranial nerves (Fig. 2–17). In some rodents, the infraorbital foramen, through which blood vessels and a branch of the fifth cranial nerve (the trigeminal) pass, is enormously enlarged in association with specializations of the masseter muscles (Fig. 10–3; p. 170). The incisive foramina, present in the palates of many mammals (Fig. 2–17), house an olfactory organ (Jacobson's organ) that allows the "smelling" of the contents of the mouth. These olfactory organs are widespread among vertebrates; a snake puts the tips of its forked tongue against this part of the palate after "testing" its immediate environment.

Sounds that cause vibration of the tympanic membrane are mechanically transmitted by the three ear ossicles (Fig. 2–12) through the air-filled chamber to the inner ear. The footplate of the stapes fills an opening into the inner ear, and, acting like a piston, transforms the movements of the ossicles to vibrations of the fluid within the cochlea. The inner ear, with the cochlea and semicircular canals, is contained by the periotic bone, which is generally covered by the squamosal bone but is exposed as the mastoid bone in some mammals (Fig. 2–17). The auditory bulla is formed by the expanded tympanic bone or by the tympanic plus the entotympanic, a bone found only in mammals. The bulbous tympanic bullae in many mammals look like structures remembered too late and hurriedly stuck to the skull, but the bullae are highly modified in some species in connection with specialized modes of life (see Fig. 10–21, p. 187, also p. 467).

The lower jaw is formed by the *dentary*. This bone typically has a coronoid process, on which the temporalis muscle inserts, a coronoid fossa, in which the masseter muscles insert, and an angular process, to which a jaw-opening muscle (the digastricus) attaches. In some herbivores, in which the masseter is enlarged at the expense of the temporalis, the coronoid process is reduced or absent and the posterior part of the dentary becomes dorsoventrally broadened (Fig. 9–1A).

Several skeletal elements in the throat region are highly modified remnants of the gill arches of fish. These structures, the *hyoid apparatus*, support the trachea, the larynx, and the base of the tongue (Fig. 2–18), and are often braced anteriorly against the auditory bullae.

TEETH. Without doubt one of the major keys to the success of mammals has been the possession of teeth. Fish, amphibians, reptiles, and mammals all have teeth, but the specialization of the dentition in mammals has gone far beyond anything found in the other groups. Only among mammals have dentitions evolved that are capable of coping with items so difficult to prepare for digestion as dry grass and large bones. So varied are the dental specializations of mammals and so closely related are they to specific styles of feeding and to patterns of adaptations of the skull, jaws, and jaw musculature, that to know in detail the dentition of any mammal is to understand much about its way of life. Most of our knowledge of the early evolution of mammals is based on studies of fossil teeth, which, because of their extreme hardness, are often the only parts of early mammals that are preserved. Teeth alone can tell us a great deal. The earliest known vertebrates, which were jawless, had bodies encased in bony plates. When some of the arches that supported the gill apparatus in primitive vertebrates became modified into jaws and jaw supports, teeth developed on the bony plates that bordered the mouth.

Although the teeth in different denti-

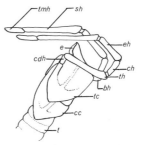

Figure 2–18. The hyoid apparatus and associated structures of the domestic cat. Abbreviations: *bh*, basihyal; *cc*, cricoid cartilage; *cdh*, chondrohyal; *ch*, ceratohyal; *e*, epiglottis; *eh*, epihyal; *sh*, stylohyal; *t*, trachea; *tc*, thyroid cartilage; *th*, thyrohyal; *tmh*, tympanohyal. (After Taylor, W. T., and Weber, R. J.: *Functional Mammalian Anatomy.* © 1951 by Litton Educational Publishing Inc. Reprinted by permission of Van Nostrand Reinhold Company.)

tions differ widely in number, structure and function, in most mammals the dentition is *heterodont*; that is to say, it consists of teeth that vary in both structure and function. In mammals, teeth occur on the premaxillary, maxillary, and dentary bones (Fig. 2–17). The anteriormost teeth, the *incisors* and *canines*, serve to gather or kill food, whereas the more specialized cheek teeth, the *premolars* and *molars*, grind or slice food in preparation for digestion. Characteristically, two sets of teeth appear in a mammal's lifetime. The *deciduous* dentition develops early and consists of incisors, canines, and premolars—but no molars. These "milk teeth" are lost and replaced by permanent teeth as the animal matures. The permanent dentition consists of a second set of incisors, canines and premolars; it also includes the molars, which have no deciduous counterparts. The deciduous dentition of some species bears little resemblance to the permanent dentition (Fig. 2–19).

The form, function, and origin of the cusp patterns of the cheek teeth, and especially of the molars, are of particular interest. The basic primitive molar, termed *tribosphenic* in reference to the basically three-cusped pattern of the occlusal surface, is found in a number of fossil mammals. (The occlusal surfaces of teeth are those that contact their counterparts of the opposing jaw—the surfaces the dentist generally attacks when putting in a filling.) A stroke of luck for functional morphologists is that the American opossum (*Didelphis virginiana*) and some other living marsupials have molars that resemble those of primitive mammals that coexisted

with dinosaurs. Opossums are omnivorous, eating insects and other small animals and soft plant material; probably many Mesozoic mammals had similar diets. Careful studies of jaw action in the opossum, therefore, can show not only how the molars of this animal function but also can probably indicate how they functioned in primitive mammals over 70 million years ago. The studies of Crompton and his coworkers (1971; Crompton and Jenkins, 1968; Crompton and Hiiemae, 1969, 1970) provide much of the basis for the following discussion of the functional morphology of molars.

In the opossum, the molars serve two functions. For up to 60 per cent of the time involved in chewing and throughout the initial stages of chewing, the high cusps of the upper and lower cheek teeth crush and puncture the food without coming together. After the food is "pulped," it is sliced by the six matching shearing surfaces shown in Fig. 2–20. This shearing is facilitated by the way in which food is trapped and steadied by the opposing molars (Fig. 2–21). Chewing occurs on one side of the jaw at a time. During cutting strokes the jaw action is not one of simple up and down movements. Precise lateral adjustments of the jaw during mastication enable opposing molars to slide against each other. As shown in Fig. 2–21, this movement involves a transverse as well as an upward component as the lower molars shear against the uppers. Attrition facets on the molars of Mesozoic mammals indicate that occlusion of the teeth during chewing has always had this transverse component (Butler, 1972). Major shearing surfaces are those designated in Fig. 2–20 as *1* and *2*, but additional cutting occurs when the surfaces on the sides of the cusps of the quadrate posterior part (the talonid) of the lower molars shear against their counterparts on the upper molars. As a result of this complex pattern of occlusion, each time the three pairs of opposing molars of one side of the jaw come together, some 18 cuts are made in the food, and food already pulped is rapidly sectioned. Natural selection has favored

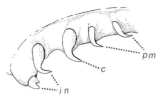

Figure 2–19. The deciduous upper dentition (of the left side) of a Neotropical fruit-eating bat (*Artibeus lituratus*). Abbreviations: *c,* canine; *in,* incisors; *pm,* premolars. (From Vaughan, 1970b.)

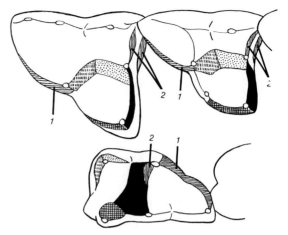

Figure 2–20. Matching shearing planes of the occlusal surfaces of the upper molars (top) and lower molars (bottom) of the Virginia opposum, *Didelphis virginiana*. The upper molars are above, the lowers are below. (After Crompton and Hiiemae, 1969.)

the evolution of "efficient" dentitions because time spent in masticating food means greater energy expenditure and less time available for food gathering. In some species, this also means a greater period of vulnerability to predators.

Two major evolutionary trends in molar structure appeared in the Mesozoic and became pronounced in the Cenozoic. In carnivores, some of the cheek teeth became bladelike and the shearing function was elaborated. In the interest of powerful sectioning of flesh, transverse jaw action was reduced in these animals and a variety of associated changes in the skull, jaws, and jaw musculature occurred (see p. 212). In herbivores, however, which must finely macerate plant material in preparation for digestion, the molars became quadrate, transverse jaw action came to be of primary importance, and distinctive features

of the skull, jaws, and jaw musculature favoring this action developed (p. 168). The dentitions of a modern carnivore, a modern herbivore, and a primitive Mesozoic mammal are compared in Fig. 2–22.

The number of teeth of each type in the dentition is designated by the *dental formula*. This is written as the number of teeth of each kind on one side of the upper jaw over the corresponding number in one lower jaw. Such a formula is incisors 3/3, canines 1/1, premolars 4/4, molars 2/3. Because the teeth are always listed in this order, the formula may be shortened to 3/3, 1/1, 4/4, 2/3. (The skull in Fig. 2–17 has this dental formula.) The dental formula lists the teeth of only one side; therefore, the total number of teeth in the formula must be doubled to give the total number of teeth in the dentition. As an example, the arrangement for man is 2/2, 1/1,

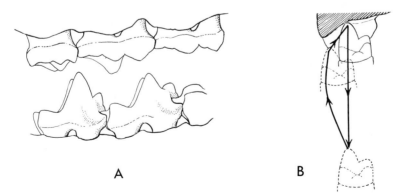

A   B

Figure 2–21. A, opposing molars of the Virginia opossum, showing the opposing cusps that steady, puncture, and crush food. B, movement of the lower teeth as they shear against the uppers. (B after Crompton and Hiiemae, 1969.)

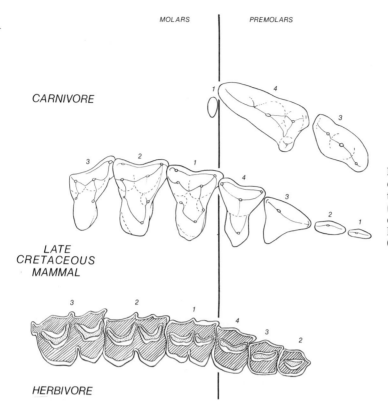

Figure 2–22. Comparisons of the occlusal surfaces of the cheek teeth of a primitive mammal (*Kennalestes*, from Upper Cretaceous), a carnivore (*Lynx*), and a herbivore (*Cervus*). (Partly after Crompton and Hiiemae, 1969.)

2/2, 3/3 × 2 = 32. The basic maximum number of teeth in placental mammals is 44 (3/3, 1/1, 4/4, 3/3), but marsupials commonly have more than this number. The number of teeth is frequently reduced, and a few placentals completely lack teeth; some specialized placentals, most notably odontocete cetaceans, have more than 44 teeth and have *homodont* dentitions (those in which all teeth are alike).

The mammalian tooth typically consists of *dentine*, a bonelike material, covered by a layer of hard, smooth *enamel*, which is largely calcium phosphate. The tooth is bound to the jaw by *cement*, and this relatively soft material may also form part of the crown of the tooth (Fig. 2–23). Most mammals have teeth that are *brachyodont*, or short-crowned, and growth ceases after the tooth is fully grown and the pulp cavity in the root closes. Many herbivores that subject their teeth to rapid wear—resulting from abrasion by silica in grass and by soil particles that adhere to plants—have *hypsodont* or high-crowned teeth (Fig. 2–23). As a further

adaptation to abrasive food, in some mammals some teeth (and in some rodents all of the teeth) retain open pulp cavities in the roots and grow continuously. These are termed *evergrowing teeth*. The roots of mammalian teeth are often divided; primitively, the upper molars have three roots and the lower molars two. Incisors and canines are single-rooted, and premo-

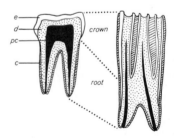

Figure 2–23. Generalized sections of mammalian teeth, showing the structure and materials. The black area is the pulp cavity (*pc*), the stippled part is dentine (*d*), the cross-hatched part is cement (*c*), and the unshaded areas are enamel (*e*). The molar on the left is similar to that of primates and is low crowned; the molar on the right is similar to that of a horse and is high crowned.

Figure 2–24. Three major types of right upper molariform teeth. Anterior is to the right; the outer edge of each tooth is toward the top. The crosshatched parts are dentine. The lophodont tooth shows an advanced stage of wear.

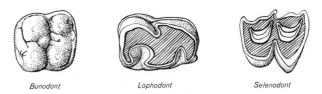

Bunodont   *Lophodont*   *Selenodont*

lars may have one or two roots. The dentitions of herbivores usually serve only two functions: the incisors, or the incisors and the canines, clip vegetation and the cheek teeth grind the food. Between these teeth there is usually a space called a *diastema* (Fig. 10–3; p. 168).

The shapes of the crowns of the molars vary in response to the demands of different diets (Fig. 2–24). In omnivores (some rodents, some carnivores, some primates, and pigs) the molars are *bunodont*; the cusps form separate, rounded hillocks that crush and grind food. In herbivores the molars may be *lophodont*, with cusps forming ridges, or may be *selenodont*, with cusps forming crescents; in these cases the teeth finely section and grind vegetation. In the dental batteries of many insectivores, bats, and carnivores are *sectorial* teeth. These have bladelike cutting edges that section food by shearing against the edges of their counterparts in the opposing jaw (as do the fourth upper premolar and the first lower molar in the dentition in Fig. 2–17).

The triangular primitive upper molar is marked by three major cusps, the *protocone*, the *paracone*, and the *metacone*, and the apex of the triangle (the proto-

cone) points inward. The lower molar has two sections, an anterior *trigonid* and a posterior *talonid*. The trigonid is triangular; the apex of the triangle, the *protoconid*, points outward, and the *paraconid* and *metaconid* form the inner edge. The talonid has two major cusps, the *hypoconid* and *entoconid*.

AXIAL SKELETON. The vertebral column has five well-differentiated sections: *cervical, thoracic, lumbar, sacral,* and *caudal.* Anterior to the sacrum, the axial skeleton is most flexible in the cervical and lumbar sections. Only the anteroposteriorly compressed thoracic vertebrae bear ribs. In some groups, such as edentates and bats, the rigidity of the ribcage is greatly enhanced by a broadening of the ribs. The first two cervicals are highly modified, the sacral vertebrae are more or less fused to support the pelvic girdle, and considerable differentiation of the vertebrae of each region is typical (Fig. 2–26). Usually from 25 to 35 presacral vertebrae are present. All mammals, with the exception of several edentates and the manatee (Sirenia), have seven cervical vertebrae.

The sternum is well developed and solidly anchors the ventral ends of the ribs, helping to form a fairly rigid ribcage. The

Figure 2–25. Basic cusp pattern of mammalian molars. A, Right upper molars; B, left lower molars. The upper pair of teeth represent the primitive cusp pattern; this was modified in some evolutionary lines by the addition of a cusp (hypocone) in the upper tooth, and the loss of a cusp (paraconid) in the lower tooth, yielding more or less quadrate teeth (lower pair) adapted to omnivorous or herbivorous diets. Abbreviations: *end,* entoconid; *hy,* hypocone; *hyd,* hypoconid; *me,* metacone; *med,* metaconid; *mel,* metaconule; *pa,* paracone; *pad,* paraconid; *pr,* protocone; *prd,* protoconid; *prl,* protoconule. (After Romer, 1966.)

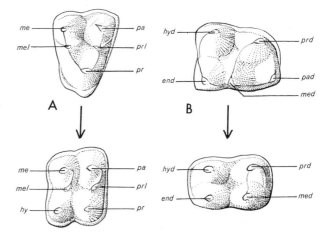

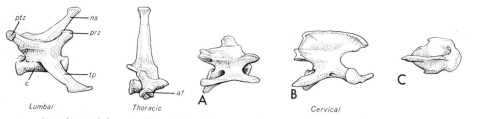

Figure 2–26. Vertebrae of the gray fox (*Urocyon cinereoargenteus*), showing the great structural variation in the parts of the vertebral column. The vertebrae are viewed from the right side; anterior is to the right. A, Fifth cervical vertebra; B, axis (second cervical); C, atlas (first cervical). Abbreviations: *af*, articular facet for the capitulum of the rib; *c*, centrum; *ns*, neural spine; *prz*, pre-zygapophysis; *ptz*, post-zygapophysis; *tp*, transverse process.

sternum is not highly variable, but in some bats departs strongly from the typical mammalian plan (Fig. 6–46, p. 115).

LIMBS AND GIRDLES. In most terrestrial mammals the main propulsive movements of the limbs are fore and aft; the toes point forward and the limbs are roughly perpendicular to the ground. In the most highly cursorial species (*cursorial* mammals are those adapted for running), the joints distal to the hip and shoulder tend to limit movement to a single plane. This allows reduction of whatever musculature does not control flexion and extension, and results in a lightening of the limbs. The mammalian pelvic girdle has a characteristic shape, with the *ilium* projecting forward and the *ischium* and *pubis* extending backward (Figs. 3–5, 3–6); these bones are solidly fused. In the shoulder girdle of placentals, the *coracoid* and *acromion* are usually reduced to small processes on the scapula, and the reptilian in-

terclavicle is gone (Fig. 2–15); the clavicle is reduced or absent in some cursorial species.

In the *manus* (hand) and *pes* (foot) of mammals there is a standard pattern of bones (Fig. 2–27), but many variations on this basic theme (some of which are described in the chapters on orders) occur among mammals with specialized types of locomotion, such as flight (bats), swimming (cetaceans and pinnipeds), or rapid running (ungulates, rabbits, some carnivores). The primitive mammalian number of digits (5), and the basic phalangeal formula of two phalanges in the thumb (pollex) and first digit of the hind limb (hallux) and three phalanges in the remaining digits (2-3-3-3-3), are retained by many mammals. Common specializations involve the loss of digits, reduction in the numbers of phalanges, or occasionally the addition of phalanges (*hyperphalangy*) as in whales and porpoises (Fig. 11–2).

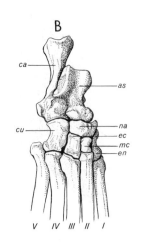

Figure 2–27. Primitive patterns of the podials (foot bones) of mammals. A, The carpus of a hedgehog (*Erinaceus europaeus*); B, the tarsus of the wolverine (*Gulo luscus*). The centrale, a carpal element that in some mammals with primitive limbs lies proximal to the trapezoid and magnum, is missing in the hedgehog. Abbreviations: *as*, astragalus; *ca*, calcaneum; *cn*, cuneiform; *cu*, cuboid; *ec*, ectocuneiform; *en*, entocuneiform; *m*, magnum; *mc*, mesocuneiform; *na*, navicular; *p*, pisiform; *r*, radius; *sc*, scapholunar (fused scaphoid and lunar); *se*, accessory sesamoid; *td*, trapezoid; *tm*, trapezium; *u*, ulna; *un*, unciform. The metacarpals and metatarsals are numbered.

MAMMALIAN ORIGINS

3

Mammals stem from an ancient reptilian lineage that arose at the start of the reptilian radiation some 300 million years ago. "The mammal-like reptiles, constituting the subclass Synapsida, were among the earliest to appear of known reptilian groups and had passed the peak of their career before the first dinosaur appeared on the earth" (Romer, 1966:173). From a late Carboniferous divergence from other reptile evolutionary lines, mammal-like reptiles became the most important reptile group in the Permian and early Triassic. (Table 3–1 lists the geological periods.) In the Triassic, however, perhaps because of competition from other reptile groups and changing climatic conditions, synapsids dwindled in importance and were nearly extinct by the close of this period. Primitive late Triassic mammals were the unspectacular descendants of the declining synapsid line. Throughout the re-

Table 3–1. GEOLOGICAL TIME SCALE FOR SPAN OF TIME SINCE LIFE BECAME ABUNDANT, SHOWING MAMMALS OR ANCESTORS OF MAMMALS TYPICAL OF EACH PERIOD

Era	Period	Estimated Time Since Beginning of Each Period (In Millions of Years)	Epoch	Mammals and Mammalian Ancestors
Cenozoic	Quaternary	2+	Recent	Modern species and subspecies; extirpation of some mammals by man.
			Pleistocene	Appearance of modern species or their antecedents; widespread extinction of large mammals.
	Tertiary	65	Pliocene	Appearance of modern genera.
			Miocene	Appearance of modern subfamilies.
			Oligocene	Appearance of modern families.
			Eocene	Appearance of modern orders.
			Paleocene	Adaptive radiation of marsupials and placentals.
Mesozoic	Cretaceous	130		Appearance of marsupials and placentals.
	Jurassic	180		Archaic mammals.
	Triassic	230		Therapsid reptiles; appearance of mammals.
Paleozoic	Permian	280		Appearance of therapsid reptiles (from which mammals evolved).
	Carboniferous	350		
	Devonian	400		
	Silurian	450		
	Ordovician	500		
	Cambrian	570		

(After Romer and Parsons, 1977.)

mainder of the Mesozoic, the "Golden Age" of reptiles, mammals persisted and continued to evolve along lines established early in synapsid history. Modest Mesozoic mammalian radiations occurred, but only after the disappearance of the ruling reptiles in the late Cretaceous did mammals begin the remarkable adaptive burst that led to their dominant position in the Cenozoic Era.

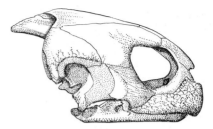

Figure 3–1. An anapsid skull (sea turtle, *Chelonia*). Note the unbroken shield of bone in the temporal region.

MAMMAL-LIKE REPTILES

GENERAL SKULL CHARACTERISTICS

The skull of mammal-like reptiles is characterized by a single opening low in the temporal part, with the postorbital and squamosal bones meeting above the opening. From a primitive reptilian skull with no temporal openings, a skull type still retained by turtles (Fig. 3–1), various patterns of perforation of the temporal part of the skull developed among early reptiles (Fig. 3–2). The openings are thought by some to have developed originally to increase the freedom for expansion of the adductor muscles of the jaw; these muscles primitively attached inside the solid temporal part of the skull. Other workers have

held that a selective advantage was gained by the reduction in the weight of the skull due to the temporal openings. According to another explanation, the stresses on the skull were greatest around the periphery of the temporal area, whereas the middle of the area was not needed for bracing of the jaw area or for muscle attachment. Therefore, the loss of bone in the middle of the temporal "shield," when solid bone in this area was no longer of adaptive importance, contributed importantly to metabolic economy by reducing the mass of bone in the skull that had to be developed and maintained (Fox, 1964).

In any case, the general trend in progressive mammal-like reptiles was clearly toward the enlargement of the temporal

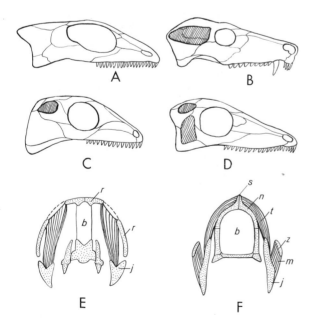

Figure 3–2. A–D, Diagrammatic views of skulls of reptiles showing the different arrangements of temporal openings. A, "Anapsid" skull (no temporal opening); B, "synapsid" skull (postorbital and squamosal meeting above opening); C, "euryapsid" skull (postorbital and squamosal meeting below the opening); D, "diapsid" skull (two temporal openings). E, F, Cross sections of skulls showing (diagrammatically) the attachments of the jaw muscles. E, Pelycosaur, with the jaw muscles originating within the remaining parts of the temporal shield (r); the sides of the braincase are cartilaginous. F, Mammal, with the jaw muscles originating on the new and completely ossified braincase (n), on the saggital crest (s), and on the zygomatic arch (z), a remnant of the original skull roof. Abbreviations: *b*, brain cavity; *j*, lower jaw; *m*, masseter muscle; *n*, new braincase formed partly by extensions of bones that originally formed the skull roof; *r*, skull roof; *t*, temporal muscles; *z*, zygomatic arch.

opening and the movement of the origins of the jaw muscles from the inner surface of the temporal shield to the braincase and to the zygomatic arch, the remnant of the lower part of the original temporal shield. In some advanced mammal-like reptiles, the postorbital bar, a vestige of the anterior part of the temporal shield, was lost.

EVOLUTION OF MAMMAL-LIKE REPTILES

The subclass Synapsida is divided into two orders, the primitive Pelycosauria, a largely Permian group, and a more advanced group, the Therapsida, of the late Permian and Triassic. The skull of *Ophiacodon* (Fig. 3–3A) illustrates several pelycosaurian features. The temporal opening is fairly small; the postorbital bar is present; the shieldlike structure of the temporal region is still evident; the teeth are not strongly heterodont; and the dentary is not greatly enlarged at the expense of other bones of the lower jaw. *Dimetrodon*, an advanced and predaceous pelycosaur, is of special interest because of the greatly elongated neural spines of the thoracic and lumbar vertebrae that formed a "sail" over the body. Skin presumably stretched between the spines. If *Dimetrodon* was able to control the blood supply to the sail, this structure might have served as an effective heat-dissipating or heat-absorbing device because of its large surface area. It may therefore represent a remarkably early and bizarre attempt at temperature regulation by pelycosaurian mammal-like reptiles.

Of more direct importance to our consideration of the origin of mammals is the order Therapsida, from which mammals arose. The therapsids are structurally diverse, and span the morphological gap between fairly primitive reptiles and animals of nearly mammalian grade. Despite the complexity introduced by a number of lines of descent and by tangential specializations, several important anatomical trends leading from the typical reptilian organization toward the mammalian pattern are apparent in therapsids. The temporal shield was reduced by an expansion of the temporal opening, and the origins of the jaw musculature were largely on the braincase and zygomatic arch; two occipital condyles replaced the single reptilian condyle. The maxillaries and palatines extended backward and toward the midline, forming a secondary palate (Fig. 3–4); the dentition became more strongly heterodont; and the dentary became progressively larger as the other (typically reptilian) jaw elements became smaller. Ribs were reduced or lost on the cervical and lumbar vertebrae; the primitive spraddle-legged limb posture was modified and the limbs tended to move beneath the body; the pectoral and pelvic girdles were strongly altered (Figs. 3–5, 3–6); and a basic pattern involving a standard phalangeal formula (2-3-3-3-3) and a simplified series of carpal and tarsal bones were established.

Without departing essentially from the

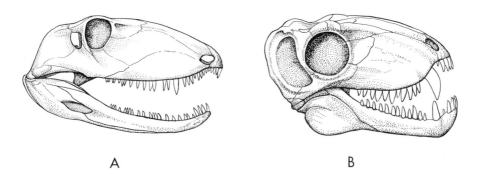

A   B

Figure 3–3. Skulls of synapsid reptiles. A, *Ophiacodon* (order Pelycosauria); B, *Phthinosuchus* (order Therapsida). (After Romer, 1966.)

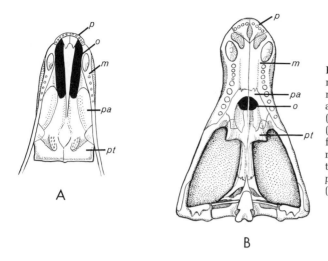

A

B

Figure 3–4. Palatal views of skulls of synapsid reptiles. A, *Scymnognathus* (order Therapsida); note that the internal nares (*o*) open into the anterior part of the mouth. B, *Cynognathus* (order Therapsida); note that the maxillaries (*m*) and palatines (*pa*) have extended medially, forming a shelf that shunts air from the external nares to near the back of the mouth. Abbreviations: *m*, maxillary; *o*, internal narial opening; *p*, premaxillary; *pa*, palatine; *pt*, pterygoid. (After Romer, 1966.)

picture presented by vertebrate paleontologists on the basis of present fossil evidence, stages in the evolution of mammals can be illustrated by considering the genera *Phthinosuchus*, *Cynognathus*, and *Probainognathus* (Figs. 3–3B, 3–7). The most primitive of these, *Phthinosuchus*, is known primarily by skull and jaw material from Permian strata in Russia, and is a therapsid reptile within the basal suborder Phthinosuchia. This type had a moderately large temporal opening and weakly heterodont dentition, but lacked such advanced features as a secondary palate. The dentary was not greatly enlarged, and the palate, as in many primitive reptiles, bore teeth. The

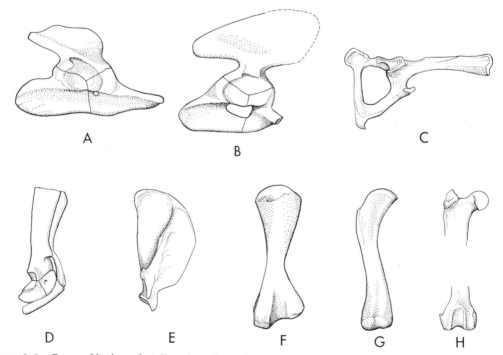

Figure 3–5. Bones of limbs and girdles of reptiles and mammals. Lateral view of left side of pelvis in: A, *Dimetrodon* (order Pelycosauria); B, *Cynognathus* (order Therapsida); C, *Erinaceus* (order Insectivora). Lateral views of the right scapulae of: D, *Kannemeyeria* (order Therapsida); E, *Lynx rufus* (order Carnivora). Anterior views of the left femora of: F, *Ophiacodon* (order Pelycosauria); G, a cynodont (order Therapsida); H, *Lynx rufus* (order Carnivora). (All drawings of reptiles after Romer, 1966.)

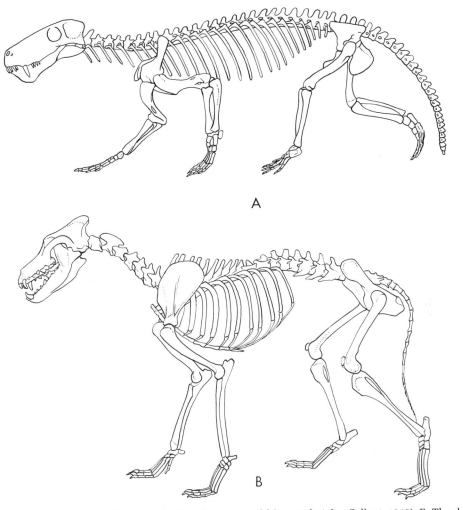

Figure 3-6.   A, The skeleton of *Lycaenops*, a Permian mammal-like reptile (after Colbert, 1949). B, The skeleton of the dire wolf (*Canis dirus*), a "modern" mammal (after Stock, 1949, courtesy of the Los Angeles County Museum of Natural History).

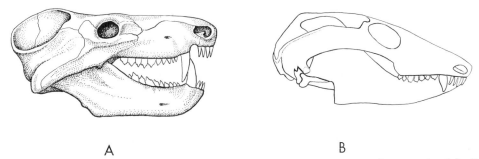

Figure 3-7.   Skulls of therapsid reptiles. A, *Cynognathus* (after Romer, 1966); B, *Probainognathus* (after Romer, A.S.: Cynodont reptile with incipient mammalian jaw articulation. *Science*, 166:881–882. Copyright 1969 by The American Association for the Advancement of Science).

limbs of this genus are poorly known, but probably the trend toward modification of the reptilian limb posture was under way.

The phthinosuchians gave rise to more advanced mammal-like reptiles included in the suborder Theriodontia, many members of which were progressive carnivores. This group, or more specifically the theriodont infraorder Cynodontia, is seemingly the basal stock from which mammals arose. Some cynodonts were over 2 m in length. The cynodont *Cynognathus*, known from the lower and middle Triassic beds of Africa, has many advanced and mammal-like features, and illustrates an early stage in the transition between reptiles and mammals. *Cynognathus* was a slender and, for the Triassic, a highly cursorial quadruped. The skull of this reptile retains a small pineal opening, a feature still present in many modern reptiles, and the postorbital bar is broad. In addition, however, a number of advanced features are evident, perhaps the most important being the secondary palate (Fig. 3–4B).

This structure is formed by an inward growth of the premaxillary, maxillary, and palatine bones, and is a plate lying beneath the original roof of the mouth and braced along the midline by a bar formed by the fused vomers. This new palate forms a chamber that shunts air from the external nares, at the front of the snout, to the internal narial openings, toward the back of the mouth. In homeotherms, such a bypass is of critical importance in allowing the animal to continue to breathe and support a high metabolic rate while masticating a mouthful of food. Another cynodont feature may have been related to homeothermy. The reduction of lumbar ribs, and the retention of a sturdy thoracic rib cage, may have been correlated with the development of a muscular diaphragm, which in turn was associated with the efficient respiration necessary for homeothermy.

Additional advanced features include a large temporal opening, a sagittal crest from which the temporal muscles must partly have originated, two occipital con-

dyles rather than the usual single reptilian condyle, and greatly enlarged dentaries in the lower jaw. The dentition is heterodont, and the cheek teeth, as in most mammals, are multicusped. Although the vertebral column is still primitive and ribs occur on all but the caudal vertebrae, the reptilian outward splaying of the legs is reduced and the limbs moved more nearly beneath the body. Correlated with the change in limb posture is an alteration of the structure of the pelvis: the ilium is shifted forward and the pubis and ischium are moved backward (Fig. 3–5). The soft anatomy and the physiology of *Cynognathus* are, of course, not known, but a trend toward homeothermy may have been well advanced in this genus and in cynodonts in general. The slender limbs of *Cynognathus* suggest that this was an active creature, and the well-developed palate reflects a need for continuous respiration during feeding, a feature critical to the maintenance of a fairly high metabolic rate.

An overly simplistic approach has frequently been taken by some in describing the differences between "reptilian" and "mammalian" limb postures. The reptilian posture has been characterized as spraddle-legged, with the humerus and femur directed horizontally, whereas mammalian limbs have been described as moving directly fore and aft and being positioned nearly vertically beneath the body. Actually, this latter posture is typical only of cursorial mammals. Jenkins (1971) found that during locomotion in a group of noncursorial species the humerus and femur functioned in postures more horizontal than vertical and at oblique angles relative to the parasagittal plane. The studies of Jenkins further demonstrate that the limb postures of terrestrial mammals are extremely diverse. Certainly a trend toward a vertical limb posture can be detected in cynodont reptiles and in early mammals, but the stereotyped picture of the vertical limb posture shared by all terrestrial mammals should be abandoned.

A handy and much used landmark in

reptilian-mammalian evolution is the structure of the jaw articulation. In reptiles, this articulation is between the quadrate bone of the skull and the articular bone of the lower jaw, but typically in mammals the squamosal and dentary bones form the articulation. A widely accepted character used in the recognition of early mammals, then, is the dependence on the squamosal-dentary articulation, with the quadrate and articular forming no part of the joint. Fossil evidence, much of which has been assembled in recent years, illustrates a succession of stages in the abandonment of the quadrate-articular jaw articulation. As might be expected, the more complete the fossil record, the more arbitrary any criterion for separating advanced mammal-like reptiles and early mammals becomes, and at present some Late Triassic fossils that have dentary-squamosal and articular-quadrate articulations side by side are regarded as mammalian. Barghusen and Hopson (1970) pointed out that a dentary-squamosal jaw articulation evolved more than once in cynodont reptiles, and suggested that additional characters be found with which to separate early mammals and their reptilian ancestors. The supplementary use of dental features in separating these animals has been proposed by Hopson and Crompton (1969).

The two mammal-like reptiles we have discussed, *Phthinosuchus* and *Cynognathus*, retain the reptilian type of articulation, but two more advanced genera, *Probainognathus* from the Middle Triassic beds in Argentina and *Diarthrognathus* from Late Triassic sandstones of South Africa, are examples of reptiles intermediate in jaw structure (Romer, 1969). Although neither is on the main evolutionary line from mammal-like reptiles to early mammals, they illustrate possible stages in this sequence. In *Diarthrognathus*, the squamosal and dentary are in contact, but a quadrate-articular joint is also retained. In *Probainognathus*, however, although a functional articulation between the quadrate (which is small and insecurely attached to the squamosal) and the reduced articular is still retained, the structure of the dentary and a depression in the squamosal that resembles an incipient glenoid fossa are clear evidence that a functional squamosal-dentary joint was established. Of particular interest is a series of contacts between the articular, quadrate, and stapes bones in *Probainognathus* that roughly foreshadow the articulations between the mammalian bones to which they gave rise: the malleus, incus, and stapes, respectively. These bones in mammals, of course, no longer function in connection with jaw support, but are situated in the middle ear and serve to transmit vibrations mechanically from the tympanic membrane to the oval window of the inner ear.

EARLY MAMMALS

For vertebrate paleontologists, much fascinating work lies ahead. The fossil record remains sadly incomplete, although field work in many parts of the world is adding to our knowledge of the evolution of mammals. The late Triassic record is sparse: Late Triassic mammals are known from Wales, continental Europe, South Africa, and China. The Jurassic is represented by fossils from North America, Europe, and Africa. Even for the Cretaceous the record is spotty: no mammals from this period are available from India, Africa, Australia, or Antarctica. A definitive outline of the early evolution of mammals is thus impossible to frame; despite the difficulties, a picture of at least the broad aspects of early mammalian evolution is emerging.

Some characteristic mammalian features of the jaws and dentition developed in relation to major structural-functional trends traceable to cynodont reptiles. In primitive therapsid reptiles, primarily the front teeth were used; the incisors and canines were powerfully built and the cheek teeth were relatively weak. The reverse was true in advanced cynodonts and early mammals, in which the cheek teeth were the more robust series. The dentary of cynodonts became increasingly larger as the jaw musculature became more massive, but excessive stresses on the jaw articulation were avoided by the development of complex jaw musculature that concentrated power in the bite of the postcanine teeth. Accompanying these muscular specializations was progressively greater control of the precision and breadth of movement of the lower jaw; in early mammals, some anteroposterior movement of the lower jaw was possible, together with transverse movement. In mammals, chewing occurs on one side of the jaw at a time. Delicate neuromuscular control of the lower jaw precludes damage to opposing cusps because of malocclusion, and precise matching of shearing planes of the cheek teeth (Fig. 3–8A) provides for increased efficiency (reduced energy expenditure) in chewing. These specializations, which were present in Late Triassic mammals, seemingly set the stage for the rapid evolution of complex molar cusp patterns (Crompton, 1974:434).

The following functional features, regarded by Crompton and Jenkins (1973) as basic to a consideration of the reptile-mammal transition and to a characterization of Late Triassic mammals, provide an appropriate summation of our discussion: (1) In mammals the lower jaw on the side involved in chewing follows a triangular "orbit" as viewed from the front (Fig. 2–19B). (2) Precise occlusion of matching shearing facets on upper and lower molars is provided for by the consistent relative positions of upper and lower molars. (3) The proper occlusal relationship between each molar and the two molars with which it occludes in the opposite jaw is ensured by the front-to-rear, serial eruption pattern

of these teeth. (4) The dentition is diphyodont. (5) Accurate alignment of the lower molars is provided for by the tongue-in-groove fit of the anterior and posterior surfaces of adjacent teeth (Fig. 3–8B). (In some advanced therapsid reptiles that were off the main evolutionary line leading to mammals, such an alignment device was present in the upper as well as the lower cheek teeth.) (6) There is a dentary-squamosal jaw articulation, although the mammalian taxa Morganucodontidae, Kuehneotheriidae and Docodonta retained a functional quadrate-articular joint.

As the fossil record of Late Triassic mammals improves, we begin to glimpse stages in the evolution of the mammalian jaw articulation. As mentioned, in some advanced therapsid reptiles the reptilian and the mammalian articulations occurred side by side. Probably both in advanced cynodonts and the first mammals, some of the small postdentary bones served to transmit sound. A fascinating possibility, and certainly a reasonable one, is that the reduction of the postdentary bones was associated primarily with selection for more efficient sound transmission by these bones rather than with enlargement of the dentary solely in the interest of more effective bracing of the jaw point (Hopson, 1966; Crompton, 1972). As the postdentary bones became smaller, they became more sensitive sound transmitters, and the dentary became progressively larger in association with the increasing size and complexity of the jaw muscles. Of the postdentary bones, the quadrate, articular, and angular were seemingly of greatest importance in sound transmission, and when freed from the lower jaw they became part

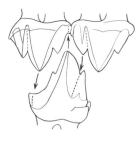

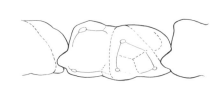

Figure 3–8. A, Shearing planes of opposing molars of a primitive, Late Triassic mammal; the shearing surfaces are outlined (after Crompton, 1974). B, Occlusal view of the lower molars of an opposum (*Didelphis virginiana*) showing the tongue-in-groove fit of the anterior and posterior surfaces of adjacent teeth.

A   B

of the ear apparatus (the incus, malleus, and tympanic ring, respectively).

The postcranial skeleton of Late Triassic mammals is known from the work of Jenkins and Parrington (1976) on fossils representing three genera. These mammals had a specialized atlas-axis complex, inherited from cynodont ancestors, that allowed freedom of rotary movement of the head. In addition, certain distinctive postcranial features first appeared in these mammals. The rather upright mammalian posture of the neck was present, and the foramina in the cervical vertebrae, through which nerves contributing to the brachial plexus passed, were enlarged, providing evidence of improvement in the neuromuscular control of the forelimbs. The recurved claws and the apparent ability of the hallux to move independently from the other digits are suggestive of climbing ability.

As presently known, Late Triassic mammals were small, with a head and body length of about 10 cm; they probably weighed some 20 to 30 gm. In general proportions they resembled a modern shrew (see Fig. 3–9), probably ate largely insects, and may have been partly arboreal.

At least briefly, we should broaden our view. Although of great interest in connection with our consideration of their evolution, these early mammals were but insignificant members of the Late Triassic vertebrate faunas, which were becoming increasingly dominated by perhaps the most spectacular vertebrates of all time, the dinosaurs. One cannot help but wonder how different would have been the sweep of vertebrate evolution, and how altered would be the face of the earth today, if the tiny Late Triassic mammals

had proved especially vulnerable to some contemporary reptilian predator and had relinquished the scene completely to the reptiles.

Current evidence points to a monophyletic origin of mammals from cynodont reptiles (Hopson and Crompton, 1969), but by Late Triassic mammals had diverged into three groups (Fig. 3–10). From one of these, the Morganucodontidae, which had basically triconodont molars (Fig. 3–11A) evolved nontherian mammals (triconodonts, docodonts, multituberculates, and perhaps the monotremes). From the second stock, the Kuehneotheriidae, with roughly triangular molars (Fig. 3–11B), arose therian mammals (symmetrodonts, pantotheres, marsupials, and eutherians). The third group, the Haramiyidae, is of uncertain affinities, but appears in the fossil record before the other two families and may have been the lineage from which multituberculates arose. Associated with differences in tooth structure between these stocks were differences in jaw action: the nontherian descendants of the Morganucodontidae retained a largely vertical jaw action, whereas transverse jaw action became progressively more important in the therian descendants of the Kuehneotheriidae. (A classification of the taxa listed in this paragraph is given in Table 3–2). Most of the Jurassic groups of mammals were seemingly experimental lines that disappeared before the end of the Mesozoic; only the monotremes (or at least vestiges of this group), marsupials, and eutherians survive today.

Among the prototherians, one of the oldest and most primitive groups is the order Triconodonta, known from the late

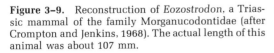

Figure 3–9. Reconstruction of *Eozostrodon*, a Triassic mammal of the family Morganucodontidae (after Crompton and Jenkins, 1968). The actual length of this animal was about 107 mm.

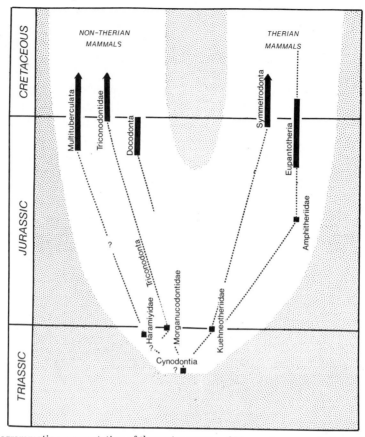

Figure 3–10. Diagrammatic representation of the major groups of Mesozoic mammals (modified from Crompton and Jenkins, 1973). Solid bars indicate the extent of the fossil record. The taxa shown in this figure appear in the classification outline in Table 3–2.

Triassic to the Lower Cretaceous. Tricono-donts were predaceous; the largest genus was nearly the size of a house cat. The dentition is heterodont, with as many as 14 teeth in a dentary. The canines are large, and typically the molars have three cusps arranged in a front-to-back row (Fig. 3–12A, B).

The order Docodonta is represented by several primitive genera known from the

Late Jurassic. Members of this group have roughly quadrate teeth, with the cusps not aligned anteroposteriorly (Fig. 3–12C, D). The braincase and postcranial skeleton seem to be on a reptilian level of development.

Among nontherian mammals, the order Multituberculata was remarkable in several ways. These were the first mammalian herbivores, and although they disappeared

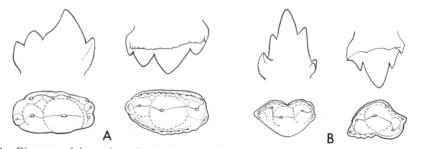

Figure 3–11. Diagrams of the molars of Triassic mammals: A, *Eozostrodon* (Morganucodontidae); B, *Kuehneotherium* (Kuehneotheriidae). In each case the lower molar is on the left and the upper molar is on the right. (Modified from Crompton, 1974.)

Table 3-2. PARTIAL CLASSIFICATION OF MAMMALS

Class Mammalia
  Subclass Prototheria
    Infraclass Eotheria
      Order Triconodonta
        Family Morganucodontidae
      Order Docodonta
        Family Docodontidae
    Infraclass Ornithodelphia
      Order Monotremata
        Family Tachyglossidae
        Family Ornithorhynchidae
    Infraclass Allotheria
      Order Multituberculata
        (?) Family Haramiyidae

  Subclass Theria
    Infraclass Trituberculata
      Order Symmetrodonta
        Family Kuehneotheriidae
      Order Pantotheria
    Infraclass Metatheria
      Order Marsupialia
    Infraclass Eutheria (including all of the orders of "placental" mammals)

(Partly after Crompton and Jenkins, 1973; and Hopson, 1970.)

in the early Tertiary and left no descendants, they were highly successful. Multituberculates appear first in the Late Jurassic, and their fossil record spans some 100 million years. These animals were widespread in the Old and New Worlds. They resembled rodents in many ways, and were clearly able to adapt in the Cretaceous to the expanding dominance of angiosperms. The lower jaw was short, deep, and strongly built, and clearly provided attachment for powerful jaw muscles; there were two incisors above and two below, and a diastema was present in front of the premolars (Fig. 3–13A). Typical of multituberculates were upper molars with three parallel rows of cuspules and remarkably specialized, bladelike, posterior lower premolars (Fig. 3–13B, C); these cheek teeth seemingly functioned to grind and section plant material.

Multituberculates persistently retained several primitive features. The olfactory lobes of the brain were large, the cerebrum was smooth, the incisive foramina were large, and the cochlea of the ear was very small as compared to that of contemporary placentals of comparable size. Considered together, these features suggest a dull-witted animal still strongly dependent on the sense of smell. Several postcranial characters indicate that multituberculates lacked, or had poorly developed, arboreal ability.

From these features, a picture emerges of a rather primitive mammal that could not remain long in competition with marsupials or placentals, but the fossil record indicates otherwise. For over 70 million years multituberculates coexisted with eutherian mammals, and for some 30 million years multituberculates and a variety of herbivorous eutherians were in competition. During the first half of the Paleocene, the multituberculates underwent an adaptive radiation, but they declined rapidly in numbers and diversity in the Late Paleocene. The decline of the multituberculates spanned some 20 million years.

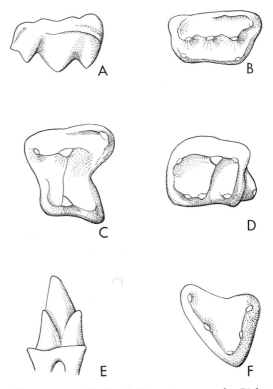

Figure 3–12. Teeth of Mesozoic mammals. Right upper molar of a triconodont (order Triconodonta): lateral view (A) and occlusal view (B). Occlusal views of right upper molar (C) and left lower molar (D) of *Docodon* (order Docodonta). Lateral view of lower left molar (E) and occlusal view of upper right molar (F) of a symmetrodont (order Symmetrodonta). (Redrawn from Romer, 1966, after Simpson, 1928.)

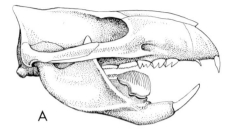

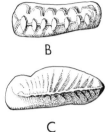

Figure 3–13. *Ptilodus* (order Multituberculata). Skull (A) and occlusal views of upper (B) and lower (C) cheek teeth. (Redrawn from Romer, 1966, after Simpson, 1937.)

The competition probably began with condylarths (ancestors of ungulates) in the Late Cretaceous, intensified when primates became common in the Paleocene, and became overwhelming in the Eocene when rodents became ubiquitous (Van Valen and Sloan, 1966). Multituberculates appear last in the Early Oligocene fossil record of Wyoming.

The therian order Symmetrodonta is known from the Late Triassic to the Upper Cretaceous. Symmetrodonts are among the oldest mammals and were probably predaceous. The molar crown pattern is marked by three fairly symmetrically situated cusps (Fig. 3–12E, F).

Because it is generally accepted that eutherian mammals evolved from the order Pantotheria, this group is of particular interest. Pantotheres occur mainly in late Jurassic beds, and have several features traceable to eutherian mammals. In the pantothere, for example, the profile of the ventral border of the dentary is interrupted by an angular process similar to that found in eutherians. The posterior "heel" on the pantothere lower molar (Fig. 3–14B) is represented in eutherians by the talonid, the posterior section of the lower molar (Fig. 3–14D). In eutherians, the talonid has a basin into which the protocone of the upper molar fits. The shape of the anterior trigonid section of the pantothere lower molar also resembles the comparable part of this tooth in some eutherians, and the triangular upper molar of pantotheres roughly resembles that of some primitive eutherians (Fig. 3–14A, C). Seemingly, the medial cusp of the upper molars in pantotheres is the paracone of eutherians.

CRETACEOUS MAMMALS

A broad view of the Cretaceous—a period of great biotic change—provides a background against which the late Mesozoic evolution of mammals can be viewed. The Late Jurassic was a time of considerable interchange of biotas between continents, as indicated by the occurrence in western Europe, East Africa, and western North America of identical or closely related species of reptiles (Colbert, 1973:175), and by intercontinental similarities between floras (Vachrameev and Akhmet'yev, 1972:419). After the earliest Cretaceous, however, dispersal between

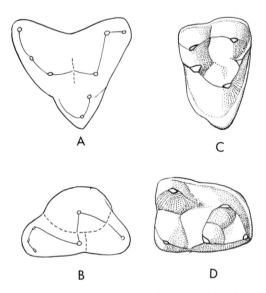

Figure 3–14. Right upper molar (A) and left lower molar (B) of *Aegialodon*, an early Cretaceous eupantothere (order Pantotheria). The upper molar is a hypothetical reconstruction. Comparable teeth (C and D) of a primitive Eocene eutherian mammal *Omomys*, a tarsier-like primate. (C and D after Romer, 1966.)

continents became sharply restricted. In the New World, a series of transgressions of a seaway from the Arctic Ocean to the Gulf of Mexico divided North America for much of the Cretaceous into two separate centers for the evolution of terrestrial plants and animals. In Eurasia, also, the dispersal of land animals was restricted: Europe was essentially an archipelago of islands during the Cretaceous, and the "Turgai Strait" Seaway separated the land faunas of Europe and Siberia. Angiosperms probably evolved in the Early Cretaceous (Muller, 1970; Krassilov, 1973), and certainly underwent a major adaptive radiation then, an event that clearly had a tremendous effect on the evolution of land faunas. Coadaptive evolution between angiosperm flowers and insects, for example, fostered a Cretaceous insect radiation.

Most dramatic were Cretaceous changes in the fortunes of the dinosaurs: throughout the Jurassic and Early Cretaceous they were diverse and abundant and dominated the terrestrial scene; in Late Cretaceous, several herbivorous groups—the ankylosaurs, ceratopsians, and hadrosaurians—diversified in association with the increasing importance of angiosperms and the decline of the gymnosperms; and by the close of the Cretaceous, dinosaurs were gone. Probably of great importance in influencing faunal changes late in the Cretaceous were climatic shifts tending toward reduced equability (greater seasonal and daily temperature extremes), fostered at least in part by the slight elevation of continents and the disappearance of most of the extensive continental seas (Axelrod and Bailey, 1968).

From the basal Late Triassic family Kuehneotheriidae several families evolved in the Jurassic, and *Aegialodon*, a descendant of one of these families, is known from the Early Cretaceous of England. This genus has been described on the basis of a single, worn, lower molar (Fig. 3–14B), the structure of which strongly suggests that *Aegialodon* belonged to the lineage that gave rise to marsupials and eutherians (see p. 39 for the classification of mammals). These "modern" mammals appear as two divergent evolutionary lines in the Early Cretaceous.

Because throughout much of the early Cretaceous intercontinental movement by land dwellers was barred by oceans and seaways, populations of mammals on different continents evolved in isolation under differing environmental pressures. This isolation of "premarsupial" and "preplacental" stocks may well have favored their differentiation; each group seemingly faced some comparable adaptive problems, but, as in the case of reproduction, to some of these problems each group developed unique solutions. Present evidence suggests that marsupials arose in North America in the Early Cretaceous (Slaughter, 1968), and until very late in the Cretaceous the North American mammalian fauna was composed largely of marsupials. Eutherians may have appeared first in the Early Cretaceous of Asia (Kielan-Jaworowska, 1975); no marsupials have yet been found there. The eutherians that appear in Late Cretaceous deposits of North America were perhaps derived from an Asian form. Kielan-Jaworowska hypothesizes that marsupials were unable to move from North America to Asia in the Late Cretaceous, but eutherians reached North America from Asia at this time. This hypothetical migration could have resulted from the rafting of eutherians from Asia to North America across the marine strait separating the two continents on currents that would not allow passage in the opposite direction (Kielan-Jaworowska, 1975).

Whereas the restricted intercontinental movement by terrestrial vertebrates in the Early Cretaceous clearly provided conditions favorable to separate mammalian adaptive radiations, this isolation cannot explain the radiations themselves. These radiations must have involved the exploitation of some newly available resources.

Considerable recent literature, including considerations of the problem by Clemens (1970:375) and Lillegraven (1974:277, 278), points to the overriding importance of the Early Cretaceous radiation of flowering plants (class Angiospermae). The

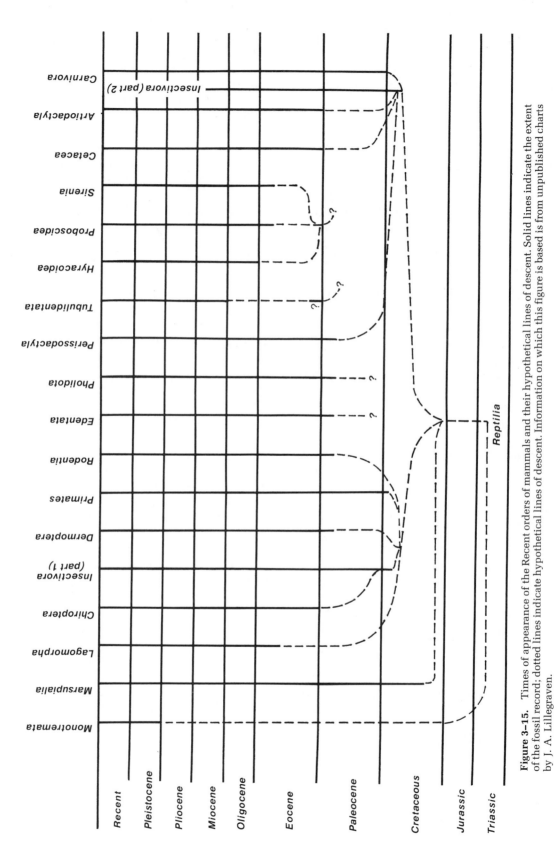

Figure 3–15. Times of appearance of the Recent orders of mammals and their hypothetical lines of descent. Solid lines indicate the extent of the fossil record; dotted lines indicate hypothetical lines of descent. Information on which this figure is based is from unpublished charts by J. A. Lillegraven.

seeds of angiosperms develop within a fruit, which is frequently edible and nutritious. Angiosperm fruits and seeds are eaten today by many mammals. Soft fruits can be dealt with effectively by many types of dentitions, and were probably of importance to Cretaceous mammals. Angiosperm pollination is made extremely efficient by the attraction of insects to flowers by nectar, and both nectar and pollen are nutritious. The Lepidoptera (moths and butterflies) appeared in the Cretaceous (MacKay, 1970), probably in response to the food offered by the flowers and leaves of angiosperms. The Isoptera (termites) also appeared at this time. Both of these insect groups are still of great importance as food for a variety of mammals, especially in tropical areas. Perhaps with widespread tropical conditions in the Cretaceous, termites and moths became equally important foods then. Mammals have a long history of insectivory: from the Late Triassic through the Jurassic, they were probably basically insectivorous. Equipped as they were for eating insects, mammals must have profited tremendously from the burgeoning populations of insects in the Cretaceous.

An interesting addendum to the story of Cretaceous mammals, and a further indication that evolutionary patterns are seldom simple, centers on the doubtful systematic positions of some fossils from Late Cretaceous deposits in North America and Asia. *Deltatheridium* (Fig. 6–2B) from Mongo-

lia, for example, has a distinctive complex of cranial and dental features and is accordingly regarded by some (Butler and Kielan-Jaworowska, 1973) as neither marsupial nor placental but as a therian of metatherian-eutherian grade. Lillegraven (1974:265) postulates that the marsupial-placental dichotomy was not yet entirely clear-cut in the Cretaceous; equivocal mammals such as *Deltatheridium* were perhaps representatives of unsuccessful evolutionary lines that were replaced before the Cenozoic by marsupials or eutherians.

For mammals, then, the Mesozoic was a time of experimentation. Natural selection, partly in the forms of predation by an imposing array of reptilian carnivores and probably some avian predators, and by competition with reptiles, birds, and other mammals, "guided" many changes in mammalian structure and function. Physiological changes not indicated by fossil material were probably of critical importance. Structural plans evolved and were workable for differing lengths of time, and many side branches from the mammalian evolutionary line proved sterile. But during the Mesozoic time of evolutionary trial and error, the basic mammalian structural plan was perfected. The extinction of the dinosaurs at the end of the Cretaceous, and the prevalence at this time of climatic conditions making homeothermy highly advantageous, set the stage for the remarkable adaptive burst of mammals in the early Cenozoic (Fig. 3–15).

4

CLASSIFICATION OF MAMMALS

Despite their remarkable success, and perhaps largely because of their greater size and their consequent inability to exploit large numbers of restricted ecological niches, mammals are much less diverse than are most invertebrate groups. Roughly 1000 genera and some 4060 species of mammals are currently recognized. These figures are insignificant by comparison with those for invertebrates. There are, for example, an estimated 750,000 species of insects, 30,000 of protozoans, and 107,000 of mollusks.

Chapters 5 through 14 consider the orders and families of mammals that are listed in Table 4–1. For each order and family, such features as size of the group, present geographic distribution, time of appearance in the fossil record, structural characters, and brief life history notes are given. Whenever appropriate, morphology is related to function. Specializations for certain modes of life or for specific locomotor abilities are frequently discussed in the sections considering the animals that possess these adaptations. Thus, adaptations for aquatic life are discussed in the chapters on cetaceans and on pinniped carnivores, adaptations for flight are con-

sidered in the section on bats, and cursorial adaptations are examined in the chapter on ungulates.

I devote considerable attention to the orders and families of mammals not because I wish to put primary stress on the taxonomic aspect of mammalogy, but rather as an attempt to provide the student with sufficient information (which, if it is not in his head, is at least at his fingertips) on the various kinds of mammals to make the discussions of the biology of mammals meaningful. The edge of a student's interest is often dulled if he must deal with information about animals with which he is completely unfamiliar. It seems pointless to me to discuss population cycles of microtines, for example, if a student has only a vague idea of what a microtine is. In the chapters on orders, then, I have tried to provide enough information, in the forms of both descriptions and illustrations, to enable the student to gain some familiarity with the orders, families, and, in some cases, subfamilies of mammals. This should serve as a background for the chapters on selected aspects of the biology of mammals.

Table 4–1. CLASSIFICATION OF RECENT MAMMALS
(Numbers of species are approximations. Geographic origins: Af, Africa; Aust, Australia; Eur, Europe; Mad, Madagascar; NA, North America; SA, South America; SE Asia, Southeast Asia.)

Classification	Common Names
Subclass Prototheria	
Order Monotremata (3 species)	
Fam. Tachyglossidae (Aust)	Spiny anteaters
Ornithorhynchidae (Aust)	Duck-billed platypuses
Subclass Theria	
Infraclass Metatheria (marsupials)	
Order Marsupialia (242 species)	
Fam. Didelphidae (NA)	Opossums
Dasyuridae (Aust)	Marsupial "mice," "rats," and "carnivores"
Notoryctidae (?)	Marsupial "moles"
Peramelidae (Aust)	Bandicoots
Caenolestidae (SA)	Rat opossums
Phalangeridae (Aust)	Cuscuses and phalangers
Petauridae (Aust)	Gliders
Burramyidae (Aust)	Pigmy possums
Tarsipedidae (Aust)	Honey possums
Phascolarctidae (Aust)	Koalas
Vombatidae (Aust)	Wombats
Macropodidae (Aust)	Kangaroos and wallabies
Infraclass Eutheria (placentals)	
Order Insectivora (406 species)	
Fam. Erinaceidae (NA)	Hedgehogs
Talpidae (Eur)	Moles
Tenrecidae (Af)	Tenrecs
Chrysochloridae (Af)	Golden moles
Solenodontidae (?)	Solenodons
Soricidae (Eur)	Shrews
Macroscelididae (Af)	Elephant shrews
Tupaiidae (SE Asia)	Tree shrews
Order Dermoptera (2 species)	
Fam. Cynocephalidae (?)	Flying lemurs
Order Chiroptera (853 species)	
Fam. Pteropodidae (Eur)	Old World fruit-eating bats
Rhinopomatidae (?)	Mouse-tailed bats
Emballonuridae (Eur)	Sac-winged bats
Craseonycteridae (SE Asia)	
Noctilionidae (?)	Bulldog bats
Nycteridae (?)	Hollow-faced bats
Megadermatidae (Eur)	False vampire bats
Rhinolophidae (Eur)	Horseshoe bats
Phyllostomatidae (?)	Leaf-nosed bats
Mormoopidae (?)	Moustached bats
Desmodontidae (SA)	Vampire bats
Natalidae (SA)	Funnel-eared bats
Furipteridae (?)	Smoky bats
Thyropteridae (?)	Disc-winged bats
Myzapodidae (?)	Sucker-footed bats
Vespertilionidae (Eur or NA)	Common bats
Mystacinidae (?)	Short-tailed bats
Molossidae (Eur)	Free-tailed bats
Order Primates (166 species)	
Fam. Lemuridae (Mad)	Lemurs
Indridae (Mad)	Indrid lemurs
Daubentoniidae (Mad)	Aye-ayes
Lorisidae (Asia)	Lorises
Galagidae (Af)	Galagos
Tarsiidae (Eur)	Tarsiers
Cebidae (SA)	New World monkeys
Callithricidae (?)	Marmosets
Cercopithecidae (Af)	Old World monkeys

Table continued on following page

Table 4–1. CLASSIFICATION OF RECENT MAMMALS (*Continued*)

Classification	Common Names
Pongidae (Af)	Great apes and gibbons
Hominidae (Af)	Man
Order Edentata (31 species)	
Fam. Myrmecophagidae (SA)	Anteaters
Bradypodidae (?)	Tree sloths
Dasypodidae (SA)	Armadillos
Order Pholidota (8 species)	
Fam. Manidae (Eur)	Scaly anteaters
Order Lagomorpha (63 species)	
Fam. Ochotonidae (Eurasia)	Pikas
Leporidae (NA)	Rabbits and hares
Order Rodentia (1690 species)	
Fam. Aplodontidae (NA)	Mountain beavers
Octodontidae (SA)	Octodonts, tuco-tucos
Echimyidae (SA)	Spiny rats
Abrocomidae (SA)	Chinchilla rats
Capromyidae (SA)	Hutias and nutrias
Chinchillidae (SA)	Chinchillas and viscachas
Dasyproctidae (SA)	Agoutis
Cuniculidae (SA)	Pacas
Heptaxodontidae (SA)	
Dinomyidae (SA)	Pacaranas
Caviidae (SA)	Guinea pigs, Patagonian ''hares''
Hydrochoeridae (SA)	Capybaras
Erethizontidae (SA)	New World porcupines
Cricetidae (?)	New World rats, mice; hamsters, muskrats, gerbils
Muridae (Eurasia)	Old World rats and mice
Geomyidae (NA)	Pocket gophers
Heteromyidae (NA)	Kangaroo rats and pocket mice
Dipodidae (Asia)	Jerboas
Zapodidae (Eur)	Jumping mice
Spalacidae (Eur)	Mole rats
Rhizomyidae (Eur)	Bamboo rats
Gliridae (Eur)	Dormice
Seleviniidae (?)	Dzhalmans
Sciuridae (Eur or NA)	Squirrels and marmots
Castoridae (NA)	Beavers
Ctenodactylidae (SA)	Gundis
Anomaluridae (?)	Scaly-tailed squirrels
Pedetidae (Af)	Springhaases
Hystricidae (Eur)	Old World porcupines
Thryonomyidae (Af)	Cane rats
Petromyidae (Af)	Dassie rats
Bathyergidae (Af)	Mole rats
Order Mysticeti (10 species)	
Fam. Balaenidae	Right whales
Eschrichtiidae	Gray whales
Balaenopteridae	Rorquals
Order Odontoceti (74 species)	
Fam. Ziphiidae	Beaked whales
Monodontidae	Norwhals and belugas
Physeteridae	Sperm whales
Platanistidae	River dolphins
Stenidae	
Phocoenidae	Porpoises
Delphinidae	Ocean dolphins
Order Carnivora (284 species)	
Fam. Canidae (NA or Eur)	Wolves, foxes, and jackals
Ursidae (Eur)	Bears
Procyonidae (NA)	Raccoons, ring-tailed cats, etc.
Mustelidae (?)	Skunks, badgers, weasels, otters, wolverines

Table 4–1. CLASSIFICATION OF RECENT MAMMALS (*Continued*)

Classification	Common Names
Viverridae (Eur)	Civets, genets, and mongooses
Hyaenidae (Asia)	Hyenas, aardwolves
Felidae (NA or Eur)	Cats
Otariidae	Sea lions and fur seals
Odobenidae	Walruses
Phocidae	Earless seals
Order Tubulidentata (1 species)	
Fam. Orycteropodidae (Eur or Asia)	Aardvarks
Order Proboscidea (2 species)	
Fam. Elephantidae (Asia)	Elephants
Order Hyracoidea (11 species)	
Fam. Procaviidae (Af)	Hyraxes
Order Sirenia (5 species)	
Fam. Dugongidae	Dugongs and sea cows
Trichechidae	Manatees
Order Perissodactyla (16 species)	
Fam. Equidae (NA)	Horses, asses, zebras
Tapiridae (Eur)	Tapirs
Rhinocerotidae (Eur)	Rhinos
Order Artiodactyla (171 species)	
Fam. Suidae (Eur)	Swine
Tayassuidae (Eur)	Javelinas, peccaries
Hippopotomidae (Asia)	Hippos
Camelidae (NA)	Camels, llamas
Tragulidae (Eur)	Chevrotains
Cervidae (Asia)	Deer, elk, moose, caribou, etc.
Giraffidae (Asia)	Giraffes, okapis
Bovidae (Eur)	Bison, antelope, gazelles, sheep, goats, cattle, etc.

5

NONEUTHERIAN MAMMALS: MONOTREMES AND MARSUPIALS

Monotremes and marsupials can conveniently be considered apart from the rest of the mammals. Monotremes and marsupials are primitive in a variety of ways, and both have a reproductive pattern different from that of other mammals: monotremes lay eggs; marsupials bear tiny and poorly developed young, and most have a choriovitelline placenta that differs from the chorioallantoic placenta of "placental" mammals (see page 398). The classification of the major groups of mammals reflects the phylogenetic isolation of the monotremes from the marsupials and placentals. Monotremes belong to the subclass Prototheria; the evolutionary line that gave rise to prototherians has been separate from that of therian mammals for some 190 million years. Both marsupials and placentals are put within the subclass Theria, but the marsupials are placed within one infraclass (Metatheria), and the placentals in another (Eutheria). The evolutionary paths of marsupials and placentals diverged from a common ancestor over 100 million years ago, in the Early Cretaceous.

ORDER MONOTREMATA

Although represented today by but three genera, each with a single species, and therefore comprising an unimportant segment of the Recent mammalian fauna, monotremes are of great interest for several reasons. Morphologically, they closely resemble no other living mammals, and they possess some features more typical of rep-

tiles than of mammals. Monotremes lay eggs and incubate them in birdlike fashion, yet they have hair and suckle their young. The order Monotremata includes the family Tachyglossidae (echidnas or spiny anteaters), occurring in Australia, Tasmania, and New Guinea, and the family Ornithorhynchidae (duck-billed platypus), restricted to eastern Australia and Tasmania.

MORPHOLOGY

Many structural features distinguish monotremes from other mammals. The monotreme skull is uniquely birdlike in appearance (Fig. 5–1); it is toothless, except in young platypuses; the sutures disappear early in life; and the elongate and beaklike rostrum (Fig. 5–2) is covered by a leathery sheath (this sheath is horny in birds). The lacrimal bones are absent and the jugals are small or absent, whereas these bones are present in most therian mammals; evidences of prefrontals and postfrontals, typically reptilian elements that are missing in therian mammals, occur on the frontals of monotremes. There is no auditory bulla, but the chamber of the middle ear is partially surrounded by oval tympanic rings.

Figure 5–1. Skull of the spiny anteater (*Tachyglossus aculeatus*). Length of skull 111 mm.

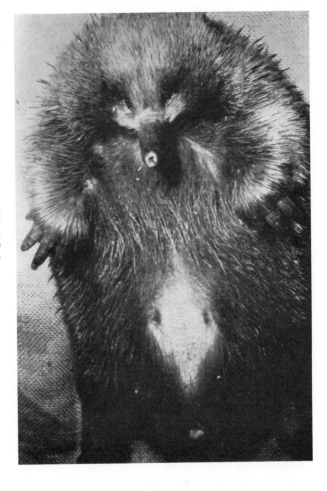

Figure 5–2. Ventral view of a live echidna (*Tachyglossus aculeatus*), showing the beak-like rostrum and the poorly developed pouch (typical of the nonbreeding season), and the tufts of hairs at the mammary "lobules." (Photograph by M. L. Augee.)

Monotreme appendages represent excellent examples of "mosaic evolution" (Crompton and Jenkins, 1973). The shoulder girdle retains a pattern of bones typical of therapsid reptiles, the forelimb has a rather reptilian posture resulting in part from fossorial specializations, and the pelvis and posture of the hind limbs are essentially therian.

The monotreme pectoral girdle contains an interclavicle, clavicles, precoracoids, and coracoids (Fig. 5–3), and provides a far more rigid connection between the shoulder joint and the sternum than that characteristic of therian mammals. Large epipubic bones extend forward from the pubes. Cervical ribs are present, and the thoracic ribs lack tubercles, processes that occur on the ribs of most other mammals and are braced against the transverse processes of the vertebrae.

As put by Howell (1944:26), no monotremes "by any strength of the imagination might be considered cursorial." Monotremes have retained a limb posture that is similar in some ways to that of a reptile. In monotremes this posture is associated with limited running ability. In the Australian echidna (*Tachyglossus aculeatus*), the humerus remains roughly horizontal to the substrate during walking (Jenkins, 1970). Rotational movements of the humerus, rather than fore and aft movements as in most mammals, are largely responsible for propulsion. In reptiles the "splayed" limb posture involves contact of the fore and hind feet with the ground well to the side of the shoulder and hip joints, respectively. The limb postures in the echidna partially depart from this pattern because the forearm angles medially and the manus (forepaw) is roughly ven-

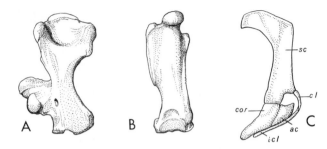

Figure 5–3. Bones of monotremes. Left humerus (A) and right femur (B) of the spiny anteater *(Tachyglossus aculeatus)*. Pectoral girdle (C) of the duck-billed platypus *(Ornithorhynchus anatinus)*. Abbreviations: *ac,* acromion; *cl,* clavicle; *cor,* coracoid; *icl,* interclavicle; *sc,* scapula. (Pectoral girdle of the platypus after Romer, 1966.)

tral to the shoulder joint; in the hind limb the foot is roughly ventral to the knee. The posture of the hind limb of the echidna resembles that of many generalized therian mammals. When the echidna is in motion, its body is elevated well above the ground in nonreptilian fashion. Despite the advances in limb posture in the echidna over the reptilian condition, locomotion is slow and appears labored and awkward.

REPRODUCTION

The monotreme reproductive system and reproductive pattern are completely unique among mammals. Eggs are laid, and these are telolecithal (the yolk is concentrated toward the vegetal pole of the ovum) and meroblastic (early cleavages are restricted to a small disc at the animal pole of the ovum), as are birds' eggs. Only the left ovary is functional in the platypus (Asdell, 1964:2), as in most birds, but both ovaries are functional in echidnas. Shell glands are present in the oviducts. A cloaca is present, to the ventral wall of which in males the penis is attached. The testes are abdominal, and seminal vesicles and prostate glands are absent. The female echidna temporarily develops a pouch during the period of incubation and caring for the young, but the platypus never develops a pouch. The mammae lack nipples, and the young suck milk from two areas (lobules) in the pouch in the echidna (Fig. 5–2), or from the abdominal fur in the platypus.

PALEONTOLOGY

The fossil record of monotremes is limited, and consists mainly of species referable to recent genera from the Pleistocene of Australia. Possible relationships between monotremes and various Mesozoic reptilian and mammalian groups have been widely discussed (Kermack and Musset, 1958; Kermack, 1963; Simpson, 1959; Romer, 1968:166, 167; Kermack and Kielan-Jaworowska, 1971). The presently widely accepted view is that monotremes arose from ancestral stock within the Late Triassic mammalian family Morganucodontidae, and that they have therefore undergone a long, independent development. The extremely specialized habits of monotremes and their isolation from competition with continental placental animals may have contributed importantly to their survival. Van Deusen (1969) pointed out, however, that *Tachyglossus* thrives over much of Australia despite fire, drought, dingos, and other placentals, and that in New Guinea *Zaglossus* is affected severely only by habitat destruction and hunting by man.

FAMILY TACHYGLOSSIDAE. Members of this group have robust bodies covered with short, sturdy spines (Fig. 5–4) that are controlled by unusually powerful panniculus carnosus muscles. *Zaglossus,* the New Guinea echidna, weighs from 5 to 10 kg, and the Australian spiny anteater *(Tachyglossus),* from about 2.5 to 6 kg. The braincase of the echidna is moderately large, and the cerebrum is convoluted. The rostrum is slender and beaklike (Figs. 5–2, 5–4), the dentary bones are slender and delicate, and the long tongue is protrusile and covered with viscous mucus secreted mostly by the enlarged submaxillary salivary glands. Food is ground between spines at the base of the tongue and adjacent trans-

Figure 5–4. Two species of monotremes: Above, the Australian spiny anteater (*Tachyglossus aculeatus*). Below, the New Guinea spiny anteater (*Zaglossus bruijni*). (The photograph of *Tachyglossus* is by M. L. Augee; that of *Zaglossus* is by Hobart M. Van Deusen.)

verse spiny ridges on the palate. The pinnae are moderately large. The limbs are powerfully built and are adapted for digging. The humerus is highly modified by broad extensions of the medial and lateral epicondyles that provide unusually large surfaces for the origins of some of the powerful muscles of the forearm (Fig. 5–3). In *Zaglossus*, the number of claws is variable; some animals have only three claws front and rear, while some have a full complement of five (Van Deusen, 1969.) In *Tachyglossus*, all digits have stout claws.

The ankles of male echidnas (and of some females) bear medially directed spurs; their function is not known.

These animals have highly specialized modes of life. They are powerful diggers and can rapidly escape predators by burrowing. Food consists largely of termites, ants, and a variety of other invertebrates. Foraging involves turning over stones and digging into termite and ant nests, and the prey is captured by the sticky tongue.

Usually one leathery-shelled egg is laid, and incubation in the temporary pouch lasts from seven to ten days. The young is helpless when hatched and remains in the pouch until the spines develop.

The Australian spiny anteater is known to be a true hibernator and will become torpid in response to cold and lack of food. During periods of torpor in experimental animals, the body temperature (5.5°C) was close to the ambient temperature (5.0°C), and the heart rate dropped to seven beats per minute (Augee and Ealey, 1968). Experimental animals were able to arouse spontaneously. One could dig slowly when its body temperature was only 10.5°C and its heart rate was seven beats per minute.

FAMILY ORNITHORHYNCHIDAE. Compared to echidnas, the duck-billed platypus is small, weighing from roughly 0.5 to 2.0 kg. Some structural features of the platypus are associated with its semiaquatic mode of life. The pelage is dense, and, as in the muskrat (*Ondatra*), the underfur is woolly. The external auditory meatus is tubular, as in the beaver (*Castor*). The eye and ear openings (pinnae are absent) lie in a furrow that is closed by folds of skin when the animal is submerged. The feet are webbed, but the digits retain claws that are used for burrowing. The web of the forefoot extends beyond the tips of the claws, and is folded back against the palm when the animal is digging or when it is on land. The ankles of the male platypus have grooved and medially directed spurs that are connected to poison glands. In a case reported by Walker (1968:6), a man spurred by a platypus suffered immediate intense pain and did not regain full use of the injured hand for months.

The braincase of the platypus, in contrast to that of tachyglossids, is small, and the cerebrum lacks convolutions. Although the young have teeth, the gums of adults are toothless and are covered by persistently growing, horny plates. Anteriorly, the occlusal surfaces of the plates form ridges that serve to chop food; posteriorly, the plates are flattened crushing surfaces. Some additional mastication is accomplished by the flattened tongue, which acts against the palate. The elongate rostrum bears a flattened, leathery bill that seems to have remarkable tactile ability.

The platypus inhabits a variety of waters, including mountain streams, slow-moving and turbid rivers, lakes, and ponds, and is primarily a bottom feeder. Aquatic crustaceans, insect larvae, and a wide variety of other animal material, and some plants, are taken during dives that are roughly a minute in duration. The platypus takes refuge in burrows dug into banks adjacent to water. Periods of inactivity, or perhaps hibernation, occur in the winter (Bourliere, 1956:195). The female digs a burrow up to 50 feet in depth, in which the eggs are laid on a nest of moist leaves. Usually two eggs are laid, and these adhere to one another. The burrow is kept plugged during the ten-day incubation period, after which the young, extremely rudimentary at hatching, are suckled for about five months.

Populations of the platypus were rapidly declining at one time and the animals appeared to be facing extinction, but, owing at least in part to protective legislation enacted by the Australian government, the platypus now seems out of danger.

ORDER MARSUPIALIA

Marsupials and placentals are representatives of two evolutionary lines that have been separate since the early Cretaceous or Jurassic (Slaughter, 1968; Clemens, 1968; Lillegraven, 1974). As a result of their long, independent histories, marsupials differ structurally from placentals in many ways. In their semiarboreal habits and omnivorous-insectivorous diet, the didelphids are seemingly the present counterparts of early therians. The brain and the reproductive pattern are more primitive in marsupials than in placentals and, perhaps largely owing to these differences, marsupials have usually been replaced by placentals whenever species of these two groups were in competition. Today, only two important strongholds for marsupials remain; the Australian region (including Australia, Tasmania, New Guinea, and nearby islands) and the Neotropics (including southern Mexico, Central America, and most of South America). When isolated from placentals for long periods of time, marsupials have undergone remarkable adaptive radiation. Most marsupials have functional counterparts among placentals.

MORPHOLOGY

The marsupial skull frequently has a small, narrow braincase, housing small cerebral hemispheres with simple convolutions. Ossified auditory bullae, when present, are usually formed largely by the alisphenoid bone, rather than in large part by the tympanic (both ectotympanic and entotympanic), petrosal, and/or basisphenoid, as occurs in most placentals. The marsupial palate characteristically has large vacuities (Fig. 5–5), and the angular process of the dentary is inflected. The dentition is unique in that there are never equal numbers of incisors above and below, except in the family Vombatidae; the cheek teeth primitively include 3/3 premolars and 4/4 molars. The primitive eutherian number of 44 teeth is often exceeded.

Marsupials often have highly specialized feet associated with specialized types of locomotion (Fig. 5–6). The unusual patterns of specialization of the hind feet of marsupials are probably a result of an arboreal heritage and the early development of an opposable first digit and an enlarged and powerfully clutching fourth digit. As in the monotremes and multituberculates, epipubic bones extend forward from the pubic bones.

REPRODUCTION

The marsupial reproductive pattern is distinctive, and is of special interest because certain features, such as the structure of the fetal membranes (described in Chapter 18), may resemble an evolutionary stage through which eutherian mammals passed. Most female marsupials have a marsupium (an abdominal pouch) or abdominal folds, within which the nipples occur. The number of nipples varies from 2, as in the family Notoryctidae and some members of the Dasyuridae, to 19, as in some members of the Didelphidae. Individual variation in the number of nipples often occurs within a species. The female reproductive tract is bifid, i.e., the vagina and uterus are double (Fig. 5–7). In all but the family Notoryctidae, which is adapted for digging, the testes are contained in a scrotum which is anterior to the penis.

The gestation period is characteristically

Figure 5–5. Ventral views of skulls of marsupials. A, New Guinea bandicoot *(Peroryctes raffrayanus*, family Peramelidae; length of skull 82 mm); B, ring-tailed possum *(Pseudocheirus corinnae*, family Petauridae; length of skull 97 mm). (After Tate and Archbold, 1937.)

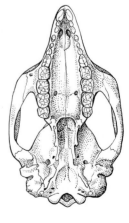

A   B

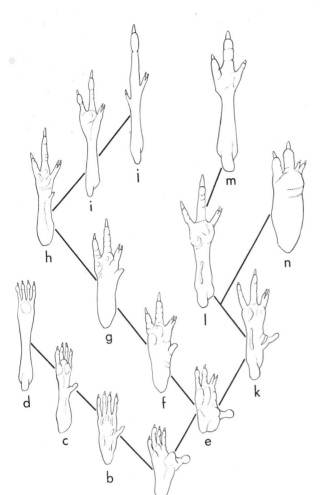

Figure 5–6. Ventral views of right hind feet of marsupials, showing patterns of specialization associated with various styles of locomotion. The presumed basic arboreal type is represented by *a*, the foot of a didelphid. The lines indicate possible evolutionary pathways leading toward greater specialization. Feet of the following kinds are shown. Dasyuridae: *b, Phascogale; c, Sminthopsis; d, Antechinomys.* Petauridae: *e, Pseudocheirus.* Peramelidae: *f, Perameles* sp.; *g, Peroryctes* sp.; *h. Peroryctes* sp.; *i, Macrotis; j. Chaeropus.* Macropodidae: *k, Hypsiprymnodon; l, Potorous; m, Macropus; n, Dendrolagus.* (After Howell, 1944.)

short (Table 18–1, page 403), and the young are tiny and rudimentary at birth. Newborn marsupials probably possess the minimal anatomical development allowing survival outside of the uterus. Organogenesis has just begun, the separation of the ventricles of the heart is incomplete, the lungs are vascularized sacs lacking alveoli, and the kidneys lack glomeruli. Also lacking are cranial nerves II to IV and VI, eye pigments, eyelids, and cerebral commissures (nerve fiber bundles connecting the cerebral hemispheres). But these naked, blind, and delicate newborn are able to

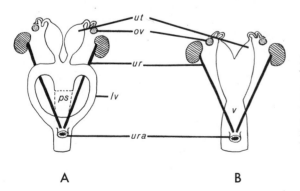

Figure 5–7. Diagrams of the female reproductive tracts of marsupials (A) and placentals (B). Abbreviations: *lv*, lateral vagina; *ov*, ovary; *ps*, pseudo-vaginal canal; *ura*, urethra; *ur*, ureter; *ut*, uterus; *v*, vagina. (After Sharman, G. B.: Reproductive physiology of marsupials. *Science,* 171:443–449. Copyright 1970 by The American Association for the Advancement of Science.)

make their way at birth from the vulva to the marsupium. Here the "embryonic" young attach to a nipple, and remain so for a period of time greatly exceeding the gestation period. The weight of the young marsupial when it leaves the pouch and that of the newborn placental are roughly the same in species of comparable full-grown size (Sharman, 1970).

Most marsupials, with the exception of the Peramelidae, have a choriovitelline placenta (see page 398).

PALEONTOLOGY

Marsupials are first known from the Early Cretaceous of North America (Slaughter, 1968), near the height of the reign of the dinosaurs. At that time the only surviving groups of primitive Jurassic mammals (discussed in Chapter 3) were the multituberculates, triconodonts, and symmetrodonts. Seemingly, the niches previously filled by triconodonts and symmetrodonts were being occupied in the late Cretaceous by the two dominant mammalian groups of today, the marsupials and the placentals. These groups probably descended from Late Jurassic pantotheres. Cretaceous marsupials and placentals were small animals with omnivorous or insectivorous feeding habits, but they provided the ancestral stock for the remarkable mammalian radiation that occurred early in the Cenozoic, after the disappearance of the ruling reptiles.

The most primitive marsupial family, and the stem group from which all other marsupials evolved, is the Didelphidae. Didelphids were abundant in North America in the Late Cretaceous, and have a nearly continuous fossil record there through the Miocene. A single didelphid genus, *Peratherium*, is known from the Eocene through the Miocene of Europe. Today, as well as during most of the Cenozoic, Australia and South America have been the two centers of marsupial abundance, and the histories of their marsupial faunas are parallel to some extent. In both continents, marsupial radiations occurred under partial or complete isolation from competition with eutherians.

Marsupials probably gained entry into Australia from their presumed North American center of origin via South America and Antarctica in the Late Cretaceous or earliest Tertiary (see p. 345). Marsupials reached Australia well before placentals, and underwent a spectacular Cenozoic radiation. Twelve living marsupial families, including some 66 genera, and two extinct families, Thylacoleonidae and Diprotodontidae, resulted from this radiation.

To one meeting for the first time the present Australian marsupial fauna, the numbers of species and the structural extremes are impressive, but, by comparison with the fauna of the Late Pleistocene, the present fauna is severely depleted. Many species of large marsupials became extinct between the Late Pleistocene and historic time, and further reductions occurred within historic times. The extinct families Thylacoleonidae and Diprotodontidae are especially noteworthy. *Thylacoleo* was a predaceous Pleistocene form roughly the size of an African lion. The third premolars were greatly elongated shearing blades and the strongly built front limbs had retractile claws (Keast, 1972:227). The Diprotodontidae was represented in the Pleistocene by several genera, of which *Diprotodon* is especially impressive. This herbivore, a huge wombat-like creature roughly the size of a rhinoceros, is the largest marsupial known. A number of very large kangaroos (Macropodidae) also became extinct before historic times. A giant among them is *Procoptodon goliah*, an extremely short-faced macropodid (Fig. 5–8) that stood some 3 m high. This huge herbivore was probably the ecological equivalent of some of today's large browsing ungulates.

Why did these imposing marsupials disappear?

This question has been considered carefully by Merrilees (1968), who stressed the possible influence of man. Aborigines probably entered Australia in the Late Pleistocene, at a time when the marsupial fauna included the large species just mentioned. With man came fire. Aborigines

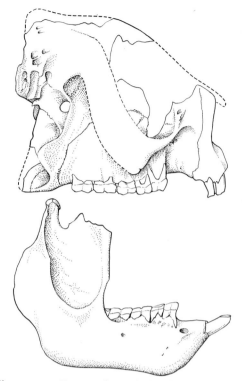

Figure 5–8. *Procoptodon goliah*, a huge, browsing, macropodid marsupial from the Pleistocene of Australia. Length of skull 218 mm. (After Tedford, 1967.)

abundance of many species and a number of extinctions (discussed by Calaby, 1971). As an example, of the some 45 species of kangaroos (Macropodidae) that occupied Australia just prior to the entry of European man, three are extinct, and roughly an additional dozen seem to have declined drastically, although some may have become rare before historic times. The destructive effectiveness of man is demonstrated by the case of the "toolache" (*Macropus greyi*), a large wallaby. This animal occupied treeless grassland in southeastern South Australia, and was still common through the early part of this century. Probably in large part because this animal was killed for its beautiful fur and because coursing these wallabies in open country became popular, their numbers declined abruptly, and no wild individuals were seen after 1924. Only some 15 years of competition with livestock and killing by man were sufficent to push the species to extinction. Fortunately, but belatedly, Australians are beginning to mourn such losses, and steps are being taken to avoid further extinctions.

Didelphid marsupials probably dispersed from North America to South America in the Late Cretaceous or earliest Paleocene. As the fossil record of South American Tertiary marsupials becomes more complete and our view of the evolution of these animals widens, we can begin to appreciate the remarkable extent of their adaptive radiation. The marsupial radiation there may well have rivalled that in Australia. Accompanying the striking structural diversification within the marsupials in response to a wealth of available habitats was a convergence by many types toward various placentals. By the middle of the Tertiary, there were marsupials that structurally (and undoubtedly functionally) resembled shrews, moles, rodents, and carnivores. In the Early Tertiary, several groups of placentals entered South America. Edentates and various ungulate groups (p. 343) seemingly "owned" the large herbivore adaptive zone throughout much of the Tertiary, to the complete exclusion of marsupials.

used fire extensively, perhaps using grass fires as an aid to hunting the large marsupials. Man may, therefore, not only have killed the large marsupials but may have made many areas less favorable for them. Merrilees suggests that post-Pleistocene aridity and man-made fires may together have tipped the balance against the large marsupials. When fauna is found in archaeological sites spanning the last 20,000 years, however, it is modern. This has led Jones (1968:203) to conclude that the major extinctions of the Australian late Pleistocene fauna took place in the period from 20,000 to 30,000 years ago, rather than in Late Pleistocene or early Recent times, as has previously been postulated.

Further reductions in the marsupial fauna began with the coming of European man to Australia. The combined effects of heavy grazing by livestock, clearing of the land for agriculture, and the introduction of the Old World rabbit (*Oryctolagus cuniculus*) caused widespread declines in the

All of the South American marsupials presumably evolved from a basal didelphid stock, and the occurrence of at least 13 genera of didelphids at a late Paleocene fossil locality near Rio de Janeiro documents an early Tertiary radiation. A number of didelphids and descendants of didelphids were seemingly omnivores, but one short-lived family (Carolameghinidae) was perhaps frugivorous, and one weasel-sized didelphid (*Sparassocynus*) was probably predaceous.

Of special interest are several groups that through their Tertiary history departed drastically from their ancestral morphological plan. Perhaps the most spectacular group is the superfamily Borhyaenoidea, which includes the doglike Borhyaenidae and the sabertooth Thylacosmilidae (Marshall, 1977a). These families are not only separable on the basis of structural features, but the two families seemingly occupied complementary adaptive zones.

The Borhyaenidae includes a number of marsupials with dentitions that suggest styles of feeding ranging from omnivorous to carnivorous. One borhyaenid, *Stylocynus*, was roughly the size of a bear and was presumably omnivorous, and a number of small and medium-sized species were also omnivores. The food of one Miocene species is not in doubt. A specimen of *Lycopsis longirostris* described by Marshall (1977b) had in its body cavity the bones of the rodent *Scleromys*. Bones next to the pelvis, that probably represent material in feces deposited at the time of death, are also those of a rodent (Fig. 5–9). *Lycopsis longirostris* is about the size of a coyote. In some borhyaenids, the first digit of the hind foot is partly opposable, a feature indicative of semiarboreal habits. All known borhyaenids are rather short legged, and the terrestrial types lack marked cursorial specializations. *Borhyaena* has a powerful dentition and a skull resembling that of a wolf in size and form (Fig. 5–10).

The family Thylacosmilidae is known from Pliocene deposits and includes five species, all of which have saber-like upper canines. The sabers of *Thylacosmilus* are long and recurved, their roots extend nearly to the occipital part of the skull, and their tips are protected by a grotesque flange on the dentary (Fig. 5–10). Although basically well adapted to a carnivorous life, both borhyaenids and thylacosmilids had one apparently maladaptive character. The canines, so important in predaceous mammals, had only a thin layer of enamel, and consequently became blunted and worn rapidly. In compensation, the roots in some forms remained open through much of life, permitting continued growth (Patterson and Pascual, 1972:262). The unusual, arched dorsal profile of the skull of *Thylacosmilus* is a contribution of the extremely long roots of the canines. Thylacosmilids probably

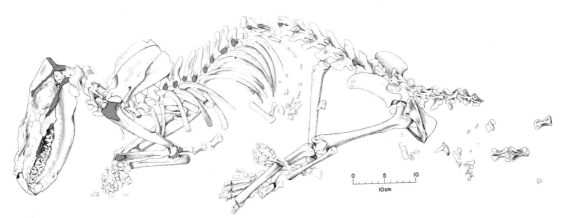

Figure 5–9. Skeleton of a borhyaenid marsupial (*Lycopsis longirostris*) from the Miocene of Colombia, South America. Note the bones of a rodent (*Scleromys*) in the body cavity area and other rodent bones (perhaps from fecal material) next to the pelvis. (After Marshall, 1976.)

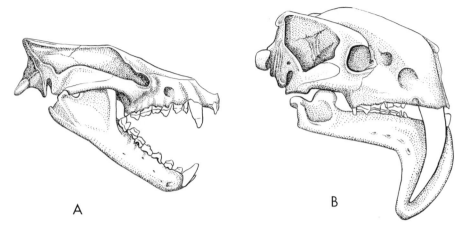

A   B

Figure 5–10. Skulls of members of the extinct marsupial family Borhyaenidae. A, *Borhyaena* (length of skull 230 mm); B, *Thylacosmilus* (length of skull 232 mm). (After Romer, 1966).

preyed on large and, in some cases, slow-moving mammals, and certainly many large mammals were their contemporaries. Ground sloths, a number of large no-toungulates, including the rhino-like *Tox-odon* (Fig. 16–9), and camel-like ungulates of the order Litopterna (Fig. 16–10) were prominent members of the late Tertiary scene. The shapes of the sabers of *Thylacosmilus* and skeletal features usually associated with powerful neck musculature indicate that the sabers were used as stabbing weapons as the predators clung to large prey with their powerful forelimbs (Marshall, 1976). This predatory style is precisely the one thought to have been used by placental saber-toothed cats (Machairodontinae) of North America and Europe, but the lack of incisors in *Thylacosmilus* and their retention and specialization in machairodonts are suggestive of different methods of taking food into the mouth. The limbs of thylacosmilids are short, and most prey was probably captured by a surprise attack rather than a chase.

The complex story of the decline and eventual extinction of the Borhyaenidae is told by Marshall (1977a). Many Early Eocene borhyaenids resembled didelphid marsupials and were of moderate size; by the Early Oligocene, however, several very large forms had evolved. But the largest of all, *Proborhyaena gigantea*, which had a

skull roughly a meter long, disappeared by the end of the Early Oligocene, and with its extinction a trend began toward reduced size in carnivorous borhyaenids. This evolutionary step moved counter to the trend toward large size that is common to many mammalian lineages. Further, the lack of cursorial adaptations in borhyaenids is unusual for a group of large carnivorous mammals. A unique adaptive interplay in the South American Tertiary between avian and marsupial predators may explain these features in borhyaenid evolution.

During the Early Tertiary in South America, the adaptive zones available to carnivores were probably "up for grabs" and, from the Early Oligocene through the Pliocene, three families of large carnivorous birds shared with borhyaenids a predatory mode of life. *Phororhacos*, a common member of the family Phororhacidae (order Gruiformes), stood some five to eight feet tall, had a heavily built, hooked bill mounted on a skull about the size of that of a horse, and was obviously predaceous. The wings were tiny and the bird was flightless, but the hind legs were long and slim and the bird must have been a swift runner. Probably the phororhacoid birds and the borhyaenid carnivores partitioned the resources available to carnivores: the phororhacoids developed and maintained large size, occupied open savanna areas, and used their cursorial ability in pursuing swift prey; the borhyaenids, on the other hand, were of more moderate size, occupied wooded country, and largely killed slow-moving prey or prey captured from ambush. Some may have been scavengers. The Oligocene extinction of the large *Proborhyaena gigantea* and the Late Tertiary reduction in the size of borhyaenids may well have been aspects of an adaptive trend toward reduction of competition with the imposing phororhacoids (Patterson and Pascual, 1972:262).

Further declines in the fortunes of borhyaenids were apparently associated with the mid-Miocene arrival in South America of members of the theretofore entirely North American raccoon family (Procyonidae). The fossil record indicates that replacement of the larger omnivorous borhyaenids by the omnivorous procyonids was complete by the Late Pliocene. Extinctions of other borhyaenids are thought by Marshall (1977a) to have been related to possible competition with marsupials of the family Didelphidae. The small to medium-sized borhyaenid omnivores declined in the Middle Pliocene and became extinct by the Late Pliocene, whereas didelphids with similar adaptations appeared in the Middle Pliocene and underwent a striking adaptive radiation in the Late Pliocene.

The history of the Borhyaenidae was thus intertwined with that of the phororhacoid birds, the immigrant South American procyonids, and the didelphid marsupials. By the Late Pliocene, the borhyaenids were gone while the three latter groups prevailed. But the Late Pliocene and Pleistocene were by no means times of faunal stability in South America and, each in its turn, the phororhacoids, the procyonids, and the didelphids suffered partial or complete competitive replacement by North American carnivores.

An event of major importance to the South American mammalian fauna occurred in Late Pliocene. At that time the land bridge connecting North and South America was established and a flood of northern placentals moved southward, while some southern mammals moved northward (see p. 345). Although the Borhyaenidae were already extinct by this time and were thus not affected by this collision of faunas, the saber-toothed thylacosmilids were still present and were decisively affected. Thylacosmilids occur in Middle and Late Pliocene deposits, but are unknown from the Pleistocene. The occurrence of placental saber-toothed cats in early Pleistocene strata immediately above Late Pliocene beds bearing thylacosmilids points toward a remarkably abrupt competitive replacement.

Carnivory was just one of a number of life styles pursued by South American marsupials. A tantalizing glimpse of one member of an unusual family of marsupials is offered by *Necrolestes* from the Middle Miocene. This small animal, the sole known representative of the family Necrolestidae, was probably insectivorous and fossorial, and may have had a mode of life similar to that of the Australian marsupial "moles" (Notoryctidae).

The family Caenolestidae, which still persists in South America in relict populations along the Andes Cordillera, began its radiation in the Oligocene. Some caenolestids were convergent toward multituberculates in a series of features, including general skull form, enlargement of the anterior incisors, and structure of the serrate, lower pair of cheek teeth (Fig. 5–11). Both multituberculates and the multituberculate-like caenolestids (subfamily Abderitinae) probably resembled rodents in feeding habits. Just as multituberculates in North America were able to persist to the Late Eocene, long after rodents became abundant, the Abderitinae lived into the Miocene of South America, well after the Early Oligocene radiation of histricomorph

Figure 5–11. Jaw of a Miocene caenolestid, showing the highly specialized, trenchant cheek tooth. (After Romer, 1966.)

rodents there (Patterson and Pascual, 1972:263).

The most rodent-like marsupials yet known are two species from the Early Tertiary of South America. These species compose the family Groeberiidae, and are remarkable in having such features as enlarged incisors with enamel only on the anterior surfaces, a sharp reduction in the number of cheek teeth, and a broad diastema.

Of special interest is the family Argyrolagidae, a supreme example of evolutionary convergence. This unique family, considered in detail by Simpson (1970b) is known from the Pliocene and Pleistocene of South America and probably diverged early from primitive didelphid ancestry. Argyrolagids do not resemble closely any other group of marsupials, but they possessed a series of morphological characters that are found today in such specialized rodents as kangaroo rats (Heteromyidae) and jerboas (Dipodidae). These rodents occupy mainly sparsely vegetated desert or semiarid areas; all are saltatorial, and all share certain distinctive morphological features. The hind limbs are long and the hind feet are modified by the loss of digits

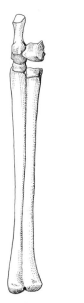

Figure 5–12. Hind foot of *Argyrolagus* (Argyrolagidae), an extinct, kangaroo rat-like marsupial. Note the appressed metatarsals of digits three and four that form a structure resembling the cannon bone of artiodactyls. (After Simpson, 1970.)

and, in some cases, by the fusion of metatarsal bones (Fig. 5–12). The long hind limbs of argyrolagids are highly specialized along similar lines: only the third and fourth digits of the foot are retained, and the metatarsals are closely appressed and resemble to some extent the cannon bone of an artiodactyl (see Fig. 14–5).

Although other arrangements occur in the kangaroo rats and jerboas (see Fig. 10–22), the trend toward lightening and lengthening of the hind foot is common to them and the argyrolagids. As in these rodents, the forelimbs of argyrolagids are small and the tail is long, and a good guess is that in life the tail was tufted. The dentition of argyrolagids was 2/2, 0/0, 1/1, 4/4, which is different from the dentitions of the rodents we are comparing them to. All of these animals have cheek teeth adapted to grinding and incisors suited for gnawing, and in both kangaroo rats and argyrolagids the cheek teeth are rootless and have a simple occlusal surface. (Among rodents, however, such an occlusal surface is not unique to kangaroo rats.)

In the animals being compared, similarities in the skulls include prominent orbits, elongation of the snout anterior to the incisors, reduction of the temporalis muscle, and inflated auditory bullae. The functional significance of several of these features may well be the same in jerboas, kangaroo rats and argyrolagids. The prominent orbits, placed far back in the skull, allow the rodents under discussion to watch for danger when the nose is close to the ground, and these features may have had the same adaptive importance in argyrolagids. Condensation of moisture on the cool nasal mucosa during exhalation is a means of reducing pulmonary water loss in kangaroo rats, and the unusual tubular extension of the nasal cavity anterior to the incisors in these animals (Fig. 10–21) is associated with improved water conservation (see p. 445). In the argyrolagids, there is an even more elongate extension of the nasal cavity (Fig. 5–13). Here, too, this specialization may have facilitated the maintenance of water balance in an arid environment.

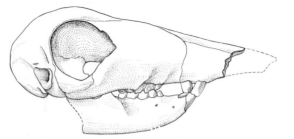

Figure 5–13. Skull of *Argyrolagus*. Note the similarity between the form of this skull and that of the kangaroo rat shown in Figure 10–21. Length of skull 55 mm. (After Simpson, 1970b).

The reduction of the temporalis muscle in argyrolagids, as indicated by a small coronoid process, was correlated with specialization of the masseter medialis: this muscle originated in the enlarged anterior part of the orbit and the eye occupied the posterior part. Similarly, in the rodents under discussion, the temporalis is reduced and the medial masseter is variously specialized. In mammals, these specialized origins are part of a system providing for the complex rotary jaw action used by herbivores. The entire form of the skull in kangaroo rats and jerboas is modified by the enormous auditory bullae (Figs. 10–21, 10–23), and in argyrolagids also the bullae are inflated. This enlargement of the bullae has been shown to be one of a remarkable series of specializations that allow kangaroo rats to detect faint, low frequency sounds made by their predators (see p. 468). Very probably the enlarged bullae of the argyrolagids served a similar end.

It seems, then, that in two separate lineages, which have been separate since at least Early Cretaceous, but which occupy (or occupied) similar dry habitats, nearly identical suites of characteristics have evolved. Because the dipodid and heteromyid rodents are structurally so amazingly closely matched by the argyrolagid marsupials, one can, with some confidence, describe some features of the argyrolagid mode of life (Simpson, 1970b). Hypothetically, the style of life was roughly as follows: Argyrolagids lived in semiarid plains or deserts with sparse or scattered vegetation, where they were able to survive at least seasonal absence of drinking water. They fed on seeds, some green vegetation, and insects, and filled their cheek pouches with food that was brought to the burrow to be eaten. When foraging, they used short bipedal or perhaps quadrupedal hops, but they could move rapidly and directly using a series of long bipedal bounds. They were able to hear the final, often nearly silent, rush of a predator, and their evasive maneuver was an explosive leap. Further evasive action involved extremely erratic (ricochetal) bounds.

One cannot help but wonder why argyrolagids did not survive until the present, for "by all rules of analogy and theories of extinction, they should have survived, as did their close ecological analogues in North America, Asia, Africa, and Australia" (Simpson, 1970b). Large mammals were the chief victims of extinction during the Pleistocene, whereas most small mammals persisted. Periods of unusual dryness should have made more habitat available to the arid land-dwelling argyrolagids, and there seems to be no period when suitable argyrolagid habitat was totally absent in South America. Finally, from the fossil record there is no indication that the extinction of argyrolagids resulted from competitive replacement: no other mammals with ecological requirements similar to those of argyrolagids have ever lived in South America. As the fossil record of South American mammals becomes more complete, information bearing on the extinction of the argyrolagids undoubtedly will come to light. At present, the mystery of the extinction of the argyrolagids remains unsolved.

FAMILY DIDELPHIDAE. Didelphids are the most primitive and generalized marsupials and are the oldest known family, dating from the early Cretaceous. As

Figure 5–14. A didelphid marsupial *(Didelphis marsupialis)*. In the United States this animal is common in the southeast and in many areas along the Pacific coast. (San Diego Zoo photograph.)

mentioned, this is the basal family of the marsupial radiation. The Didelphidae includes 12 Recent genera with 66 species, and occurs from southeastern Canada, with the American opossum *(Didelphis virginiana;* Fig. 5–14), to southern Argentina, with the Patagonian opossum *(Lestodelphis halli).*

In these New World opossums, the rostrum is long (Fig. 5–15), the braincase is usually narrow, and the sagittal crest is prominent. The dental formula is 5/4, 1/1, 3/3, 4/4 = 50. The incisors are small and unspecialized, and the canines are large. The upper molars are basically tritubercular with sharp cusps, and the lower molars have a trigonid and a talonid (Fig. 5–16).

Except for the opposable and clawless hallux in all species, a feature probably inherited from arboreal ancestral stock, and the webbed hind feet in the water opossum *(Chironectes minimus),* the feet

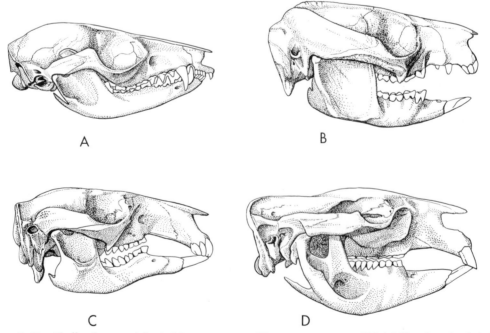

Figure 5–15. Skulls of marsupials. A, Mouse opossum *(Marmosa canescens,* Didelphidae; length of skull 35 mm); B, brush-tailed possum *(Trichosurus vulpecula,* Phalangeridae; length of skull 87 mm); C, wallaby *(Wallabia bicolor,* Macropodidae; length of skull 135 mm); D, wombat *(Vombatus ursinus,* Vombatidae; length of skull 180 mm).

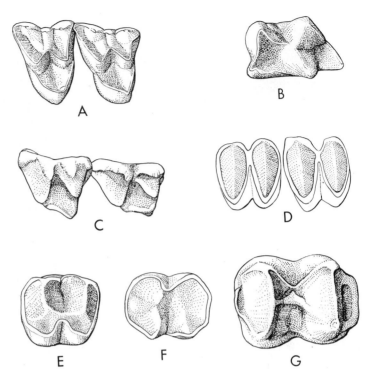

Figure 5–16. Occlusal views of molars of marsupials. A, Second and third right upper molars and B, third lower left molar of a mouse opossum (*Marmosa canescens*, Didelphidae); C, second and third upper right molars of the New Guinea bandicoot (*Peroryctes raffrayanus*, Peramelidae; after Tate and Archbold, 1937); D, second and third right upper molars of a wombat (*Vombatus ursinus*, Vombatidae); E, second upper right molar and F, third lower left molar of a phalanger (*Trichosurus vulpecula*, Phalangeridae); G, second upper right molar of a wallaby (*Wallabia bicolor*, Macropodidae).

are unspecialized, with no loss of digits or syndactyly. (Syndactyly is the condition in which two digits are attached by skin, as shown by a number of examples in Figure 5–6.) The foot posture is plantigrade. A marsupium is present in some didelphids, but is represented by folds of skin protecting the nipples in others, and is absent from some. The tail is long and is usually prehensile.

Although they occupy a wide range of habitats, didelphids are primarily inhabitants of tropical or subtropical areas, where they are often locally abundant. Most didelphids are partly arboreal and are omnivorous. The water opossum, however, is largely aquatic, is an accomplished diver, and is carnivorous. In this species, both sexes have a marsupium. In the male, the scrotum lies in this pouch; in the female, a sphincter can tightly close the entrance to the marsupium, allowing the animal to swim while the young remain in the pouch.

The small mouse opossum (*Marmosa*), a widespread neotropical didelphid, is one of the most abundant small mammals in some parts of Mexico. Although mouselike in general appearance, it seems to be largely insectivorous in some areas, at least during the summer (Smith, 1971). Poorly defined folds of skin protect the nipples of *Marmosa*, and the young simply hang on to the nipples and the mother's venter as best they can.

The American opossum has been introduced on the Pacific Coast of the United States, where it has thrived under suburban conditions. In many areas, however, it has been unable to occupy wild sections away from human habitation.

FAMILY DASYURIDAE. The Dasyuridae (Fig. 5–17) probably evolved from didelphid ancestry, but dasyurids are more progressive than didelphids, both dentally and with regard to limb structure. The fossil record is too incomplete to give clear evidence of the lines of descent of Australian marsupials, but the dasyurids are thought to be the most direct descendants of the marsupials that originally colonized Australia, probably in the late Cretaceous. Although the earliest known dasyurid is from the Australian middle Tertiary, the family must have arisen at a far earlier time. Recent members of this family in-

Figure 5-17. Four members of the family Dasyuridae. A, A marsupial "rat" (*Dasyuroides byrnei*); B, a marsupial "mouse" (*Sminthopsis crassicaudata*); C, a marsupial "mouse" (*Antechinus stuartii*); D, the native "cat" (*Dasyurus viverrinus*). (The photographs of *Dasyuroides* and *Antechinus* are by Jeffrey Hudson; those of *Sminthopsis* and *Dasyurus* are by Anthony Robinson.)

clude 20 genera and 50 species, and the geographic range includes Australia, New Guinea, Tasmania, the Aru Islands, and Normanby Island (Van Deusen and Jones, 1967:61).

Many of the major characters of dasyurids are shared by other marsupials, but several features are diagnostic of the former. The dental formula is 4/3, 1/1, 2-4/2-4, 4/4 = 42-50. The incisors are usually small and either pointed or bladelike, the canines are large and have a sharp edge, and the upper molars have three sharp cusps adapted to an insectivorous and carnivorous diet. The skulls of some dasyurids resemble rather closely those of didelphids (Fig. 5-18). The forefoot has five digits, and the hind foot has four or five digits. The hallux is clawless and usually vestigial, and is absent in some cursorial genera (Figs. 5-6, 5-19). There is no syndactyly. The foot posture is plantigrade in many species, but the long-limbed jumping marsupials, such as *Antechinomys* and the cursorial and carnivorous native "cat" (*Dasyurus viverrinus*; Fig. 5-17) and Tasmanian "wolf" (*Thylacinus cynocephalus*; Fig. 5-20), are digitigrade. The marsupium is often absent; when present, it is often poorly developed. The tail is long and well furred, conspicuously tufted in some species, and never prehensile. Dasyurids' size ranges from that of a shrew (*Planigale*) to that of a medium-sized dog (*Thylacinus*).

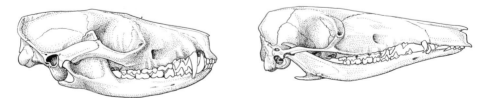

Figure 5-18. Skulls of marsupials. Left, native "cat" (*Dasyurus viverrinus*), Dasyuridae; length of skull 72 mm. Right, bandicoot (*Perameles* sp.), Peramelidae; length of skull 81 mm.

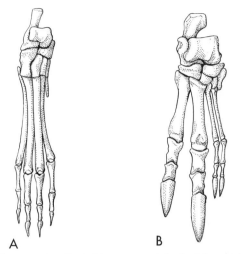

Figure 5–19. Feet of two marsupials. A, A terrestrial species, the Australian native "cat" (*Dasyurus viverrinus*, Dasyuridae); B, an arboreal species, a tree kangaroo (*Dendrolagus* sp., Macropodidae). (After Marshall, 1972.)

A wide variety of terrestrial habitats are occupied by dasyurids, and a few species are arboreal. A remarkably diverse array of marsupials are grouped within the Dasyuridae. The smaller species fill the feeding niche occupied in Eurasia and North America by shrews (Soricidae), and resemble these animals in the possession of long-snouted heads and unspecialized limbs. A group of rat-sized types seems adapted to preying on insects and small vertebrates, and the desert-dwelling genus *Antechinomys* has long slender limbs and a long tufted tail, and uses a rapid, bounding, quadrupedal gait. Another group, the native "cats," consists of somewhat civetlike dasyurids that weigh from roughly 0.5 to 3 kg and prey on a variety of small vertebrates. These "cats" are agile and effective predators, and, although primarily terrestrial, are capable climbers. The largest marsupial carnivores are the Tasmanian devil (*Sarcophilus harrisii*) and the Tasmanian "wolf." The Tasmanian devil is a stocky, short-limbed dasyurid, weighing from roughly 4.5 to 9.5 kg; it is now restricted to Tasmania. It is a persistent scavenger, but will also kill a wide variety of small vertebrates. The Tasmanian wolf is doglike in both size and general build (Fig. 5–20); it has long limbs and a digitigrade foot posture.

Although it is treated here as a dasyurid,

Figure 5–20. The Tasmanian "wolf" *Thylacinus cynocephalus*, Dasyuridae), which may now be extinct. (Photograph courtesy of Edwin H. Colbert.)

Thylacinus has been separated from the dasyurids and put in the family Thylacinidae by Ride (1970:226). Now extinct in New Guinea, on the Australian mainland, and very possibly in Tasmania, *Thylacinus* is able to prey on such large animals as the larger species of wallabies.

The most divergent dasyurid is the numbat or banded anteater—a small, long-snouted animal considered by some to be the sole representative of the family Myrmecobiidae. The teeth are small and widely spaced in the long tooth row, and the long protrusile tongue is used in capturing termites. This animal was formerly widespread in eucalyptus forests, in which fallen branches and logs provided lush populations of termites. But with clearing of these forests by man have come severe restrictions in the range of the numbat.

FAMILY NOTORYCTIDAE. This remarkable family is represented by a single species of marsupial "mole" that inhabits arid parts of northwestern and south central Australia. Many of the diagnostic characters of these mouse-sized animals are adaptations for fossorial life. The eyes are vestigial, covered by skin, and lack lenses and, as indicated by the specific name *Notoryctes typhlops* (*typhlops* means blind in Greek), are not functional. The ears lack pinnae. The nose bears a broad cornified shield, and the nostrils are narrow slits. The dental formula is usually 4/3, 1/1, 2/3, 4/4 = 44, but the incisors vary in number. The incisors, canines, and all but the last upper premolar are unicuspid; the paracone and metacone of the upper molars form a prominent single cusp, and the lower molars lack a talonid. As an adaptation serving to brace the neck when the animal forces its way through the soil, the five posterior cervical vertebrae are fused. The forelimbs are robust, and the claws of digits three and four are remarkably enlarged and function together as a spade, whereas the other digits are reduced. The central three digits of the hind feet have enlarged claws, the small first digit has a nail, and the fifth digit is vestigial. The marsupium is partially divided into two compartments, each with a single nipple.

The fur is long and fine textured, and varies in color from silvery white to yellowish red. There is no fossil record of notoryctids.

These animals use their powerful forelimbs and the armored rostrum to force their way through soft sandy soil. When the "mole" forages near the surface, the soil is pushed behind the animal and no permanent burrow is formed. The food is predominantly invertebrate larvae, but a variety of animal material and some vegetable material have been eaten by notoryctids in captivity. In contrast to moles (Order Insectivora; Family Talpidae), which characteristically inhabit moist soils, notoryctids inhabit sandy desert regions supporting scattered vegetation.

FAMILY PERAMELIDAE (Fig. 5–21). Members of this family are called bandicoots, and are characterized in general by an insectivore-like dentition and a trend toward specialization of the hind limb for running or hopping. Eight Recent genera, represented by 22 Recent species, are known, mainly from Australia, Tasmania, and New Guinea. Some species of bandicoots have been extirpated or have become uncommon over parts of their former range owing, apparently, to the grazing of livestock, to brush fires, and to the introduction of various placental mammals.

The smaller bandicoots are rat-sized; the largest species weighs roughly 7.0 kg. The dental formula is 4-5/3, 1/1, 3/3, 4/4 = 46 or 48. The incisors are small, and the molars are tritubercular (Fig. 5–16) or quadritubercular. The rostrum is slender (Fig. 5–18), and the ears of some species resemble those of rabbits. The marsupium is present and opens to the rear, and bandicoots, alone among marsupials, have a chorioallantoic placenta (see page 398). Although often long, the tail is not prehensile. The fourth digit of the hind foot is always the largest, and the remaining digits are variously reduced (Fig. 5–22). The hind foot posture is usually digitigrade, and the hind limbs are elongate. The opposable hallux, probably inherited by peramelids from an arboreal ancestral stock, is rudimentary or may be lost. The

Figure 5–21. Two peramelid marsupials. A, A New Guinea bandicoot (*Peroryctes raffrayanus*; photograph by Stanley Oliver Grierson); B, a "rabbit" bandicoot (*Macrotis lagotis*; photograph by Anthony Robinson).

second and third digits of the hind foot are joined (syndactylous) as far as the distal phalanges by an interdigital membrane, and the muscles of these digits are partially fused, allowing them to act only in unison (Jones, 1924). An extreme degree of cursorial specialization occurs in the pig-footed bandicoot (*Chaeropus ecaudatus*), in which the forelimb is functionally didactyl and the hind foot is functionally monodactyl during running. The

second and third digits of the forelimb are large and clawed; the first and fifth are absent, and the fourth is vestigial. The curious appearance of the forefoot is the source of the animal's common and generic names (*Chaeropus* means pig-footed in Greek).

The structure and function of the specialized peramelid hind foot are unique and have been described in detail by Marshall (1972). In mammals, extreme reduction

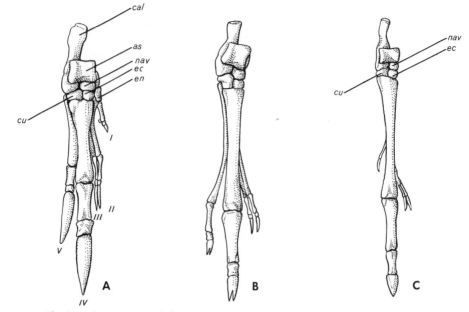

Figure 5-22. The feet of some peramelid marsupials; the least specialized foot is on the left and the most specialized is on the right. A, A long-nosed bandicoot (*Perameles* sp.); B, a rabbit bandicoot (*Macrotis* sp.); C, the pig-footed bandicoot (*Chaeropus ecaudatus*). Abbreviations: *as*, astragalus; *cal*, calcaneum; *cu*, cuboid; *ec*, ectocuneiform; *en*, entocuneiform; *nav*, navicular. The digits are numbered in A. (After Marshall, 1972.)

in the number of digits is usually associated with good running ability. Most highly cursorial ungulates have retained only the third digit (in the case of the horse) or digits three and four (in the case of some antelope). A similar trend occurs in peramelids. Probably partly due to an early development of syndactyly involving the second and third digits and the use of these digits for grooming, the general trend in the cursorial peramelids is toward the reduction of all digits but the fourth, with a great enlargement of this digit (Fig. 5–22). These specializations are accompanied by an alteration in the structure and function of the tarsal bones. The ectocuneiform bone makes broad contact with the proximal end of the fourth metatarsal and partially supports this digit, a character unique to peramelids. The mesocuneiform is lost, and the weight of the body is borne mainly by the cuboid, ectocuneiform, navicular, and astragalus bones. The calcaneum does not serve a major weight-bearing function, but of course serves as a point of insertion for the extensors of the foot, muscles of great importance in locomotion.

Horses, antelope, and peramelids provide beautiful examples of a similar functional problem, refining running ability, that is solved by different structural means. In the horse, only the third digit is retained and it is supported largely by the ectocuneiform, the navicular, and the astragalus (Fig. 14–4, page 241). In the pronghorn antelope, only two digits are retained and the cannon bone (the fused third and fourth metatarsals) is supported largely by the fused cuboid and navicular and the fused mesocuneiform and ectocuneiform (Fig. 14–5); the calcaneum is no longer a weight-bearing element. In the most cursorial peramelid (*Chaeropus*), the fourth digit is greatly enlarged and is supported, as outlined above, by the cuboid, navicular, ectocuneiform, and astragalus (Fig. 5–19). Marshall (1972) points out, however, that the structure of the hind limb of peramelids is not entirely modified for running, perhaps because of the burrowing tendencies of these animals. The fibula is large and movement at the ankle joint is not restricted to a single plane, as it is in most cursorial mammals. (Cursorial adaptations are discussed in more detail on page 238.)

Bandicoots are largely insectivorous, but small vertebrates, a variety of invertebrates, and some vegetable material are also eaten. Some species take refuge in nests that they build of sticks and other plant debris; all species of *Macrotis* dig burrows in which they hide during the day. *Chaeropus* is reported by Jones (1924:171) to squat like a jackrabbit beneath saltbushes (*Atriplex*) in semiopen areas; in jackrabbit fashion, *Chaeropus* depends on its speed to escape enemies, but seeks shelter in hollow logs when chased by dogs. The limb structure of *Chaeropus* is considerably more specialized in some ways than that of rabbits, but the styles of locomotion and methods of escape are similar. Unfortunately, *Chaeropus* seems never to have been observed by a modern biologist and is probably extinct. It was last seen in about 1926.

FAMILY CAENOLESTIDAE. Recent members of this family (three genera and seven species) bear the common name of rat opossum, and are restricted to parts of the Andes Mountains of northern and western South America. The earliest known caenolestids are from the Eocene of South America, and in the Oligocene and Miocene a diverse group of caenolestids, including some highly specialized types (Fig. 5–11), appeared. The Recent caenolestids are representatives of a conservative evolutionary line that dates at least from the early Eocene. There are three separate and seemingly relict populations of caenolestids: the northern population occupies parts of western Venezuela, Colombia, and Ecuador; the intermediate population occurs in a section of the Andes in southern Peru; and the southern population occupies Chiloé Island and a nearby coastal area of south central Chile.

The rat opossums are rat- or mouse-sized marsupials that resemble shrews because of their elongate heads and small eyes (Fig. 5–23). The skull is elongate and the brain is primitive; the olfactory bulbs are large and the cerebrum lacks fissures. The dental formula is 4/3-4, 1/1, 3/3, 4/4 = 46 or 48; the first lower incisors are large, and the remaining lower incisors, the canine, and the first premolar are unicuspid. The atlas bears a movable cervical rib. The feet are five-toed and unspecialized. The tail is long but not prehensile, and there is no marsupium.

Caenolestids appear to be ecological homologues of the shrews, but, in contrast to shrews, the rat opossums may be a declining group. They are largely terrestrial, but are agile climbers, eat mostly insects, and occur primarily in mesic, forested areas. The limbs and style of locomotion are unspecialized, as in shrews.

PHALANGEROID MARSUPIALS: PHALANGERIDAE; PETAURIDAE; BURRAMYIDAE; TARSIPEDIDAE; PHASCOLARCTIDAE

The phalangeroid marsupials are considered here to include the Australian families listed, which are referred to by such names as possum, ringtail, cuscus, glider, noolbender, and koala. Until recently, all of the structurally and behaviorally diverse phalangeroid marsupials were in-

Figure 5–23. A caenolestid marsupial (*Lestoros inca*). This individual was taken in Peru, at an elevation of 3530 m. (Photograph by John A. W. Kirsch.)

cluded in a single family, Phalangeridae, but Ride (1970) divided this family into the families listed above; I am following his system of classification.

The phalangeroid marsupials occupy a geographic range that includes much of Australia and many of its coastal islands, Tasmania, New Guinea and nearby coastal islands, Celebes, Timor, and Ceram. To the east of New Guinea, they inhabit the Bismarck and Louisiade archipelagos, D'Entrecasteaux Group, Solomon Islands, and, by introduction, New Zealand. These are the most widely distributed Australian marsupials, a fact probably explained by their arboreal habits and the high probability of their "island hopping" on rafts of vegetation. Phalangeroids are first known from the Late Oligocene of Tasmania (Tedford, et al.; 1975).

The phalangeroid skull is usually broad and somewhat flattened, and the lower tooth row is usually interrupted by a diastema (Figs. 5–15B, 5–24A). The dental formula is 2-3/1-2, 1/0, 1-3/1-3, 3-4/3-4 = 24-40. At most, three upper incisors are present. The first lower incisor is long and robust, and often projects forward; the second and third lower incisors are absent or vestigial. The molars are either moderately flat-crowned and tubercular or have ridges (Figs. 5–16E, F; 5–24B); in the single case of *Tarsipes*, they are small and peglike.

The limbs are adapted for climbing. The hands have five clawed digits; the first and second digits in some species are opposable to the remaining digits. The first digit of the hind foot is large, opposable, and clawless; digits two and three are syndac-tylous, and digits four and five are robust (Fig. 5–6E). The tail is long (with the exception of the koala, *Phascolarctos cinereus*, in which it is vestigial) and is often prehensile. The marsupium is present, and in all but the koala it opens anteriorly.

FAMILY PHALANGERIDAE. In this family are the possums and cuscuses, a group of primarily arboreal animals. Three genera and about 10 species are known. The brush-tailed possum (*Trichosurus vulpecula*) is one of the most familiar of Australian mammals, for it frequently maintains resident populations in suburban areas, where it often seeks shelter in roofs of houses and feeds on cultivated plants.

These marsupials are of moderate size, ranging in weight from approximately 1 to 6 kg. The skull is broad and has deep zygomatic arches (Fig. 5–15B). The molars are bilobed with rounded cusps (Fig. 5–16E, F). As adaptations to arboreal life, the hands and feet are large and have a powerful grasp, and the tail is prehensile. The cuscuses have short ears, woolly fur, and an odd, teddy bear–like appearance (Figs. 5–25, 5–26).

Members of this family mostly inhabit wooded areas, but the adaptable brush-tailed possum also occupies treeless areas, where it takes refuge in rocks or in the burrows of other mammals. This animal is locally destructive to plantations of introduced pines. Phalangerids are omnivorous, and are known to take a wide variety of plant material as well as insects, young birds, and birds' eggs. The brush-tailed possum is solitary and has a sternal scent

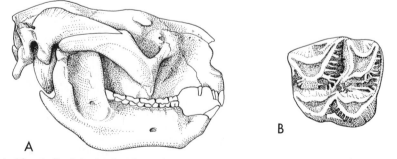

A   B

Figure 5–24. A, The skull of the koala (*Phascolarctos cinereus*, Phascolarctidae; length of skull 132 mm). B, Occlusal view of the second molar, upper right tooth row, showing the crescentic areas of dentine, exposed by wear, and the complex pattern of furrows.

Figure 5-25. Two views of a New Guinea cuscus *(Phalanger maculatus,* Phalangeridae). Note the prehensile tail with the "traction ridges" on the bare distal part of its ventral surface. (Photographs by Stanley Oliver Grierson.)

gland, considerably larger in males, which produces a musky smell that is used in the scent marking of objects within the animal's territory. This marsupial is one of Australia's most valuable fur bearers. During 1959, some 107,500 brush-tailed possum skins were marketed in Victoria.

FAMILY PETAURIDAE (Fig. 5-27). This family includes the ringtails, so named because of their prehensile tails, and the greater and lesser gliding possums, some of the handsomest and most remarkable of all marsupials. Eight genera and some 25 species are currently recognized.

Most members of this family are fairly small; weights range from about 100 gm to 1.5 kg. The skull is broad, and the four-cusped molars have fairly sharp outer cusps forming a roughly W-shaped ectoloph. The tail is prehensile in some petaurids and long and bushy in others. Some species are strikingly marked (Fig. 5-27B, C). The gliders *(Petaurus* and *Schoinobates)* have furred membranes that extend between the limbs and function as lifting surfaces for gliding. In these gliders, the claws are sharp and recurved, like those of a cat, and increase the ability of the animal to cling to the smooth trunks and large branches of trees.

The petaurids are nocturnal and arboreal creatures and occur in wooded areas. (Figure 5-28 shows a habitat of the gliders.) Ringtails are nocturnal and are strictly herbivorous, eating both leaves and fruit. They make conspicuous nests ("dreys") of leaves and twigs in the dense scrub of eastern and southeastern Australia. Curious specializations, similar to those of the primate *Daubentonia,* occur in two petaurid genera of "striped possums." In *Dactylopsila,* and to a more advanced degree in *Dactylonax,* the fourth digit of the hand is elongate and slender, and its claw is recurved. In addition, the incisors are robust and function roughly as do those of rodents. Striped possums tear away tree bark with their incisors, and extract insects from crevices and holes in the wood with the specialized fourth finger and the tongue. The conspicuous striped color pattern of *Dactylopsila* (Fig. 5-27C) is of interest because it is associated, as in skunks, with a powerful, musky scent.

Figure 5–26. Two New Guinea cuscuses (Phalangeridae). A, *Phalanger vestitus* (photograph by Stanley Oliver Grierson); B, *P. orientalis* (photograph courtesy of Hobart M. Van Deusen).

Figure 5–27. Three members of the family Petauridae. A, A ring-tailed possum *(Pseudocheirus forbesi)*; B, a gliding possum *(Petaurus breviceps)*; C, a striped possum *(Dactylopsila trivirgata)*. (Photographs by Stanley Oliver Grierson.)

The gliders are strikingly similar to flying squirrels *(Glaucomys)* in gliding style and ability and some can glide over 100 m. *Schoinobates,* the greater glider, is remarkable in having perhaps the most specialized marsupial diet: its food is entirely leaves and blossoms, chiefly those of eucalyptus trees. Sugar gliders *(Petaurus breviceps;* Fig. 5–27B) live in family groups, and scent marking plays an important role in the social organization of the group. Each individual has a particular odor recognized by the others. The cohesion of the group is also aided by mutual scent marking, for all members of the group become permeated with the scent of the group's dominant males (Schultze-Westrum, 1964).

FAMILY BURRAMYIDAE. The type genus of this family was known for many years only from Pleistocene fossil material; finally, in 1966, at a ski lodge on Mt. Hotham in Victoria, a representative of the genus was found alive. More recently, it has been found at other localities. This family contains five genera and eight species of small mouselike marsupials, called pigmy possums.

These diminutive marsupials are from about 60 to 120 mm. in head and body length, are delicately built, and have large eyes and mouselike ears (Fig. 5–29). The tail is long and prehensile in all species, and has a lateral fringe of hairs in the pigmy glider *(Acrobates pygmaeus).* This species has a narrow gliding membrane that is bordered by a fringe of long hairs. Traction between the digits and the trunks and branches of trees is increased in the pigmy glider by expanded pads at the tips

Figure 5–28. Rain forest in eastern Victoria, Australia. This community is inhabited by two species of lesser gliding possums (*Petaurus*) and by greater gliding possums (*Schoinobates volans*). (Photograph by Diana Harrison.)

of the fingers and toes; the surfaces of the pads have "traction ridges" that further increase the clinging ability of these animals.

Members of this family are restricted to wooded areas. They are apparently insectivorous-omnivorous, but the feeding habits of some members of the group are not known. As in some small placental mammals, in some burramyid marsupials the ability to become torpid during cold weather is well developed. Pigmy gliders become torpid in their nests on cold days, and the tails of the members of two genera of pigmy possums (*Cercartetus* and *Eudromicia*) become greatly enlarged with fat as winter approaches and these animals undergo periods of torpor.

FAMILY TARSIPEDIDAE. This family contains but one species, the highly spe-

Figure 5–29. A pygmy possum (*Cercartetus concinnus*, Burramyidae). (Photograph by Anthony Robinson.)

cialized, slender-nosed honey possum or noolbender *(Tarsipes spencerae)*. This remarkable animal's many specializations obscure its relationships to other marsupials, and its taxonomic position has long been uncertain.

Tarsipes is small, only about 15 to 20 gm in weight, and has a long prehensile tail. The pelage is marked by three longitudinal stripes on the back. The rostrum is long and fairly slim, and the dentary bones are extremely slender and delicate. The cheek teeth are small and degenerate, and only the upper canines and two medial lower incisors are well developed. The snout is long and slender and the long tongue has bristles at its tip (these specializations are similar to those of some nectar-feeding bats of the family Phyllostomatidae). All digits but the syndactylous second and third digits of the hind feet have expanded terminal pads resembling to some extent those of the primate *Tarsius*.

Honey possums occur in forested and shrubby areas. Like the hummingbirds and nectar-feeding bats, honey possums feed on nectar, pollens, and, to some extent, on small insects that live in flowers. The long protrusile tongue is used to probe into flowers. *Tarsipes* can climb delicately over even the insecure footing of clusters of flowers at the ends of branches, and often clings upside down to flowers while feeding. Although the animal is still common in some areas today, the expansion of agriculture in southwestern Western Australia is restricting the honey possum's range.

FAMILY PHASCOLARCTIDAE. The familiar koala or native "bear" *(Phascolarctos cinereus)* is the sole member of this family. This highly specialized herbivore is restricted to some wooded parts of southeastern Australia.

The tufted ears, odd looking naked nose, and chunky tailless form make the koala one of the most distinctive of Australian marsupials (Fig. 5-30). These are fairly large marsupials; the adults range from 8 to 10 kg in weight. The skull is broad and sturdily built, and the dentary bones are deep and robust (Fig. 5-24A). The roughly quadrate molars have crescentic ridges (Fig. 5-24B), and there is a diastema in both the upper and lower tooth rows between the cheek teeth and the anterior teeth. Branches are grasped between the

Figure 5-30. The koala *(Phascolarctos cinereus,* Phascolarctidae). (Photograph by Diana Harrison.)

first two and the last three fingers of the hand, and between the clawless first digit and the remaining digits of the foot; the long curved claws aid in maintaining purchase on smooth branches.

Koalas are fairly sedentary and feed on only a few species of smooth-barked eucalyptus trees. Maturation of a koala takes considerable time. A single young is born and is carried in the pouch for six months, after which it rides on its mother's back for a few more months. The young koala is dependent on its mother for a year, and sexual maturity is not reached until three or four years of age. According to Ride (1970:88), koalas grunt in piglike fashion when feeding at night, and when alarmed make continuous wails. Koalas lived in southwestern Western Australia during the late Pleistocene, but no longer occur there even though suitable habitat is present.

FAMILY VOMBATIDAE. This family is represented by two living genera and four species. Known as wombats, these animals are completely herbivorous, and show remarkable structural convergence toward rodents. Because of the efforts of man, wombats have become scarce or absent over much of their former range, and now are restricted to parts of eastern and southern Australia, Tasmania, and the islands between Australia and Tasmania.

Wombats are stocky animals with small eyes and rodent-like faces (Fig. 5–31), and attain over 35 kg in weight (Troughton, 1947:144). The skull and dentition bear a striking resemblance to those of some rodents (Figs. 5–15D, 5–16D). The skull is flattened, the rostrum is relatively short, and the heavily built zygomatic arches flare strongly to the sides. The area of origin of the anterior part of the masseter muscle is marked by a conspicuous depression in the maxillary and jugal that is similar to the comparable depression in the maxillary and premaxillary of the beaver (*Castor*; see Fig. 10–13). The dental formula is 1/1, 0/0, 1/1, 4/4 = 24; all teeth are rootless and evergrowing. Only the anterior surfaces of the incisors bear enamel, and the incisors and the first premolars are separated by a wide diastema. The molars are bilophodont (Fig. 5–16D). As in rodents, the coronoid process of the dentary is reduced and the masseter muscle, rather than the temporalis, is the major muscle of mastication.

The limbs are short and powerful, and the foot posture is plantigrade. The forefeet have five toes; all digits have broad, long claws. The hallux of the hind foot is small and clawless, but the other digits have claws. Digits two and three of the hind feet are syndactylous. The tail is vestigial. The marsupium opens posteriorly and contains one pair of mammae.

Figure 5–31. A wombat (*Vombatus ursinus*, Vombatidae). (Photograph by Anthony Robinson.)

This family is first known from the late Tertiary (Gill, 1957), and both Recent genera have fossil species. The Pleistocene trend toward large size that is apparent in many other mammalian groups also occurred in the Vombatidae, as evidenced by the huge Pleistocene "wombat" *Phascolonus*.

Wombats are powerful burrowers. Their burrows are extensive networks of tunnels and are sufficiently wide to admit a small person. As a boy, Peter Nicholson (1963) studied wombats in Victoria by crawling into their burrows. He found that almost without exception they were amiable and inquisitive, and that they seemed sociable and visited each other's burrows. Young wombats learned to burrow within their mother's burrow system by digging small subsystems, but abandoned the maternal burrows about four months after leaving their mother's pouches. Wombats dig burrows in the open or beneath rock piles as do marmots (*Marmota*), their North American rodent counterparts. Level or mountainous terrain supporting dry or moist sclerophyll forests or grassland is inhabited, and wombats are able to go for long periods without drinking water. Their food is largely herbs and grass, but includes bark, roots, and fungi (Troughton, 1947:139, 140). The widespread destruction of wombats in settled areas is not desirable from a biologist's viewpoint, but seems inevitable because of "conflict of interests" with man. The openings to wombat burrows are hazardous to large livestock, and wombats are locally destructive to crops. Some encouragement comes from Blanchetown in South Australia: a 3000-acre reserve for wombats near this town has been purchased by public subscription (Ride, 1970:92).

FAMILY MACROPODIDAE (Fig. 5–32). Members of this familiar marsupial group, which includes the kangaroos, euros, and wallabies, are the ecological equivalents of such ungulates as antelope. Both macropodids and ungulates are cursorial, and both have highly specialized limbs, but rather than using quadrupedal locomotion, as do the ungulates, the macropodids are primarily bipedal. Further, both groups are herbivorous, and have skulls and dentitions specialized for this mode of feeding. Even some specializations of the digestive system are similar in these two groups. Ride (1970:50) cites studies by Professors H. Waring and A. R. Main of the University of Western Australia on digestion in the tammar (*Macropus eugenii*), euro (*M. robustus*) and quokka (*Setonix brachyurus*). These animals have intestinal bacteria that digest the cell walls of plants; these (and probably other) macropodids can thus utilize the digestible inner parts of the cells, can utilize the byproducts of the bacterial digestion, and are able to digest the bacteria. The ruminant ungulates also depend on bacteria to increase the efficiency of their utilization of vegetation (see p. 244).

The present distribution of the 19 Recent genera and approximately 47 Recent species of macropodids includes New Guinea, Bismarck Archipelago, the D'Entrecasteaux Group, Australia, and, by introduction, some islands near New Guinea and New Zealand. The family Macropodidae appears first in the middle Tertiary (late Oligocene or early Miocene) of Australia. Wallabies and a kangaroo are known from the late Tertiary, and in the Pleistocene unusually large macropodids occurred.

Living macropodids vary tremendously in size and structure. The musky rat kangaroo (*Hypsiprymnodon moschatus*) weighs only 500 g, whereas the great gray kangaroo (*Macropus giganteus*), the largest living marsupial, reaches 2 m in height and approximately 90 kg in weight. The marsupium is usually large and opens anteriorly. The macropodid skull is moderately long and slender, and the rostrum is usually fairly long (Fig. 5–15C). The dental formula is 3/1, 1-0/0, 2/2, 4/4 = 32 or 34. The upper incisors have sharp crowns with their long axes oriented more or less front-to-back. The tips of the procumbent lower incisors are held against a leathery pad just behind the upper incisors when the animals gather vegetation. This specialized arrangement serves a cropping

Figure 5–32. Three kinds of macropodid marsupials. A, Great gray kangaroo (*Macropus giganteus*); B, pademelon (*Thylogale billardierii*); C, red-necked wallaby (*Macropus rufogriseus*). (Photographs by Diana Harrison.)

function, similar to that of the lower incisors and the premaxillary pad in the artiodactylan ungulates lacking upper incisors. There is a broad diastema between the macropodid incisors and the premolars. The molars are quadritubercular and bilophodont (Fig. 5–16G). In many macropodids, the last molar does not erupt until well after the animal becomes adult. A unique situation occurs in the little rock wallaby *(Peradorcas concinna)*, in which nine molars may erupt in succession. Usually four or five molars are functional at one time, and replacement is from the rear as the molars are successively lost from the front.

Macropodids are highly specialized for jumping. The forelimbs are five-toed and usually small; they are used for slow movement on all fours or for food handling (Frith and Calaby, 1969). The hind limbs are elongate, especially the fourth metatarsal. The hallux is missing in all but *Hypsiprymnodon*. Digits two and three are small, syndactylous, and used for grooming; the fourth is the largest digit; and the fifth is often also robust (Fig. 5–33). The unusual pattern of digital reduction and the dominance of the fourth digit in the most highly cursorial Australian marsupials are perhaps due to their arboreal ancestry. In these ancestors, the foot was five-toed and the hallux was opposable; the fourth was the longest remaining digit, and the foot was adapted to grasping branches. With specialization of the foot for running or hopping, the hallux was lost and the longest toe, the fourth, became the most important digit. In most macropodids the foot is functionally two-toed during rapid locomotion, which is characteristically bipedal, but in *Megaleia* the foot

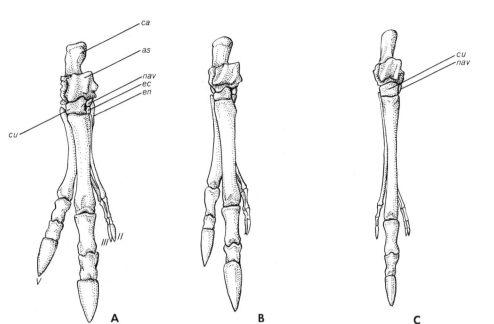

Figure 5-33. The feet of some macropodid marsupials; the least specialized foot is on the left, and the most specialized is on the right. A, A scrub wallaby *(Thylogale* sp.); B, a kangaroo *(Macropus* sp.); C, red kangaroo *(Megaleia rufa)*. Abbreviations: *as*, astragalus; *ca*, calcaneum; *cu*, cuboid; *ec*, ectocuneiform; *en*, entocuneiform; *nav*, navicular. The digits are numbered in A.

is functionally one-toed (Fig. 5–33C). In the macropodid tarsus there is no contact between the ectocuneiform and the fourth metatarsal (in contrast to the arrangement in the Peramelidae; see Fig. 5–22). Because the hind limb posture of macropodids is basically plantigrade, the calcaneum is an important weight-bearing element of the tarsus (which it is not in the digitigrade Peramelidae). The macropodid tail is usually long and robust, and functions in the more specialized species as a balancing organ and as the posterior "foot" of the tripod formed by the plantigrade hind feet and tail, on which the animal can sit when not in motion.

Several macropodid genera depart from the familiar structural pattern of kangaroos and from the grazing or browsing habit. *Hypsiprymnodon,* a muskrat-sized inhabitant of rain forests and riparian situations, has a tail of modest length and retains all of the digits of the hind foot. The hind limbs are not greatly elongate, and the animal uses quadrupedal rather than saltatorial locomotion. Animal material forms a

large share of the food of this seemingly primitive macropodid. The tree kangaroos *(Dendrolagus,* Macropodinae) spend considerable time on the ground, but frequently use their arboreal ability to escape from danger. This mode of life is reflected by the large and robust forelimbs with strong recurved claws; by the hind limbs, which are not strongly elongate; and by the short, broad hind foot (Fig. 5–19). Saltation, typical of terrestrial kangaroos, has not been completely abandoned by tree kangaroos; not only are these animals agile climbers, but also they leap between trees and from trees to the ground. Their food is large fruit and leaves.

The running ability of the larger kangaroos *(Macropus)* is impressive. Speeds on level terrain of roughly 50 km per hour are attained, and leaps covering distances of 8.5 m and heights of 3 m have been reported (Troughton, 1947:213). The solitary hare wallabies *(Lagorchestes),* which are roughly the size of a large rabbit, also are renowned for their great speed. These animals have a jackrabbit-like style of escape. They hide beneath bushes or clumps of grass, burst out suddenly when frightened, and run away at high speed. The highly developed jumping ability of macropodids allows these animals to move easily for long distances between scattered sources of water or forage and to escape enemies by erratic leaps. These abilities, rather than the capacity for great speed, are perhaps of primary adaptive importance. Saltation may have been developed by small forms ancestral to kangaroos as a means of erratic escape in open areas. This ricochetal style of locomotion is known in a number of desert-dwelling rodents. According to Howell (1944:247), the kangaroo's "method of traveling by saltation was hardly begun for the purpose of ultimate speed. Rather has it built speed into the locomotor pattern that was already established, probably for some other purpose." The locomotion of rock wallabies (*Petrogale*) is adapted to the rocky country they inhabit. According to Jones (1924:231), their movements are spectacular: "There seems to be no leap it will not take, no chink between boulders into which it will not hurl itself."

INSECTIVORES, DERMOPTERANS, AND BATS

6

This chapter deals with insectivores and some of their descendants. An insectivore ancestry is generally accepted for bats and, on the basis of limited fossil material and anatomical evidence based on Recent species, the flying lemurs (Dermoptera) are also thought to have an insectivore derivation. The flying lemurs are customarily put in a separate order, as by Simpson (1945: 53–54). But Miller (1907) classified them as bats, whereas Romer (1968:179) saw little reason for separating flying lemurs from insectivores and listed the Dermoptera as a suborder of the Insectivora (1966: 380). I have followed recent tradition in considering the Dermoptera to be a separate order, but because the chiropterans and dermopterans probably arose from insectivore stock, I feel justified in discussing these groups in the same chapter.

ORDER INSECTIVORA

A firm introductory warning about the reality and limits of this order is appropriate. Because the most primitive placental mammals were seemingly insectivores and

their descendants have retained dentitions that are still adapted to an insect diet although often highly specialized, the tendency is to include a number of primitive types together with all modern descendants in the Insectivora. As an example, the families Tupaiidae and Macroscelididae, currently included in the order, may well be living members of a primitive group that should most appropriately be separated from the Insectivora, but the fossil record is too sparse to allow a resolution of this problem.

Taxonomic assignment of some equivocal fossil types has also been difficult and controversial. Members of the Cretaceous Asiatic family Deltatheriidae (Fig. 6–1A), for example, have been regarded by some as insectivores important in the evolution of such groups as carnivores; but, on the basis of more complete fossil material, these animals now appear to be members of a sterile evolutionary line, separate from both the marsupials and eutherians, that died out in Late Cretaceous (Lillegraven, 1974:265). The present classification of the unwieldly assemblage of "insectivores" may actually reflect phylogenetic relationships, but we

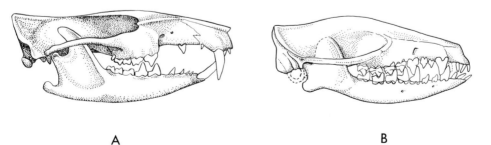

A   B

Figure 6–1. Skulls of insectivore-like mammals from the Cretaceous. A, *Zalambdalestes* (length of skull 50 mm; after Romer, A. S.: *Vertebrate Paleontology*, 3rd ed., The University of Chicago Press, 1966). B, *Kennalestes* (length of skull 27 mm; after Kielan-Jaworowska, 1975).

lack sufficient evidence to be sure that it does. The problem in any case is complex, and one must remember that because the order Insectivora has long been used as a convenient grab bag repository for taxa of doubtful affinities, it should be viewed with suspicion.

Insectivores comprise today the third largest order of mammals, containing roughly 77 genera and 400 species. They have an unusually wide geographic distribution: they occur throughout much of both hemispheres, but are absent from most of the Australian region, all but the northernmost part of South America, and the polar regions.

The order Insectivora is ancient, and insectivores are probably the most direct living descendants of primitive eutherian mammals. Zofia Kielan-Jaworowska, of the Polish Academy of Sciences, led a number of important paleontological expeditions between 1963 and 1971 to the Gobi Desert of Mongolia, where excellent late Cretaceous specimens of mammals were found. Some of these fossils are regarded as eutherian (Kielan-Jaworowska, 1969b). These fossils include the small eutherian *Kennalestes* (Fig. 6–1B), which can here serve as an example of an early insectivore. *Kennalestes* may ultimately have provided the basal stock for the establishment of eutherians in North America in the Late Cretaceous. Kielan-Jaworowska suggests (1975) that the earliest known North American eutherian evolved from the Asiatic *Kennalestes*. As previously mentioned, the marsupials apparently originated in the Early Cretaceous in North America and never reached Asia, whereas the eutherians may have originated in the Old World and dispersed to the New World in Late Cretaceous. Future paleontological work will perhaps support this premise, but, whether it does or not, the basal eutherian stock seems to be the Insectivora.

MORPHOLOGY

Because of the broad structural diversity that occurs among the mammals included in the Insectivora, an anatomical diagnosis of the group is difficult to frame. The order Insectivora has been separated by many authorities into two major divisions (or suborders): the lipotyphlans, including the families Erinaceidae, Talpidae, Tenrecidae, Chrysochloridae, Solenodontidae, and Soricidae, and the menotyphlans, including the Macroscelididae and Tupaiidae. This division is based on a series of morphological differences that indicate a long separation of the lines of descent represented by these two groups.

Because of resemblances between primitive prosimian primates, such as lemurs, and the family Tupaiidae, some authors have placed this family under the order Primates (Simpson, 1945:61; Walker, 1968:395). Evidence based on studies of the nervous system, however, indicates no close tupaiid-primate relationship (Campbell, 1966), and paleontological evidence as interpreted by Szalay (1968) indicates that tupaiids were derived from the Insectivora and not from the Primates. The taxonomic scheme used here—the retention of the menotyphlan families within the Insectivora—is upheld by Findley (1967:103), who states that "acceptance of the menotyphlans as a natural taxon and the grouping of them with insectivores seem justified as a matter of practical expediency."

Questions of taxonomy aside, the structural contrasts between the lipotyphlans and menotyphlans indicate that these groups occupy contrasting adaptive zones. Broadly speaking, the lipotyphlans have primitive brains and depend more on olfaction than on vision; they usually have specialized dentitions; their limbs are generally unspecialized and they pursue a generalized quadrupedal locomotion. Menotyphlans, in contrast, have more progressive brains and acute vision; the dentitions are specialized along less insectivorous lines; and locomotion is less generalized (macroscelidids hop and many tupaiids climb).

For all Recent lipotyphlous insectivores, the following characters are reasonably diagnostic: the tympanic bone is annular, no auditory bulla is present, and the ento-tympanic bone is absent; the tympanic cavity is often partially covered by processes from adjacent bones; the olfactory bulbs are longer than the rest of the brain and are largely interorbital; the eyes and the optic foramina are usually small; the jugal is reduced or absent and the zygomatic arch is incomplete in some groups; the orbitosphenoid is mainly anterior to the braincase; the teeth have sharp cusps, and usually the crown pattern of primitive placentals is recognizable; the anterior dentition is often modified by the enlargement and specialization of the incisors and the reduction of the canines; the limbs (except those of the fossorial groups) are usually unspecialized and are never adapted to saltation.

Menotyphlans are characterized by the following features: the auditory bulla is complete and the entotympanic is large; the olfactory bulbs are shorter than the rest of the brain and do not extend between the orbits; the eyes and the optic foramina are enlarged; the jugal is large and the zygoma is complete; the orbitosphenoid forms part of the braincase; the hind limbs are greatly elongate and adapted to hopping (Macroscelididae; Fig. 6-2C) or are slightly elongate, with the manus and pes enlarged, and are adapted to climbing (Tupaiidae; Fig. 6-2B).

VENOMOUS INSECTIVORES

Except for the duck-billed platypus (Order Monotremata), the only mammals known to be venomous are shrews. Over 350 years ago there were reports on the effects on man of being bitten by the short-tailed shrew of North America (*Blarina brevicauda*), and work in the present century has confirmed that this animal has venomous saliva. It has also been demonstrated that the European water shrew (*Neomys fodiens*) and the Haitian solenodon (*Solenodon paradoxus*) are venomous. The salivas of other insectivores closely related to the two shrews listed above have been studied but are apparently not toxic. In some people, the bites of *Suncus murinus* cause minor aches, and hypersensitivity and reddening of the skin, especially at the finger joints (G. L. Dryden, personal communication).

Both *N. fodiens* and *B. brevicauda* have similar adaptions for delivering venom and the effects of the venoms are similar. (Venomous insectivores and their venoms are discussed by Pournelle, 1968, and Pucek, 1968.) In these shrews, the first lower incisors have concave medial surfaces, forming a crude channel, and the ducts from the venom-producing submaxillary salivary glands open near the bases of these teeth. *Neomys fodiens* salivates copiously during attacks on prey, and saliva is seemingly channeled to wounds via the two first lower incisors. Pearson (1942) showed that mice injected with extracts of the submaxillary glands of this shrew were strongly affected; the activity of the mice was reduced rapidly by what seemed to be a neurotoxic action (a neuro-

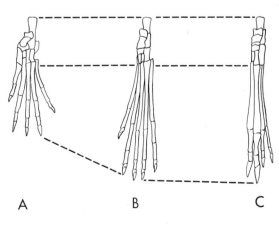

Figure 6-2. Left hind feet of several insectivores, showing differing degrees of elongation of the metatarsals and phalanges. A, Gymnure (*Echinosorex*, Erinaceidae); B, a tree shrew (*Tupaia*, Tupaiidae); C, an elephant shrew (*Rhynchocyon*, Macroscelididae). (After Evans, 1942.)

A   B   C

toxin impairs normal function of the nervous system). Frogs bitten by *N. fodiens* were partially immobilized, and when forced to move seemed uncoordinated. Laboratory mice injected with a homogenate of these salivary glands immediately developed paralysis of the hind limbs.

The toxic saliva of *Solenodon* is produced by the submaxillary salivary glands, is carried by ducts to the bases of the large and deeply channeled posterior surfaces of the second lower incisors, and presumably enters a wound by capillary action. The saliva of this insectivore is similar in effect to the salivas of the venomous shrews.

Perhaps of greatest interest is the functional importance of venom to insectivores. *Blarina brevicauda* can kill mice considerably larger than itself, and Eadie (1952) reported that meadow voles (*Microtus pennsylvanicus*) were an important fall and winter food of this shrew. Frogs and small fish are known to be preferred foods of *N. fodiens*. Both of these shrews attack prey from behind and direct bites at the neck and base of the skull, an area where neurotoxic venom might be readily introduced into the central nervous system. The adaptive importance to a very small predator of making its relatively large prey helpless would seem to be great, and one wonders why more shrews are not venomous.

FAMILY ERINACEIDAE. Members of the family Erinaceidae, the hedgehogs, are the most primitive lipotyphlans. The family is represented today by 10 genera and 14 species; they occur in Africa, Eurasia, southeastern Asia, and the island of Borneo. Erinaceids are first known from the Oligocene, and fossil material is known from the Oligocene to the Pliocene in North America, and from the Oligocene to the Recent in the Old World. The family Adapisoricidae, containing primitive relatives of the hedgehogs, is represented from the Cretaceous to the Oligocene in both hemispheres.

Erinaceids vary from the size of a mouse to that of a small rabbit (1.4 kg). The eyes and pinnae are moderately large, and the snout is usually long. The zygomatic arches are complete. The dental formula is 2-3/3, 1/1, 3-4/2-4, 3/3 = 36-44. The first upper and, in some species, the first lower incisors are enlarged, but the front teeth (Fig. 6–3) never reach the degree of specialization typical of shrews. In hedgehogs, the upper molars have simple nonsectorial cusps, with the paracone and metacone near the outer edge; the hypocone completes the quadrate form of the upper tooth. Both the trigonid and talonid of the lower molars are well developed (Fig. 6–4B). The molars are thus better adapted to an omnivorous than to an insectivorous diet. The feet retain five digits in all but one genus, and the foot posture is plantigrade. An obvious specialization is the possession of spines in members of the subfamily Erinaceinae (Fig. 6–5). In these animals, the sheet of muscle beneath the skin (panniculus carnosus) is greatly enlarged and controls the pulling of the skin around the body and the erection of the spines.

In various parts of their wide range,

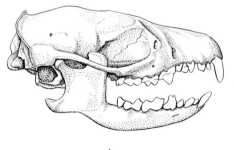

A

Figure 6–3. Skull of insectivore. A hedgehog (*Erinaceus* sp., Erinaceidae: length of skull 32 mm).

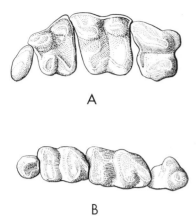

A

B

Figure 6-4. Cheek teeth of a hedgehog (*Erinaceus* sp., Erinaceidae). Fourth premolar and three molars of the upper right (A) and lower left (B) tooth rows.

hedgehogs occupy deciduous woodlands, cultivated land, and tropical and desert areas. Hedgehogs are omnivorous, but animal food seems to be preferred and a wide variety of invertebrates is taken. An African hedgehog (*Erinaceus albiventris*) that my family kept in captivity for a time in Kenya voraciously attacked beetles; he quickly chewed up medium-sized chrysomelid beetles, but had considerable difficulty with large scarabeid beetles.

Kingdon (1974:32) reports that hedgehogs will attack and kill small snakes, and that during the attacks hedgehogs direct their spines forward, leaving only a small part of their body exposed to the snakes' strikes. Hedgehogs seem remarkably resistant to snake venom. Some members of this family protect themselves by rolling into a tight ball with the spines erected.

Members of the subfamily Erinaceinae are probably heterothermic. (Heterothermic animals can regulate their body temperature physiologically, but temperature is not regulated precisely or at the same level at all times.) Hibernation occurs in the widespread genus *Erinaceus*, and estivation is practiced by the desert species *Parachinus aethiopicus*. A related species from India, *P. micropus*, has survived in captivity for periods of from four to six weeks without food or water (Walker, 1968:133), and this species and *Hemiechinus auritus* are known to have winter periods of dormancy in India. In Kenya, East Africa, *Erinaceus albiventris* disappears and apparently hibernates through the long dry season from May to September or October. In this instance, the animals are probably responding primarily to food shortages, for temperatures remain moderate through this period. The ability to undergo periods of torpor, involving a lowering of body temperature (hypothermia) and a lowered metabolic rate, may be developed in all members of the subfamily Erinaceinae as a means of achieving metabolic economy during critical periods of stress due to cold, heat, or shortages of food.

FAMILY TALPIDAE. This family includes a group of small rat- or mouse-sized animals usually referred to as moles. These predominantly burrowing insectivores (15 genera and 22 species) occur in parts of North America, Europe, and Asia. The European fossil record of talpids begins in the late Eocene; talpids are known first in the New World from the Oligocene. Apparently, the anatomical modifications typical of Recent fossorial genera were attained early, for the Recent European genus *Talpa* is first known from the Miocene.

The head and forelimbs of most talpids are modified for fossorial (burrowing) life (Fig. 6–6). The zygomatic arch is complete, the tympanic cavity is not fully enclosed by bone, and the eyes are small and often lie beneath the skin. The snout

Figure 6-5. European hedgehog (*Erinaceus europaeus*) with young. (Photograph by G. L. Dryden.)

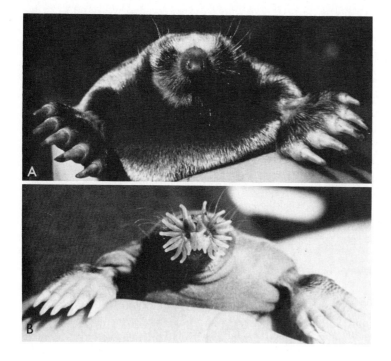

Figure 6–6. Heads and forelimbs of two moles (Talpidae). A, Hairy-tailed mole (*Parascalops breweri*). B, Star-nosed mole (*Condylura cristata*); the unique "star" of finger-like structures is found in no other mammal. (Photographs by G. L. Dryden.)

is long and slender, the ears usually lack pinnae, and the fur is characteristically lustrous and velvety. The dental formula is 2-3/1-3, 1/0-1, 3-4/3-4, 3/3 = 34-44. The first upper incisors are inclined backward (Fig. 6–7B), and the upper molars have W-shaped ectolophs (Fig. 6–8). In the fossorial species, the forelimbs are more or less rotated from the usual orientation typical of terrestrial mammals in such a way that the digits point to the side, the palms face backward, and the elbows point upward (Fig. 6–9B). In addition, the phalanges are short, the claws are long, and the clavicle and humerus are unusually short and robust. The scapula is long and slender (Fig. 6–9A) and serves both to

anchor the forelimb solidly against the axial skeleton and to provide advantageous attachments for some of the powerful muscles that pull the forelimb backward. The anteriormost segment of the sternum (the manubrium) is greatly enlarged and extends forward to beneath the base of the skull. These specializations serve both to increase the area for attachment of the large pectoralis muscles, and to move the shoulder joint forward and allow the forepaws to remove or loosen soil beside the snout.

The clavicle is short and broad, and provides a large secondary articular surface for the humerus (Fig. 6–9B). The double articulation of the shoulder joint,

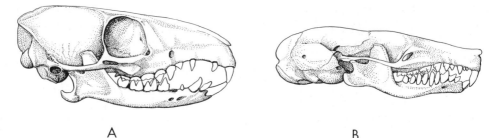

A   B

Figure 6–7. Skulls of insectivores. A, A tree shrew (*Tupaia* sp., Tupaiidae; length of skull 52 mm); B, the eastern mole (*Scalopus aquaticus*, Talpidae; length of skull 37 mm).

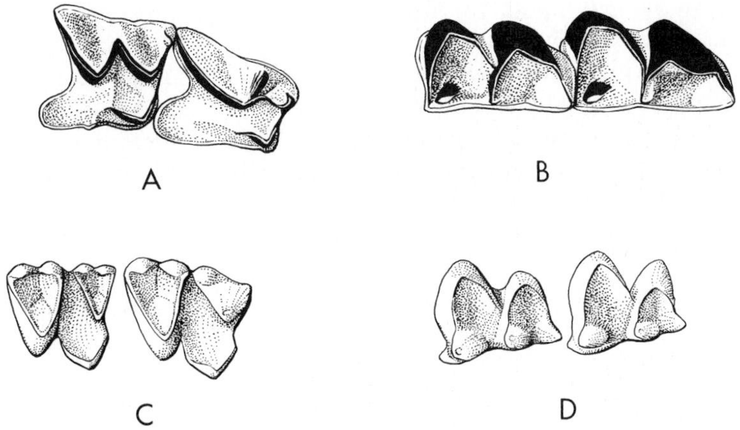

Figure 6–8. Cheek teeth of insectivores. The fourth upper right premolar and first molar (A) and the first two lower left molars (B) of the vagrant shrew (*Sorex vagrans*, Soricidae). The pigmented parts of the teeth are shown in black. The first and second upper right molars (C) and the comparable lower left molars (D) of the eastern mole (*Scalopus aquaticus*, Talpidae).

with articular contacts between the humerus and the scapula and clavicle, provides an unusually strong bracing for this joint during the powerful rotation of the

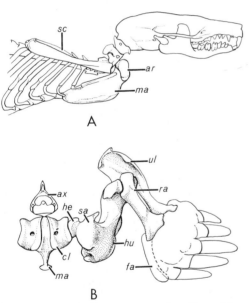

Figure 6–9. Details of the pectoral girdle and forelimb of the eastern mole (*Scalopus aquaticus*, Talpidae). A, Side view of the pectoral girdle, B, Anterior view of part of the pectoral girdle and the forelimb, with the shoulder joint slightly disarticulated to show the head of the humerus and its secondary articular surface. The head of the humerus articulates with the glenoid fossa of the scapula, and a second articulation (involving considerably larger surfaces) occurs between the secondary articular surface of the humerus and an articular surface of the clavicle. Abbreviations: *ar*, articular surface of the clavicle; *ax*, axis; *cl*, clavicle; *fa*, position of the falciform bone; *he*, head of the humerus; *hu*, humerus; *ma*, manubrium of the sternum; *ra*, radius; *sa*, secondary articular surface of the humerus; *sc*, scapula; *ul*, ulna.

humerus that accompanies the digging stroke of the forelimb. In some genera, the falciform bone is large and serves to increase the breadth of the forepaw and to brace the first digit (Fig. 6–9B).

One talpid, the Asiatic shrew mole (*Uropsilus*), lacks fossorial or aquatic specializations and resembles a shrew in general form. The remarkable subfamily Desmaninae inhabits the Pyrenees Mountains, and some mountains in Portugal, southeastern Europe, and parts of Russia. These animals are adapted to semiaquatic life; they have webbed forefeet, and the greatly enlarged hind feet are webbed and bear a fringe of stiff hairs that increase the effectiveness of the hind feet as paddles. These animals also have strange flexible snouts that have an extremely highly developed sense of touch and smell, and enable the animals to detect underwater prey.

Fossorial talpids occur typically in moist and friable soils in forested, meadow, or streamside areas, and feed largely on animal material. A species that occurs in the eastern United States (*Scalopus aquaticus*), however, locally penetrates the moderately dry sandhill prairies of eastern Colorado, where the characteristic ridges of soil appear only during wet weather. In most areas, these ridges are a common evidence of the presence of moles, and are made by the animals as they travel just beneath the surface by forcing their way through the soil. Soil from deep burrows is deposited on the surface in more or less conical "mole hills." The burrows of pocket gophers (*Geomys bur-*

sarius) are used to some extent by *Scalopus*, as indicated by the occasional capture in Colorado of moles in traps set for pocket gophers. The semiaquatic Old World desmanines live along the banks of lakes, ponds, or streams, and feed largely on aquatic invertebrates. Their burrows open beneath the surface of the water and extend upward to a nest chamber above the water level. Moles are not known to hibernate or estivate.

FAMILY TENRECIDAE. The tenrecs are a group of insectivores that vary widely in structure and in habits. Various tenrecs bear a general resemblance to such diverse mammals as shrews, hedgehogs, muskrats, mice, and otters. The family includes 11 genera and 23 species, and inhabits Madagascar, the Comoro Islands, and west central Africa. The meager fossil record of tenrecs indicates little about their evolution. The oldest fossil records are from the Miocene of East Africa, and Pleistocene fossils are known from Madagascar.

Tenrecs vary from roughly the size of a shrew to that of a cottontail rabbit. The snout is frequently long and slender. The jugal is absent, the eye is usually small, and the pinnae are conspicuous. The tympanic bone is annular and the squamosal forms part of the roof of the tympanic cavity. The anterior dentition varies between species; the first upper premolars are never present, and the molars are 3/3 in all but *Tenrec* (4/3) and *Echinops* (2/2). The upper molars have crowns that are triangular in occlusal view, and only in one genus (*Potamogale*) is a W-shaped ectoloph present in these teeth.

An unusually broad array of adaptive types occurs within the Tenrecidae. *Tenrec* roughly resembles a tailless, coarse-pelaged, long-snouted opossum (Fig. 6–10), and has spines interspersed with soft hairs. It is omnivorous. *Echinops* (Fig. 6–11), *Hemicentetes* (Fig. 6–12), and *Setifer* are also spiny, and the latter two genera resemble hedgehogs (Erinaceidae) closely. In these two genera, the panniculus carnosus muscle is powerfully developed, and enables the animals to erect the spines. It also contributes to the ability of these animals to roll into a ball. The feet and head are tucked beneath the body during this protective movement, and "the sphincter muscles running around the body at the junction of the spiny dorsum and the hairy venter permit the spiny dorsal skin to be drawn together, thus enclosing the animal in an impregnable shield of spines" (Gould and Eisenberg, 1966). These authors found that newborn *Echinops* and *Setifer* reacted to being disturbed by rolling into a ball. *Hemicentetes* has a group of 14 to 16 specialized quills on the middle of the back, the "stridulating organ," that rub together when un-

Figure 6–10. A tenrec (*Tenrec ecaudatus*, Tenrecidae). (Photograph by J. F. Eisenberg and Edwin Gould.)

Figure 6–11. A Madagascar "hedgehog" (*Echinops telfairi*, Tenrecidae). (Photograph by J. F. Eisenberg and Edwin Gould.)

derlying dermal muscles are twitched to produce sounds in a variety of patterns of repetition. Differences in these sounds depend on differences in associated behavior of the animals (Gould, 1965) and may be used in intraspecific communication.

The subfamily Potamogalinae includes animals that in many ways are the most remarkable tenrecids. The potamogales or otter shrews occur in west central Africa; they are the only living members of a primitive lineage, and have probably survived because of their highly specialized, semiaquatic style of life. Although the giant otter shrew (*Potamogale velox*) has been known to scientists since 1860, the genus to which the dwarf otter shrews belong (*Micropotamogale*) was not described until 1954 (Heim de Balsac, 1954).

The giant otter shrew is quite large for an insectivore, measuring some 600 mm in length and weighing about 1 kg, and is highly specialized for the life of a miniature otter. The body is long and stream-lined, the limbs are rather short and stocky, and the large tail is laterally compressed. Propulsion beneath the water is controlled by lateral movements of the flattened tail, and a number of features are associated with this locomotor style. The caudal vertebrae have high neural spines, transverse processes, and most have chevron bones (bones that form an inverted arch beneath the centra and occur on the caudal vertebrae of some mammals). These unusual caudal vertebrae provide attachments for the powerful tail musculature, which is aided by the greatly enlarged gluteal muscles. The posterior parts of the gluteal muscles, which in quadrupedal mammals serve to move the hind limbs, attach to the muscles overlying roughly the first five caudal vertebrae and move the tail. In the sinuous motion of the back and tail, and even in overall body form, *Potamogale* resembles a large salamander. Giant otter shrews live in permanent streams and rivers and in coastal

Figure 6–12. A streaked tenrec (*Hemicentetes semispinosus*, Tenrecidae). (Photograph by J. F. Eisenberg and Edwin Gould.)

swamps, and, although they rely partly on fish, they seem to prefer freshwater crabs to other food (Kingdon, 1974a:17). The habits of potamogales remain poorly known and provide fascinating opportunities for the resourceful biologist.

Some tenrecs are known to become torpid under natural conditions during seasons of food shortage (Eisenberg and Gould, 1970) or in the laboratory for unknown reasons (G. L. Dryden, personal communication).

FAMILY CHRYSOCHLORIDAE. Another variation on the fossorial insectivore theme is typified by chrysochlorids, the golden moles. These animals resemble "true" moles (Talpidae), but even more closely resemble, in fossorial adaptations and in function, the marsupial "moles" (Notoryctidae). The five genera and roughly 11 species comprising the family Chrysochloridae occur widely in southern Africa, where they occupy forested areas, savannas, and sandy deserts. The earliest fossil chrysochlorids from the Miocene of East Africa resemble Recent species, and these and Pleistocene fossil material give no firm evidence of the derivation of the group. Butler (1969) suggests that the Tenrecidae and the Chrysochloridae may be related.

Golden moles have modes of life similar to those of the fossorial members of the Talpidae and possess some parallel adaptations, as well as some contrasting structural features. The ears of golden moles lack pinnae, and the small eyes are covered with skin. The pointed snout has a leathery pad at its tip (Fig. 6–13A). The zygomatic arches are formed by elongate processes of the maxilla, and the occipital area includes bones, the tabulars, not typically found in mammals. The skull is rather abruptly conical instead of being flattened and elongate as in many insectivores. An auditory bulla is present and is formed largely by the tympanic bones; the malleus is enormously enlarged (Fig. 6–13B). The dental formula is usually 3/3, 1/1, 3/3, 3/3 = 40. The first upper incisor is enlarged, and the molars are basically tritubercular and

lack the stylar cusps and the W-shaped ectoloph typical of talpids. The permanent dentition of golden moles emerges fairly late in life. The forelimbs are powerfully built and the forearm rests against a concavity in the rib cage. The fifth digit of the hand is absent, and digits two and three usually have huge picklike claws. A "third bone" is present in the forelimb (Fig. 6–13B). The forelimbs are not rotated as are those of talpids, but more or less retain the usual mammalian posture, with the palmar surfaces downward.

Golden moles are adept burrowers. Bateman (1959) studied golden moles in the laboratory and found that a 60 g golden mole could push up a 9 kg weight covering its cage; this amounts to exerting a force equal to some 150 times the animal's weight. During digging, the wedge-shaped head butts upward as the claws of the powerful forelimbs sweep downward and backward (Kingdon, 1974a:24, 25), and when the animal is close to the surface a ridge marks the course of the burrower's progress. Both deep and shallow burrows are constructed; the depth of the burrows may depend on the amount of soil moisture. The roofs of shallow burrows in sandy soil frequently collapse, leaving a furrow in the sand as a trace of the former burrow. The diet of golden moles consists mostly of invertebrates; two desert-dwelling genera (Cryptochloris and Eremitalpa) also eat legless lizards. In sandy deserts Eremitalpa occasionally forages for insects blown by the wind into furrows in the sand (Eloff, 1967).

FAMILY SOLENODONTIDAE. Represented today by but two genera and two species, the solenodons seem to be relict types that are unable to survive in competition with other placentals recently introduced into their ranges. Solenodons occurred in sub-Recent and Recent times in Cuba, Haiti, and Puerto Rico, but are now restricted to Haiti (Solenodon paradoxus) and to Cuba, where a declining and endangered population of Atopogale cubanus occurs. The extinct genus Nesophontes occupied the West Indies at

Figure 6–13. The cape golden mole *(Chrysochloris asiatica;* Chrysochloridae). A, Note the leathery nose pad and the greatly enlarged claws of digits two and three. B, Note the enormously enlarged malleus. This specialization may aid in conduction of sound through the bones of the skull by increasing the inertia of the malleus and may improve hearing underground. The "third bone" of the forelimb replaces the flexor muscle of the third digit. (Photograph of external view by Jennifer U. M. Jarvis; X-ray photograph by E. N. Keen.)

least until the arrival of the Spaniards. The introduction by man of the house rat *(Rattus),* the mongoose *(Herpestes),* and dogs and cats into the West Indies and the extensive clearing of land for agricultural purposes combined to cause the rapid decline of the solenodons. These animals are now rare in most areas, and hopes for their survival under natural conditions seem dim.

Solenodons are roughly the size of a muskrat, and have the form of an unusually large and big-footed shrew. The five-toed feet and the moderately long tail are nearly hairless. The snout is long and slender, the eyes are small, and the pinnae are prominent. The zygomatic arch is incomplete, no auditory bulla is present, and the dorsal profile of the skull is nearly flat. The dentition is 3/3, 1/1, 3/3, 3/3 = 40. The first upper incisor is greatly enlarged and points backward slightly; the second lower incisor has a deep lingual groove that may function to transport the toxic saliva that empties from a duct at the base of this tooth. The upper molars lack a W-shaped ectoloph, and are basically tritubercular. A sharp and bladelike (trenchant) ridge is formed by a high crest at the outer edge of each molar.

Solenodons are generalized omnivorous feeders that prefer animal material. They often find food by rooting with their snouts or by uncovering animals with their large claws. Solenodons are rather archaic creatures that seem to have little competitive ability. Their distribution on islands is probably the key to their continued, if tenuous, survival.

FAMILY SORICIDAE. Members of this family, the shrews, are among the smallest and least conspicuous of mammals. In many areas they are the most numerous insectivores, however, and they have the widest distribution of any insectivorous family. The family Soricidae is represented today by some 24 genera and 291 species, and occurs throughout the world except in the Australian area, most of South America, and the polar areas. Soricids appear in the Oligocene in both Europe and North America. Because soricids are rare as fossils, their early evolution is obscure; they may have evolved from an early erinaceid ancestral stock (Dawson, 1967a:20).

Shrews are small: the smallest weighs only 2 g (the smallest living mammal), and the largest weighs roughly 180 g, the size of a rat. The snout is long and slim, the eyes are small, and the pinnae are usually visible. The feet are five-toed; except for fringes of stiff hairs on the digits in semiaquatic species, and enlarged claws in semifossorial forms, they are unspecialized. The foot posture is plantigrade or semiplantigrade. The narrow and elongate skull usually has a flat dorsal profile (Fig. 6–14B); there is no zygomatic arch or tympanic bulla, and the tympanic bone is annular (Fig. 6–14A). The specialized dentition consists of from 26 to 32 teeth; the dental formula of *Sorex* is 3/1, 1/1, 3/1, 3/3 = 32. In the subfamily Soricinae, the teeth are pigmented; the first upper incisor is large, hooked, and bears a notch and projection resembling those on the upper mandible of a falcon (Fig. 6–14B). Behind the first upper incisor is a series of small unicuspid teeth (presumably incisors, a canine and premolars); P⁴ is large and has a trenchant ridge; and the upper

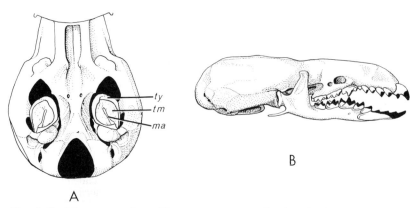

Figure 6–14. The skull of the vagrant shrew (*Sorex vagrans*, Soricidae; length of skull 17 mm). A, Ventral view of the basicranial region, showing the annular (ringlike) tympanic bone (*ty*), the tympanic membrane (*tm*), and the malleus (*ma*). B, Side view, showing the pincer-like anterior incisors. The pigmented parts of the teeth are shown in black.

molars have W-shaped ectolophs (Fig. 6–8A). Both the trigonid and talonid of the lower molars are well developed (Fig. 6–8B), and the first lower incisor is greatly enlarged and procumbent (leaning forward).

Although they require rather special care and feeding, breeding colonies of some species of shrews have been maintained for considerable periods in the laboratory. Dryden (1975) has maintained *Suncus murinus*, Fons (1974) has kept *S. etruscus*, and Hellwing (1971) has maintained *S. russula*.

Because they are unusually small, shrews can exploit a unique mode of foraging. Many shrews patrol for insects in spaces beneath logs, beneath fallen leaves and other plant debris, and in the narrow spaces and crevices beneath rocks. Surface runways of rodents and burrows of rodents may also be used as feeding routes. Due to their style of foraging, shrews are seldom observed, even in areas where they are common. Although the shrew is typically associated with moist conditions, some species, such as the gray shrew (*Notiosorex crawfordi*) of the southwestern United States and the piebald shrew (*Diplomesodon pulchellum*) of southern Russia, inhabit desert areas. Aquatic adaptations occur in some species that are capable of diving and swimming and feed mainly on aquatic invertebrates. One of the most aquatic species is the Tibetan water shrew (*Nectogale elegans*), which inhabits mountain streams and feeds primarily on fish. In this species, the streamlined shape is enhanced by the strong reduction of the pinnae, and the digits and feet have fringes of stiff hairs that greatly increase the effectiveness of the appendages as paddles. The distal part of the tail is laterally compressed and the edges bear lines of stiff hairs.

Innate following behavior in some shrews of the subfamily Crocidurinae results in the formation of "caravans." When a female and her litter are moving, the first young grabs her tail, the next young takes the first young's tail in its mouth, and so on, forming a chain of young that under some conditions is dragged by the mother. This behavior is known only in some crocidurine shrews.

FAMILY MACROSCELIDIDAE. This menotyphlan group, comprising the elephant shrews, is alone among insectivores in being specialized for hopping. The common name of elephant shrew refers to the vague resemblance between the trunk of an elephant and the long, flexible snouts of these shrews. The present distribution includes Morocco, Algeria, and the part of Africa south of the southern end of the Red Sea. There are five genera represented by 28 living species. The sparse fossil record of elephant shrews is limited to Africa. One extinct species of the living genus *Rhynchocyon* is known from the Miocene, and an extinct genus is recorded from the Pliocene.

Elephant shrews vary from the size of a mouse to that of a large rat, and have large eyes and ears and remarkably long snouts (Figs. 6–15, 6–16). The distal segments of the limbs are long and slender, the hind limbs being somewhat longer, and the tail is moderately long. The forelimbs and hindlimbs have four or five digits; in *Rhynchocyon*, the fore feet are functionally tridactyl, with the first digit absent and the fifth much reduced, whereas the hind feet have four digits. This more extreme loss of digits in the forelimb than in the hindlimb is most unusual in mammals. Elephant shrews have the cranial features listed for menotyphlans at the beginning of this chapter. In contrast to the Tupaiidae, the large orbits are never bordered by a complete postorbital bar (Fig. 6–17). The dental formula is 1-3/3, 1/1, 4/4, 2/2-3 = 36-42; the last upper premolar is the largest molariform tooth. The upper molars are quadrate and have four major cusps.

A considerable array of habitats is occupied by elephant shrews, including open plains, savannas, thornbush, and tropical forests. *Rhynchocyon*, the giant elephant shrew, is primarily a forest dweller, and in some places, as along the coast of Kenya, occupies what appear to be relict strips of evergreen forest. *Rhynchocyon chrysopygus* (Fig. 6–16) has been stud-

Figure 6-15. Two species of elephant shrews (Macroscelididae). A, Spectacled elephant shrew (*Elephantulus rufescens*); B, four-toed elephant shrew (*Petrodromus tetradactylus*). (Photographs by Galen Rathbun.)

Figure 6-16. Golden-rumped elephant shrew (*Rhynchocyon chrysopygus*). (Photograph by Galen Rathbun.)

ied by Galen Rathbun (1973, and personal communication) in the coastal forests of Kenya, where these animals are diurnal foragers. They feed on a wide variety of invertebrates, many of which are dug from the leaf litter by the long claws of the front feet. This species is strikingly colored: a dark chestnut brown, with almost a purplish cast, covers the back and flanks, against which the bright yellow rump patch stands out in sharp contrast. Rathbun has found that this species typically lives in pairs that occupy stable territories, and that scent marking by both sexes is an important territorial behavior. A resident pair will chase conspecific intruders, with males chasing males and

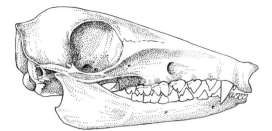

Figure 6–17. Skull of golden-rumped elephant shrew *(Rhynchocyon chrysopygus;* length of skull 67 mm).

Figure 6–18. A tree shrew *(Tupaia longipes,* Tupaiidae). (Photograph by M. W. Sorenson.)

females chasing females. Rathbun postulates that the conspicuous yellow rump functions as a target during aggressive encounters; he has found that the skin beneath this patch is thicker than skin over the rest of the body, and that scars and cuts are concentrated beneath the patch.

Studies by Rathbun (1976) on the spectacled elephant shrew (Fig. 6–15), *Elephantulus rufescens,* in thornbush country in southern Kenya, indicate that this species may be somewhat more advanced behaviorally than is the giant elephant shrew. The territories of brush-dwelling *E. rufescens* contain intricate patterns of trails, many of which are interrupted at points where the animals regularly leap over obstacles Some 24 per cent of the daylight behavior of these diurnally active animals is devoted to cleaning the trails; the front feet are used to sweep away leaf and twig litter. This species spends its life above ground, never seeking refuge in burrows, usually forages within 1 m of trails, and uses trails as escape routes. When startled, these animals bound along the trails at amazing speed. Dung is deposited at specific sites on the borders of and within the home range, and scent marking with the sternal glands is concentrated in border areas. Rathbun found that the bulk of the diet is termites and ants.

FAMILY TUPAIIDAE. Members of this family, called tree shrews, roughly resemble small, long-snouted squirrels (Fig. 6–18), and occur from India, through Burma, to the islands of Sumatra, Borneo, and the Philippines. This family is represented solely by its Recent members (five genera and 15 species). In addition to the

cranial features listed for menotyphlan insectivores at the beginning of this chapter, tree shrews are characterized by well-developed postorbital processes that join the zygoma (Fig. 6–7). The dental formula is 2/3, 1/1, 3/3, 3/3 = 38; the upper incisors resemble canines, and the upper canine is reduced. The upper molars have trenchant, W-shaped ectolophs (Fig. 6–19A), and the lower molars retain the basic insectivore pattern (Fig. 6–19B). The limbs are pentadactyl and the digits have strongly recurved claws. The long tail is heavily furred in 12 species, is tufted in one species, and is covered with short hairs in two species.

Tree shrews occupy deciduous forests and forage both in the trees and on the ground. They are opportunistic feeders, and utilize a variety of foods, but animal material and fruit are preferred. These animals are diurnal and are characteristically highly vocal. The mountain tree shrew occurs in social groups in which a rigid dominance hierarchy is apparent, whereas in parts of Borneo other species occupying lowland areas do not form social groups (Sorenson and Conaway, 1968). These authors report an interesting breeding be-

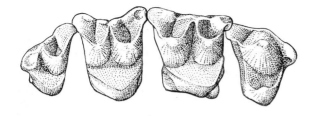

A

Figure 6-19. Cheek teeth of a tree shrew (*Tupaia* sp., Tupaiidae). Fourth premolar and molars of upper right (A) and lower left (B) tooth rows.

B

havior in *Tupaia montana*. The male emits a shrill call when ejaculation occurs. This call is probably an advertisement to other males of the receptivity of a female in estrus, and results in the sharing of a receptive female by males that rank high in the dominance hierarchy.

ORDER DERMOPTERA

Members of this order are generally called flying lemurs because of their lemur-like faces and their ability to glide between trees. One family, Cynocephalidae, with but one genus (*Cynocephalus*) and two species, represents the order. The distribution includes tropical forests from southern Burma and southern Indochina, Malaya, Sumatra, Java, Borneo, and nearby islands, to southern Mindanao and some of the other southern islands of the Philippine group. An extinct dermopteran family (Plagiomenidae) is recorded in North America from the late Paleocene and early Eocene, but the family Cynocephalidae is known only from the Recent.

Members of the family Cynocephalidae are of modest size (roughly 1 to 1.75 kg), and have large eyes and faces that resemble those of Old World fruit bats or some prosimian primates. The brownish, chest-

nut, or gray pelage is irregularly blotched with white. The molars have retained a basically three-cusped insectivore pattern, and the type of occlusion between upper and lower cheek teeth suggests that shearing rather than grinding is the main masticatory action, an unusual situation considering that dermopterans are largely herbivorous. The anterior dentition is specialized: the lateral upper incisor is caniniform, and the first two lower incisors are broad and pectinate (comblike). The unusual lower incisors are used to groom the fur, but also function to scrape leaves during feeding. The dental formula is 2/3, 1/1, 2/2, 3/3 = 34.

A broad furred membrane extends from the neck, starting just behind the ear, to near the ends of the fingers, from the hand and forelimb to the body and to near the end of the toes, and connects the hind foot and the tail to its tip, forming a tail membrane (uropatagium) that resembles that of a bat (Fig. 6-20). The hands and feet retain five digits which bear needle-sharp, curved claws that serve to clutch branches. As in bats, the neural spines of the thoracic vertebrae are short, the sternum is keeled, the ribs are broad, the radius is long, and the distal part of the ulna is strongly reduced. The great lengthening of the intestine typical of herbivorous mam-

Figure 6–20. A, A flying lemur (*Cynocephalus volans*) gliding between trees in a tropical forest in Mindanao (Philippine Islands). B, A flying lemur with a young animal clinging with needle-sharp claws to the bare skin where the gliding membrane joins the body. (Photographs by Charles H. Wharton, © 1948 National Geographic Society.)

mals is well illustrated by flying lemurs. *Cynocephalus*, which has a head plus body length of only about 410 mm, has an intestinal tract approaching 4 m in length, some nine times the head-and-body length (Wharton, 1950:272). The cecum, a blind diverticulum at the proximal end of the colon, is greatly enlarged (to about 48 cm in length) and is divided into compartments. This chamber harbors microorganisms that help break down cellulose and other relatively indigestible carbohydrates. Cecal enlargement is usually associated with an herbivorous diet (as in many rodents).

Flying lemurs are nocturnal and are slow but skillful climbers, but they are nearly helpless on the ground. They perform glides of distances well over 100 m in traveling to and from feeding places (Fig. 6–20). The diet includes leaves, buds, flowers, and fruit. Winge (1941:145) reports that the enlarged tongue and specialized lower incisors are used in cowlike fashion in picking leaves. According to Wharton (1950), flying lemurs seek refuge during the day in holes in trees, and several individuals may occupy the same den. He reports that, while traveling along branches and feeding, these animals invariably remain upside down. The distribution of flying lemurs is being restricted in some areas by the clearing of forests for agriculture by man, and in some regions the animals are hunted for their meat and their fur.

ORDER CHIROPTERA

Because of their apparent secretiveness, their emergence in gathering darkness during Shakespeare's "very witching time of night," their ability to fly, and their unusual form that for the casual observer sets them apart from the more familiar groups of animals, bats have long been central figures in superstitions and shadowy forms in poetry.

Bats, and an uneasy creeping in one's scalp

As the bats swoop overhead!
Flying madly.

Pipistrello!
Black piper on an infinitesimal pipe.
Little lumps that fly in air and have voices indefinite, wildly vindictive;

Wings like bits of umbrella.

Bats!

Creatures that hang themselves up like an old rag, to sleep;
And disgustingly upside down.
Hanging upside down like rows of disgusting old rags
And grinning in their sleep.
Bats!

This is the bat of D. H. Lawrence, but the biologist's bat has taken on substance. Accelerated research on bats in recent years has revealed some of the fascinating aspects of the lives of these animals: extraordinarily complex social behavior, involving the maintenance of harems by males and the use of an array of vocal communication signals; a coordinated assemblage of neuromuscular and behavioral adaptations allowing bats to perceive in detail their prey and their environment by the use of sound; and an ability unsurpassed in other mammals to conserve energy daily or to survive through periods of stress by drastic reductions in the metabolic rate. Biologists have come to realize that bats not only deserve respect as remarkably specialized products of some 65 million years of evolution, but that they merit our protection for their importance in terrestrial ecosystems as efficient predators of insects.

Bats are a remarkably successful group today, and comprise the second largest mammalian order (behind the Rodentia); approximately 168 genera and 853 species of living bats are known. Bats are nearly cosmopolitan in distribution, being absent only from arctic and polar regions and some isolated oceanic islands. Although bats are frequently abundant members of temperate faunas, they reach their highest densities and greatest diversity in tropical and subtropical areas. In certain Neotropical localities, for example, there are more species of bats than of all other kinds of mammals together. Bats occupy a number of terrestrial environments, including temperate, boreal and tropical forests, grasslands, chaparral, and deserts. Because man-made structures often afford excellent roosting sites and agricultural areas provide high insect populations, bats are doubtless more abundant in some areas now than they were before these areas were occupied by European man. But in some areas, the molesting of colonies of bats by man has caused alarming declines in populations, and the poorly controlled use of insecticides presents another threat.

Two sharply differentiated suborders of bats are recognized. The suborder Megachiroptera includes the family Pteropodidae, the Old World fruit bats, and the suborder Microchiroptera includes all the other 17 families of bats. Microchiropterans are nearly cosmopolitan in distribution and are largely insectivorous. Two functional contrasts between the megachiropterans and the microchiropterans are of particular importance. Megachiropterans are not known to hibernate, and maintain their body temperatures within fairly narrow limits by physiological and behavioral means, whereas many microchiropterans are heterothermic, and some hibernate for long periods. In addition, whereas microchiropterans use echolocation as their primary means of orientation and can fly and capture insects in total darkness, most megachiropterans use vision and therefore are helpless in total darkness. One exception is the megachiropteran *Rousettus*, in which the ability to echolocate perhaps evolved independently. *Rousettus* uses clicks made by the tongue as the basis for its acoustical orientation (Novick, 1958). All microchiropterans use ultrasonic pulses produced by the larynx.

Echolocation, a means of perceiving the environment even in total darkness (see Chapter 21), and flight, allowing great motility, have been two major keys to the success of bats. These abilities enable bats to occupy at night many of the niches filled by birds during the day. In addition, the remarkably maneuverable flight of bats facilitates a mode of foraging for insects that birds have never exploited. Heterothermy, allowing bats to hibernate or to operate at a lowered metabolic output during part of the diel cycle, has enabled these animals to occupy areas only seasonally productive of adequate food and to utilize an activity cycle involving only nocturnal or crepuscular foraging periods. The metabolic economy resulting from hibernation and from lowered metabolism during part of the diel cycle has affected the longevity of some bats. For their size, some microchiropteran bats are remarkably long-lived. *Myotis lucifugus*, a small bat weighing roughly 10 gm, may live as long as 24 years (Griffin and Hitchcock, 1965).

MORPHOLOGY

Many of the most important diagnostic features of bats are adaptations for flight. The bones of the arm and hand (with the exception of the thumb) are elongate and slender (Fig. 6–21), and flight membranes extend from the body and the hind limbs to the arm and the fifth digit (plagiopatagium), between the fingers (chiropatagium), from the hind limbs to the tail (uropatagium), and from the arm to the occipitopollicalis muscle (propatagium, Fig. 6–22B). In some species, the uropatagium is present even when the tail is absent. The muscles bracing the wing membranes are often well developed and serve to anchor a complex network of elastic fibers (Fig. 6–22A). Rigidity of the outstretched wing during flight is partly controlled by the specialized elbow and wrist joints, at which movement is limited to the anteroposterior plane.

In most microchiropteran species, the enlarged greater tuberosity of the humerus locks against the scapula at the top of the upstroke (Fig. 6–23), allowing the posterior division of the serratus anterior muscle, which tips the lateral border of the

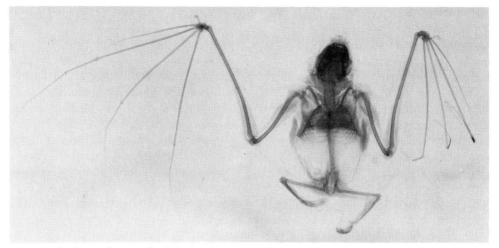

Figure 6–21. An X-ray photograph of the big fruit-eating bat (*Artibeus lituratus*, Phyllostomatidae), showing the great elongation of the bones of the arm and hand. (From Vaughan, 1970b.)

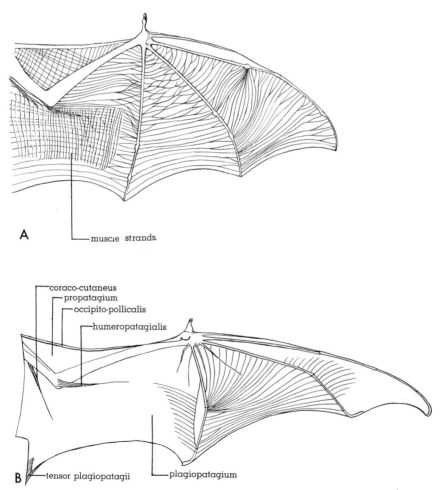

Figure 6–22. Ventral views of the wings of two bats, showing the parts of the wing and muscles and elastic fibers that brace the membranes. A, The big fruit-eating bat (*Artibeus lituratus*, Phyllostomatidae); note the muscle strands that reinforce the plagiopatagium and the system of elastic fibers. This broad-winged bat does not remain on the wing for long periods. B, The western mastiff bat (*Eumops perotis*, Molossidae). This narrow-winged bat is a fast and enduring flier. (*Artibeus* from Vaughan, 1970a. *Eumops* from Vaughan, 1970b.)

94

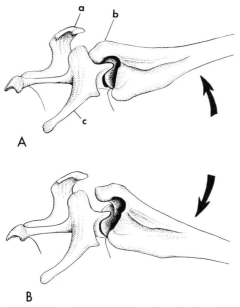

A

B

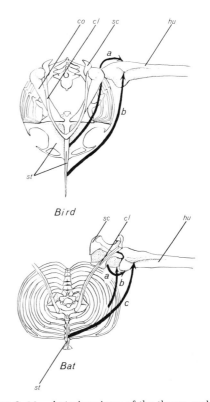

Bird

Bat

Figure 6–23. Anterior view of the left shoulder joint of a free-tailed bat (*Molossus ater*) at the top of the upstroke of the wing (A) and during the downstroke (B). The greater tuberosity of the humerus (*b*) locks against the scapula at the top of the upstroke, transferring the responsibility for stopping this stroke to the muscles binding the scapula to the axial skeleton. During the downstroke the greater tuberosity of the humerus moves away from its locked position. This type of action and this type of shoulder joint also occur in the Vespertilionidae and other advanced families of bats. The acromion process (*a*) and the coracoid process (*c*) of the scapula are shown. (From Vaughan and Bateman, 1970.)

Figure 6–24. Anterior views of the thorax and part of the left forelimb of a bird and a bat, with some of the major muscles controlling the wing-beat cycle shown diagramatically. In the bird the supracoracoideus muscle (*a*) raises the wing and the pectoralis muscle (*b*) powers the downstroke; both muscles originate on the sternum. In the bat the downstroke is primarily controlled by three muscles, the subscapularis (*a*), serratus anterior (*b*), and pectoralis (*c*). Only the pectoralis originates on the sternum. Many muscles power the upstroke in bats. Abbreviations: *cl*, clavicle; *co*, coracoid; *hu*, humerus; *sc*, scapula; *st*, sternum. (From Vaughan, 1970a.)

scapula downward, to help power the downstroke of the wing (Fig. 6–24). The adductor and abductor muscles of the forelimb raise and lower the wings and are therefore the major muscles of locomotion; a contrasting arrangement occurs in terrestrial mammals, in which the flexors and extensors provide most of the power for locomotion. The distal part of the ulna is reduced in bats, and the proximal section usually forms an important part of the articular surface of the elbow joint (Fig. 6–25). The clavicle is present and articulates proximally with the enlarged manubrium and distally with the enlarged acromion process and enlarged base of the coracoid process (Fig. 6–24). The hind limbs are either rotated to the side 90° from the typical mammalian position and have a reptilian posture during quadrupe-

dal locomotion, or they are rotated 180°, have a spider-like posture, and are used primarily to suspend the animal upside down from a horizontal support. The fibula is usually reduced, and support for the uropatagium, the calcar (see Fig. 6–26), is usually present.

The evolution of the muscular control pattern of the wing-beat cycle typical of microchiropteran bats has seemingly been strongly influenced by their use of echolocation. Highly maneuverable flight is essential for these bats, because objects are perceived in detail only at fairly close range. In contrast, birds use vision for more long-range perception of their envi-

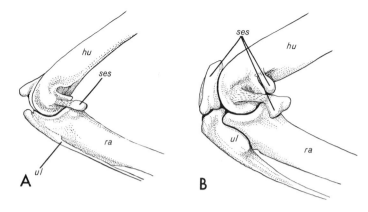

Figure 6–25. Lateral view of the right elbow of a myotis (*Myotis volans*, Vespertilionidae) and of a free-tailed bat (*Molossus ater*, Molossidae). Abbreviations: *hu*, humerus; *ra*, radius; *ses*, sesamoid; *ul*, ulna. (From Vaughan, 1970.)

ronment, and have relatively little need for extremely maneuverable flight. In both groups, similar trends toward rigidity of the axial skeleton and lightening of the wings occur, but many of the muscular and skeletal specializations that enable these animals to control their wings differ between the groups. The pectoral girdle in birds is braced solidly by a tripod formed by the clavicula and coracoids, anchored to the sternum, and by the nearly bladelike scapula, resting almost immovably against the ribcage. The pectoralis and supracoracoideus muscles, both of which originate on the sternum, supply nearly all of the power for the wing beat.

In bats, nearly the reverse mechanical arrangement occurs: the scapula is braced against the axial skeleton by the clavicle alone, and the job of powering the wing beat is shared by many muscles (Fig. 6–24). This division of labor is made possible partly by the freedom of the scapula to rotate on its long axis. The pectoralis, the subscapularis, the posterior division of the serratus anterior, and the clavodeltoideus muscles control the downstroke of the wings; only the pectoralis originates on the sternum. The muscles of the deltoideus and trapezius groups and the supraspinatus and infraspinatus muscles largely power the upstroke.

A morphological trend of critical importance to bats and all other flying animals is toward the reduction of wing weight. Propulsion is obtained in all flying animals by movements of the wings, and the kinetic energy produced by such movements depends upon the speed and weight of the wing. The amplitude of a stroke and its speed are progressively greater toward the wing tip. Consequently, reduction of the weight of the distal parts of the wing results in a reduction of the kinetic energy developed during a wing stroke. A considerable advantage in metabolic economy is thus gained, for as less kinetic energy is developed during each stroke, less energy is necessary to control the wings. In addition, light wings can be controlled with speed and precision during the extremely rapid maneuvers used when bats chase flying insects. Reduction of the weight of the wings had been furthered in bats by many specializations. Movement at the elbow and wrist joints is limited to one plane, thus eliminating musculature involved in rotation and bracing at these joints. In addition, the work of extending and flexing the wings is transferred from distal muscles (of the forearm and hand) to large proximal muscles (pectoralis, biceps, and triceps), thereby allowing a reduction in the size of the distal musculature. Certain forearm muscles are made nearly inelastic by investing connective tissue and, because of this modification and specializations of their attachments, these muscles "automatically" extend the chiropatagium with extension at the elbow joint, or flex the chiropatagium with flexion at this joint (Fig. 6–27).

The hind limbs of bats are generally quite thin, but are not drastically reduced in length, because of their importance in supporting the trailing edge of the plagiopatagium and the lateral edge of the uropatagium. According to Howell and Pylka

Figure 6–26. A fishing bat *(Noctilio labialis)*, showing several stages in the wing-beat cycle. A, Top of the up-stroke; B, midway through the downstroke; C, the end of the downstroke; D, midway through the upstroke. (Photographs by J. Scott Altenbach, courtesy of Carl Brandon.)

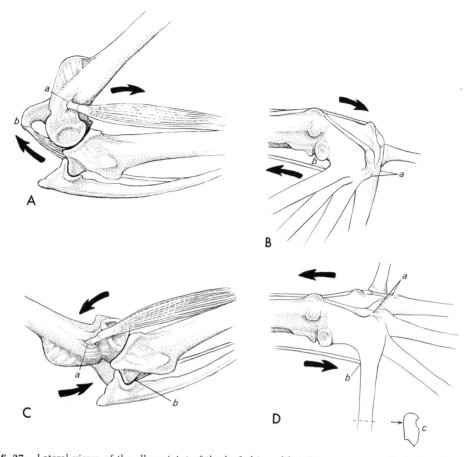

Figure 6–27. Lateral views of the elbow joint of the leaf-chinned bat (*Mormoops megalophylla*), showing the "automatic" flexion and extension of the fingers caused by certain forearm muscles in many advanced bats. Flexion of the elbow joint (A) moves the origins of the extensor muscles (a) *toward* the wrist and the origin of one flexor muscle (b) *away* from the wrist. Because the flexor muscle is largely inelastic, with flexion at the elbow (B) the distal tendon of the flexor (b) pulls on the fifth digit and tends to flex the fingers. With extension at the elbow joint (C), the origin of the extensor muscles (a) is moved away from the wrist and the origin of the flexor muscle (b) moves toward the wrist. This action results in pulling the extensor tendon (a) toward the elbow (D) and releasing tension on the flexor tendon (b), thus extending the fingers. In part D the complex cross-sectional shape of the fifth digit (c) is shown. (From Vaughan and Bateman, 1970.)

(1976) the fact that bats generally hang upside down is related to the thinness of the hind legs, which probably evolved under selective pressures favoring reductions in weight. The hanging position was perhaps necessitated by the inability of the delicate femur to tolerate compression stresses associated with other roosting postures.

FLIGHT

The three modern groups of flying animals—insects, birds, and bats—are each highly successful. Viewing the terrestrial scene, there are more flying than non-flying species of animals, but each flying group has evolved a different type of wing: that of birds is formed of feathers braced by a simplified forelimb skeleton along the leading edge; insect wings are membranous sheets of chitin braced by intricate patterns of chitinous veins; and bat wings are sheets of skin braced by the five-digited forelimb and elastic connective tissue. Flight styles also differ, with birds depending on rather fast and not especially maneuverable flight, insects usually using extremely rapid wing beats, a variety of flight speeds, and often a remarkable ability to hover, and bats utilizing chiefly slow, highly maneuverable flight. As might be expected, diverse and complex mechanical and aerodynamic problems are faced by these groups of fliers, and animal flight remains incompletely understood. Considerations of bird and insect flight can (thankfully!) be left to ornithologists and entomologists, but inasmuch as some 22 per cent of terrestrial mammals are bats, we must come to grips with flight in bats.

Most students have been introduced at least once to basic aspects of aerodynamics; this topic can therefore be treated briefly. Because the wings of animals usually provide both the thrust and the lift necessary for sustained flight, whereas in aircraft the wings provide only the lift, flight in animals presents special problems.

Lift is generated when an airstream sweeps over a wing with an asymmetrical cross section. The profile of the cross section of a wing, the airfoil, varies widely between species of flying animals, but characteristically in birds and bats it has an arched dorsal surface and a concave ventral surface (Fig. 6–28B). The tendency is for the parts of the airstream flowing over the opposite surfaces of the wing to arrive at the trailing edge simultaneously; this necessitates faster movement of air over the dorsal than the ventral surface. The more rapidly the air moves over a surface, the less pressure it exerts, which is a relationship described by Daniel Bernoulli in 1738 and exploited by flying vertebrates for over 150 million years. The unequal forces on opposing wing surfaces creates lift, a force opposing the force of gravity on a flying animal.

Lift is also created when a surface is presented at an angle of attack to the airstream (angle of attack is the angle that the chord line of the airfoil makes with the plane of motion of the wing; Fig. 6–28B). Within limits, lift can be increased by raising the angle of attack of a wing. When, however, the angle of attack of the wing becomes so great that the air moving over the upper surface breaks away from the wing and forms turbulent eddies, the lift produced by the wing abruptly falls as the drag sharply rises, and stalling occurs. (Drag is the force exerted by air on an object in motion and in a direction opposite to that of the motion.)

Wing performance is under the influ-

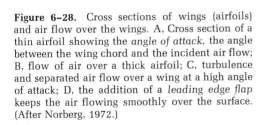

Figure 6–28. Cross sections of wings (airfoils) and air flow over the wings. A, Cross section of a thin airfoil showing the *angle of attack*, the angle between the wing chord and the incident air flow; B, flow of air over a thick airfoil; C, turbulence and separated air flow over a wing at a high angle of attack; D, the addition of a *leading edge flap* keeps the air flowing smoothly over the surface. (After Norberg, 1972.)

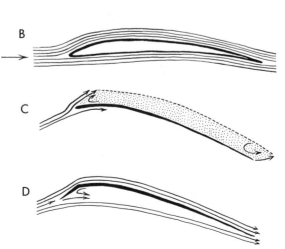

ence of a series of variables. Lift increases directly as the surface area of the wing, but so does drag; lift is increased (within limits) as the camber of an airfoil is increased (camber is the curvature or arching of an airfoil), but this also increases drag; lift increases as the square of the speed, as does drag. Intuitively, then, one might expect that some of the constraints forcing modifications of wing design on fast fliers are of relatively little importance in slow-flying bats. This seems to be true, and leads us to a consideration of the unique structure and function of the chiropteran wing.

The wings of bats form very thin airfoils of high camber. Several important features enhance the performance of these wings under the conditions of low-speed flight typical of most bats (Norberg; 1969, 1972). Thin airfoils, essentially cambered plates, are more effective in producing high lift at low speeds than are conventional airfoils with some thickness (Fig. 6–28). Of further importance is the ability of the bat to vary the camber of the wing in the inter-

est of producing high lift at low speeds. Camber of the bat wing is largely under the control of the occipitopollicalis muscles, the flexors of the thumb, the inclination of the dactylopatagium minus (Fig. 6–26), the fifth digit, and the hind limbs.

Compared to birds, bats have low wing loadings (Table 6–1). Wing loading is an expression of the ratio of body weight to wing area (W/S); in general, the lower the wing loading, the slower an animal can fly and still maintain adequate lift to remain airborne. Most bats also have broad wings with low aspect ratios. The aspect ratio is the relationship of the length of a wing to its mean breadth, and for wings of irregular shape it is expressed as the ratio of the span squared to the wing area (b^2/S). Some bats that fly rapidly and remain in flight for long periods have long, narrow, high-aspect ratio wings (Fig. 6–22B). Compared to broad wings, narrow wings suffer less loss of lift owing to air spillage from the area of high pressure on the ventral surface to the area of low pressure on the dorsal surface of the wing tip.

Table 6–1. Comparisons of Wing Loadings (Grams of Body Weight/Square Centimeter of Wing Area) of Bats and Birds

Species	Weight (gm)	Wing Area (square cm)	Wing Loading
Phyllostomatid bats			
Artibeus phaotis	9.9	101.74	0.098
A. lituratus	59.6	380.36	0.180
Phyllostomus discolor	42.2	261.60	0.162
Glossophaga soricina	10.6	99.29	0.106
Desmodontid bats			
Desmodus rotundus	27.8	198.94	0.161
Vespertilionid bats			
Myotis nigricans	4.2	67.58	0.062
Rhogeessa tumida	3.9	55.91	0.070
Molossid bats			
Molossus sinaloae	23.8	133.33	0.179
M. molossus	16.1	95.15	0.169
Birds			
Ruby-throated Hummingbird	3.0	12.4	0.242
House Wren	11.0	48.4	0.227
Chimney Swift	17.0	104.0	0.163
Redwinged Blackbird	70.0	245.0	0.286
Mourning Dove	130.0	257.0	0.364
Peregrine Falcon	1222.5	1342.0	0.911
Common Loon	2425.0	1358.0	1.786
Golden Eagle	4664.0	6520.0	0.715
Canada Goose	5662.0	2820.0	2.007

(Data from Lawlor, 1973; Poole, 1936.)

Wings that are strongly tapered toward the tip minimize this spillage and loss of lift, and are typical of fast-flying bats.

Another feature that differs strongly between bats with contrasting flight styles is the breadth of the membranes that form the leading edge of the wing, the propatagium, the dactylopatagium brevis, and the dactylopatagium minus (Fig. 6–22). These membranes can be canted downward to form "leading edge flaps," structures that greatly improve the efficiency of cambered-plate, low-speed airfoils (Fig. 6–28C). The leading edge flap can be seen to be larger in the low-speed wing than in the high-speed wing. Flaps of several sorts are used in aircraft to avoid stalling at low speeds, and in birds the alula (the small "bastard wing" formed by feathers attached to the thumb) and the slotting between feathers at the wing tip serve this function. By tending to avoid the separation of air flow from flight surfaces behind these slots, they improve low speed performance at high angles of attack. Leading edge flaps are especially effective in airfoils with thin sections and a thin leading edge (Abbott and von Doenhoff, 1949), precisely the type of airfoil formed by the wings of bats.

To produce lift, an airfoil must move through the air, and this requires a means of propulsion. In animals, propulsion is created by movements of the wings, and photographs of the wing-beat cycle in bats in level flight indicate that the downstroke is the power stroke whereas the upstroke is largely a recovery stroke (Fig. 6–26). During the downstroke, the wings are fully extended and the powerfully braced fifth digit and the hind limbs maintain the plagiopatagium at a fairly constant angle of attack. But toward the wing tip, the air pressure against the membranes becomes progressively greater as the speed of the wing becomes increasingly higher. This increase in pressure, coupled with the elasticity of the membranes between the digits, causes the trailing edges of the chiropatagium to lag behind the well-braced leading edge—in effect, the wing tip is twisted into a propeller-like shape and

serves a propeller-like function. As the wing tip sweeps rapidly downward, it tends to force air backward, resulting in forward thrust of the animal. The membrane between the third and fourth digits (dactylopatagium longus) is probably of primary importance in producing thrust. During the upstroke, or recovery stroke, the wing is partly flexed, the stroke is directed upward and to some extent backward, and the force of the air stream partially aids the movement. Judging in part from the large muscles that power the downstroke and the relatively small muscles that control the upstroke, one would expect that the latter demands relatively little power and energy.

What about the function of the proximal segment of the wing, the plagiopatagium, during the wing-beat cycle? This segment retains an angle to attack appropriate to the production of lift through much of both the downstroke and the upstroke, and it seems to function primarily as a lifting surface. In many bats, a series of muscular and skeletal specializations insure that the fifth digit is braced against forces tending to cause extension beyond the point at which its angle of attack is proper for the development of lift.

Some bats can fly very slowly and some can hover, and during these types of flight the action of the wings departs from that used in level flight. In hovering by the nectar-feeding bat *Leptonycteris sanborni*, studied by Altenbach (1977), the downstroke is directed largely forward and the upstroke, backward (Fig. 6–29). The posture of the wings during the downstroke is similar to that in level flight, but because the stroke is largely horizontal, vertical thrust is developed. The upstroke, however, is complicated by a reversal of the usual posture of the wing tip: the tip turns over in such a way that the dorsal surface of the chiropatagium faces downward and the leading edge of the wing still leads in this stroke, but is posterior to the trailing edge (Fig. 6–29). Toward the end of the upstroke, the reversed wing tip is flipped rapidly backward and produces considerable upward thrust; and at the

Figure 6–29. A nectar-feeding bat *(Leptonycteris sanborni)* in flight, showing some positions of the wings during slow or hovering flight. (Photographs by J. Scott Altenbach.)

start of the downstroke the wing tip swings into its normal posture. This powerful flip probably demands considerable energy, but the vertical thrust that it develops strongly augments the thrust resulting from the downstroke and enables the bat to remain nearly stationary in the air. Probably because of the high energy cost of hovering, it is generally used only briefly by bats.

Although they appear to interrupt awkwardly the otherwise smooth chiropatagium, the arm and the fingers enhance low-speed flight performance by serving as turbulence generators. At the low speeds and high angles of attack common to the wings of many bats, a turbulent boundary layer (the boundary layer is the thin layer of air very near the surface of the wing that is affected by friction with the surface) is better able to remain against the wing and continue to allow the development of lift than is a laminar boundary layer, which

tends to separate completely from the wing surface with a consequent loss of lift and rise in drag (Fig. 6–30). The turbulent boundary layer formed behind turbulence generators (such as the arm or digits of bats) serves as a transition layer within which there is exchange of momentum between the fast-flowing outer layers of air and the inner layers that are slowed by friction with the surface of the wing. The turbulent boundary layer can thus transfer kinetic energy to the surface layers from the rapid outer layers and can retain laminar flow in the lift-producing outer layers. In small slow fliers, which most bats are, the type of air flow induced by turbulence generators is especially important in allowing lift to be maintained at high angles of attack.

During the early evolution of the bat wing, selection seemingly favored refinements in design that allowed the development of high lift at low speeds. But later,

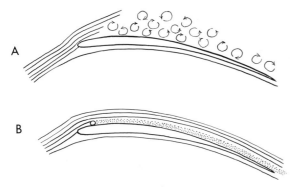

Figure 6–30. Effect of a turbulence generator. A, Separation of air flow over a wing and loss of lift; B, the boundary layer of air flowing over the wing changes from laminar to turbulent flow after encountering a protruding structure (such as the digit of a bat) and the lift produced by the airfoil is retained. (After Norberg, 1972.)

perhaps in the Eocene and Oligocene, bats underwent an adaptive radiation involving, in part, exploitation of various styles of flight. The wings of some bats (members of the family Molossidae and some members of the Emballonuridae, for example) developed characteristics advantageous during rapid flight. Because lift varies as the square of the speed of an airfoil, it would seem that rapid-flying bats could afford the luxury of higher wing loadings because of the greater lift developed per unit of wing area at higher speeds; but, because drag also increases as the square of the speed, during rapid flight a reduction in wing surface area, angle of attack, and camber would be highly advantageous. Wing design in rapid-flying bats is clearly the result of a series of evolutionary compromises and not all have wings that are alike; nonetheless, a number of bats have evolved roughly the same type of high-speed wing. (High speed is used here only in a relative sense, for probably few bats achieve speeds in level flight above 80 km. per hour.) This wing is narrow and often has an elongate and strongly tapered tip, the membranes that form the leading edge flap are narrow, and the camber is low. Together these features serve to reduce drag and to favor efficient (in terms of energy outlay), rapid, and enduring flight. In the case of free-tailed bats (Molossidae), the wing membranes contain unusually large numbers of elastic fibers that brace the membranes against the force of the airstream at higher speeds.

PALEONTOLOGY

Because of their small size, delicate structure, nonterrestrial habits, and occurrence mostly in tropical areas where fossilization seldom occurs, bats are rare as fossils. Consequently, the evolution of bats is poorly known. The earliest undoubted fossil bat (*Icaronycteris index*, Fig. 6–31) is from early Eocene beds in Wyoming. On the basis of this beautifully preserved specimen, Jepsen (1966) described the extinct family Icaronycteridae. Although this bat has several primitive features, such as claws on the first two digits of the hand and fairly short, broad wings, its basic limb structure is that of modern bats. The upper molars of *Icaronycteris* have the W-shaped ectoloph typical of most insectivorous bats, and this bat has been put in the suborder Microchiroptera. Late Eocene and early Oligocene deposits in France have yielded the earliest records of the modern microchiropteran families Emballonuridae, Megadermatidae, Rhinolophidae, and Vespertilionidae. Megachiropterans appear first in the Oligocene of Italy.

The Eocene appearance of a bat clearly beautifully adapted for flight, the Oligocene appearance of many modern families, and the assignment by paleontologists of fossils from the late Eocene to the still-living genus *Rhinolophus* indicate an early origin of bats. The antiquity of bats is further indicated by the fact that some genera were essentially as they are now in the Oligocene, a time when horses were three

Figure 6–31. A beautifully preserved early Eocene bat (*Icaronycteris index*). (From Jepsen, 1970.)

bat of the still-living genus *Tadarida*, for example, were found by Sige (1971) to be nearly identical to those of present day members of this family. Although no clear fossil evidence bears on the matter, a Paleocene origin of bats seems probable, and a late Cretaceous divergence of chiropteran ancestors from primitive insectivore stock seems possible. Jepsen, in his excellent discussion of the evolution of bats (1970:22), summed up the status of our knowledge: "At present bat history has a completely open end, in the distant past, that only more fossils can close."

SUBORDER MEGACHIROPTERA

Inasmuch as only one family represents this suborder today, the descriptions given below for the family Pteropodidae characterize the suborder Megachiroptera.

FAMILY PTEROPODIDAE. Most pteropodids are fruit eaters, and many species are called flying foxes because of their foxlike faces and large size. These bats are abundant and often conspicuous members of many tropical biotas in the Old World. This family is represented by 40 Recent genera and 149 Recent species. Pteropodids occur widely in tropical and subtropical regions from Africa and southern Eurasia to Australia, and on many South Pacific islands as far east as Samoa and the Carolines.

Members of this family are often large, up to 150 cm (nearly 5 feet!) in wing

toed and no bigger than sheep and bears and antelope had not yet appeared on the scene. The wings of an Oligocene molossid

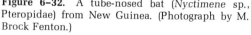

Figure 6–32. A tube-nosed bat (*Nyctimene* sp., Pteropidae) from New Guinea. (Photograph by M. Brock Fenton.)

spread, and differ from members of the suborder Microchiroptera in many ways. The face is usually foxlike, with large eyes, usually a moderately long snout with a simple unspecialized nose pad, and simple ears lacking a tragus (Fig. 6–32). (The tragus, a fleshy projection of the anterior border of the ear opening, may be seen on a member of the Vespertilionidae in Fig. 6–33E). The orbits are large, and are bordered posteriorly by well developed postorbital processes which may meet to form a postorbital bar. The rostrum is never highly modified (Fig. 6–34). The dental formula is 1-2/0-2, 1/1, 3/3, 1-2/2-3 = 24-34. The molars are never tuberculosec-

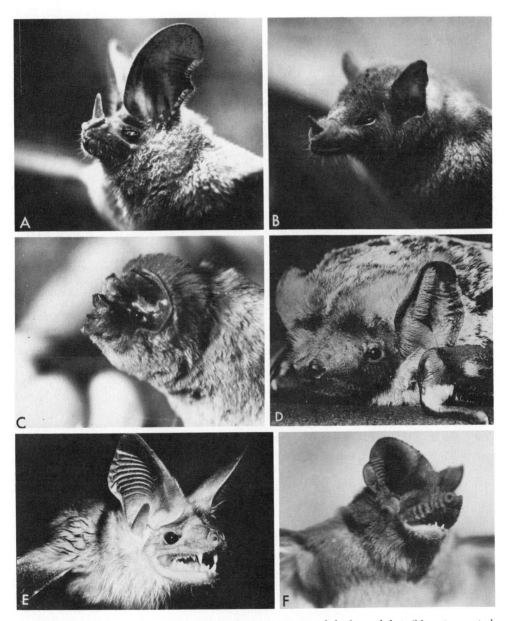

Figure 6–33. Faces of some microchiropteran bats. A, Big-eared leaf-nosed bat (*Macrotus waterhousii,* Phyllostomatidae), an omnivore; B, long-nosed bat (*Leptonycteris sanborni,* Phyllostomatidae), a nectar and pollen feeder. The following bats are insectivorous: C, leaf-chinned bat (*Mormoops megalophylla,* Mormoopidae); D, hoary bat (*Lasiurus cinereus,* Vespertilionidae); E, pallid bat (*Antrozous pallidus,* Vespertilionidae); F, Brazilian free-tailed bat (*Tadarida brasiliensis,* Molossidae). (Photographs of *Lasiurus cinereus* and *Antrozous pallidus* by Patricia Brown.)

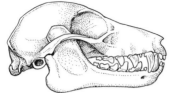

Figure 6–34. The skull of a megachiropteran bat (*Pteropus* sp., Pteropodidae). Length of skull 62 mm.

torial with W-shaped ectolophs, as in most microchiropterans, but are low, moderately flat crowned, more or less quadrate, and lack stylar cusps (Fig. 6–35). The teeth are adapted basically to crushing fruit. The wing is primitive in having two clawed digits, and the greater tuberosity of the humerus is not enlarged (Fig. 6–36) to make contact with the scapula at the top of the upstroke. The tail is typically short or rudimentary.

Broadly speaking, pteropodids utilize two types of food. Most members of the subfamily Pteropodinae are fruit eaters, whereas members of the subfamily Macroglossinae eat mostly nectar and pollen. The fruit eaters as a rule are large bats with fairly robust or moderately reduced dentitions. The jaws in these species are usually fairly long, or, in some species that presumably eat hard fruit, the jaws are shorter and the teeth and dentary bones are unusually robust. The fruit bats often roost in trees in large colonies (Fig. 6–37); in the case of the African genus *Eidolon*, as many as 10,000 have been observed roosting together. Fruit bats occasionally travel long distances during their nocturnal

foraging, and *Pteropus* regularly flies at least 15 km. from roosting sites to feeding areas (Breadon, 1932; Ratcliffe, 1932). The fruit eaters are usually not particularly maneuverable fliers, but have a steady, direct style of flight. They are adroit at clambering in vegetation, where the clawed first and second digits of the wing come into play.

Hypsignathus monstrosus, the hammer-headed bat, is unique among mammals in the fantastic degree to which specializations of the vocal apparatus have been carried in the males. This large frugivorous pteropodid, with a wingspread approaching 1 m, occupies tropical forests in much of central Africa. Communal displays by males in courtship areas enter importantly into the breeding cycle of *Hypsignathus* (see page 373). The males on the courtship arena, which is called a lek, use a penetrating call, described by Kingdon (1970a:170) as "guttural, explosive and blaring," to attract females. The remarkable specializations of the vocal apparatus clearly evolved in association with the importance of loud vocalizations during breeding displays.

Externally, the most striking feature of the male is the strange hammer-headed appearance (Fig. 6–38). This is due in part to the enlarged and elevated nasal bones, but is accentuated by a large pouch that encloses the rostrum and extends back over the cranium. These features enhance the resonance of the calls, and pharyngeal sacs in the throat are probably also resonators. Equally impressive are internal features at-

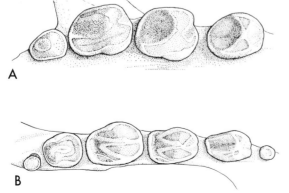

A

B

Figure 6–35. The cheek teeth of a megachiropteran bat (*Pteropus* sp.). A, Right upper tooth row, showing two molars and two premolars. B, Lower left tooth row, showing three premolars and three molars. (From Vaughan, 1970b.)

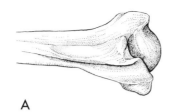

A

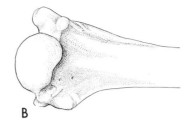

B

Figure 6–36. The proximal end of the right humerus in three bats. Anterior views are on the left and posterior views are on the right. A and B, *Pteropus* sp (Pteropodidae); C and D, *Myotis lucifugus* (Vespertilionidae); E and F, *Molossus ater* (Molossidae). (From Vaughan, 1970b.)

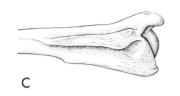

C

D

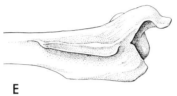

E

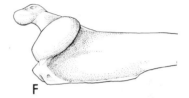

F

Figure 6–37. Part of a flying fox "camp" in northeastern Australia. (Photograph by Roger E. Carpenter.)

Figure 6–38. Two views of head of a male hammer-headed bat *(Hypsignathus monstrosus)*, showing the highly modified lips and inflated rostrum. These specializations are associated with an ability to produce very loud, resonant sounds. (Photographs by R. L. Peterson.)

tending the massive enlargement of the larynx. This structure, which contains huge vocal chords, has moved into the thorax, where it occupies most of the space filled by the heart and lungs in other mammals. As a result of this migration, the large trachea lies against the diaphragm and curves sharply craniad to the lungs, which are also forced against the diaphragm. The thoracic cavity thus serves largely as a container for the huge larynx in male *Hypsignathus*, with a drastic sacrifice in lung capacity. Kingdon has called this animal a flying loud-speaker; this characterization seems especially apt when one considers the enlarged lips of males, which the animals can form into almost perfect megaphones. The anatomy of the larynx is described by Schneider, et al. (1967), and early studies on the morphology of the vocal apparatus of this bat are by Matschie (1899) and Lang and Chapin (1917).

The pteropodids that eat nectar and pollen are small by comparison with their fruit-eating relatives, and have long slender rostra, strongly reduced cheek teeth, and delicate dentary bones. The tongue is long and protrusile and has hairlike structures at its tip to which pollen and nectar adhere. Pollen, which adheres to the fur (Fig. 6–39) and is ingested when the bats groom themselves, is probably an essential source of protein to nectar-feeding bats (see page 303). Some species roost in groups in caves, and some roost solitarily in vegetation. Flight is slow and maneuverable.

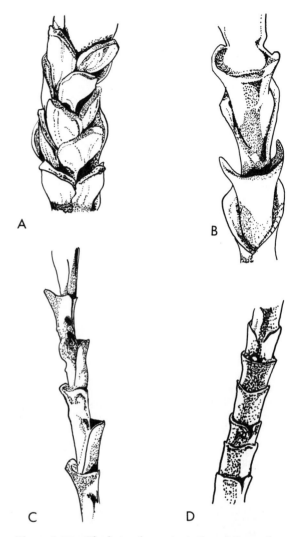

Figure 6–39. The hairs shown in A, B, and C are of nectar-feeding bats and are adapted to catching pollen; the hair shown in D is that of an insectivorous bat. A, *Epomophorus* sp.; B, *Glossophaga* sp.; C, *Musonycteris harrisoni*; D, *Pteronotus davyi*. (After photographs by D. J. Howell.)

SUBORDER MICROCHIROPTERA

Recent members of this suborder are usually small. The eyes are often small, the rostrum is usually specialized, and the nose pad and lower lips may be modified in a variety of ways (Figs. 6–33, 21–2). The ears have a tragus in all but members of the family Rhinolophidae, are usually complex, and are frequently large. The postorbital process is usually small. Dentitions vary tremendously, but most microchiropterans (except the Desmodontidae, and some members of the Phyllostomatidae) have tuberculosectorial molars; the upper molars have a W-shaped ectoloph with strongly developed stylar cusps, and in the lower molars the trigonid and talonid are roughly equal in size (Fig. 6–40A). In many insectivorous species and in some frugivorous members of the Phyllostomatidae, one or more premolars above and below are caniniform, and in some insectivorous species the premaxillae are separate (Fig. 6–41).

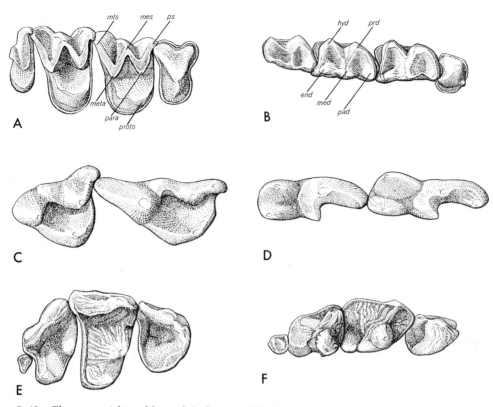

Figure 6–40. The upper right and lower left cheek teeth of three bats. A and B, an insect-eating vespertilionid (*Lasiurus cinereus*); A, the upper fourth premolar and three molars and B, the comparable lower teeth, C and D, a nectar-feeding phyllostomatid (*Leptonycteris sanborni*): C, second and third upper molars and D, the comparable lower teeth. E and F, a fruit-eating phyllostomatid (*Artibeus jamaicensis*): E, upper fourth premolar and three molars and F, the comparable lower teeth. Abbreviations: *end*, entoconid; *hyd*, hypoconid; *med*, metaconid; *mes*, mesostyle; *meta*, metacone; *mts*, metastyle; *pad*, paraconid; *para*, paracone; *prd*, protoconid; *proto*, protocone; *ps*, parastyle.

The flight apparatus of the microchiropterans is more progressive than that of the megachiropterans. In microchiropterans, the second digit does not bear a claw and lacks a full complement of phalanges, and its tip is connected by a ligament to the joint between the first and second phalanges of the third digit. During flight this connection allows the second digit to brace the third digit, which forms much of the leading edge of the distal part of the wing, against the force of the airstream (Norberg, 1969). The greater tuberosity of the humerus is usually enlarged and locks

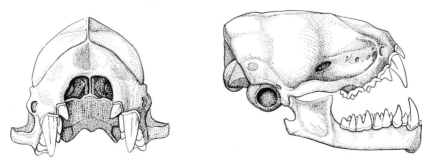

Figure 6–41. The skull of the hoary bat (*Lasiurus cinereus*, Vespertilionidae): left, anterior view, showing the emarginate front of the palate; right, side view, showing the shortened rostrum typical of some insect-feeding bats. Length of skull 17 mm.

against a facet on the scapula at the top of the upstroke of the wings (Fig. 6–23). The size of the tail and uropatagium are variable (Figs. 6–26, 6–29). The shape of the wing varies according to foraging pattern and style of flight. In general, slow, maneuverable fliers have short broad wings, whereas rapid enduring fliers have long narrow wings (Fig. 6–22).

Since their divergence from primitive insectivore stock, perhaps in the Cretaceous or early Paleocene, microchiropteran bats have undergone a remarkable adaptive radiation. Seventeen Recent families of microchiropterans and approximately 128 genera are now recognized. This large number of families and genera reflects the great structural diversity and widely contrasting modes of life that occur within this suborder.

FAMILY RHINOPOMATIDAE. Members of this small family, containing but one genus with three species, occur in northern Africa and southern Asia east to Sumatra. These animals are called mouse-tailed bats because of the long tail that is largely free from the uropatagium. No fossil representatives of the family are known.

These bats are considered to be the most primitive members of the Microchiroptera. The premaxillaries resemble those of megachiropterans in being separate from one another, and their palatal portions are much reduced. The second digit of the hand, in contrast to the arrangement in all other microchiropterans, retains two well-developed phalanges. Perhaps the clearest indication of the primitiveness of these bats is the structure of the shoulder joint. In contrast to the situation in most microchiropterans, the greater tuberosity of the humerus is small and does not lock against the scapula at any point in the wing-beat cycle. Other rhinopomatid features include laterally expanded nasal chambers, no fusion of cervical, thoracic or lumbar vertebrae, and a complete fibula.

The dentition is adapted to an insectivorous diet. The molars are tuberculosectorial; the upper molars have W-shaped ectolophs of the usual microchiropteran type. The dental formula is 1/2, 1/1, 1/2,

3/3 = 28. These are fairly small bats (the length of the head and body is up to 80 mm) with slender tails whose length approaches that of the head and body. The eyes are large, and the anterior bases of the large ears are joined by a fold of skin across the forehead.

Mouse-tailed bats are insectivorous and typically occupy hot, arid areas. They roost in a wide variety of situations, including fissures in rocks, houses, ruins, and caves; one species roosts in large colonies in some Egyptian pyramids. Although locally common, mouse-tailed bats over much of their range are outnumbered by other types of bats, and compared to other microchiropteran families are not an important group today. Rhinopomatids perhaps hibernate in some areas. Large deposits of subcutaneous fat occur in the abdominal area and around the base of the tail in individuals from some localities. These bats tolerate body temperatures as low as 22° C and can spontaneously rewarm themselves (Kulzer, 1965). The structure of the rhinopomatid kidney suggests a remarkable ability to concentrate urine in the interest of water conservation, a specialization of considerable adaptive importance for a group that inhabits arid areas.

FAMILY EMBALLONURIDAE. This family contains a variety of bats that are frequently called sac-winged or sheath-tailed bats. These bats range from small to large; *Taphozous peli*, an African emballonurid, is among the largest of the insectivorous microchiropterans, with a wingspread of nearly 70 cm. Twelve genera and about 44 species are currently recognized, and the wide geographic range of emballonurids includes the Neotropics (much of southern Mexico, Central America, and northern South America), most of Africa, southern Asia, most of Australia, and the Pacific Islands east to Samoa. The earliest fossil emballonurid is from the Eocene or Oligocene of Europe.

These small bats combine a number of primitive features with several noteworthy specializations. In the possession of postorbital processes and reduced premaxillaries that are not in contact with one another, emballonurids resemble pteropodids. In addition, the shoulder joint and elbow joints are primitive. An advancement over the rhinopomatids is the retention of only the metacarpal in the second digit; the flexion of the proximal phalanges of the third digit onto the dorsal surface of the third metacarpal is a specialization also found in some advanced families of bats. External obvious specializations include a glandular sac in the propatagium in some genera and the emergence of the tail from the dorsal surface of the uropatagium. The nose is simple; that is to say, it lacks leaflike structures or complex patterns of ridges and depressions. In addition to the more common gray and dark brown species of emballonurids, some species of one genus (Saccopteryx) have handsome whitish stripes on the back, and members of the genus Diclidurus are white.

These insectivorous bats typically inhabit tropical or subtropical areas, where they use a great variety of roosting sites. Emballonurids occupy houses, caves, culverts, rock fissures, hollow trees, vegetation, or the undersides of rocks and dead trees for daytime retreats, and usually roost in colonies. These bats are often fairly tolerant of well-lighted situations. In East Africa, Taphozous mauritianus often roosts on the trunks of large trees such as baobab trees (Adansonia digitata; Fig. 6–42). In some areas, emballonurids probably forage mainly over water. Some members of the genus Taphozous have long narrow wings, are swift and dashing fliers, and often forage in clearings and above the canopies of tropical forests.

A distinctive feature of some emballonurids is the glandular sac in the propatagium. Recent work on one species of Neotropical emballonurid (Saccopteryx bilineata) has shown that this sac, especially well developed in males, is used in ritualized displays during the breeding season (see page 371).

FAMILY CRASEONYCTERIDAE. This family, recently described by Hill (1974), is recorded only from Thailand. As far as is known, only one species (Craseonycteris thonglongyai) represents the family.

Figure 6–42. A baobab tree (Adansonia digitata) in Kenya, East Africa. Cavities in these trees are used as daytime retreats by various kinds of mammals, including some species of bats.

Craseonycteris is delicately built and is among the smallest of all bats, with small eyes and large ears. The premaxillae are not fused to adjacent bones, a feature that may increase the mobility of the upper lip, and the much reduced coronoid process of the dentary probably allows a wide gape of the jaws. The dental formula is 1/2, 1/1, 1/2, 3/3 = 28, and is of the usual insectivorous type with W-shaped ectolophs on the upper molars. The greater tuberosity of the humerus extends beyond the head of the humerus and may serve as a locking device, the second digit of the wing has only one very short phalanx, and the wing is broad. The pelvis and axial skeleton are highly specialized: the last three thoracic vertebrae and all but the last two lumbars are fused, and the sacral vertebrae are fused, whereas the pelvis is delicately built. The hind limbs are slender, and the fibula is threadlike. These bats resemble members of the Rhinopomatidae and Emballonuridae in several ways, and Hill (1974) considers the Craseonycteridae to be rather closely related to these families.

Nothing has been published on the habits of these tiny bats. They have been captured in a cave, and the vertebral fusions and slender hind limbs are probably adaptations to a style of roosting involving hanging pendant from the ceiling of a cave. Hill (1974) suggests that these bats may glean insects from leaves.

FAMILY NOCTILIONIDAE. Although this family is not important in terms of numbers of species (it contains but two species of one genus), it is of special interest because one species is structurally and behaviorally highly specialized for eating fish. Noctilionid bats are often referred to as bulldog bats or fishing bats. They oc-

cupy the Neotropics from Sinaloa, Mexico, and the West Indies to northern Argentina in South America. There is no fossil record of noctilionids.

Both structurally and in general external appearance, noctilionids are distinctive. They are fairly large (from roughly 20 to 75 gm in weight and up to two feet in wingspread), and the heavy lips, somewhat resembling those of a bulldog, the pointed ears, and the simple nose make the face unmistakable (Fig. 6–43). The dorsal pelage varies in color from orange to dull brown, and a whitish or yellowish stripe is usually present from the interscapular area to the base of the tail. The hind limbs and feet are remarkably large, especially in *Noctilio leporinus*, and the feet have sharp recurved claws. The premaxillae are complete, and in adults the two maxillae are fused together and are fused with the premaxillae, forming a strongly braced support for the enlarged upper medial incisors. The dental formula is 2/1, 1/1, 1/2, 3/3 = 28. The teeth are robust, and the molars are tuberculosectorial.

Figure 6–43. The face of a fishing bat (*Noctilio labialis*, Noctilionidae). (Photograph by N. Smythe and F. Bonaccorso.)

The seventh cervical vertebra is not fused to the first thoracic, the shoulder joint and elbow joint are primitive, and the second digit of the hand has a long metacarpal and a tiny vestigial phalanx. The pelvis is powerfully built, with the ischia strongly fused together and fused to the posterior part of the laterally compressed, keel-like sacrum. The tibia and hind foot of N. leporinus have a series of unusual specializations to be considered with the mode of foraging of the animal.

The feeding habits of the two species of Noctilio differ (Hooper and Brown, 1968; Fleming et al., 1972). Noctilio labialis eats largely insects, which it seems to catch over water. Howell and Burch (1974) report that several individuals taken in June in Costa Rica had fed largely on the pollen of a tropical tree (Brosimum); perhaps the animals scooped floating pollen from the surface of the water. N. leporinus, however, is a markedly atypical microchiropteran in that it eats largely fish. The style of foraging of this species is now known to involve the use of the hind claws as gaffs (Bloedel, 1955). This bat recognizes concentrations of small fish or single fish immediately beneath the surface of the water by detecting (by means of echolocation) the ripples or breaks in the surface that these fish create (Suthers, 1965, 1967). The bat skims low over the water and drags the feet in the water, with the limbs rotated so that the hooklike claws are directed forward. (This involves rotation of the hind limbs 180° from the typical mammalian position.) When a small fish is "gaffed," it is brought quickly from the water and grasped by the teeth. From 30 to 40 small fish were captured in this fashion per night by a N. leporinus under laboratory conditions. A series of modifications of the hind limb are clearly advantageous in allowing this animal to pursue efficiently its specialized style of foraging. The long calcar, which is roughly as long as the tibia, the calcaneum, the digits and claws, and the distal part of the tibia are all strongly compressed so that they are streamlined with respect to their direction of movement when they are dragged through the water. During the foraging sweeps the short tail is raised; the bladelike calcar is pulled craniad and is clamped against the flattened side of the tibia. In this way the large uropatagium is brought clear of the water and the streamlined calcar and tibia knife through the water, producing a minimum of drag.

Noctilionids roost during the day in groups in hollow trees and rock fissures, caves, and occasionally in buildings. N. leporinus is seemingly most common in tropical lowland areas, frequently occurring along coasts where they forage along rivers or streams, over mangrove-lined marshes and ponds, or over the sea. In western Mexico in the dry season, I have taken individuals as they foraged over small disconnected ponds in a nearly dry stream bed. These ponds supported large numbers of small fish.

FAMILY NYCTERIDAE. Members of this small family (including 13 species of one genus) are called hollow-faced bats. These bats occur in Madagascar, Africa, the Arabia-Israel area, the Malay Peninsula, and parts of Indonesia, including Sumatra, Java, and Borneo. No fossil nycterids are known.

Externally, these fairly small bats can be recognized by their large ears (Fig. 6–44), very small eyes, moderate or small size, and distinctive "hollow" face. The skull has a conspicuous interorbital concavity (Fig. 6–45) that is probably associated with the "beaming" of the ultrasonic pulses used in echolocation. This concavity is connected to the outside by a slit in the facial skin. The dental formula is 2/3, 1/1, 1/2, 3/3 = 32, and the molars are tuberculosectorial. Postcranially, these bats combine primitive and specialized features. The shoulder joint and elbow joint are fairly primitive, but the retention of only the metacarpal of the second digit of the hand, and the reduction of the number of phalanges of the third digit to two, are obvious specializations. The pectoral girdle is modified in the direction of enlargement and strengthening of the bracing of the sternum, a pattern parallel to the trend in birds toward the strengthening of the pec-

Figure 6–44. Faces of two African bats. A, A slit-faced bat (*Nycteris thebaica*, Nycteridae); B, African false vampire bat (*Cardioderma cor*, Megadermatidae).

toral girdle. The sternum in nycterids is robust and the mesosternum is strongly keeled; the manubrium is broad, the first rib is unusually strongly built, and the sev-

enth cervical and first thoracic vertebrae are fused.

This general pattern also occurs in the family Megadermatidae (a family to which

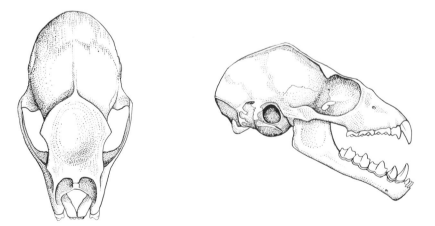

Figure 6–45. The skull of a slit-faced bat (*Nycteris thebaica*, Nycteridae): left, dorsal view, showing the depression in the forehead; right, side view, showing the flattened profile. Length of skull 19 mm. (After Hill and Carter, 1941.)

the nycterids are closely related), and reaches its most extreme development in the family Rhinolophidae (Fig. 6–46). Because the specializations of the pectoral girdle in these bats parallel to some extent the roughly similar modifications of this girdle in birds, they might be thought to be associated with a progressive structural trend in bats. Actually, it is doubtful that this is the case. Some of the most advanced and successful families of bats have less birdlike pectoral girdles than those in *the microchiropteran families listed above, but have modifications of the shoulder and elbow joints and forelimb musculature that provide for efficient flight. Perhaps the nycterid–megadermatid–rhinolophid pectoral girdle is associated with a foraging style typified by short intervals of flight. In any case, this style of pectoral girdle seems to be a divergent type and does not represent a progressive morphological trend common to most "advanced" microchiropterans.

Hollow-faced bats inhabit tropical forests and savanna areas, and seem to feed largely on arthropods that are picked from vegetation or from the ground. Nycterids are amazingly delicate and maneuverable fliers, and when foraging often seem to drift effortlessly around the trunks of large trees and near foliage. Flying insects form part of the diet, but orthopterans and flightless arachnids such as spiders and scorpions are also important food items. These bats remain on the wing only for fairly short intervals, for they retire to a resting place to eat their larger prey.

Nycterids roost in a variety of situations, and some are even known to occupy burrows made by porcupines and aardvarks. In Kenya, East Africa, in some remote safari camps, the pits dug as essential parts of privies ("longdrops" in local parlance) are occasionally used as daytime retreats by *Nycteris thebaica,* to the consternation of the uninitiated users of these toilets.

FAMILY MEGADERMATIDAE. This is not a large family, consisting of but four genera and five species. These bats are known as false vampires, an inappropriate title as they neither resemble vampires nor feed on blood. They occur in tropical areas in East Africa, southeastern Asia including Indonesia, the Philippines, and Australia. The fossil record of megadermatids is scanty; the earliest fossil is from the Eocene or Oligocene of Europe.

These are fairly large, broad-winged bats. The largest species has a wingspread approaching 1 m. The ears are large and

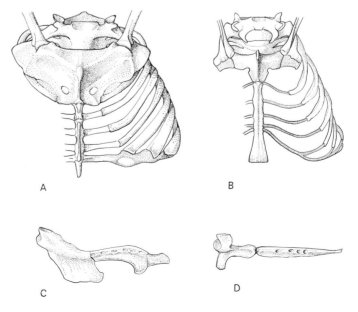

Figure 6–46. Ventral views of the thorax and lateral views of the sternum of a rhinolophid (*Hipposideros commersoni;* A and C) and a vespertilionid (*Myotis yumanensis;* B and D). Note the highly specialized sternum of *Hipposideros,* to which the first two ribs are fused. (From Vaughan, 1970b.)

A   B

C   D

are connected across the forehead by a ridge of skin. The snout bears a conspicuous "nose leaf," and the eyes are large and prominent (Fig. 6–44). The premaxillae and upper incisors are absent; the upper canines project forward and have a large secondary cusp (Fig. 6–47). The molars are tuberculosectorial; the dental formula is 0/2, 1/1, 1-2/2, 3/3 = 26 or 28. In *Megaderma*, and to a still greater extent in *Macroderma*, the W-shaped ectoloph of the upper molars is modified by the partial loss of the commissures connecting the mesostyle to the paracone and metacone. This trend is toward the development of an anteroposteriorly aligned cutting blade, and may be associated with the carnivorous habits of these genera. The shoulder and elbow joint are primitive; the second digit of the hand has one phalanx, and the third has two phalanges. The pectoral girdle has specializations similar to those of the nycterids, but the strengthening of the pectoral girdle is carried further in megadermatids. The manubrium of the sternum is relatively broader in megadermatids than in the nycterids, and is fused with the first rib and the last cervical and first thoracic vertebrae into a robust ring of bone. The megadermatid sternum is moderately keeled. The tail is very short or absent.

These bats occur in tropical forests and savannas, often near water, and utilize a variety of foods. Of the five species of megadermatids, three are known to be carnivorous and two are mostly insectivorous. The Australian ghost bat (*Macroderma gigas*), an unusually large, pale colored megadermatid, feeds on a variety of small vertebrates. In some areas it seems to feed largely on other bats. The ghost bat and related species in southeast Asia frequently consume their prey while hanging from the ceilings of spacious covered porches or verandas of large homes, and detract from the gracious atmosphere by littering the floors with feet, tails, and other discarded fragments of frogs, birds, lizards, fish, bats, and rodents. *Megaderma lyra* of India was observed by Prakash (1959) to eat bats of the genera *Rhinopoma* and *Taphozous*, a gecko, and a large insect, and in stomach contents of these bats he found bones of amphibians and fishes. The carnivorous species of megadermatids may hunt partly by sight. Another megadermatid, the insectivorous African *Lavia frons*, uses a style of foraging similar to that of a flycatcher. This partially diurnal bat hangs from a branch and makes short flights to capture passing insects.

Cardioderma cor of Africa has the most sedentary style of foraging known for any insectivorous bat (Vaughan, 1976). This entirely nocturnal bat perches in low vegetation when foraging and at some seasons regularly gives loud calls audible to man that seem to be territorial announcements. The body revolves through approximately 360° as the hanging bat meticulously "scans" the ground, listening for sounds made by terrestrial invertebrates, such as large beetles and centipedes. When prey is detected, the bat flies directly to the ground, snatches up the food, and returns

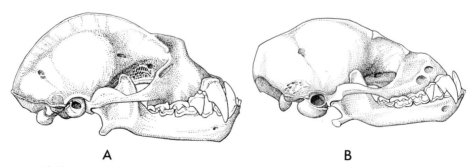

             A   **B**

Figure 6–47. Skulls of two African bats that eat large beetles. A, Giant leaf-nosed bat (*Hipposideros commersoni*, Rhinolophidae; length of skull 32 mm); B, African false vampire bat (*Cardioderma cor*, Megadermatidae; length of skull 27 mm).

to the same perch to consume it. When insects are abundant during the wet seasons, *Cardioderma* spends very little time in flight: on some nights this bat perches for periods averaging nearly 11 minutes and spends less than 1 per cent of its foraging time in flight. But in the dry season, when prey abundance declines, flights from perch to perch are more frequent and more time is spent in flight, although flights after prey average only 3 seconds. Considering all seasons, flights after prey average only 5 seconds. In *Cardioderma*, and perhaps in all megadermatids, the technique of searching for prey has departed markedly from that of most insectivorous bats, whereas the style of flight and the morphology of the forelimb has remained generalized.

Megadermatids roost in many types of places, from hollow trees (Fig. 6–42), caves, and buildings, in the case of most species, to sparse, occasionally sunlit vegetation, in the case of *Lavia*. This bat has been found by Wickler and Uhrig (1969) to occupy fairly small foraging territories and to have several calls audible to man during social interactions.

FAMILY RHINOLOPHIDAE. This is a large and successful Old World family, with 10 genera and approximately 127 species. Its members are often called horseshoe bats because of the complex and basically horseshoe-shaped cutaneous ridges and depressions on the nose (Fig. 6–48). Some rhinolophids are quite small, but *Hipposideros commersoni* of Africa, at the other extreme, is the largest insectivorous bat, some individuals reaching weights of over 100 g. The geographic distribution includes much of the Old World from western Europe and Africa to Japan, the Philippines, Indonesia, Melanesia, and Australia. This may be an extremely ancient family, for some late Eocene fossils from Europe have been assigned to the living genus *Rhinolophus*.

Because of the unique and complex face, rhinolophids are one of the most unmistakable groups of bats. The ears are usually large, but lack a tragus, and the eyes are small and inconspicuous. The tail is of

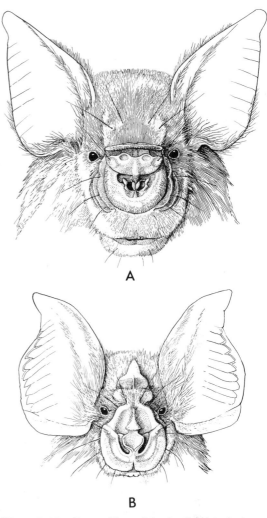

Figure 6–48. Faces of bats of the family Rhinolophidae. A, Giant leaf-nosed bat *(Hipposideros commersoni)*; B, *Rhinolophus landeri*.

moderate length in some species, but is small or rudimentary in others. The pectoral girdle is remarkable because it represents the extreme development of the trend (that occurs also in the Nycteridae and Megadermatidae) toward powerful bracing and enlargement of the sternum. In the most extreme manifestation of this trend, the seventh cervical vertebra, the first and second thoracics, the first and most of the second rib, and the enormously enlarged and shieldlike manubrium of the sternum are fused into a powerfully braced ring of bone (Fig. 6–46). The shoulder joint has a moderately well-developed locking device. In some rhinolophids, all but the last two lumbar vertebrae are fused; a similar specialization occurs in the Natalidae (Fig. 6–49). The pelvis is uniquely modified by enlargement of the anterior parts and an accessory connection between the ischium and the pubis. These unusual pelvic specializations may be responses to the mechanical stresses imposed on the hind limbs and pelvis by the repeated take offs and landings that occur during foraging in some of these bats. When these bats roost, they often hang upside down, and the hind limbs are rotated 180° from the usual mammalian posture so that the plantar surfaces of the feet face forward. The extreme adaptations for strengthening the pectoral and pelvic girdles that are typical of rhinolophids occur to a comparable degree in no other family of bats.

Horseshoe bats are common in many areas, and in Germany the "Hufeisennase" is a familiar inhabitant of attics and church steeples. These bats have wide environmental tolerances; various species inhabit temperate, subtropical, tropical, and desert regions. Rhinolophids hibernate in some parts of their range, and characteristically rest or hibernate with the body enshrouded by the wing membranes. Several species in East Africa are migratory. The food is largely arthropods, and the style of foraging resembles that of the nycterids and some megadermatids. Horseshoe bats pick spiders and insects from vegetation or capture flying insects in midair, and *Rhinolophus ferrumequinum* was observed to alight on the ground and capture flightless arthropods (Southern, 1964). When foraging, *Hipposideros com-*

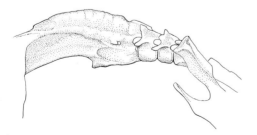

Figure 6–49. Lateral view of the left side of the fused lumbar vertebrae of a funnel-eared bat (*Natalus stramineus*, Natalidae). (From Vaughan, 1970b.)

mersoni of Africa hangs fairly high in trees, uses echolocation to detect large and straight-flying beetles at distances up to 20 m, and makes brief and precise interception flights that last an average of but 5.1 seconds (Vaughan, 1976). This bat returns to the perch to consume prey, which consists of very large beetles up to 60 mm in length. Like *H. commersoni*, a number of rhinolophids make short foraging flights and do not remain continuously on the wing while foraging. Perhaps the wing membranes are important in some species in aiding in the capture of insects. Webster and Griffin (1962) demonstrated photographically that one species of rhinolophid is able to capture insects in the chiropatagium.

In contrast to many bats that emit pulses from the open mouth in echolocation, rhinolophids keep the mouth closed during flight; the ultrasonic pulses used in echolocation are emitted through the nostrils and are "beamed" by the complex nasal apparatus (Mohres, 1953). A remarkable series of coordinated behaviors is associated with the highly specialized rhinolophid style of echolocation (see page 461).

Most horseshoe bats are colonial, but some are solitary. Many kinds of roosting sites are used; caves, buildings, and hollow trees are generally preferred, but foliage and the burrows of large rodents are used by some species.

FAMILY PHYLLOSTOMATIDAE. This is the most diverse family of bats with respect to structural variation, and contains more genera than does any other chiropteran family. Forty-four genera and some 120 species are included in the Phyllostomatidae. These Neotropical "leaf-nosed bats" are so named because of the conspicuous leaflike structure that is nearly always present on the nose (Figs. 6–33, 21–2). These bats have exploited the widest variety of foods used by any family of bats. Some leaf-nosed bats have retained insectivorous feeding habits, but some are carnivorous and eat small vertebrates, including rodents, birds, and lizards; some eat nectar and pollen, and some are frugivorous. Phyllostomatids are the most important bats in the Neotropics, and occur from the southwestern United States and the West Indies south to northern Argentina. These bats can be traced back to the Miocene of Colombia, South America.

The great structural variation that occurs in the Phyllostomatidae is largely associated with an adaptive radiation into a wide variety of feeding niches. Within the family are some fairly small bats (*Choeroniscus* has a wingspread of roughly 220 mm) as well as the largest New World bat (*Vampyrum* has a wingspread of over 1000 mm). In most species, the nose leaf is conspicuous and is spear shaped, but in a few species the nose leaf is rudimentary or is highly modified (Fig. 6–33B). The ears vary from extremely large to small, and a tragus is present. The tail and uropatagium are long in some species, with many stages of reduction and the absence of the tail and uropatagium being represented by various species. Some species have a uropatagium but lack a tail; only *Sturnira* has completely lost the uropatagium. The wings are typically broad; the second digit has one phalanx and the third has three phalanges. The shoulder joint has a moderately well-developed locking device formed between the greater tuberosity of the humerus and the scapula, but the elbow joint and forearm musculature are primitive, and the forelimb is, for a bat, generalized. Probably all phyllostomatids, whatever their feeding habits, remain on the wing only for short periods during foraging.

The forelimbs are not used only for flight but are important in many species in food handling as well as in climbing over and clinging to vegetation (as in the fruit-eating species). The importance of such use of the forelimbs has probably favored the retention in phyllostomatids of limbs more generalized than those of many strictly insectivorous groups of bats. The seventh cervical and first thoracic vertebrae are not fused, and no fusion of elements to form the sturdy "pectoral ring" characteristic of rhinolophoid bats occurs in phyllostomatids. In some leaf-nosed bats, however, the sternum is strongly keeled. The ventral parts of the pelvis are lightly built in most species, but the ilia are robust and are more or less fused to the sacral vertebrae. These vertebrae are fused into a solid mass that becomes laterally compressed posteriorly. The acetabulum is characteristically directed dorsolaterally; the hind limbs are rotated 180° from the usual mammalian orientation and have a spider-like posture. Because of this position of the hind limbs, some phyllostomatids are unable to walk on a horizontal surface and use the hind limbs only for hanging upside down.

All of the Recent leaf-nosed bats probably evolved from an ancestral type that had tuberculosectorial teeth adapted to a diet of insects. Of the six Recent subfamilies, however, only one has retained this dentition, and in some species there is no trace of the ancestral pattern. The noteworthy adaptive radiation of phyllostomatids will be traced by considering the dentitions and foraging habits of each subfamily.

The subfamily Phyllostomatinae deviates least from the ancestral structural plan, and some species retain insectivorous feeding habits. This subfamily contains all of the leaf-nosed bats with tuberculosectorial teeth of the ancestral type; however, in some species (*Chrotopterus* and *Vampyrum*, for example) the W-shaped ectoloph of the upper molars is distorted by the reduction of the stylar cusps and the closeness of the protocone, paracone, and metacone. Most members of this subfamily are insectivorous, and some species are known to pick insects either from vegetation or from the ground. On the other hand, a few of the largest phyllostomatines resemble their Old World look-alike ecological counterparts, the megadermatids, in their carnivorous habits. The large phyllostomatine species *Phyllostomus hastatus, Trachops cirrhosus, Chrotopterus auritus,* and *Vampyrum spectrum* are known to feed on small vertebrates. Beneath the roosts of *V. spectrum*, feathers and the tails of rodents and geckos frequently give indications of feeding preferences. The means by which these carni-vorous-omnivorous bats perceive small vertebrates is not known. They may well hear the faint sounds made as their prey moves, and the large eyes of these bats indicate that hunting may also involve the use of vision. Bats of this type generally have large ears, however, suggesting highly discriminatory echolocation.

Nectar feeding is popular among tropical vertebrates (as indicated by the presence of over 300 species of hummingbirds in the American tropics), and has also been adopted by bats of the subfamily Glosso-phaginae of Mexico and Central and South America, and by bats of the subfamily Phyllonycterinae of the West Indies. These bats feed on the nectar and pollen of a great variety of plants and have many structural features associated with this mode of life. The tongue is long and pro-trusile, and has a brushlike tip (Fig. 6–50); the rostrum is elongate and the dentaries are slender (Fig. 6–51). The cheek teeth have largely lost the tuberculosectorial pattern (Fig. 6–40). The hairs of at least some of these nectar feeders have divergent scales (Fig. 6–39) that catch pollen as the bat feeds on nectar. This pollen is swallowed when the bats groom their fur

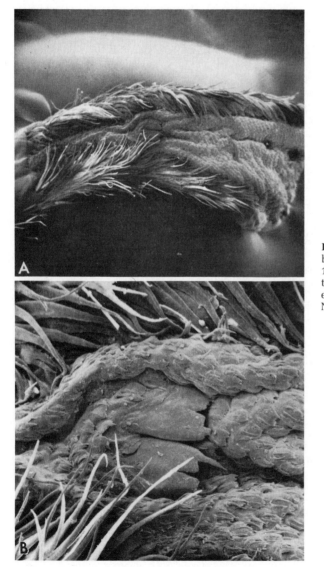

Figure 6–50. The tongue of a nectar-feeding bat (*Leptonycteris sanborni*) under 20× (A) and 100× (B) magnification. The tip of the tongue is to the left. (Photographs taken with a scanning electron microscope by Donna J. Howell and Norman Hodgkin.)

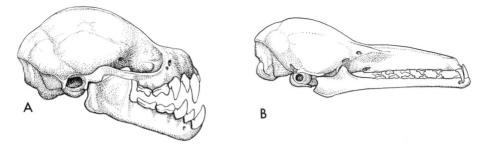

Figure 6–51. Skulls of leaf-nosed bats (Phyllostomatidae). A, A fruit-eater (*Artibeus phaotis*; length of skull 19 mm); B, a nectar-feeder (*Choeronycteris mexicana*; length of skull 30 mm).

and provides a protein supplement without which the animals could not survive (see page 303).

In nectar-feeding bats, the wings are usually broad and the uropatagium is reduced. Although they probably are not able to remain on the wing for long periods of time, as can some insectivorous bats, nectar feeders can maneuver delicately through dense tropical vegetation and can hover. Flowers are seemingly located by the sense of smell, and these bats feed by hovering briefly and thrusting their long tongues into the flowers. The pollination of many night-blooming Neotropical plants is accomplished by nectar-feeding phyllostomatids, just as many plants in the Old World tropics are pollinated by nectar-feeding pteropodids. Selective forces determined by this method of pollination have probably been important in the evolution of flower structure and the timing of pollen and nectar production in these plants (see page 302). Glossophagine bats are clearly not so tightly restricted to a nectar and pollen diet as they were formerly thought to be (Howell and Burch, 1974). A dietary continuum from heavy reliance on insects (in *Glossophaga*) to virtual total dependence on nectar and pollen (in *Leptonycteris*) occurs in the subfamily.

The members of three subfamilies of phyllostomatids—Carolliinae, Sturnirinae and Stenoderminae—are frugivorous. The success of these groups and the richness of this food source in the neotropics is indicated by the fact that within the Phyllostomatidae, the largest and most abundant group of neotropical bats, half of the species (approximately 60 out of 121) are basically fruit eaters. Several variations on this fruit-eating theme can be recognized. Members of the subfamily Carolliinae have reduced molars with the original tuberculosectorial pattern largely obliterated. These bats apparently prefer ripe, soft fruit, and are known to eat a great variety of it. The second frugivorous subfamily, Sturnirinae, is composed of fairly small, often brightly colored bats that have robust molars with no trace of the basic tuberculosectorial pattern. Indeed, their molars strongly resemble those of New World monkeys (Cebidae). Sturnirines eat small and often hard fruit, such as the fruits of low-growing species of nightshade (*Solanum* sp.).

The third frugivorous subfamily, Stenoderminae, contains bats with robust teeth that are highly modified for crushing fruit. The upper molars have lost the stylar cusps, and the inner portion is much enlarged and is marked by complex rugosities (Fig. 6–40E, F). The rostrum is short and the coronoid process of the dentary is fairly high in many species, conferring considerable mechanical advantage for powerful jaw action to the large temporal muscles. Large species of the stenodermine genus *Artibeus* are remarkably abundant in some neotropical areas, and their piercing calls are characteristic sounds of the tropical nights. Often many *Artibeus* of several species concentrate on a single fig tree (*Ficus* sp.) with abundant fruit. In central Sinaloa, Mexico, two students and I camped beneath such a fig tree—but only for one night. The activities of dozens of *A. lituratus, A. hirsutus* and *A. jamaicensis* caused a nearly continuous rain of fruit and bat excrement through part of the night, and with sunrise came herds of aggressive local pigs to gather from the ground the night's fallout of figs. Stenodermines often eat unripe and extremely hard fruits, and it is perhaps as an adaption to this type of food that the robust teeth and powerful jaws evolved.

A final caution regarding the classification of phyllostomatid bats is in order. Great diversity in echolocation ability and in karyotypes (numbers, sizes, and shapes of chromosomes) occurs within the Glossophaginae (Howell, 1974; Baker, 1967). These differences indicate that this subfamily, as now conceived, may be polyphyletic; that is to say, it may contain bats derived from separate phylogenetic lines. The validity of this taxon is thus in doubt.

FAMILY MORMOOPIDAE. The three genera and eight species that comprise this family were traditionally considered as members of the family Phyllostomatidae (Miller, 1907:118; Simpson, 1945:57), but recent studies have shown that they differ so markedly from the phyllostomatids that they merit recognition as a separate family (Forman, 1971; Smith, 1972; Vaughan and Bateman, 1970). All mormoopids can appropriately be called leaf-chinned bats, because in all species a conspicuous, leaf-like flap of skin occurs on the lower lip. These bats are largely tropical in distribution, and occur from the southwestern United States and the West Indies south to Brazil.

Leaf-chinned bats are fairly small, and have several distinctive, externally visible specializations. The snout and chin always have cutaneous flaps or ridges (that reach their most extreme form in Mormoops, Fig. 6–33C), but a nose leaf is never present. The ears are moderately large, have a tragus, and vary in shape, but always have large ventral extensions that curve beneath the fairly small eyes. The tail is short and protrudes from the dorsal surface of the fairly large uropatagium. The rostrum is tilted more or less upward (this feature is most extremely developed in Mormoops), and the floor of the braincase is elevated. The coronoid process of the dentary is

reduced, allowing the jaws to gape widely. The teeth are of the basic insectivorous type; the dental formula is 2/2, 1/1, 2/3, 3/3 = 34.

The number of phalanges in the hand is as in the phyllostomatids, but the shoulder and elbow joints differ markedly from the phyllostomatid pattern. The greater tuberosity of the humerus in mormoopids does not form a well-developed locking device with the scapula; the head of the humerus is more or less elliptical, perhaps favoring a specialized wing-beat cycle. The elbow joint is specialized in all species and, in Mormoops, modifications of the distal end of the humerus and the forearm musculature provide for a highly efficient "automatic" flexion and extension of the hand. The musculature of the hand is reduced and simplified; this serves to lighten the hand and probably favors maneuverability and endurance. The hind limbs do not have the spider-like posture typical of phyllostomatids, but have a reptilian posture that allows leaf-chinned bats to crawl on the walls of caves with considerable agility.

Leaf-chinned bats are among the most abundant bats in many tropical localities, where they are seemingly the major chiropteran insectivores. They are most common in tropical forests, but occur also in some desert areas. Some species appear early in the evening; their insect-catching maneuvers resemble those of their temperate zone counterparts, the vespertilionids. Leaf-chinned bats usually roost in caves or deserted mine shafts, and may concentrate in large numbers. A colony of Mormoops observed by Villa-R. (1966:187) in Neuvo Leon, Mexico, contained more than 50,000 bats, and a colony of four species of mormoopids in Sinaloa was estimated to contain between 400,000 and 800,000 bats (Bateman and Vaughan, 1974). When the bats from the latter colony emerged in the evening, they swept down the nearby arroyos and trails in such numbers and at such speeds that one hesitated to move across their path. When they form large colonies, these bats seem to disperse several miles from their roosting site to forage

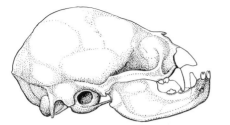

Figure 6–52. The skull of a vampire bat (*Desmodus rotundus*, Desmodontidae). Length of skull 24 mm.

at night, and remain continuously on the wing for several hours. Their impact on tropical ecosystems must be great, for the bats in the Sinaloan colony probably consume over 1400 kg (about 1½ tons) of insects per night. It is not surprising that the bats must disperse over a wide area to forage.

FAMILY DESMODONTIDAE. This family contains the vampire bats, the only mammals that feed solely on blood. Only three genera, each with a single species, comprise this group, but they are widely distributed from northern Mexico southward to northern Argentina, Uruguay, and central Chile. No extinct genera have been found, but *Desmodus* is known from the Pleistocene.

Vampires are fairly small bats. Mexican specimens of the largest and most common species, *Desmodus rotundus*, usually weigh from 30 to 40 gm. The skull and dentition are highly specialized. The rostrum is short and the braincase is high (Fig. 6–52), and in all species the cheek teeth are reduced in both size and number. In *D. rotundus*, the most specialized species, the dental formula is 1/2, 1/1, 2/3, 0/0 = 20. The upper incisors are unusually large and are compressed and bladelike, as are the upper canines. These teeth have remarkably sharp cutting edges. The cheek teeth are tiny. Except for the canine, the lower teeth are small. The thumb in *Desmodus* is unusually long and sturdy and contributes an additional segment with three joints to the forelimb during quadrupedal locomotion. The hind limbs are large and robust, and the fibula is not reduced. The proximal part of the femur and the tibia and fibula are curiously flattened and ridged; these irregularities provide large surfaces for the attachment of the powerful hind limb musculature. Vampires can run rapidly and easily and can even jump short distances (Fig. 6–53). Their flight is strong and direct and not highly maneuverable.

The feeding habits of this family are of particular interest. Vampires begin foraging after complete darkness, and have one foraging period per night. *Diaemus* and *Diphylla* prefer the blood of birds, but *Des-*

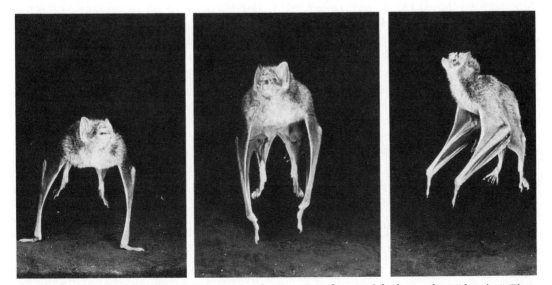

Figure 6–53. A vampire bat (*Desmodus rotundus*) leaping. Note the use of the long, robust "thumbs." (Photographs by J. Scott Altenbach.)

modus feeds on mammals. *Desmodus* alights on the ground near its chosen host, usually a cow, horse, or mule, and climbs up the foreleg to the shoulder or neck. The bat uses its upper incisors and canines to make an incision several millimeters deep from which blood is "lapped" by the tongue. Vampires occasionally feed on the feet of cattle, at which time their ability to jump quickly may enable them to avoid injury when the host animal moves its feet. In *Desmodus*, the ingestion of blood is facilitated by an anticoagulant in the saliva that retards the clotting of blood. It has been estimated that each bat takes a meal of blood each night that amounts to over 50 per cent of the fasting weight of the bat; a vampire weighing 34 gm, then, takes roughly 18 gm of blood per night (Wimsatt, 1969a). Because of the nightly vampire-caused drain of blood from cattle in certain localities, and because vampires transmit rabies and other diseases, these bats are of great economic importance in many Neotropical areas. Occasionally vampires feed on man.

Renal function in vampires is of great interest. These animals begin urinating soon after they begin to feed, and rapidly lose much of the water taken in with the blood meal. This enables the bats to fly back to their roosts with less expenditure of energy and at less risk from predation than if they were burdened by the full weight of the ingested blood. Back in the roost, the bats continue digesting the now partially dehydrated blood and are faced with a problem of excreting large amounts of nitrogenous wastes without losing excessive amounts of water. At this time, rather than freely excreting water, as was done earlier in the foraging-digestion cycle, the kidney exerts a remarkable ability to concentrate wastes, and highly concentrated urine is excreted. In this notable instance, a tropical mammal that lives in environments that seldom lack accessible water has evolved a kidney surpassing that of many desert mammals in its ability to concentrate urine and conserve water (Horst, 1969).

FAMILY NATALIDAE. This small Neo-

tropical family includes a single genus with five species. These bats are commonly referred to as funnel-eared bats, and occur from Baja California, northern Mexico and the West Indies, southward to Colombia, Venezuela, and Brazil. The only fossils are those of the living genus from the Pleistocene and Recent.

These small bats weigh from roughly 5 to 10 gm, and have slender, delicate-looking limbs, broad wing membranes, and a large uropatagium that encloses the long tail. The funnel-shaped ears with a tragus, the simple nose lacking any sort of nose leaf, and the long soft pelage that is frequently yellowish or reddish in color are characteristic. The skull has a long wide rostrum with complete premaxillaries, and the braincase is high. The teeth are tuberculosectorial; the dental formula is 2/3, 1/1, 3/3, 3/3 = 38. The humeroscapular locking device is well developed, reduction of the phalanges of the hand is well advanced (the second digit lacks a phalanx and the third has two), and the manubrium of the sternum is unusually broad and has a well developed keel. Some of the most distinctive natalid features, however, are those of the axial skeleton that serve to reduce its flexibility: the thoracic vertebrae are anteroposteriorly compressed and fit tightly together; the ribs are broad and the narrow intercostal spaces are largely spanned by sheets of bone; all except the last two lumbar vertebrae are fused into a solid, laterally compressed, dorsally and ventrally keeled mass (Fig. 6–49); and the sacral vertebrae are mostly fused. As a result of these specializations, the strongly arched thoracolumbar section of the vertebral column is nearly rigid, with movement between this and the sacral section of the column allowed only by the "joint" formed by the last two lumbar vertebrae. These specializations seem to brace and cushion the vertebral column against shock transmitted to it by the hind limbs when this bat alights on the ceilings of caves.

Funnel-eared bats are insectivorous and their foraging flight is slow, delicate, and maneuverable. Individuals released in dense vegetation are amazingly adroit at flying slowly through small openings between the interlacing branches of trees and shrubs. These bats inhabit tropical and semitropical lowlands and foothills, and typically roost in groups in warm, moist, and deep caves or mines. These are handsome little bats; groups of *Natalus stramineus* scattered over the ceiling of a cave look like bright orange jewels in the beam of a flashlight.

FAMILY FURIPTERIDAE. This small family contains but two genera, each with one species. These bats occur in northern South America south to southern Brazil and northern Chile, and in Trinidad. Furipterids, known as smoky bats because of their grey pelage, are seemingly closely related to the Natalidae, Thyropteridae, and Myzopodidae. All these groups share certain structural similarities. No fossil furipterids are known.

Externally, furipterids resemble natalids in the structure of the ears and in their slender build. The shoulder joint and the fused lumbar vertebrae are also similar in these families. Furipterids differ from natalids in minor features of the skull and dentition, such as partially cartilaginous premaxillaries and reduced canines. The furipterid dental formula is 2/3, 1/1, 2/3, 3/3 = 36. The thumb of smoky bats is greatly reduced and is functionless.

These bats apparently are not common, and their habits are poorly known. They are insectivorous, and have been found in caves and buildings. Most of the area inhabited by smoky bats is tropical, but *Amorphochilus* occurs in arid coastal sections of northwestern South America.

FAMILY THYROPTERIDAE. Two small Neotropical species of bats comprise this family. These bats are known as discwinged bats because of the remarkable sucker discs that occur on the thumbs and feet. These animals, and the one member of the family Myzopodidae, are the only bats and, except for two genera in the order Hyracoidea, the only mammals that have true suction cups. Disc-winged bats occur in southern Mexico, Central America, and South America as far as Peru

and southern Brazil. No fossil thyropterids have been recorded.

In general appearance and in many skeletal details, these small, delicately formed bats resemble natalids, but the lumbar vertebrae are not fused as in the latter. The skulls of natalids and thyropterids are similar and the dental formulae are the same. The thumb is reduced but retains a small claw, and its first phalanx has a sucker disc. The second digit is short, being represented by only a rudimentary metacarpal, and as a result the membrane between digits two and three is unusually small. The third digit has three bony phalanges. The digits of the feet have only two phalanges each, the third and fourth digits are fused, and the metatarsals bear a suction disc. The discs have a complex structure that allows them to act as suction cups; the bats can cling to smooth surfaces and can even climb a vertical glass surface. A fibrocartilaginous framework braces each disc; the rim of the disc consists of 60 to 80 chambers, each supplied by a sudoriferous gland (sweat gland). These glands improve the tightness of contact with the substrate by insuring that the face of the disc is constantly moistened. The disc itself lacks muscles, but specialized forearm muscles produce suction by cupping the middle of the face of the disc, and release suction by lifting a section of the rim of the disc (Wimsatt and Villa-R., 1970). These are the most elaborate and efficient suction discs known in mammals.

Disc-winged bats are insectivorous and are restricted to tropical forests. The roosting habits of these bats are highly specialized. They roost only in the young, slightly unfurled leaves of certain tropical plants that are partially or completely shaded by larger trees. Such a roosting site is provided by the "platanillo" (*Heliconia* sp.), which resembles the banana plant. While a young leaf of this plant is beginning to unroll, it forms a tube roughly four feet long and an inch or so in diameter with a small opening at its tip. Several discwinged bats may occupy such a tubular leaf in a head-to-tail row, heads upward, with the sucker discs anchoring them to the slippery surface of the smooth leaf. Because the leaf soon unfurls, it is suitable for occupancy for only about 24 hours and the bats move periodically to new and more suitable leaves. Findley and Wilson (1974) found that these bats usually roost in social groups of six or seven, that the bats of a given group always roost together, and that each group occupies an exclusive area within which it roosts in the daytime.

FAMILY MYZOPODIDAE. The only species representing this family is *Myzopoda aurita*, the sucker-footed bat, a species restricted to Madagascar. No fossils of this family are known.

This bat is probably related to the Natalidae, Furipteridae, and Thyropteridae, as indicated by the structure of the shoulder joint. The lumbar vertebrae are not fused as they are in natalids. The cheek teeth are of the standard tuberculosectorial-insectivorous type, and the dental formula is 2/3, 1/1, 3/3, 3/3 = 38. The ears are very large and the ear opening is partly covered by an unusual mushroom-shaped structure of a sort found in no other bat. The claw of the thumb is rudimentary, and the thumb bears a sucker disc. Only the metacarpal of the second digit is bony; the third digit has three ossified phalanges. The foot bears a sucker disc on its sole and, as in thyropterids, each digit has only two phalanges. In *Myzopoda*, the metatarsals are fused and all the toes fit tightly against one another.

Myzopoda appears to be rare, and its life history is unknown. Its dentition indicates insectivorous feeding habits.

FAMILY VESPERTILIONIDAE. This is the largest family of bats in terms of numbers of species, and is the most widely distributed. Thirty-three genera and approximately 280 species are included in this family, and in temperate parts of the world these are usually by far the most common bats. In the New World, vespertilionids occur from the tree line in Alaska and Canada southward throughout the United States, Mexico, and Central and South America. All of the Old World is inhabited north to the tree line in northern Europe and Asia. Most islands, with the exception of some that are remote from large land masses, support vespertilionids. As can be inferred from their geographic distribution, these bats occupy a wide variety of habitats, from boreal coniferous forests to barren sandy deserts. In the Neotropics, however, they are greatly outnumbered by bats of other families, particularly by leaf-nosed bats (Phyllostomatidae).

Perhaps because of the diversity of habits and structure represented within the Vespertilionidae, no common name for this group is in general use; they are usually simply called vespertilionid bats. This family can be traced back to the middle Eocene in both Europe and North America, but it apparently did not reach Africa and South America until the Pleistocene. The genus *Myotis* is remarkable for its broad geographic distribution, which includes roughly the entire area occupied by the Vespertilionidae, and for its long fossil record, which begins in the middle Oligocene of Europe.

Vespertilionids are rather plain-looking bats that lack the distinctive facial features characteristic of many families. A nose leaf is rarely present, nor do complex flaps or pads occur on the lower lips. The eyes are usually small. The ears are of moderate or large size, and the tragus is present, but differs in shape markedly between species. These bats are usually small, weighing from 4 to 45 gm. The wings are typically broad, and the uropatagium is large and encloses the tail. The shoulder joint is of an advanced type, and provides for a locking of the large greater tuberosity of the humerus (Fig. 6–36C, D) against the scapula at the top of the upstroke of the wing. The elbow joint is also advanced, and the spinous process of the medial epicondyle, which is well developed in many species, enables certain forearm muscles to "automatically" extend and flex the hand (Fig. 6–26). The shaft of the ulna is vestigial, but the proximal portion forms an essential part of the elbow joint. The second digit of the hand has two bony phalanges, and the third digit has three. The fibula is rudimentary. The manubrium of the sternum has a keel, but the body of the sternum has at best a slight ridge. Except in one genus (*Tomopeas*), all presacral verte-brae are unfused.

The teeth are tuberculosectorial, and the W-shaped ectoloph of the upper molars is always well developed. The dental for-mula varies from 1/2, 1/1, 1/2, 3/3 = 28 to 2/3, 1/1, 3/3, 3/3 = 38. The skull lacks post-orbital processes; the palatal parts of the premaxillaries are missing and the front of the palate is emarginate (Fig. 6–41). In gen-eral, vespertilionids are mostly small plain bats that are characterized by refinements of the flight apparatus that make them ef-ficient, maneuverable fliers.

Most vespertilionids are insectivorous and in their ability to capture flying in-sects they are unexcelled. Most children in Europe and North America gain their first experience with bats by watching vesperti-lionid bats, silhouetted against the twilight sky, making abrupt turns and sudden dives while pursuing insects. The most commonly used vespertilionid foraging technique is probably also the most de-manding: it involves the pursuit and cap-ture of flying insects by bats that remain on the wing throughout most of their foraging periods. The insects are perceived by echolocation (Chapter 21). The bats emit ultrasonic pulses, and locate and fol-low insects by utilizing the reception and interpretation of echoes of the pulses from the bodies of the insects. Insects are usually followed in their erratic flight by a series of intricate maneuvers by the bat, and are either captured in the mouth or, in the case of some species of bats, are trapped by a wingtip or by the uropa-tagium (Webster and Griffin, 1962). This type of foraging demands highly maneu-verable flight, and this is the type of flight to which vespertilionids seem best adapted.

Styles of foraging vary between different vespertilionids The tree-roosting bats (*Lasiurus*) remain on the wing throughout their foraging, while others alight to eat large prey. Some species snatch insects or arachnids from leaves or pounce on them on the ground. The pallid bat (*Antrozous pallidus*), a common species in the south-western United States, feeds on such large terrestrial arthropods as scorpions, Jerusa-lem crickets (*Stenopelmatus*), and sphinx moths (Sphingidae). This bat often uses porches, shallow caves, or abandoned buildings as places to rest and eat its prey. Often there are accumulations of discarded legs, fragments of exoskeletons, and wings beneath these roosts. Some vespertilionids capture insects from the surface of the water, and several species of *Myotis* cap-ture fish or crustaceans from the water, probably by gaffing the prey with the claws of the feet. Some vespertilionids are known to have an early evening and a predawn foraging period. Kunz (1973b) found that in Iowa several temporal pat-terns of foraging occurred among six spe-cies of vespertilionid bats: these species had major periods of foraging within five hours after sunset and some of them had a well-marked second foraging period, whereas some had only a minor second period or none at all. The nocturnal activ-ity patterns of most vespertilionids remain unknown, but in all probability the times of peak activity of some bats coincide with times of peak activity of their insect prey.

A wide variety of roosting places is uti-lized by vespertilionids. They adapt well to urban life, and frequently roost during the day in attics of churches or houses, in spaces between rafters of barns or ware-houses, or behind shutters or loose boards. Crevices in rocks, spaces beneath rocks or behind loose bark, caves, mines, holes in trees, and foliage are also utilized. Often these bats are colonial, frequently with nursery colonies of females with young oc-cupying one roost and adult males using another; but many species, such as the foliage-roosting bats, roost singly or in small groups. Some species rest for part of the night beneath bridges or in porches or buildings, often in places never used as daytime retreats.

In temperate regions, many vesperti-lionids hibernate. Although the hiberna-tion sites of some species are not known, some well known species hibernate in caves and mines or buildings, and some species migrate fairly long distances to reach favorite hibernacula. Excellent reviews of hibernation (Davis, 1970) and migration in bats (Griffin, 1970) are available. Two small European species are known to migrate over 1000 miles from Russia to Bulgaria (Krzanowski, 1964). Some temperate zone vespertilionids hibernate for short periods and may be at least intermittently active in the winter, but others in colder areas hibernate throughout the winter except for occasional short arousals. In tropical areas, vespertilionids may respond to seasonal changes in insect abundance by local migrations. This is the case in southern Kenya, East Africa. The weight of much observational evidence points toward migration in foliage-roosting bats in the United States. However, some red bats *(Lasiurus borealis)* remain in cold regions in the central United States throughout the winter and may be active on warm days (Davis and Lidicker, 1956).

FAMILY MYSTACINIDAE. One rather aberrant species, *Mystacina tuberculata*, the short-tailed bat, is the sole member of this family. This species occurs only in New Zealand, and the family has no fossil record.

Mystacina has some characteristics of vespertilionids and some of free-tailed bats (Molossidae), but in general is sufficiently distinct from either group to merit recognition as a member of a separate family. Vespertilionid characteristics of *Mystacina* include the advanced locking shoulder joint, one phalanx in the second digit and two in the third, and the lack of fusion of presacral vertebrae. The skull is roughly like that of vespertilionids, but there is no anterior palatal emargination. The teeth are tuberculosectorial, and the dental formula is 1/1, 1/1, 2/2, 3/3 = 28. The limbs of *Mystacina* resemble in some ways those of molossids: the wing membranes and uropatagium are tough and leathery, the first phalanx of the third digit folds back on the dorsal surface of the metacarpal, and the hind foot is unusually broad; the fibula is complete and the hind limb is robust. Unlike vespertilionids or molossids, the tail of *Mystacina* is short and protrudes from the dorsal surface of the uropa-

tagium. Each of the claws of the thumb and foot has a secondary talon at its ventral base.

This unusual bat is insectivorous and captures insects in midair or from vegetation. The wing can be folded compactly, owing to the unique pattern of flexion of the third digit, and during quadrupedal locomotion it is partially protected by the leathery proximal part of the plagiopatagium. The limbs are seemingly well adapted to quadrupedal locomotion. These bats are agile and rapid runners and are known to chase insects over the branches of trees. *Mystacina* lives in forested areas, where it roosts during the day in caves and in tree hollows.

FAMILY MOLOSSIDAE. Members of this family, the free-tailed bats, are important components of tropical and subtropical chiropteran faunas throughout much of the world. Eleven genera and 88 species are included in the family. Molossids occupy the warmer parts of the Old World, from southern Europe and southern Asia southward, and they inhabit Australia and the Fiji Islands. In the New World, they occasionally occur as far north as Canada, but the main range begins in the southern and southwestern United States and the West Indies and extends southward through all but the southern halves of Chile and Argentina.

Structurally, this is a peripheral group of bats; the most extreme manifestations of many of the typically chiropteran adaptations for flight occur in the Molossidae. The greater tuberosity of the humerus is large (Fig. 6–36E, F), and the locking device between it and the scapula is highly developed. The origins of the extensor carpi radialis longus and brevis and flexor carpi ulnaris muscles are well away from the center of rotation of the elbow joint and probably act more effectively than in any other bats as "automatic" extensors and flexors of the hand. The wing is typically long and narrow (Fig. 6–22B), with the fifth digit no longer than the radius, and the membranes are leathery because they are reinforced by numerous bundles of elastic fibers. In many species, refinements of structure that favor high-speed flight occur. In many Neotropical molossids, for example, the arrangement of the muscles of the forearm is such that the forearm is flattened and streamlined with respect to the airstream during flight. In the interest of rigidity of the outstretched wing during flight, movement at the wrist and elbow joint is strictly limited to one plane. The muscles that brace the fifth digit and maintain an advantageous angle of attack of the plagiopatagium during the downstroke of the wings are large and unusually highly specialized. Except for fusion of the last cervical and first thoracic vertebrae, the presacral vertebrae are unfused. The body of the sternum is not keeled.

The general appearance of molossids is distinctive. The tail extends well beyond the posterior border of the uropatagium when the bats are not in flight, and the fur is usually short and velvety. (In one genus, *Cheiromeles*, the fur is so short and sparse that the animal appears naked.) The muzzle is broad and truncate, and the thick lips are wrinkled in some species (Fig. 6–33F). Typically, the ears are broad, project to the side, and are like short wings. As viewed from the side, the pinnae are arched and resemble an airfoil of high camber. The ears are frequently braced by thickened borders and are connected by a fold of skin across the forehead. Because of the unique design of the ears, in most species they do not directly face the force of the airstream during flight, an adaptation probably of considerable importance to these fast-flying bats.

The skull is broad, the teeth are tuberculosectorial, and the dental formula varies from 1/1, 1/1, 1/2, 3/3 = 26 to 1/3, 1/1, 2/2, 3/3 = 32. Several characteristically molossid features are associated with the well-developed quadrupedal locomotion typical of these bats. The first phalanges of digits three and four flex against the posterodorsal surfaces of their respective metacarpals, providing for the chiropatagium to be folded into a compact bundle, no longer than the forearm, that is manageable when the animals run. The feet are broad, and have sensory hairs along the outer edges of the first and fifth toes. The fibula is not reduced, and the short hind limbs are stoutly built. Within the structural limits of the basic chiropteran plan, these bats have seemingly made the best of two types of locomotion. The highly specialized wings are clearly adapted to fast, efficient flight, whereas the primitive hind limbs have not lost their ability to serve in rapid quadrupedal locomotion.

These insectivorous bats are remarkable for their speedy and enduring flight. Whereas most bats fly fairly close to the ground or to vegetation when foraging, many molossids fly high and may move long distances during their nightly foraging. Some populations of Brazilian free-tailed bats *(Tadarida brasiliensis)*, the bats that occur in great numbers in Carlsbad Caverns and other large caves in the southwestern United States, fly at least 50 miles to their foraging areas each night (Davis, et al., 1962). Observations on these bats in Texas were made with radar and helicopters, and dispersal flights were tracked (Williams et al., 1973). Dispersing bats were recorded at elevations of over 3000 m, and masses of bats moved at an average speed of 40 km per hour. The western mastiff bat *(Eumops perotis)* forages over broad areas, and in southern California may on occasion fly more than 650 m above the ground (Vaughan, 1959:22). Because of the temperature inversions that frequently prevail for many nights in this area, in the winter these high-flying bats may be surrounded by air warmer than that at the ground, and may be catching insects that are flying in the warm "strata." Some molossids remain in flight for much of the night; foraging periods of at least six hours have been recorded for some species.

The flight of many molossids is unusually fast, and some species rival swifts and swallows in aerial ability. In some areas of Mexico, early-flying mastiff bats *(Molossus ater)* mingle with late-flying flocks of migrating swallows, and the bats seem at least the equals of the swallows in speed and maneuverability. One gets the impression, in fact, that the swallows are hastened to

Figure 6–54. The roosting place of mastiff bats (*Eumops perotis*, Molossidae) in a granite cliff. The animals occupy the space beneath the tongue-shaped slab of rock at the upper right. (From Vaughan, 1959.)

their roosts by the sudden appearance in abundance of their chiropteran counterparts.

Some molossids make spectacular dives when returning to their roosts. The western mastiff bat, for example, often makes repeated high-speed dives and half loops past the roosting site. It returns to its roost in a cliff by diving toward the base of the cliff, pulling sharply upward at the last instant, and entering the crevice with momentum to spare. Several other molossids are known to return to their roosting places by similar maneuvers. Because the wings of many molossids are narrow and have relatively small surface areas relative to the weights of the bats, these animals must attain considerable speed before they can sustain level flight. As a result, some species roost high above the ground in cliffs (Fig. 6–54), buildings, or palm trees, in situations where they can dive steeply downward for some distance in order to gain appropriate flight speed. These species are unable to take flight from the ground.

At least some species of molossids seem to forage in groups, the cohesion of which is insured in part by loud vocalizations audible to man. It is common knowledge among field workers that often when molossids are heard overhead or are captured in nets over water, single individuals are the exception and groups are the rule. To widely dispersing bats that take daytime shelter in communal roosts and seek concentrations of insects (perhaps the molossid strategy), group foraging with communication between members of the group would be advantageous.

Most molossids inhabit warm areas. Migration, therefore, is not characteristic of molossids in general. The Brazilian free-tailed bat, however, is known to make extensive migrations from the United States to as far south as southern Mexico (Villa-R. and Cockrum, 1962). The tremendous deposits of guano in some large caves inhabited by molossids attest to the effect that large colonies of these bats must have on insect populations in some areas.

ORDER PRIMATES

7

Primates have been most successful in tropical and subtropical areas, where today they pursue mostly arboreal modes of life. Some anthropoids, such as baboons and chimpanzees, have become partly or mostly terrestrial, but only man has become fully bipedal. Approximately 47 genera and 166 species of primates are living today, of which 16 genera and 63 species are in the New World. The primitive primates, such as lemurs, lorises, and tarsiers, are included in the suborder Prosimii; the more progressive monkeys, apes, and man are in the suborder Anthropoidea.

The evolution of the principal anatomical features of primates was influenced strongly by arboreal life. Stereoscopic vision, the dextrous, grasping hand, and the remarkable agility and muscular coordination of primates are seemingly the result of arboreal life. These features were in large part responsible for the primate trend toward enlargement of the brain. A good general description of primates is given by Anderson (1967a:151): primates are "eutherian mammals having generalized limb structure, primitively arboreal habits, omnivorous diet and comparatively unspecialized teeth; grasping with mobile digits and possessing freely movable limbs; phylogenetically replacing claws with nails and developing enlarged and sensitive pads on digits; reducing nose and sense of smell, enlarging eyes and improving vision, enlarging brain, and progressively improving placentation."

Most primates are omnivorous, and many species seem to be largely opportunistic feeders. Because soft foods are usually taken, the molars of primates are largely bunodont and brachyodont, and have the quadrate form typical of molars of

generalized feeders. Early in the evolution of primates, a hypocone was added to the upper molar, and the paraconid of the lower molar disappeared, leaving a basically four-cusped pattern (Fig. 7–1). The primate trend toward shortening of the rostrum is probably related to the importance of stereoscopic vision and the lack of importance of the sense of smell, rather than to a need for greater mechanical advantage for jaw musculature.

Primates comprise one of the oldest eutherian orders, dating from the late Cretaceous (Van Valen, 1965a:743). The fossil record of primates has considerable gaps, some of which result from the primate preference for tropical or subtropical enviornments. In these areas conditions are generally not suitable for the fossilization of animals. Primates arose from primitive insectivores, and some early primate families departed so little from the structural plan of insectivores that ordinal assignment to either Insectivora or Primates is difficult. The primate divergence from insectivores involved a trend toward a fruit-

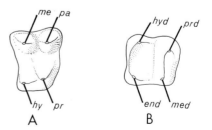

Figure 7–1. A diagrammatic representation of the basic four-cusped crown pattern of primates. A, a right upper molar; B, a left lower molar. Abbreviations: *end*, entoconid; *hy*, hypocone; *hyd*, hypoconid; *me*, metacone; *med*, metaconid; *pa*, paracone; *pr*, protocone; *prd*, protoconid.

eating or leaf-eating habit and was accompanied by changes in dentition that favored thorough mastication and partial oral digestion of food (Szalay, 1968). The Eocene, Old World family Adapidae (Fig. 7–2) may have been the basal group that gave rise to most prosimians, but the lack of an adequate fossil record for most prosimian families makes any conclusions on prosimian evolution tentative. An origin of the anthropoids from some prosimian group is assumed by many paleontologists, but is not documented by fossil evidence. Cebids and callithricids appear to be strictly New World types, whereas the other anthropoids arose in the Old World and all but hominids have remained there except where introduced by man.

Suborder Prosimii

The five families included in this suborder contain an assemblage of mostly arboreal mammals that in some cases bear only a marginal resemblance to the more "standard" primates (monkeys, great apes, and man) comprising the suborder Anthropoidea. The prosimians are primitive primates and, in the case of several families, occupy restricted geographic areas and pursue specialized modes of life. Even among prosimians, however, the importance of vision, manual dexterity, and vocal communication is apparent.

FAMILY LEMURIDAE. The lemurs inhabit Madagascar and the nearby Comoro Islands. Among the 15 Recent species, belonging to five genera, some are arboreal, some are semiarboreal, and some are largely terrestrial. These are the most primitive living primates. The fossil record of lemurids is from the Eocene, Pleistocene, and sub-Recent deposits in Madagascar. One extinct giant of presumably arboreal habits had an elongate skull 30 cm in length (roughly one foot!). The survival of lemurs is perhaps related to their insular distribution; they have been isolated on Madagascar since the early Cenozoic and have never been in competition with progressive primates (except man).

In contrast to most primates, the cranium of lemurs is elongate and the rostrum is usually of moderate length, giving the faces of some lemurs a foxlike appearance (Fig. 7–3). In more typical primate fashion, the lemurid braincase is large, crests for the origin of the temporal muscles are inconspicuous, and the foramen magnum is directed somewhat downward. The largest lemurs are roughly the size of a house cat; the smallest are the size of a mouse. The dental formula is 0-2/2, 1/1, 3/3, 3/3 = 32-36. The upper incisors are usually reduced or absent, and between those of the two sides is a broad diastema; the lower canine is incisiform and the first lower premolar is caniniform. The molars are basically tritubercular. The pollex and hallux are more or less enlarged and are opposable in all genera (Fig. 7–4A). The pelage is woolly, the tail is long and heavily furred, the limbs are usually slim, and the tarsal bones are not greatly elongated (Fig. 7–5A). Conspicuous color patterns occur in some species.

Lemurs are variously omnivorous, insectivorous, or herbivorous-frugivorous and, depending on the species, are diurnal or

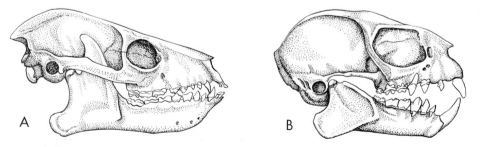

Figure 7–2. Skulls of fossil primates. A, An Eocene lemur (*Notharctus*, Adapidae; length of skull 75 mm); B, an Eocene tarsier-like primate (*Tetonius*, Anaptomorphidae; length of skull 46 mm). (After Romer, 1966.)

Figure 7–3. A lemur (*Lemur fulvus*, Lemuridae). (Photograph by D. Schmidt; San Diego Zoo photo.)

nocturnal. They are agile climbers, and the hands are used both for climbing and for food handling. Some species make great leaps from branch to branch. Some lemurs store fat in preparation for esuvation during the dry season. Large social groups occur in some species and, as in higher primates, vocalization seems important in

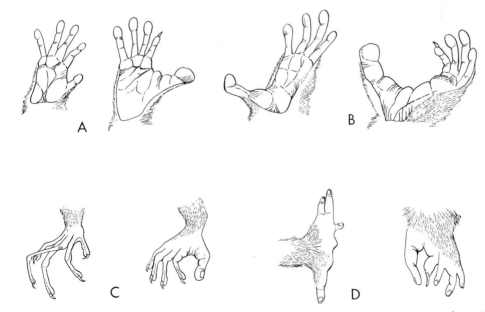

Figure 7–4. Hands and feet of some prosimian primates (the hand is on the left in each pair). A, A lemur (*Lemur mongoz*, Lemuridae); B, an indrid (*Propithecus diadema*, Indridae); C, an aye-aye (*Daubentonia madagascariensis*, Daubentoniidae); D, a potto (*Arctocebus calabarensis*, Lorisidae).

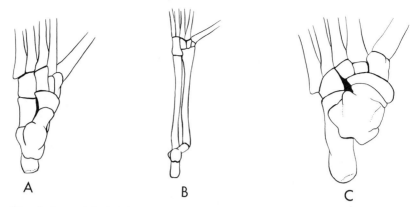

Figure 7–5. Dorsal views of the left feet of several primates, showing the tarsal bones. A, A lemur (*Lemur* sp., Lemuridae); B, a tarsier (*Tarsius*, Tarsiidae); C, a gorilla (*Gorilla gorilla*, Pongidae). Note the remarkable elongation of the calcaneum and the navicular in the tarsier. (Redrawn by permission of Quadrangle Books from *The Antecedents of Man* by W. E. LeGros Clark, © 1959, 1962, 1971 by W. E. LeGros Clark.)

maintaining contact between members of a group. (Lemur social behavior is discussed on p. 374).

FAMILY INDRIDAE. The indrids, often called woolly lemurs, are not a diverse group, including but three genera and four species, and are restricted to Madagascar. There are Pleistocene and sub-Recent records of indrids from Madagascar, and matching the extinct huge lemurids are extinct indrids that are also of great size.

These animals are fairly large (up to 900 mm in head-and-body length); two genera have shortened rostra and monkey-like faces; the snout is fairly long in the other genus. The dental formula is 2/2, 1/0, 2/2, 3/3 = 30. The upper incisors are enlarged and the first lower premolar is caniniform. The hands and feet are highly modified for grasping branches during climbing (Fig. 7–4B). The pelage is conspicuously marked in some species, and the tail is long in three species.

These primates are largely herbivorous. Their leaf-eating habits resemble those of the Neotropical howler monkey (*Alouatta*; Cebidae) and the African colobus monkey (*Colobus*; Cercopithecidae). Indrids are typically fairly slow, deliberate climbers. The hind limbs are long relative to the front limbs, and when traveling on the ground these primarily arboreal and diurnal animals proceed by a series of hops. The hands are used for climbing and for handling food, but manual dexterity seems

limited and food is often picked up in the mouth. One genus is solitary or occurs in pairs; the other two genera typically live in small bands. A specialized laryngeal apparatus enables *Indri* to produce loud resonant calls. These howls are given with greatest frequency in the morning and evening, as are the calls of the howler monkey, and perhaps function in maintaining territorial boundaries between neighboring bands. All indrid species are vocal to some extent.

FAMILY DAUBENTONIIDAE. This family is represented by only one highly specialized Recent species (*Daubentonia madagascariensis*) with the common name of aye-aye. This nocturnal animal occurs locally in northern Madagascar, where it is restricted to dense forests and stands of bamboo. The fossil record of the Daubentoniidae consists of sub-Recent fossils from Madagascar of an extinct species that was larger than the surviving aye-aye.

Aye-ayes weigh approximately 2 kg; they have prominent ears and a long bushy tail. The skull and dentition are remarkably specialized, and depart strongly from the usual primate plan. The skull is short and moderately high. The orbit is prominent and faces largely forward; the postorbital bar and zygomatic arch are robust, and the rostrum is short and deep (Fig. 7–6A). The dentition differs from the basic primate type both in the extensive loss of teeth and in the strong specialization of

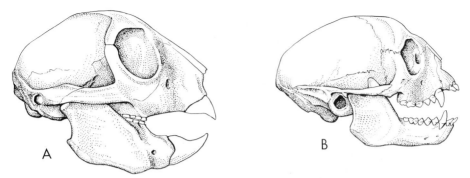

Figure 7–6. Skulls of primates. A, An aye-aye (*Daubentonia madagascariensis*, Daubentoniidae; length of skull 90 mm); B, a marmoset (*Saguinus geoffroyi*, Callithricidae; length of skull 51 mm). (*Saguinus* is after Hall, E. R., and Kelson, K. R.: *The Mammals of North America*, © 1959 by the Ronald Press Company, New York.)

the teeth that are retained. The dental formula is 1/1, 0-1/0, 1/0, 3/3 = 18 or 20. The canine is often absent, and the cheek teeth have flattened crowns with no clear cusp pattern. The laterally compressed incisors are greatly enlarged, wear to a sharply beveled edge because only the anterior surfaces are covered with enamel (as in rodents), and are ever-growing. Because of the shape of the teeth and the presence of a diastema between the incisors and the cheek teeth, *D. madagascariensis* was first described as a rodent. The hand is unique among primates. The digits are clawed, and all but the nonopposable pollex are long and slender; the third digit is remarkably slender (Fig. 7–4C). In the hind paw, the hallux is opposable and bears a nail, but the other digits are clawed.

Aye-ayes are nocturnal and are mainly insectivorous. They are arboreal, capable of making graceful leaps between branches. Their foraging technique is noteworthy. The elongate third finger is used to tap on wood harboring wood-boring insects; the aye-aye then listens carefully for insects within the wood, and the remarkable third digit is used for removing adult and larval insects from holes or fissures in the wood. When necessary, the powerful incisors tear away wood to enable the third digit to reach insects in deep burrows. Surprisingly, this strange mode of foraging is shared by two Australian genera of marsupials of the family Petauridae. In *Dactylonax*, the most specialized of these marsupials, the front incisors are modified

and the manus is specialized along lines parallel to those in the hand of *Daubentonia*, except that the fourth rather than the third digit is the probing finger.

As is the case with many mammals that occupy limited areas, the future of the aye-aye seems dim. Because *Daubentonia* is restricted to continuous, heavy forests, its continued survival depends largely on the extent to which land is not cleared for agriculture in northern Madagascar.

FAMILY LORISIDAE. The lorises are more widely distributed than are the primitive primates of Madagascar, and are locally common. Lorises occur in Africa south of the Sahara, in India, Ceylon, and southeast Asia, and in the East Indies. The fossil record of lorisids is scanty, but it suggests that these animals evolved in the Old World and have never occurred elsewhere. A Pliocene form is known from Asia.

The eyes face forward in the lorisids (Fig. 7–7), rather than more or less to the side as in the lemurids, and the rostrum is short. Lorisids are arboreal, and their locomotion usually involves methodical hand-over-hand climbing. Lorises vary from the size of a rat to that of a large squirrel. The braincase is globular, the facial part of the skull is often short and ventrally placed, and the anteriorly directed orbits are separated by a thin interorbital septum. The dental formula is 1-2/2, 1/1, 3/3, 3/3 = 34 or 36. The upper incisors are small, the lower canine is incisiform, and the molars are basically quadritubercular.

Figure 7-7. A loris (*Loris tardigradus*, Lorisidae). (Photograph by Ron Garrison; San Diego Zoo photo.)

The manus and pes are specialized in a variety of ways for clutching branches. In the genus *Arctocebus*, an odd, pincer-like hand has been developed by the reduction of digits two and three and a change in the postures of the remaining digits; the first digit of the pes is opposable and is frequently greatly enlarged (Fig. 7-4D). Circulatory adaptations in the appendages provide for an increased blood supply to the digital flexor muscles that are used in gripping branches during extended periods of contraction. These same circulatory modifications, involving the formation of a *rete mirabile*, are also important in this and many other mammals in conserving body heat. (A *rete mirabile* is a complex meshwork of small arteries and veins that are intertwined so that the warm blood passing to an appendage is cooled by the cooler blood coming from the appendage, and the cooler blood from the appendage is warmed by the arterial blood. One result of this system is the avoidance of much of the energy loss that would accompany the warming of drastically cooled blood from a poorly insulated limb as the blood entered the general blood stream.)

These nocturnal primates are insectivorous and carnivorous, and prey is usually captured by the hands after a stealthy approach. The specialized lorisine genus *Arctocebus* spends considerable time upside down, and is reported to sleep in this position.

FAMILY GALAGIDAE. Galagos are handsome animals that reach the size of a large squirrel and have very large eyes, expressive ears that resemble those of some bats, and a remarkable ability in some species to make prodigious arboreal leaps. Seven species of galagos are recognized by Hill and Meester (1971). The tail is long and well furred and is used as a balancing organ during leaping. In contrast to lorises, galagos have unusually long hind limbs with powerful thigh muscles.

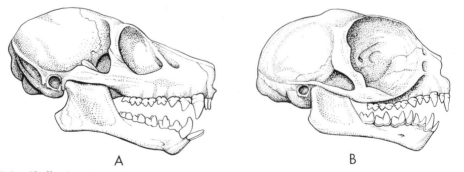

Figure 7-8. Skulls of prosimian primates. A, A galago (*Galago* sp., Lorisidae; length of skull 65 mm); B, a tarsier (*Tarsius spectrum*, Tarsiidae; length of skull 36 mm). (Redrawn by permission of Quadrangle Books from *The Antecedents of Man* by W. E. LeGros Clark, © 1959, 1962, 1971 by W. E. LeGros Clark.)

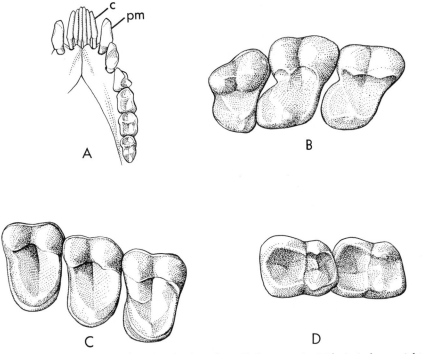

Figure 7–9. Teeth of prosimian primates. Teeth of a galago (*Galago* sp., Lorisidae): A, lower right tooth row, showing the incisiform canine (*c*) and the caniniform premolar (*pm*); B, upper right molars. Teeth of a tarsier (*Tarsius spectrum*, Tarsiidae): C, upper right molars; D, first and second lower left molars. (Redrawn by permission of Quadrangle Books from *The Antecedents of Man* by W. E. LeGros Clark, © 1959, 1962, 1971 by W. E. LeGros Clark.)

The skull has a long rostrum (for a primate) and the dental formula is 2/2, 1/1, 3/3, 3/3 = 36; specialized lower incisors and canines are procumbent (Fig. 7–8A) and form a comblike structure (Fig. 7–9A) used in grooming the fur and in feeding on resin. The specialized hands and feet are well adapted to grasping: the thumb and

Figure 7–10. Traction patterns on the palm of the hand of the greater galago (*Galago crassicaudatus*).

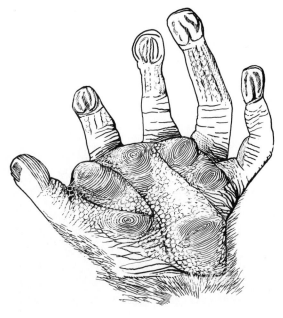

hallux are both large and opposable, the fourth digits are unusually long, and the distalmost pads of the digits have well developed traction ridges (Fig. 7-10) of importance during climbing. The second digit of the hind foot is short and bears a claw used for grooming, but all other digits bear flattened nails. The foot segment of the hand limb is long in association with the leaping ability; the elongation involves the tarsus, but is not as extreme as that in *Tarsius* (Fig. 7-5B).

Galagos are common in many sections of Africa, and the arboreal leaps of *Galago senegalensis* are a fascinating part of the twilight scene in some parts of East Africa. This species (Fig. 7-11) usually takes refuge by day in family groups in holes in such trees as the baobab (Fig. 6-42), and the evening dispersal from these retreats often involves the use of repeatedly followed pathways through the trees. During our 14-month stay at a camp in southern Kenya, East Africa, my family and I often observed a galago that used the roof of our cabin and adjacent trees as part of its pathway. Many evenings after sunset, this animal could be heard landing on our thatched roof; it then took six or so leaps to travel through a large umbrella tree (*Acacia tortilis*), after which it made an enormous leap to a bush, spanning about 12 feet horizontally and 8 feet vertically. It then dropped to the ground and hopped kangaroo-fashion some 20 m to a group of trees where it resumed its arboreal travel. A leap of 7 m by a galago is mentioned by Kingdon (1971:318).

Galagos have an extremely varied diet,

but seem to prefer insects. When beetles and other insects are abundant during the rainy season, galagos depend on this food source. A precisely judged leap terminating in a quick grab with one hand is a common style of capturing insects. Kingdon (1971:318-319) reports that *G. senegalensis* feeds also on seeds and small vertebrates and that it takes nectar from the large flowers of baobab trees. Of particular interest is the habit of eating the resin (gum) of trees such as acacias. In some areas with long dry seasons and periodic food shortages, this food may be of major survival value.

Galagos have a variety of vocalizations that serve as warnings, communication signals between the mother and young, and perhaps as appeasement during intraspecific encounters. *Galago senegalensis* has a "vocabulary" of about ten basic sounds (Andersson, 1969), and the loud and raucous calls of the greater galago (*G. crassicaudatus*) are an impressive addition to the chorus of night sounds in wooded parts of East Africa.

FAMILY TARSIIDAE. This family is represented today by three species of the genus *Tarsius*, and occurs in jungles and secondary growth in Borneo, southern Sumatra, in some East Indian islands, and in some of the Philippine Islands. Tarsiids are known from the Eocene of Europe, but there are no fossils representing the remainder of the Cenozoic.

The tarsier is roughly the size of a small rat and, with its large head, huge eyes, long limbs, and long tail, has a distinctive appearance (Fig. 7-12). The most conspic-

Figure 7-11. Bushbabies or lesser galagos (*Galago senegalensis*). A, A group of five that slept together on a *Commiphora* branch during the day; B, an individual preparing to leap from a *Commiphora* branch. Photographs taken in southern Kenya, East Africa. (Photograph of the group by Thomas R. Huels; photograph of the single galago by Richard G. Bowker.)

Figure 7–12. Tarsiers (*Tarsius* sp., Tarsiidae). (Photograph by Ron Garrison; San Diego Zoo photo.)

uous cranial features are the enormous orbits, which face forward and have expanded rims and a thin interorbital septum (Fig. 7–8B). The eye of the tarsier is apparently adapted entirely to night vision, for it lacks cones in the retina. The dental formula is 2/1, 1/1, 3/3, 3/3 = 34. The medial upper incisors are enlarged, the premolars are simple, the crowns of the upper molars are roughly triangular, and the lower molars have large talonids (Fig. 7–9D). The neck is short, a characteristic of many saltatorial vertebrates. All but the clawed second and third pedal digits have flat nails, and all digits have disclike pads (Fig. 7–13A). The limbs, especially the hind ones, are elongate; the tibia and fibula are fused.

The trend toward jumping ability that is apparent in galagos is developed to an extreme degree in the family Tarsiidae. As in all highly specialized jumpers, the hind foot is elongate, but in the tarsier the elongation has been unique. It involves two tarsal bones (hence the name *Tarsius*) rather than metatarsals, as in such jumpers as elephant shrews (Fig. 6–2) and kangaroo rats (Fig. 10–22E). In *Tarsius*, the calcaneum and navicular are greatly elongate (Fig. 7–5B), whereas the metatarsals are not unusually long in relation to the phalanges (Fig. 7–13A). An important functional end is achieved by this unusual system of foot elongation: because the elongation has occurred in the tarsus, the dexterity and grasping ability of the digits themselves (the metatarsals and phalanges) have not been sacrificed. A reduction of dexterity would have accompanied an elongation of the metatarsals and the resulting functional reduction of the number of digit segments. In elephant shrews and kangaroo rats, dexterity and gripping ability of the hind foot are not important, and a more "direct" means of elongation — lengthening of the already somewhat elongate metatarsals — occurred.

Tarsiers are primarily arboreal and nocturnal and feed largely on insects which they pounce upon and grasp with the hands. Fogden (1974:161) observed tarsiers quietly watching and waiting at a low perch and leaping down to the ground to capture insects. Although more highly adapted to leaping than any other primate, tarsiers can walk and climb quadrupedally, can hop or run on their hind legs on the ground, and can slide down branches (Sprankel, 1965). Tarsiers and some species of galagos share the ability to leap long distances with great precision, and in both of these types of primates the landing from a leap is largely bipedal. In association with jumping ability, much of the weight of the tarsier is concentrated in the hind limbs, which together comprise 21

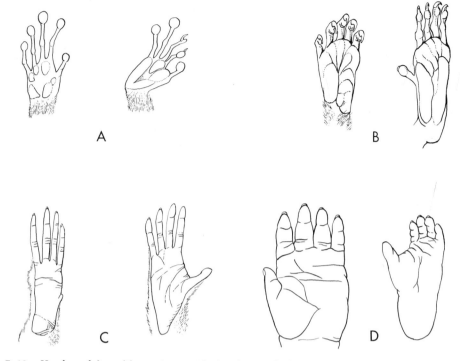

Figure 7–13. Hands and feet of four primates (the hand is on the left in each pair.) A, A tarsier (*Tarsius spectrum*, Tarsiidae); B, a marmoset (*Callithrix sp.*, Callithricidae); C, a woolly spider monkey (*Brachyteles arachnoides*, Cebidae); D, a gorilla (*Gorilla gorilla*, Pongidae). (A, B, and D redrawn by permission of Quadrangle Books from *The Antecedents of Man* by W. E. LeGros Clark, © 1959, 1962, 1971 by W. E. LeGros Clark.)

per cent of the total weight of the animal; the musculature of the thighs alone equals 12 per cent of the body weight, largely owing to great enlargement of the quadriceps femoris (Grand and Lorenz, 1968), a powerful extensor of the shank. Some work has shown that tarsiers usually live in pairs, but Fogden (1974:162–163) observed single animals most frequently, suggesting that only large, dominant males associate with females. Perhaps the relative silence of tarsiers reflects their limited social life.

SUBORDER ANTHROPOIDEA

The five families within this suborder are "higher" primates, and have many progressive features not typical of members of the suborder Prosimii. Anthropoids—monkeys, marmosets, apes, and man—are the most familiar primates and are vastly more important in terms of taxonomic diversity and adaptability than are the relatively primitive prosimians.

FAMILY CEBIDAE. The New World monkeys all belong to this family, which includes 29 Recent species of 11 genera. Cebids range from southern Mexico, through Central America, to southern Brazil. Cebids first appear in the early Oligocene of South America. Primitive primates ancestral to the cebids may have entered South America from Central America on logs or debris that floated across the stretch of water that separated these land masses during much of the Tertiary.

These are usually small, slim monkeys: the largest cebid is the howler monkey (*Alouatta*), which weighs up to 9 kg. New World Monkeys are arboreal and have elongate limbs and curved nails on the digits. The pollex is not opposable and is reduced in some species (Fig. 7–13C), but the hand usually has considerable dexterity. The hallux is strongly opposable. The tail is long in all but one genus and is prehensile in four of the eleven genera. The skull is more or less globular, with a high braincase, and the rostrum is typically short. The orbits face forward, and

Figure 7–14. A howler monkey (*Alouatta palliata*, Cebidae). (San Diego Zoo photo.)

the nostrils are separated by a broad internarial pad and face to the side (Fig. 7–14), a condition termed platyrrhine. The dental formula is 2/2, 1/1, 3/3, 3/3 = 36, and the lateral pair and medial pair of cusps of the molars are separated by a central, anteroposteriorly aligned depression (Fig. 7–15). Brightly colored and bare patches of skin on the rump (ischial callosities) do not occur in cebids as they often do in Old World monkeys (family Cercopithecidae).

Cebids typically occur in tropical forests, and most are diurnal. They are basically vegetarians; fruit is often preferred, but a wide variety of plant and animal material is eaten. The night monkey (*Aotes trivirgatus*), an inhabitant of Central and South America, is apparently insectivorous and carnivorous and may feed to some extent on bats. Cebids are active, intelligent animals and are adroit climbers; some species move with amazing speed through the trees. For dazzling arboreal ability, the Neotropical spider monkey (*Ateles*) is probably only surpassed by the Old World gibbons (*Hylobates*, Pongidae).

Most cebids are vocal to some extent, and several species have loud, penetrating calls. Outstanding among these is the howler monkey (Fig. 7–14), in which the hyoid apparatus is enlarged into a resonating chamber. The males emit loud "roaring" sounds that carry for long distances through the tropical rain forest. These sounds are seemingly important in keeping the members of a troop together. These troops may include up to 40 individuals.

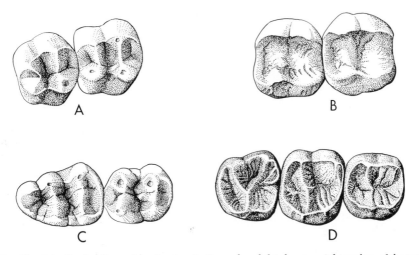

Figure 7–15. Cheek teeth of anthropoid primates. A, Second and third upper right molars of the orange-crowned mangabey (*Cercocebus torquatus*, Cercopithecidae); B, first and second upper right molars of a saki monkey (*Pithecia monachus*, Cebidae); C, second and third lower left molars of a baboon (*Papio* sp., Cercopithecidae); D, upper right molars of the orangutan (*Pongo pygmaeus*, Pongidae). Note the cross lophs on the teeth of the mangabey and the baboon, and the extra posterior cusp (hypoconulid) on the third lower molar of the baboon. (*Pongo* redrawn by permission of Quadrangle Books from *The Antecedents of Man* by W. E. LeGros Clark, © 1959, 1962, 1971 by W. E. LeGros Clark.)

The territories of troops are probably announced and partly maintained by the loud vocalizations. Most cebids are gregarious, the most common social aggregation consisting of a family group. Some species form unusually large troops; the squirrel monkey (*Saimiri*) occurs in bands of up to 100 animals.

FAMILY CALLITHRICIDAE. Members of this family are called marmosets, and are small, often conspicuously marked primates that inhabit tropical forests from Panama and the northern part of South America south to southern Brazil. Four genera and 33 species are known. These are somewhat squirrel-like primates that vary in size from that of a mouse to that of a squirrel. Marmosets are seemingly more closely related to cebids than to any other primate group. The earliest record of marmosets is in the Upper Oligocene.

Marmosets are primarily arboreal, but they do not grasp branches as do many other primates and never use brachiation (a mode of arboreal locomotion involving using the hands in swinging from branch to branch). Instead, they use their clawed hands and feet for climbing quadrupedally, and they bound through the trees in squirrel-like fashion. The marmoset skull has a short rostrum with prominent and forward-directed orbits (Fig. 7–6B). The dental formula is 2/2, 1/1, 3/3, 2/2, or 3/3 = 32 or 36. The medial incisors are chisel-like and frequently project forward; the upper molars are approximately triangular and usually have sharp cusps. The body and limbs are slender and the tail is always long. The hallux has a flat nail; all other digits bear laterally compressed claws. The hand resembles that of a squirrel and is slender, and the pollex is not opposable (Fig. 7–13B). Conspicuous ruffs or tufts of fur occur on the heads of some species, and can be erected.

Marmosets are omnivorous. The diet consists mostly of fruit and insects; lizards and small birds and their eggs may be important foods for some species. Marmosets are social and typically live in small family groups. In some Neotropical areas, they are the most common primates. These animals are often extremely vocal and emit a variety of high-pitched sounds and piercing "alarm" calls, some of which resemble those of birds that live in the same areas.

FAMILY CERCOPITHECIDAE. These are the Old World monkeys, and are the most successful primates in terms of numbers of species (60 Recent species of 11 genera). They occupy a wide range, including Gibraltar, northwest Africa, Africa south of the Sahara, southern Arabia, much of southeastern Asia east to Japan, Indonesia east to Timor, and the Philippine Islands. Among nonhominid primates, cercopithecids have the greatest tolerance for cold climates; some of these primates occupy high forests in Tibet, and others live in northern Honshu, Japan, where winter snows occur. Cercopithecids first appear in the Oligocene of Egypt and, like the living species, fossil forms are known only from the Old World. Just as in many other groups, some Pleistocene cercopithecids reached large sizes; an extinct South African baboon of this epoch reached the size of a gorilla.

In weight, cercopithecids range from 1.5 kg to over 50 kg, and some species are stocky in build, quite unlike most cebids. The nostrils are close together and face downward (Fig. 7–16), a condition termed catarrhine. The skull is often robust and heavily ridged and, compared to cebids, the rostrum is long (particularly in the baboons). The dental formula is 2/2, 1/1, 2/2, 3/3 = 32, as in the apes (Pongidae) and man (Hominidae). The medial upper incisors are often broad and roughly spoon shaped; the upper canines are usually large and in some species are tusklike. When the jaws are closed, the lower canine rests in a diastema between the upper canine and the last incisor. The first lower premolar is enlarged and forms a shearing blade that rides against the sharp posterior edge of the upper canine (Fig. 7–17A). Most of the molars have four cusps, the outer pair connected to the inner pair by two transverse ridges producing a bilophodont tooth. The last lower molar has an additional posterior cusp, the hypoconulid (Fig. 7–15C).

Figure 7–16. Two cercopithecid monkeys. A, Japanese macaques (*Macaca fuscata*); B, mona guenon (*Cercopithecus mona*). Note the dense fur of the macaques; these animals live in northern Honshu, Japan, where snow falls in the winter. (Photograph of *Macaca* by Carl B. Koford; photograph of *Cercopithecus* is a San Diego Zoo photo.)

All of the digits have nails, and the pollex and hallux are opposable except in the strongly arboreal, leaf-eating genus *Colobus*, in which the pollex is vestigial or absent. The tail is vestigial in some species, but long in others. Ischial callosities are well developed in many species, and the bare rump skin is frequently bright red. The conspicuous patch is used in conjunction with ritualized postures, as a means of communication between members of a social group. Bare facial skin

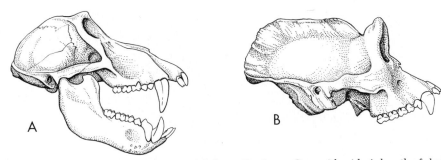

Figure 7–17. Skulls of anthropoid primates. A, A baboon (*Papio* sp., Cercopithecidae); length of skull 200 mm. B, A gorilla (*Gorilla gorilla*, Pongidae); length of skull 320 mm.

may also be red, but is bright blue in the mandrill (*Papio sphinx*). These patches of skin are more brightly colored in the male than in the female in some species. The olfactory epithelium is greatly reduced in cercopithecids, and apparently their sense of smell is rudimentary. The facial muscles are well developed and produce a wide variety of facial expressions. Some cercopithecids are brightly or conspicuously marked. For example, the variegated langur of Indochina (*Pygathrix nemaeus*) has a bright yellow face, a chestnut strip beneath the ears, black and chestnut limbs, a gray body, and a white rump and tail.

Although most cercopithecids are probably largely omnivorous, some are adapted to an herbivorous diet. Members of the subfamily Colobinae (the arboreal langurs and colobus monkeys) are herbivorous, frugivorous, and some species seem to feed primarily on leaves. The baboons (*Papio, Theropithecus*) are the most successful terrestrial cercopithecids, and one species, in areas where suitable trees are not available, assembles in large groups (up to 750 individuals) on cliffs (Kummer, 1968a:310). The remarkably complicated social behavior of baboons and of certain other Old World monkeys is reasonably well known (see Jay, 1968; De Vore, 1965), and is discussed on page 374.

Interesting contrasts between the behavior of the baboons and the equally terrestrial patas monkey (*Cercopithecus patas*) are discussed by Hall (1968:114). Savanna baboons are highly vocal, live in fairly large troops controlled by several dominant males, and are prone to noisy, rough, and aggressive interactions. In contrast, the patas monkey usually maintains "adaptive silence," but has a repertoire of soft calls, lives in small troops, each with a single adult male that serves as a watchdog, is rarely aggressive, and never fights. The patas monkey has a slim greyhound-like build and is the fastest runner of all primates, having been timed at a speed of 55 km per hour. Adaptations for speed in this animal include elongation of the limbs, carpals, and tarsals, shortening of the digits, reduction of the pollex and hallux, and the development of palmar and plantar pads. This remarkably cursorial primate has a quiet mode of life, usually attempts to escape detection, and depends on its speed to escape danger. In these respects the patas monkey is the primate counterpart of the small antelope. The noisy baboon troop, however, frequently depends on its aggressive dominant males to confront and discourage a predator, and terrestrial locomotion is relatively unimportant as a means of escaping enemies.

Sexual dimorphism is pronounced in both the baboons and the patas monkey, as it is in many primates. The male baboon of South and East Africa weighs roughly 33 kg, the female 16.5 kg; the male patas monkey averages 13 kg, and the female, 6.5 kg (Hall, 1968:114). Probably all cercopithecids are basically social, and vocalizations and facial expressions play central roles in social interactions. The life span of these monkeys is long: a Chacma baboon (*Papio ursinus*) lived in captivity for 45 years, and life spans of 20 or 25 years in the wild may be common (Walker, 1968:447).

FAMILY PONGIDAE. The gibbons and great apes are included in this family, members of which occur in equatorial Africa, southeast Asia, Java, Borneo, Sumatra, and the Mentawi Islands. Eight Recent species of four genera are known; all species are restricted to tropical forests. The pongid record begins in the Oligocene of Egypt, when the gibbon-like genus *Propliopithecus* appears. Judging from the fossil record, pongids evolved in Africa, reached Europe in the Miocene and Asia in the Pliocene, and never occurred in the New World.

The family Pongidae includes two subfamilies that differ markedly in structure and modes of life. The gibbons, subfamily Hylobatinae, are arboreal and are the most rapid and spectacular climbers and brachiators of all mammals. Gibbons are relatively small, weighing from 5 to 13 kg, have remarkably long arms, and use the hands like hooks rather than as grasping structures during brachiation. The great apes, subfamily Ponginae, vary from 48 to 270 kg (nearly 600 pounds!) in weight, and have robust bodies and powerful arms. The hands and feet are similar to those of man, but the hallux is opposable (Fig. 7–13D). The pongid skull is typically robust, and in older animals is marked by bony crests and ridges; it is long relative to its width (Fig. 7–17B). The teeth are large. The dental formula is 2/2, 1/1, 2/2, 3/3 = 32, as in cercopithecids and man. The incisors are broad and the premaxillae and anterior parts of the dentaries are broadened to accommodate them. The canines are large and stoutly built, but are never tusklike. The upper molars are quadrangular and basically four-cusped, and the lower molars have an additional posterior cusp (hypoconulid). In contrast to cercopithecids, a trend toward elongation of the molars does not occur in pongids, and the molars lack well-defined cross ridges (Fig. 7–15D). The tooth rows are parallel and the mandibular symphysis is braced by a bony shelf (the "simian shelf").

The forelimbs are longer than the hind limbs and the hands are longer than the feet; all digits bear nails. Pongids have no tails. The thorax is wide and the scapula has an elongate vertebral border. Adaptations allowing advantageous muscle attachments during erect or semierect stances include lengthening of the pelvis and enlargement and lateral flaring of the ilium. Regarding structural details, locomotor ability, brain size, and level of intelligence, the great apes are closer to man than are any other mammals.

Pongids are largely vegetarians, but some are occasionally carnivorous. The chimpanzee (*Chimpansee*), for example, occasionally catches and eats the colobus monkey. Arboreal locomotion in pongids involves brachiation in some species. The gorilla (*Gorilla*) and chimpanzee are mostly terrestrial and, although capable of bipedal stance and limited bipedal locomotion, are mostly quadrupedal. The behavior of pongids and cercopithecids has been studied intensively in recent years. Good general references on primate behavior include De Vore (1965) and Jay (1968). Owing primarily to the efforts of man, who acts almost as if his survival and prosperity were threatened by his next of kin, some of the great apes are dangerously close to extinction. Destruction of habitat and killing of the animals themselves, fostered by man's anachronistic feeling that his position as the dominant form of life justifies any form of exploitation of his environment, has led to serious reductions of some primate populations. Perhaps no more than 2500 wild orangutans (*Pongo pygmaeus*) survive today; these are restricted to parts of the islands of Sumatra and Borneo. As a sad stroke of irony, an important drain on the declining populations of orangs resulted from their capture and exportation to European and American zoos, institutions dedicated in part to the preservation of vanishing species.

FAMILY HOMINIDAE. Man is the only living member of the family Hominidae. In man, the skull has a greatly inflated cranium, housing a large cerebrum, and the rostral part of the skull is virtually absent. The foramen magnum is beneath the skull, a feature associated with an upright stance. The dentition is not as robust as in the pongids: the incisors are less broad, the canines typically rise but slightly above adjacent teeth, and the cheek teeth are less heavily built. The premolars are usually bicuspid. The upper molars have four cusps; the first lower molar has five cusps, the second has four, and the third has five. The dental formula 2/2, 1/1, 2/2, 3/3 = 32 occurs in most individuals, but one or more of the posterior molars (the "wisdom teeth") may not appear. The tooth rows are not parallel, as they are in pongids, nor is the simian shelf present in the mandible. The pollex, but not the hallux, is opposable. With a change in the posture and use of the forelimbs, the thorax has become broad and the scapulae have come to lie dorsal to the ribcage, as in bats, rather than lateral to the ribcage, as in most mammals. As in the case of many primates, in humans the males are considerably larger than the females.

If the same standards used in the classification of other mammalian orders were applied to the primates, man would be included with the gibbons and great apes in a single family. "Differences of no greater magnitude than those separating the hominids and the pongids characterize subfamilies in some other orders of mammals" (Anderson, 1967a:177). The ancestry of hominids can seemingly be traced back to the Miocene.

The astounding growth of the populations of man throughout much of the world is leading to progressively more acute problems that threaten, if not the very survival of man, at least his present style of life. It is obviously essential that men of different races learn to accept and appreciate living and working together as equals, and that effective birth control measures be followed in most parts of the world in order to halt population growth. Man is clearly creating an environment in which he is ill suited, both psychologically and physically, to live. In his attempt to improve his lot, man is actually changing his environment in ways that make it less able to support life over a long period. Our present exploitation and modification of the environment cannot long continue. The adaptive ability of man is being put to more critical tests today than ever before; the future of man may well be determined by choices he makes within the next few decades.

EDENTATES; SCALY ANTEATERS; THE AARDVARK

The orders of mammals considered in this chapter (Edentata, Pholidota, Tubulidentata) are not brought together because they share a common ancestry; indeed, they are probably rather distantly related to one another. They also seem to be isolated phylogenetically from other orders of mammals. But as a result of convergence (not closeness of relationship!), there are some striking structural similarities between these three orders, one of the most obvious being the loss or reduction of teeth. Primarily as a matter of convenience, these aberrant and unrelated orders are discussed in the same chapter.

ORDER EDENTATA

Although the edentates are not of great importance today (the order Edentata contains only 14 living genera and 31 species), they are remarkably interesting animals because of their unique structure, their large and bizarre fossil types, and because of the probable influence of the Tertiary separation of South America and North America on their evolution.

The living members of the order Edentata share a series of distinctive morphological features. Extra zygapophysis-like (xenarthrous) articulations (Fig. 8–1) brace the lumbar vertebrae. The incisors and canines are absent; the cheek teeth, when present, lack enamel, and each has a single root. The tympanic bone is annular; the brain is small and the braincase is usually long and cylindrical; the coracoid process is unusually well developed and the clavicle is present. The ischium is variously expanded and specialized (Fig. 8–2) and

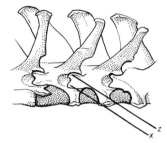

Figure 8–1. Three lumbar vertebrae (viewed from the left; anterior is to the left) of the nine-banded armadillo (*Dasypus novemcinctus*), showing the "xenarthrous" articulations (x) supplementing the normal articulations between zygapophyses (z).

usually forms an ischiocaudal, as well as an ischiosacral, symphysis. The hind foot is typically five-toed, and the forefoot has two or three predominant toes with large claws. Major edentate structural trends are toward reduction and simplification of the dentition, specialization of the limbs for such functions as digging and climbing, and rigidity of the axial skeleton. Perhaps the use of the powerful forelimbs when the animals are in the bipedal posture, which is adopted frequently during feeding or defense, has influenced the evolution of the rigid axial skeleton and the bracing of the pelvis.

PALEONTOLOGY

Edentates, strictly a New World group, originated in North America, where members of the primitive suborder Palaeanodonta are known from the Paleocene. Palaeanodonts were probably descendants of the stock that gave rise to the more specialized suborder of edentates, the Xenarthra, and disappeared from the

A

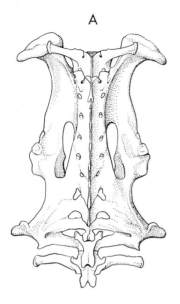

B

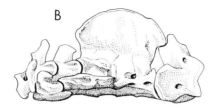

Figure 8–2. Parts of the skeleton of the nine-banded armadillo (*Dasypus novemcinctus*). A, Dorsal view of the pelvic girdle, showing the great degree of fusion of vertebrae with the ilium and ischium; B, lateral view of the cervical vertebrae (anterior is to the right). The axis and cervicals three, four, and five are fused.

fossil record of North America in the Oligocene, apparently without reaching South America. Edentates seemingly reached South America in the Late Cretaceous or Early Paleocene. The earliest fossils from this region are from the Late Paleocene, and consist of scutes from the armor of members of the family Dasypodidae (armadillos), the most successful surviving family. A fossil armadillo from the Early Eocene had features—such as vestiges of enamel on the teeth—that are suggestive of palaeanodont ancestry.

A diverse array of forms resulted from a mid-Tertiary radiation of South American dasypodids: a large Pliocene type (*Macroeuphractus*) had enlarged canine-like teeth and was probably a scavenger; the Miocene armadillo *Stegotherium* had a long snout and was most likely a termite eater; another Miocene form (*Peltephilus*) and its relatives had specialized scutes that formed pointed horns on the front of the snout; and enormous, rhinoceros-sized armadillos of the subfamily Pampatheriinae lived in the Pleistocene of both North and South America.

Although our knowledge of the modes of life of the extinct South American armadillos is incomplete, their great structural diversity indicates that they exploited a variety of foods and together formed an important part of the South American Tertiary biotas. The early dasypodids were armored with ossified dermal scutes, as are all modern species, and are probably the basal group from which other xenarthrans evolved.

Dispersal of mammals between North and South America was restricted from the Paleocene until Late Pliocene (see page 342); during this interval not only dasypodids but also other xenarthrans underwent a radiation in South America partly free from competition with the other mammalian orders that were becoming established in North America. Several now extinct evolutionary lines of xenarthrans arose. The Glyptodontidae appeared in the Late Eocene and represent one line that probably evolved from dasypodid stock. These ponderous creatures, some of which were nine feet long, had unusual deep skulls (Fig. 8–3A). Many of the unique structural features of the glyptodonts are associated with their development of a nearly impregnable, turtle-like carapace composed of many fused polygonal scales. These are the most completely armored vertebrates known. The limbs are distinctive and highly specialized, and the last two thoracic vertebrae and the lumbar and sacral vertebrae are fused into a massive arch that, together with the ilium, support the carapace. Patterson and Pascual (1972:267) suggest that the post-Miocene

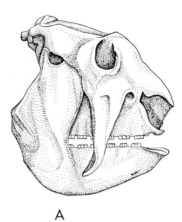

Figure 8-3. Skulls of extinct edentates: A, *Glyptodon*, length of skull 560 mm; B, *Paramylodon*, length of skull 510 mm. (After Romer, 1966.)

A

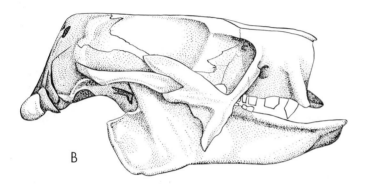

B

diversification of the glyptodonts was favored by the spread of pampas (grassland) in South America.

Additional evolutionary lines are represented by various ground sloths of several extinct families. Ground sloths first appear in South American Oligocene deposits. These animals were herbivores; their teeth lacked enamel and were ever-growing. The family Megatheriidae appears in the Oligocene of South America, and includes *Megatherium*, a massive ground sloth larger than an elephant, as well as smaller types four to eight feet in length. These animals were covered with hair, lacked the upper canine-like tooth, and walked on the outer edges of the unusually specialized hind feet. A second family of ground sloths, the Megalonychidae, is closely related to the megatheriids, but differs from that group in having the anteriormost cheek teeth modified into "canines." *Megalonyx*, a Pleistocene genus that reached the size of a cow, was widely distributed in North America. The remains of

smaller species of megalonychids have been found in the West Indies in association with human artifacts. These sloths seemingly survived into the Recent. A third family of ground sloths, the Myodontidae, appeared in the Oligocene, and is characterized in part by the development of upper "canines" (Fig. 8-3B) and remarkably robust limbs. Protection was afforded some members of this group by round dermal ossicles embedded in the presumably thick skin. A trend toward large size is apparent in the mylodonts, as in other ground sloths.

The glyptodonts, megatheriids, megalonychids, and mylodonts underwent much of their evolution in the Tertiary in isolation from the progressive North American mammalian fauna. Nevertheless, when the land bridge between the Americas was reestablished in the Pliocene, the glyptodonts and ground sloths were remarkably successful in invading North America. By some means of chance dispersal, ground sloths reached North America before the

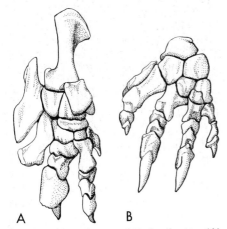

Figure 8–4.   The right pes of *Nothrotherium* (A) and the right manus of *Paramylodon* (B). (After Romer, 1966.)

joining of this land mass and South America, and one genus (*Megalonyx*) seemingly evolved in North America.

The plains-dwelling mylodont *Paramylodon* was widespread in North America and is the most common edentate in the Pleistocene deposits of Rancho La Brea in Los Angeles. This edentate had large claws on digits two and three (Fig. 8–4B), an arrangement similar to that in the living armadillo (Fig. 8–5B).

The common megatheriid *Nothrotheriops* (Fig. 8–6) occurred in North America and South America in the Pleistocene, and its remains from Gypsum Cave in Nevada include bones, skin, and hair. *Nothrotheriops* probably walked on the sides of its highly modified hind feet (Fig. 8–4A), as did other large ground sloths. The powerful claws of the forelimbs were probably used to grasp and tear down vegetation in preparation for ingestion. Dung of this animal is well preserved in dry caves in Nevada and Arizona. Dung from Rampart Cave in Arizona contains such plants as Mormon tea (*Ephedra*) and globe mallow (*Sphaeralcea*) (Long, Hansen, and Martin, 1974); these plants remain common today in dry parts of the southwestern United States. This ground sloth persisted into the Recent in this area, and probably did not disappear until some 11,000 years ago. Thus ended a most fascinating and spectacular cycle of edentate evolution.

Highly developed protective devices (as in Dasypodidae) and narrow, specialized feeding habits (Bradypodidae and Myrmecophagidae) have perhaps been important in allowing Recent edentates to survive

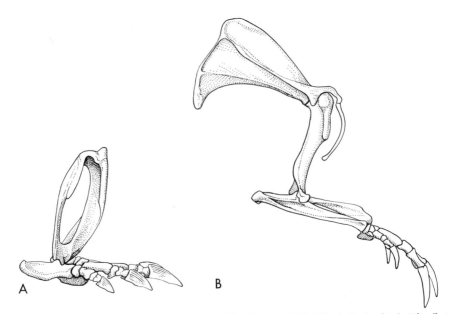

Figure 8–5.   The right limbs of the nine-banded armadillo. A, Part of hind limb; B, forelimb. The flattening of the bones of the forearm and shank increases the surface area for attachment of muscles, and the elongation of the olecranon and the calcaneum give added mechanical advantage for power to the muscles that insert on them.

Figure 8–6. How *Nothrotheriops* might have looked. This extinct, megatheriid ground sloth survived until some 11,000 years ago in the southwestern United States.

under competitive pressure from more "advanced" eutherians. It is probably significant that the edentate stronghold is in tropical areas; these regions often provide last refuges for declining groups that were formerly more widespread and abundant.

FAMILY MYRMECOPHAGIDAE. Members of this group, the anteaters, are highly specialized for feeding on ants and termites. Anteaters occur in tropical forests of Central and South America south to Argentina, and are represented by four Recent species of three genera. Although not important in terms of numbers of species, anteaters are often common in suitable habitats and appear to "own" their narrow feeding niche in the Neotropics.

The most obvious structural features of anteaters are associated with their ability to capture insects, to dig into or tear apart insect nests, and to climb. The skull is long and roughly cylindrical (Fig. 8–7A), the zygoma are incomplete, and the long rostrum contains complex, double-rolled turbinals. Teeth are absent, and the dentary is long and delicate. The long, vermiform tongue is protrusile, and is covered with sticky saliva secreted by the fused submaxillary and parotid salivary glands. These glands are situated in the neck and are enormously enlarged. The forelimbs are powerfully built; the third digit is enlarged and bears a stout, recurved claw, and the remaining digits are reduced. The giant anteater (*Myrmecophaga*) walks on its knuckles with its toes partly flexed, whereas the other anteaters (*Cyclopes* and *Tamandua*), which are fully

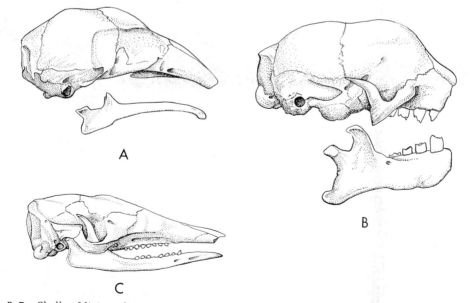

A

C

B

Figure 8–7. Skulls of living edentates. A, Two-toed anteater (*Cyclopes didactylus*, Myrmecophagidae; length of skull 46 mm); B, three-toed sloth (*Bradypus griseus*, Bradypodidae; length of skull 76 mm); C, nine-banded armadillo (*Dasypus novemcinctus*, Dasypodidae; length of skull 95 mm). (*Cyclopes* and *Bradypus* after Hall, E. R., and Kelson, K. R.: *The Mammals of North America*, © 1959 by The Ronald Press Company, New York.)

or partly arboreal, walk on the side of the manus with the claws toed inward. It is of interest to observe that the unusual posture of the manus of *Myrmecophaga* during terrestrial locomotion is essentially the same as that used by *Nothrotheriops*, a megatheriid ground sloth discussed earlier.

In anteaters, the pes has four or five clawed digits and has a plantigrade posture. The size of anteaters ranges from that of a squirrel (in *Cyclopes*; 350 g) to that of a large dog (in *Myrmecophaga*; 25 kg). *Myrmecophaga* is covered with long coarse fur, and the nonprehensile tail has long hairs that hang downward, but in the other anteaters, the fur on the body and tail is shorter and the tail is prehensile.

Anteaters use the powerful forelimbs to expose ants and termites by tearing apart their nests; the insects are captured by the long tongue, are swallowed whole, and are ground up by the thickened pyloric portion of the stomach. All of the anteaters pursue a slow and fairly awkward type of terrestrial locomotion. In the two genera that climb trees, the claws of the manus are used as grappling hooks or are used for grasping as the animals travel along branches in a hand-over-hand fashion. *Cyclopes* is nocturnal and entirely arboreal, and forages for insects high in the trees. The two species of *Tamandua* are largely arboreal but are terrestrial to some extent, and are mostly nocturnal. *Myrmecophaga* is entirely terrestrial and seems largely diurnal. The defensive behavior of these animals is unusual. When threatened, anteaters stand on the hind limbs and use the tail or the back braced against a support to form a solid tripod; they slash or grasp at an enemy with the claws of the manus. The power of the forelimbs and the unusual size of the claws provide these animals with a formidable defense.

Of particular interest is the fact that, as in the Recent anteaters, the large members of the extinct families Mylodontidae, Megatheriidae, and Megalonychidae had large claws on the strong forelimbs; the hind limbs and pelvic girdles were specialized in directions indicating considerable bipedal ability; and the tail was sturdy and may have functioned as a brace during bipedal stance. These large edentates may well have defended themselves against saber-toothed cats (*Smilodon*) and other large Pleistocene predators much as Recent anteaters protect themselves from today's less imposing predators.

FAMILY BRADYPODIDAE. The tree sloths are strange animals (Fig. 8–8) that are so highly modified for a specialized form of arboreal locomotion that they have nearly lost the ability to move on the ground. The six Recent species of tree sloths belong to two genera, and range from Central America (Honduras) through the northern half of South America to northern Argentina. These animals primarily inhabit tropical rain forests.

The adaptive zone of tree sloths is quite different from that of the anteaters and involves strictly arboreal habits and an entirely herbivorous diet. The bradypodids differ strongly from the myrmecophagids, especially in skull characteristics. The tree sloth skull is short and fairly high, with a strongly reduced rostrum. The zygomatic arch is robust but incomplete, and its jugal portion bears a ventrally projecting jugal process similar to that present in many extinct edentates (Fig. 8–7B). The premaxil-

Figure 8–8. A two-toed tree sloth (*Choloepus hoffmanni*). (San Diego Zoo photo.)

laries are greatly reduced and the turbinals are complexly rolled, as in myrmecophagids. Five maxillary and four or five mandibular teeth are present; the anteriormost teeth in *Choloepus* are caniniform and are kept sharp by abrasion between the posterior surface of the upper tooth and the anterior surface of the lower tooth. The persistently growing teeth are roughly cylindrical, and have central cores of soft dentine surrounded successively by hard dentine and cement.

A departure from the usual mammalian pattern of seven cervical vertebrae occurs in the bradypodids, in which from six to nine occur, the number differing between species and in some cases even between individuals of the same species. Xenarthrism (Fig. 8–1) is strongly developed in the thoracic and lumbar vertebrae, and, as in some extinct ground sloths, the coracoid and acromion processes of the scapula are united. The externally visible digits do not exceed three in number and, except for the long and laterally compressed claws, are syndactylous (bound together). Tree sloths are not large, weighing from 4 to 7 kg, and are covered with long coarse hair. This fur provides a habitat for algae, which grow in the flutings on the surfaces of the hairs during the rainy season and tint the fur green. In addition, the adults of two genera of moths (*Bradypodicola* and *Cryptoses*; Pyralididae, Microlepidoptera) hide in large numbers in the dense pelage. In tree sloths, the tail is rudimentary or is short.

These remarkably specialized animals are strictly herbivorous, eating primarily leaves and seldom descending to the ground. Climbing is done in an upright position by embracing a branch, or by hanging upside down and moving along hand-over-hand. Tree sloths spend considerable time hanging upside down, and at times sleep in this position. In both genera the forelimbs are greatly elongate and are considerably longer than the hind limbs, and when the animals are on the ground the limbs splay out to the side. The animals are unable to support their weight on their limbs and progress on the ground by slowly dragging their bodies forward with the forelimbs. In some tropical basins, where the ground is inundated during part of the rainy season, tree sloths disperse by swimming.

Although tree sloths occupy tropical environments, in which temperature extremes virtually never occur, they possess some adaptations typical of animals that are subjected to cold stress in boreal areas. In tree sloths, and in anteaters and armadillos, the limbs have retia mirabilia (see page 136). By performing countercurrent heat exchange, these specialized vascular bundles allow the limbs, which are unusually long in the case of tree sloths, to become considerably cooler than the body without causing a serious cooling of the body or an unusual metabolic drain. In addition, tree sloths are insulated by long, fairly dense pelage, consisting of long guard hairs and short, fine-textured underfur, a type of pelage characteristic of inhabitants of cold climates.

What is the adaptive importance to tropical animals of such specializations for the retention of body heat? Apparently edentates, and perhaps tree sloths especially, are extremely sensitive to cold. Tree sloths are probably heterothermic, with body temperatures varying from 28° to 35°C; when they are inactive, they have difficulty maintaining a constant body temperature. Tree sloths have been observed to shiver at ambient temperatures as high as 27°C (80°F)! The specializations in tree sloths for the retention of body heat, therefore, enable these extremely cold-sensitive animals to maintain nonlethal temperatures in the body during cool Neotropical nights. It is not surprising that tree sloths range northward no further than the tropical forests of Honduras.

FAMILY DASYPODIDAE. Members of this family, which include all of the armadillos, are remarkable for the protective armor that occurs in all species. In terms of numbers of species and breadth of distribution, this is the most successful edentate family. Twenty-one Recent species of armadillos of nine genera are known, and the geographic range includes Florida (by introduction), much of the south central United States as far north as Kansas, Mexico, Central America, and nearly all of South America to near the southern end of Argentina. Armadillos are common in many areas and occupy tropical forests and savannas, semideserts, and temperate plains and forests.

The most obvious and unique structural feature of armadillos is the jointed armor. This consists of bony scutes covered by horny epidermis (Fig. 8–9). The scutes form plates that occur in a variety of patterns, but always present are a head shield and a series of plates protecting the neck and much of the body. Sparse hair usually occurs on the flexible skin between the plates and on the limbs and the ventral surface of the body. Individuals of some species can curl into a ball so that their limbs and vulnerable ventral surfaces are largely protected by the armor. The largest species, the giant armadillo (*Priodontes giganteus*), weighs up to 60 kg; the smallest, the pigmy armadillo (*Chlamyphorus truncatus*), is roughly the size of a small rat (120 g).

The skull is often elongate and is dorsoventrally flattened; the zygomatic arch is complete and the mandible is slim and elongate (Fig. 8–7C). The teeth are borne only on the maxillary (except in one species), are nearly cylindrical, and vary from 7-9/7-9 to 25/25. The teeth are frequently partially lost with advancing age. The axial skeleton is fairly rigid and is partially braced against the carapace; the second and third cervical vertebrae, and in four species other cervicals as well, are fused (Fig. 8–2B); 8 to 13 sacral and caudal vertebrae form an extremely powerfully braced anchor for the pelvis (Fig. 8–2A); xenarthral articulations between thoracic and lumbar vertebrae (Fig. 8–1) produce a rigid vertebral column; and elongate metapophyses from the lumbar vertebrae brace but do not contact the carapace.

Bracing of the carapace by the skeleton is remarkably extensive in some species, but the skeleton does not actually contact the carapace. For example, in the familiar nine-banded armadillo (*Dasypus novemcinctus*), the species that enters the United States, the carapace is partially supported by prominent dorsolateral processes from the ilium and ischium (Fig. 8–2A), and by the modified tips of the neural spines of all the thoracic and lumbar vertebrae. In this species, the powerfully developed panniculus carnosus muscles and a broad ligament from the expanded dorsolateral flange of the ilium attach to the inner surface of the carapace and bind it to the body, but no bony contact is made between the carapace and the skeleton. The ribcage is made rigid by a broadening of the ribs, by heavy intercostal muscles, and by ossification of the greatly enlarged parts of the ribs that in most mammals are the costal cartilages.

Figure 8–9. A naked-tailed armadillo (*Cabassous centralis*). (Photograph by Lloyd G. Ingles.)

In all armadillos the limbs are power-fully built, and the fore and hind feet bear large, heavy claws (Fig. 8–5). The feet are five-toed in all but one genus, and the foot posture is usually plantigrade. The tibia and fibula are fused proximally and dis-tally and are highly modified for giving or-igin to powerful muscles of the shank (Fig. 8–5A). The femur has a prominent third trochanter. Retia mirabilia occur in the limbs.

Compared to other edentates, armadillos are more generalized in their feeding and their locomotion. Most species feed pri-marily on insects, but a variety of inver-tebrates, small vertebrates, and vegetable material is also taken. All armadillos are at least partly adapted for digging, and some species are highly fossorial. In some spe-cies, the forelimbs are unusually powerful and the manus is somewhat like that of a mole. Such a fossorial creature is the pygmy armadillo (*Chlamyphorus trun-catus*), which utilizes a style of digging seemingly unique among mammals. The soil is dug away and pushed beneath the animal by the long claws of the forepaws, and the hind feet rake the soil behind the animal. The pelvic scute is then used to pack the soil behind the body. During the packing, the front limbs push the animal backward and the hind quarters vibrate rapidly from side to side (Rood, 1970). No permanent burrow is formed. Running is limited in armadillos by a structural pat-tern that achieves power but no speed; some species can be run down by a man. The "pichi" (*Zaedyus pichiy*) was ob-served in southern Argentina by Simpson (1965a:201), who found this armadillo to depend heavily on its sense of smell. A cap-tive individual "seemed to pay practically no attention to anything she saw or heard but lived by and for her nose." The pichi cannot curl into a ball, and escapes from enemies either by burrowing beneath a thorn bush or by clutching a solid surface with the claws and pulling the carapace down against the ground.

As indicated by their wide range, arma-dillos are more resistant to cold than are tree sloths, but the possession of retia mirabilia in the limbs suggests that arma-dillos share the general edentate sensitiv-ity to cold. As a further indication, arma-dillos in the northern parts of their range have been found to suffer as high as 80 per cent winter mortality when prolonged cold spells occurred (Fitch, Goodrum, and Newman, 1952). Despite such mortality, armadillos have extended their range into the south-central United States in the last 90 to 100 years (see Fig. 16–14. They were formerly restricted to extreme southern Texas.

ORDER PHOLIDOTA

The pangolins, or scaly anteaters, com-prise a single family, Manidae, and are represented today by a single genus (*Manis*) with eight species. Pangolins occur in tropical and subtropical parts of the southern half of Africa and in much of southeast Asia. Their fossil record is poor, and while it documents the probable Oli-gocene occurrence of these animals in Eur-ope, it contributes little to our knowledge of the evolution or relationships of the group. The paucity of fossils and the lim-ited number of Recent forms suggest that pangolins have never been abundant or diverse.

Pangolins are strange-looking creatures that at a glance seem more reptilian than mammalian (Fig. 8–10). They are of moder-ate size, weighing from about 5 to 35 kg. Although pangolins and South American anteaters (Myrmecophagidae) are not close-ly related, they share some anatomical features associated with their insect-eating habits. The skull of the pangolin is conical and lacks teeth; the dentaries are small and lack angular and coronoid processes; the tongue is extremely long and vermi-form, and originates on tremendously specialized parts of the ribcage rather than from the hyoid. The last pair of cartilag-inous ribs have lost their vertebral connec-tions and have formed an enormously elongate posterior extension of the sternum. This extension passes into the posterior part of the abdominal cavity, curves up-

Figure 8–10. A pangolin (*Manis gigantea*, Pholidota). (Photograph, courtesy of the American Museum of Natural History.)

ward, and ends near the kidneys. The fused and spatulate end of these ribs gives origin to the accessory musculature of the tongue. The tongue and its accessory musculature is therefore longer than the head and body together, allowing the tongue to be extremely protrusile. The scales are the most distinctive feature; these cover the dorsal surface of the body and the tail and are composed of agglutinated hair. The skin and scales of pangolins account for a large share of the weight of these animals: Kingdon (1971:356) reports that these parts constitute one third to one half of the weight of the ground pangolin (*Manis temmincki*). The manus and pes have long recurved claws; the pes has five toes and the manus is functionally tridactyl. The walls of the pyloric part of the stomach are thickened. This part of the stomach usually contains small pebbles and seems to function to grind food as does the gizzard of a bird.

The food of pangolins is mostly termites, but ants and other insects are also taken. The insects are located by smell, and pangolins seem highly selective in their choice of food. Sweeney (1956) found that only rarely would a pangolin dig for the "wrong" species of ant or termite. Pangolins have been observed to tear apart termite-infested wood while lying on their backs; the termites that fall onto their chests are captured deftly by darting movements of the tongue. The pangolin rolls into a ball when disturbed, erects the scales, flails the tail, and moves the large, sharp-edged scales in a cutting motion. Foul-smelling fluid is sprayed from the anal glands of some species as further protection. Pangolins are fairly awkward on the ground, but some are capable climbers.

ORDER TUBULIDENTATA

This order, the aardvarks, includes but one family (Orycteropodidae). The one Recent species (*Orycteropus afer*) is an anomalous-looking creature (Fig. 8–11) that inhabits Africa south of the Sahara Desert. The earliest record of tubulidentates is from the Miocene of Africa, and in the Pliocene an extinct member of the Recent genus occupied parts of Europe and Asia. Skeletal similarities between the two groups indicate that tubulidentates may have evolved from condylarths (primitive ungulates; order Condylarthra).

Many structural parallels occur between tubulidentates and certain edentates, probably not because the groups are closely related, but because of similarities in habits. Both tubulidentates and some edentates are powerful diggers and feed on

Figure 8–11. A juvenile aardvark (*Orycteropus afer*). (San Diego Zoo photo.)

ants and termites. The aardvark weighs up to roughly 82 kg, and the thick, sparsely haired skin provides protection from insect bites. The skull is elongate, and the dentary is long and slender (Fig. 8–12). In the adult dentition, incisors and canines are lacking; the cheek teeth are 2/2 premolars and 3/2 molars. Each tooth is rootless and consists of many (up to nearly 1,500) hexagonal prisms of dentine, each surrounding a slender, tubular pulp cavity. The columnar teeth lack enamel but are surrounded by cement. The anteriormost teeth erupt first, and are often lost before the posterior molars are fully erupted. The slender tongue is protrusile.

Olfaction is used in finding insects; the olfactory centers of the brain are unusually well developed, and the turbinal bones are remarkably large and complex. The nostrils are highly specialized in a fashion unmatched by any other mammal. Fleshy tentacles, which presumably have an olfactory function, occur on the nasal septum (Fig. 8–13), and dense hair surrounds the nostrils and can seal them when the aardvark digs. The pollex is absent, and the hind foot is five-toed; the robust claws are flattened and blunt.

The powerful forelimbs are used in burrowing and in dismantling termite and ant nests, and the hind limbs thrust accumu-

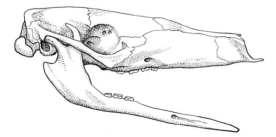

Figure 8–12. The skull of the aardvark (*Orycteropus afer*). (After Hatt, 1934.)

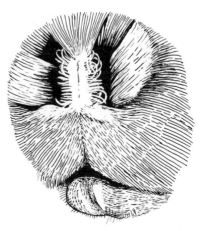

Figure 8–13. The complex nose of an aardvark (*Orycteropus afer*); note the fleshy tentacles on the nasal septum and the dense tracts of hair that can seal the nostrils. (After Kingdon, 1971.)

lated soil from the burrow. Although the foot posture is digitigrade, aardvarks are slow runners and can be run down by a man. Burrows dug by aardvarks are numerous in some areas: Walker (1968:1318) reports some 60 burrow entrances in an area 100 by 300 m. These burrows are used as retreats by a variety of mammals, including the warthog *(Phacochoerus africanus).*

ORDER
LAGOMORPHA

9

Although lagomorphs—the rabbits (Leporidae) and pikas (Ochotonidae)—are not a diverse group, including but 10 genera with 63 Recent species, they are important members of many terrestrial communities and are nearly worldwide in distribution. Considering large land masses, lagomorphs were absent only from the Australian region and from southern South America before recent introductions by man. Lagomorphs occupy diverse terrestrial habitats from the arctic to the tropics, and, in many temperate and boreal regions, rabbits are subject to striking population cycles marked by periods of great abundance alternating with times of extreme scarcity. In such regions, population cycles of many carnivores are influenced strongly by changes in population densities of rabbits.

Many important diagnostic features of Recent lagomorphs are related to their herbivorous habits and, in the case of leporids, to their cursorial locomotion. Lagomorphs have fenestrated skulls, a feature highly developed in some leporids (Fig.

9–1A). The anterior dentition resembles that of a rodent, but whereas rodents have 1/1 incisors, rabbits have 2/1 incisors; the second incisor is small and peglike, and lies immediately posterior to the first (Fig. 9–1A). As in rodents, the lagomorph incisors are ever-growing. A long postincisor diastema is present in lagomorphs, and the canines are absent. The cheek teeth are hypsodont and rootless, and the crown pattern features transverse ridges and basins (Fig. 9–2A and 9–3B). The distance between the upper tooth rows is greater than that between the lower tooth rows, allowing occlusion of upper and lower cheek teeth only on one side at a time and requiring a lateral or oblique jaw action. The masseter muscle is the primary muscle of mastication; the temporalis is small and the coronoid process, its point of insertion, is rudimentary.

The clavicle is either well developed (Ochotonidae) or rudimentary (Leporidae), and the elbow joint limits movement to a single anteroposterior plane (Fig. 9–2B, C). The tibia and fibula are fused dis-

A

B

Figure 9–1. A, Skull of the antelope jackrabbit *(Lepus alleni)*; note the highly fenestrated maxillary and occipital bones. B, Anterior part of the skull of the arctic hare *(L. arcticus)*; note the procumbent incisors and the receding nasals. These specializations are associated with this animal's habit of using the incisors to scrape away ice and snow to reach food. *(L. arcticus* after Hall, E. R., and Kelson, K. R.: *The Mammals of North America,* © 1959 by The Ronald Press Company, New York.)

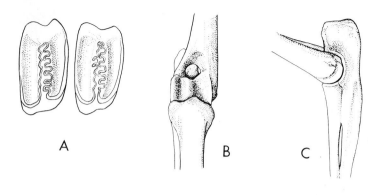

Figure 9–2. Leporid structural features. (The drawings are of the antelope jackrabbit, *Lepus alleni.*) A, Occlusal view of upper right premolars three and four. B, Anterior view of the right elbow joint; movement is limited to a single (anteroposterior) plane by this "tongue and groove" articulation. When full extension of the forearm is reached, a process on the olecranon of the ulna locks into the conspicuous hole in the humerus and braces the joint. C, Medial view of the right elbow, showing the tight fit between the articular surface of the humerus and the articular surfaces of the radius and ulna. The radius and ulna are partially fused.

tally; the front foot has five digits and the hind foot has four or five digits. The soles of the feet, except for the distalmost toe pads in *Ochotona*, are covered with hair. The foot posture is digitigrade during running but is plantigrade during slow movement. The tail is short, and in *Ochotona* is not externally evident.

The lagomorphs are a group of mammals with no clear relationships to other Recent eutherian orders. The first fossil record of mammals with lagomorph-like characters is from the late Paleocene of Asia, when the primitive family Eurymylidae appears. (But eurymylids may not be lagomorphs, for they have such nonlagomorph features as 1/1 incisors.) Although rodents and rabbits were once thought to be closely related, the fossil evidence indicates no such relationship. It appears, rather, on the basis of cusp patterns, that the earliest rodents (Paramyidae) and the earliest lagomorphs were derived from separate eutherian ancestry. The family Leporidae probably originated in Asia, but under-

went most of its early (Oligocene and Miocene) evolution in North America. Leporids became well established in the Old World in the Pliocene, and the advanced subfamily Leporinae arose there. The pikas appeared first in the Oligocene of Eurasia, and spread in the Pliocene to Europe and to North America. The Recent genus *Ochotona* is known from the early Pliocene. In contrast to the leporids, which have remained widespread since the Pliocene, the ochotonids reached their greatest diversity and widest distribution in the Miocene, when they occupied Europe, Asia, Africa, and North America (Dawson, 1967b:305), and have declined since. In North America, pikas are now of local occurrence in high mountains north of Mexico. They occur more widely in the Old World, where they inhabit eastern Europe and much of northern and central Asia.

The environmental tolerances, or perhaps the competitive success, of ochotonids has seemingly changed in the New

Figure 9–3. Skull (A) and occlusal view of the upper right fourth premolar and first molar (B) of the pika (*Ochotona princeps*).

World since the Miocene. For example, in the Miocene an ochotonid occupied riparian communities in the Great Plains, Rocky Mountain, and Great Basin regions of North America (Wilson, 1960:7); riparian situations in these areas no longer support pikas. Factors influencing the striking post-Miocene decline of the ochotonids are poorly understood.

Why, although they are an old and thriving group, have lagomorphs not undergone a greater adaptive radiation? Perhaps their conservatism is related to the limitations of their functional position as "miniature ungulates." Competition with members of the larger and more diverse order Artiodactyla, a group highly adapted to a herbivorous diet and to cursorial locomotion, may have limited lagomorphs to the exploitation of but a single limited adaptive zone, although this zone was occupied with great success over broad areas. Of interest in this regard are the scarcity and local occurrence of lagomorphs in many parts of East Africa, where there is an extremely rich ungulate fauna.

FAMILY OCHOTONIDAE. The pikas are represented today by 1 genus with roughly 14 species. Pikas are less progressive with regard to cursorial adaptations than are the rabbits, and they usually venture only short distances from shelter. Pikas occur in the mountains of the western United States and south central Alaska, and occur over a wide area in the Old World, including eastern Europe and much of Asia south to northern Iran, Pakistan, India, and Burma.

In contrast to rabbits, pikas are small, weighing about 100 to 150 gm; they have short, rounded ears, short limbs, and no externally visible tail (Fig. 9–4). The ear opening is guarded by large valvular flaps of skin that may protect it during severe weather. The skull is strongly constricted between the orbits and lacks a supraorbital process; the rostrum is short and narrow. The skull is less strongly arched in ochotonids than in leporids (Fig. 9–3A), and the angle between the basicranial and palatal axes is lower. The maxilla has a large fenestra. The dental formula is 2/1, 0/0,

Figure 9–4. A pika (*Ochotona princeps*) at a lookout point on a rock, with a mouthful of plant material for its hay pile. (Photograph by O. D. Markham.)

3/2, 2/3 = 26. The third lower premolar has more than one re-entrant angle, and the re-entrant enamel ridges of the upper cheek teeth are straight (Fig. 9–3B). The anal and genital openings are enclosed by a common sphincter, and males have no scrotum.

In North America, pikas typically occupy talus slopes in the high mountains. These herbivorous animals are characteristic of boreal or alpine situations, and they occur from near sea level in Alaska to the treeless tops of some of the highest peaks in the Rocky Mountains and Sierra Nevada–Cascade chain. In North America, habitat requirements usually include fairly extensive areas of large rocks or irregular boulders adjacent to growths of forbs and grasses. When frightened, pikas seek shelter in the labyrinth of spaces and crevices between rocks, and seldom forage far from such shelter. Large "hay piles" are built each summer in the shelter of large, usually flat-bottomed boulders, and the dried but green material is eaten during the winter when snow covers the ground and little vegetation growth occurs. In Eurasia, pikas occupy an extensive geographic range and a wide range of habitats, including talus, forests, rock-strewn terrain, and open plains and desert-steppe areas. Unusually large hay piles, weighing up to 20 kg, are made by pikas inhabiting dry areas in southern Russia (Formozov, 1966).

FAMILY LEPORIDAE. The rabbits are a remarkably successful group in terms of ability to occupy a variety of environments over broad areas. Rabbits are now nearly cosmopolitan. Their distribution before introductions by man included most of the New and Old Worlds, and rabbits have been introduced into New Zealand, Australia, parts of southern South America, and various oceanic islands in both the Atlantic and Pacific. Nine Recent genera represented by 49 Recent species are known.

Several major leporid evolutionary trends in structure are recognized by Dawson (1958:6). The cheek teeth have become hypsodont, some of the premolars have become molariform, and the primitive crown pattern has been modified into a simple arrangement in which most traces of the primitive cusp pattern have been lost. These changes resemble those in some groups of strictly herbivorous ro-

dents. The skull has become arched, and the angle between the basicranial and palatal axes has increased. The changes are associated with a posture involving a greater angle between the long axis of the skull and the cervical vertebrae than that typical of primitive leporids. Trends in limb structure leading to increased cursorial ability include elongation of the limbs and specializations of articulations so that movement is limited to one plane.

The leporid skull (Fig. 9–1A) is more or less arched in profile, and the rostral portion is fairly broad. The maxillae, and often the squamosals, occipitals, and parietals, are highly fenestrated, and a prominent supraorbital process is always present. The auditory bullae are globular and the external auditory meatus is tubular. The dental formula is usually 2/1, 0/0. 3/2, 3/3 = 28; the re-entrant enamel ridges of the upper cheek teeth are usually crenulated (Fig. 9–2A). The clavicle is rudimentary and does not serve as a brace between the scapula and the sternum. The limbs, especially the hind limbs, are more or less elongate; movement at the elbow

Figure 9–5. A white-tailed jackrabbit (*Lepus townsendii*) in its partially white winter pelage. This animal is from Colorado, but in more northerly parts of its range this jackrabbit is almost entirely white in winter, (Photograph by George D. Bear.)

joint is limited to the anteroposterior plane (Fig. 9–2B, C). The tail is short. The ears have a characteristic shape: the proximal part of the ear is tubular and the lower part of the ear opening is well above the skull (Fig. 9–5). The testes become scrotal during the mating season. In some species that inhabit regions with snowy winters, the animals molt into a white winter pelage in the fall (Fig. 9–5) and into a brown summer pelage in the spring. Wild leporids weigh from 0.3 to roughly 5 kg.

Leporids inhabit a tremendous array of habitats, from arctic tundra and treeless and barren situations on high mountain peaks to coniferous, deciduous, and tropical forests, open grassland, and deserts. Some species, such as *Sylvilagus palustris* and *S. aquaticus* of the southeastern United States, are excellent swimmers and lead semiaquatic lives. Leporids are entirely herbivorous and utilize a wide variety of grasses, forbs, and shrubs. Several species are known to reingest fecal pellets and are thought to obtain essential nutrients (proteins and some vitamins) from material as it passes through the alimentary canal for a second time.

Habitat preference and cursorial ability differ markedly among different species of leporids and are strongly interrelated. Broadly speaking, species with relatively poor cursorial ability, such as *Brachylagus idahoensis* and *S. bachmani* of the western United States, scamper short distances to the safety of burrows or dense vegetation when disturbed, and typically occur in stands of big sagebrush *(Artemesia tridentata)* or dense chaparral, respectively. Other cottontails, such as *S. floridanus* of the eastern and *S. audubonii* of the western United States, are intermediate in cursorial ability and typically inhabit areas with scattered brush, rocks, or other cover, and do not run long distances to reach a hiding place. Representing the extreme in cursorial specialization among lagomorphs are the North American jackrabbits *(Lepus californicus, L. townsendii,* and *L. alleni* and its relatives). These animals have greatly elongate hind limbs, have adopted a bounding gait, and occupy areas with

limited shelter, such as deserts, grasslands, or meadows, where they take shelter in "forms" (Fig. 9–6). Instead of taking cover at the approach of danger, they depend for escape on their running ability. (Adaptations that contribute to running ability are discussed on page 238). Jackrabbits, and other similarly adapted members of the genus *Lepus*, are extremely rapid runners for their size; some attain speeds of up to 70 km per hour. This speed allows some species to occupy open areas with little cover, where safety depends upon outrunning predators. So strongly entrenched is the habit of seeking safety by running that some speedy leporids only take to shelter as a last resort when injured or exhausted.

Although rabbits are seemingly peaceful and nonaggressive types, they are strong competitors and are remarkably adaptable. In some parts of Australia, the extinction or near extinction of certain marsupials is due primarily to competition with introduced European rabbits *(Oryctolagus)*. In addition, these prolific rabbits have caused

Figure 9–6. A white-tailed jackrabbit *(Lepus townsendii)* in its "form," a hollowed-out hiding place beneath a bush or other (often scanty) shelter. This animal is in its brown summer pelage. (Photograph by George D. Bear.)

great damage to crops and rangeland, and at various times have been a primary agricultural pest in many parts of Australia as well as in New Zealand, where they were also introduced. The range of environmental conditions to which leporids have adapted is tremendous. Populations of rabbits *(Lepus arcticus)* along the arctic coasts of Greenland use their protruding incisors (Fig. 9–1B) to scrape through snow and ice to reach plants during the long arctic winters, whereas far to the south, in the deserts of northern Mexico, jackrabbits *(Lepus alleni)* maintain their water balance through hot dry periods by eating cactus and yucca.

ORDER RODENTIA

10

Rodents comprise the largest mammalian order, including some 34 living families, 354 genera, and roughly 1700 species. Rodents are a spectacularly successful group; they are virtually cosmopolitan in distribution and are important members of nearly all faunas. They are also a remarkably complicated group with respect to morphological diversity, lines of descent, and parallel evolution of similar features in different groups. Because of these complexities, zoologists often have not agreed on the relationships between families and other taxa. As a result, superfamily and subordinal groupings remain in dispute; therefore numbers of species and genera listed here are approximations.

The terms *sciuromorph, myomorph,* and *hystricomorph* have been used repeatedly to designate major taxonomic divisions of the rodents. These terms, which are based on important differences in the structure of the skull and masseter muscles, are entrenched in the literature on mammals. Although they have frequently been given subordinal rank (Sciuromorpha, Myomorpha, Hystricomorpha), there is no general agreement about their use in the taxonomic scheme of the rodents. The classification proposed by Wood (1965) is used here (Table 10–1). This system recognizes but three suborders, and does not assign to suborders those families of uncertain relationships. In my judgment, Wood's classification is useful not only because it indicates the relationships of some families, but also because it identifies the families whose relationships are poorly known.

Rodents are fascinating in large part because of the very features that make them difficult to classify. Their complex pat-

Table 10–1. A Classification of Recent Rodents

Order Rodentia
  Suborder Protrogomorpha
    Superfam. Aplodontoidea
      Fam. Aplodontidae
  Suborder Caviomorpha
    Superfam. Octodontoidea
      Fam. Octodontidae
      Fam. Echimyidae
      Fam. Abrocomidae
      Fam. Capromyidae
    Superfam. Chinchilloidea
      Fam. Chinchillidae
      Fam. Dasyproctidae
      Fam. Cuniculidae
      Fam. Heptaxodontidae
      Fam. Dinomyidae
    Superfam. Cavioidea
      Fam. Caviidae
      Fam. Hydrochoeridae
    Superfam. Erethizontoidea
      Fam. Erethizontidae
  Suborder Myomorpha
    Superfam. Muroidea
      Fam. Cricetidae
      Fam. Muridae
    Superfam. Geomyoidea
      Fam. Geomyidae
      Fam. Heteromyidae
    Superfam. Dipodoidea
      Fam. Dipodidae
      Fam. Zapodidae
    Superfam. Spalacoidea
      Fam. Spalacidae
      Fam. Rhizomyidae
    Superfam. Gliroidea
      Fam. Gliridae
      Fam. Seleveniidae

Lineages of uncertain relationships and not assigned to suborders:
      Fam. Sciuridae
      Fam. Castoridae
      Fam. Ctenodactylidae
      Fam. Anomaluridae
      Fam. Pedetidae
      Fam. Hystricidae
      Fam. Thryonomyidae
      Fam. Petromyidae
      Fam. Bathyergidae

(Based on Wood, 1965.)

terns of evolution, repeated and striking parallelism, intricate systems of resource allocation, and highly refined adaptations to such demanding environments as the arctic and the deserts make them an unusually rewarding group to study. No apologies for the complicated systematic arrangement used in this chapter are in order. Systematics should reflect the phylogenetic relationships between the organisms considered, and it is obvious that no simple system of classification could indicate the relationships between the diverse and incompletely studied array of living rodents.

MORPHOLOGY

Recent members of the order Rodentia share a series of distinctive cranial features. The upper and lower jaws each bear a single pair of persistently growing incisors, a feature developed early in the evolution of rodents and one that committed them to a basically herbivorous mode of feeding, but permitted the exploitation of such often abundant foods as insects. Because only the anterior surfaces are covered with enamel, the incisors assume a characteristic beveled tip as a result of wear. The occlusal surfaces of the cheek teeth are often complex and allow for effective sectioning and grinding of plant material. The dental formula seldom exceeds 1/1, 0/0, 2/1, 3/3 = 22, and a diastema is always present between the incisors and the premolars. The incisors and canines are always 1/1, 0/0.

The glenoid fossa of the squamosal is elongate, allowing anteroposterior and transverse jaw action. The mandibular symphysis has sufficient "give" in many species to enable the transverse mandibular muscles to pull the ventral borders of the rami together and spread the tips of the incisors. The masseter muscles are large and complexly subdivided, and provide most of the power for operating the lower jaw. By comparison with the masseters, the temporal muscles are usually small and their point of insertion, the coronoid process, is reduced, particularly in some

hystricomorphous rodents (Fig. 10–1). In general, rodents have undergone little specialization other than in the head area. Notable exceptions occur in some rodents that inhabit barren arid regions and are adapted for jumping, in some that burrow, and some that glide.

PALEONTOLOGY

Rodents are an ancient group: the earliest fossil records are from the late Paleocene of North America. The most primitive known rodents, the Paramyidae (Fig. 10–2A), may be the basal group from which all rodents evolved. The structure of the paramyid skull and dentary indicates that the temporalis muscle was large and that the masseter muscles were not highly specialized and originated entirely from the zygomatic arch (Fig. 10–3A). The dental formula is 1/1, 0/0, 2/1, 3/3, and the cheek teeth are brachyodont.

The structure of early rodents and their efficiency as small herbivores was apparently highly adaptive, for the Eocene paramyids were abundant in Eurasia and North America. According to two of the leading authorities on the evolution of rodents (Lavocat, 1974; Wood, 1974), the New World Caviomorpha and the African families Thryonomyidae, Petromyidae, and Bathyergidae evolved from closely related or identical Northern Hemisphere ancestors. There is disagreement as to the ancestry of the Hystricidae, but Lavocat would derive this family from the ancestry just mentioned. A number of lines of evidence, such as close similarities in jaw musculature, skull and jaw structure, and nearly identical patterns of the cephalic ar-

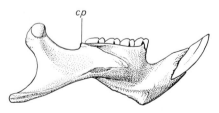

Figure 10–1. Dentary bone of the nutria (*Myocastor coypu*, Capromyidae); note the strongly reduced coronoid process (*cp*). Compare this dentary to that of *Aplodontia*, shown in Figure 10–2.

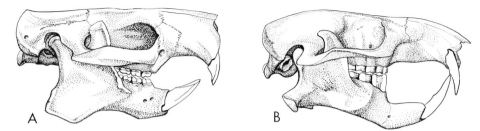

Figure 10–2. Skulls of primitive rodents. A, *Paramys* (Paramyidae), a primitive late Paleocene and early Eocene sciuromorph. Length of skull 89 mm. (After Romer, A. S.: *Vertebrate Paleontology*, 3rd ed., The University of Chicago Press.) B, Mountain beaver (*Aplodontia rufa*, Aplodontidae), the most primitive living rodent. Length of skull 68 mm.

teries (Bugge, 1974) point to this common origin of the Caviomorpha and the African families.

The caviomorphs, represented today by a large and thriving group of Neotropical rodents, first reached South America in the Late Eocene or Early Oligocene, when this continent was isolated by a seaway from North America and by the South Atlantic Ocean from Africa. These earliest South American rodents must have immigrated via rafting, but it is uncertain whether dispersal was from Africa to South America at a time when the continents had not yet drifted as far apart as they are today, as postulated by Lavocat, or was from North America to South America, as held by Wood. The occurrence in the Late Eocene or early Oligocene of Texas and central Mexico of rodents that have most characters that would be expected of a caviomorph ancestor suggests that dispersal from North America may have resulted in the initial establishment of rodents in South America. In any case, the Caviomorpha and their African relatives have been isolated from each other since at least Early Oligocene. In view of this isolation and the environmental contrasts between Africa and South America, it is not surprising that these groups, while retaining certain basic similarities, have followed divergent evolutionary paths.

The early evolution of the rodents was rapid, and by Late Eocene, when they were forcing to extinction their closest competitors (the multituberculates), some rodents had abandoned the primitive arrangement of jaw musculature. Seemingly, the time between the Late Eocene and Middle Oligocene was one of accelerated rodent evolution (Wilson, 1972), and it was then that the major patterns of specialization of the jaw musculature typical of modern rodents were established. A number of Recent families, such as the Sciuridae, Cricetidae and Erethizontidae, appeared in the Oligocene. The Miocene spread of grassland in both the New and Old World provided new adaptive zones for rodents, and some families that are adapted to semiarid conditions, such as the Geomyidae and Pedetidae, appeared at this time. In contrast to the marsupials, edentates, carnivores, and ungulates, rodents seemingly suffered no major Pleistocene or post-Pleistocene extinctions, and today we see a rich and abundant rodent fauna.

EVOLUTION OF SPECIALIZATIONS OF THE JAW MUSCULATURE AND SKULL

Rodents have occupied an extraordinarily diverse array of habitats and, although primarily vegetarians, have exploited a wide variety of food, including insects, fish, reptiles, birds, mammals, and above- and underground parts of trees, grasses, and forbs. Morphological diversity in rodents has resulted from selective pressures favoring a variety of feeding habits, styles of locomotion, and techniques of burrowing. Rodents are variously terrestrial, arboreal, fossorial, saltatorial, and semiaquatic, and some are skilled gliders.

Despite their diversity and success in adapting to contrasting environments and different styles of life, rodents have rather consistently followed certain basic trends in the evolution of the jaw muscles, the bones from which they take origin or insertion, and the teeth. Even in the early stages of their evolution, selective pressures apparently favored forward migration of the jaw muscles. As pointed out by Wood (1959:171), however, there were seemingly a very limited number of ways in which this could occur, given the structural constraints imposed by the skull design and jaw musculature of rodents. Accordingly, the trend in the evolution of the jaw musculature seems to have been toward "myriad detailed variations on a limited number of basic patents. . . ." (Wood, 1974:50). This variation, involving repeated and complex parallelism and intricate variation on the basic theme, has obscured both the relationships between rodents and their evolutionary patterns.

Since their appearance in Late Paleocene, rodents have had dentitions featuring a division of labor between the incisors and the cheek teeth. This dentition serves chiefly two functions. The incisors serve as chisels with which food is gnawed, vegetation is clipped, or, in some fossorial forms, soil and rocks are dug away. These teeth are subject to heavy wear, and became evergrowing early in the evolution of rodents. The cheek teeth, separated from the incisors by a broad diastema, perform a different function, that of mastication of food. A complicated and rather rotary jaw action allows the lower cheek teeth to move transversely against the upper teeth, producing a crushing and grinding action. Not only are the gnawing and grinding functions performed by different teeth, but they must be performed separately. When the cheek teeth are in position for grinding, the tips of the incisors do not meet; the lower jaw must therefore be moved forward for the incisors to be in position for gnawing.

This division of labor between incisors and cheek teeth clearly "guided" the evolution of the rodent jaw musculature. During gnawing, muscles that attach far forward on the jaw and skull are advantageous because they confer great power on the jaw action through increased mechanical advantage. Furthermore, because forward movement of the lower jaw is a prerequisite for gnawing, selection probably favored attachments of the jaw muscles that were far forward on the rostrum and that caused a forward shift of the jaw during their contraction. During the grinding of food by the cheek teeth, jaw muscles with mechanical advantage for power are also important, but the complex jaw action associated with the grinding movements demands jaw musculature with precise control over anterior, posterior, and transverse jaw movements, as well as the simpler opening and closing action of the jaws. The specializations of the jaw musculature and the skulls of rodents to be considered can best be put in functional perspective if the importance of complex rotary grinding movements and powerful forward movements of the jaw are kept in mind.

In the rodent literature, the terms sciuromorph, hystricomorph, and myomorph have been widely used to refer to different patterns of specialization of the jaw musculature, skull, and lower jaw. Not every rodent can readily be fitted into one of these groups, however, nor do experts agree on how the types evolved. The myomorph pattern, for example, could have evolved from either sciuromorph or hystricomorph types, but the weight of evidence favors a hystricomorph origin (Klingener, 1964:76, 77). Wood (1965) dealt with these types of specialization in his consideration of grades, or morphological stages of specialization, and stressed that for the most part each grade has been reached from a preceding grade by separate lineages of rodents evolving along parallel morphological lines. Thus, although a grade contains animals that have attained a similar level of morphological development, it does not necessarily include animals that are closely related. An understanding of the terms listed above and an appreciation for some of the complexities of rodent evolution can be gained by considering the work of Wood.

Primitive rodents from the Paleocene and Eocene are included by Wood in grade one. In these rodents, the incisors were evergrowing, the cheek teeth were reduced to 2/1 premolars and 3/3 molars, and the angular process of the lower jaw was in roughly the same vertical plane as that of the rest of the jaw. The temporalis muscle was large, had an extensive area of origin, and inserted on the high coronoid process of the dentary (Fig. 10–3A). All of the divisions of the masseter muscle originated on the zygomatic arch. The medial division seemingly passed almost vertically downward from the zygomatic arch to the dentary and pulled the jaw straight upward. The cheek teeth of the early members of grade one were low crowned and had a crown pattern based on cusps of no more than four crests; some later members, such as the living *Aplodontia rufa*, had high-crowned or ever-growing cheek teeth. Relationships between members of this grade seem quite close.

Presumably as a result of competition between the rapidly diversifying rodents, as early as Late Eocene some phylogenetic lines further specialized the skull and jaw musculature in a way that increased the effectiveness of gnawing and grinding. These specializations involved primarily the masseter lateralis and the masseter medialis and their areas of attachment. In some rodents, the insertion of the anterior part of the masseter lateralis shifted onto the anterior surface of the zygomatic arch and the adjacent side of the rostrum. This pattern, which can be termed *sciuromorphous*, occurs in the living families Sciuridae, Castoridae, Heteromyidae, and Geomyidae. In these animals, the temporalis muscle is highly reduced in some and large in others, and the angular process of the dentary is in roughly the same plane as that of the rest of the den-

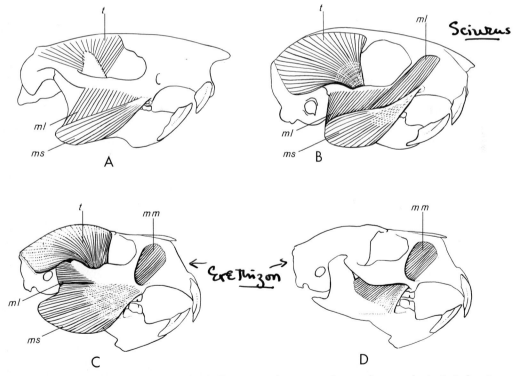

Figure 10–3. Patterns of specialization of the skulls, jaws, and jaw musculature of some rodents. A, *Ischryotomus* (Paramyidae), a primitive Eocene rodent; the jaw muscles are restored. (After Wood, 1965.) Note that the masseter muscles originate entirely on the zygomatic arch. B, Abert's squirrel (*Sciurus aberti*, Sciuridae); the anterior part of the masseter lateralis originates on the rostrum and the zygomatic plate. C and D, Porcupine (*Erethizon dorsatum*, Erethizontidae); the anterior part of the masseter medialis originates largely on the rostrum and passes through the greatly enlarged infraorbital foramen. (The temporalis muscle, typically reduced in size in hystricomorphous rodents, is unusually large in porcupines.) Abbreviations: *ml*, masseter lateralis; *mm*, masseter medialis; *ms*, masseter superficialis; *t*, temporalis.

tary. Associated in the Castoridae with specializations of the powerful anterior part of the masseter lateralis is a conspicuous flattening of the anterior part of the zygomatic arch and a depression in the side of the rostrum just anterior to the arch (Fig. 10–28B). In other rodents of grade two the anterior part of the large masseter medialis no longer takes origin from the zygomatic arch but passes through the enlarged infraorbital foramen to an often extensive area of origin on the side of the rostrum (Fig. 10–3C, D). Rodents that have this specialization of the masseter medialis can be termed *hystricomorphous*, and include the suborder Caviomorpha and the families Dipodidae, Zapodidae, Theridomyidae, Petromyidae, Thryonomyidae, Anomaluridae, Ctenodactylidae, Hystricidae, and Pedetidae. Among these families, the temporalis is often strongly reduced, and in some the angular process flares sharply outward. In both hystricomorphous and sciuromorphous rodents, the specialized divisions of the masseter are angled forward and confer a strong anterior pull to the lower jaw.

Members of the African fossorial family Bathyergidae have enormous dentaries and large masseter muscles, but lack either of the types of specialization described above. Lavocat (1973) believes that bathyergids have secondarily reduced the infraorbital foramen and the muscle that passed through it. Perhaps, then, the specialized bathyergids passed through grade two some time in the past.

Some rodents, also assigned to stage two by Wood, have utilized both the sciuromorphous and hystricomorphous types of masseteric specializations (Fig. 10–4). These rodents can be termed *myomorphous*, and include the superfamilies Muroidea, Spalacoidea and Gliroidea. The temporalis in these rodents is typically reduced, and the angular process does not flare outward. The myomorphous masseter pattern allows for a very strong anterior pull on the lower jaw, and the tremendous success of muroid rodents indicates that it is a highly adaptive arrangement.

The grade two level of masseteric specialization was clearly reached by a number of independent evolutionary lines that often differ strikingly from one another in such features as the structure of the lower jaw or the crown pattern of the cheek teeth.

Some rodents that Wood includes in grades one and two achieved a further level of specialization, grade three. Rodents of this grade evolved cheek teeth that are either extremely high crowned or are ever-growing. Such teeth occur in a variety of rodents of diverse ancestry; many of these rodents are entirely vegetarians or fossorial, or both. At least some members of the following families have reached this grade: Aplodontidae, Geomyidae, Hetero-

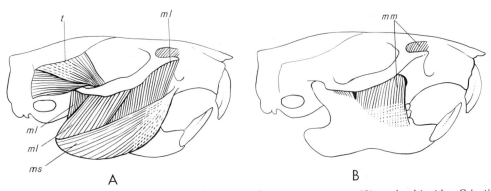

Figure 10–4. Zygomasseteric pattern in myomorphous rodents. A, A cotton rat (*Sigmodon hispidus*, Cricetidae); the masseter superficialis originates on the rostrum, and the anterior part of the masseter lateralis originates on the anterior extension of the zygomatic arch. B, The superficial muscles have been removed; the masseter medialis originates partly on the rostrum and passes through the slightly enlarged infraorbital foramen. (After Rinker, 1954.) Abbreviations are the same as in Figure 10–3.

myidae, Castoridae, Thryonomyidae, Bathyergidae, Ctenodactylidae, Pedetidae, Spalacidae, Rhizomyidae, Cricetidae (Microtinae), and most members of the suborder Caviomorpha (with the exception of the Erethizontidae). In fossorial rodents, or in rodents that eat such abrasive plants as grasses, ever-growing cheek teeth seem to be highly advantageous.

SUBORDER PROTROGOMORPHA

Rodents assigned by Wood (1965) to this suborder include the basal family Paramyidae, five other extinct families, and the surviving but relict family Aplodontidae. All protrogomorphs share the primitive zygomasseteric arrangement of the paramyids (Fig. 10–3A), a pattern common to the animals that evolved during the earliest known rodent radiation in the Eocene. Derived from aplodontid ancestry, the highly specialized family Mylagaulidae appeared in Early Miocene. These extinct woodchuck-sized rodents were fossorial, and are renowned for their possession of prominent horns.

FAMILY APLODONTIDAE. This family is of interest primarily because of the unique primitive morphological features that characterize its one living member. *Aplodontia rufa*, the mountain "beaver," is restricted to parts of the Pacific Northwest. This animal is roughly the size of a small rabbit, and has a robust, short-legged form. *Aplodontia* is generally regarded as the most primitive living rodent. Its zygomasseteric arrangement is close to that of the ancestral family Paramyidae, with the masseters having an entirely zygomatic origin. The skull is flat and the coronoid process of the dentary is large (Fig. 10–2B). The cheek teeth are ever-growing (a specialized feature) and have a unique crown pattern (Fig. 10–5A). The dental formula is 1/1, 0/0, 2/1, 3/3 = 22.

The earliest records of aplodontids are from the Oligocene of western North America. The aplodontids spread later to Europe and Asia, but since the middle Pliocene have lived only in the moist, forested parts of the Pacific slope of North America, where they occur today from central California to southern British Columbia. Widespread late Tertiary aridity in North America may have restricted the aplodontids to their present range.

Aplodontia occurs in small colonies, favors moist areas supporting lush growths of forbs, and often builds its burrows next to streams. The diet includes a variety of forbs and the buds, twigs, and bark of such riparian plants as willow *(Salix)* and dogwood *(Cornus)*. On occasion *Aplodontia* builds "hay piles" of cut sections of forbs (Grinnell and Storer, 1924:157). Although usually terrestrial, this animal is known to climb to some extent in search of food.

SUBORDER CAVIOMORPHA

Caviomorph rodents have hystricomorphous skulls; typically, the origin of the anterior division of the masseter medialis occupies much of the side of the rostrum (Fig. 10–3C, D). The recognition of relationships between the members of this assemblage of rodents rests partly on the structure of the lower jaw, which is *hystrognathous*: the angular process flares

Figure 10–5. Crowns of first two right upper molars of two rodents. A, Mountain beaver (*Aplodontia rufa*, Aplodontidae); note the simplified and unique crown pattern. B, Merriam's kangaroo rat (*Dipodomys merriami*, Heteromyidae); note the highly simplified crown pattern. The outer border of the tooth is above; anterior is to the right. The unshaded part is enamel and the stippled part is dentine.

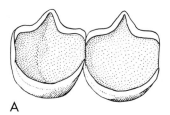

A   B

conspicuously outward and is not in the same plane as that of the alveolus of the lower incisors. A division of the specialized masseter lateralis originates from this part of the dentary. The incisors have "multiserial" enamel; this complex type of enamel structure is considered by Wilson (1972) to be important in improving the strength of the incisors, and in his opinion its development set the stage for the forward movement of the masseter medialis and more powerful jaw action. ("Uniserial" enamel, another complicated type, occurs in many other rodents.) In most caviomorphs, the malleus and incus of the middle ear are fused, but in a few species this is not always the case. The pattern of branching of the carotid arteries in most caviomorphs is nearly identical. Members of the families Hystricidae and Bathyergidae share with the caviomorphs some of the characters mentioned above, and Lavocat (1974) and Wood (1974) agree that these groups may share a common ancestry.

When one considers this ancestry, the North American fossil record is of importance. *Franimys*, the earliest rodent known by more than isolated teeth, appears in the Late Paleocene of North America. The incipient development of a hystricognathous lower jaw in this rodent, a feature typical of Caviomorpha, Hystricidae and Bathyergidae, documents a very early trend toward this distinctive pattern of specialization. Wood (1975) hypothesizes that animals similar to *Franimys* were among the early (perhaps Late Paleocene) rodent immigrants to Eurasia and that, given an adequate fossil record, the hystricids and the bathyergids could be traced back to this early stock, which he names Franimorpha. Wood believes that the caviomorph lineage and those of the hystricids and bathyergids have evolved separately since perhaps the Late Paleocene, and that those morphological features shared by these groups, but not present in the Franimorpha, are the result of independent, but parallel, evolution. The occurrence of a fully hystricognathous rodent in the Eocene of Texas at a time when such

rodents are unknown elsewhere is interpreted by Wood (1972) as evidence that South American caviomorphs were derived from a stock that dispersed from North America, rather than from the Old World.

Except for the New World porcupines (Erethizontidae), all caviomorph rodents almost certainly have a common ancestry within an immigrant stock that entered South America in the early Tertiary. Competition from other orders of mammals was apparently not intense, for the caviomorphs rapidly radiated; among fossils from the Early Oligocene (Deseadan), when rodents first appear in South America, six of the living caviomorph families can be distinguished. The Miocene changes in climate in South America that were accompanied by an expansion of grasslands strongly affected the fortunes of the caviomorphs. The Octodontidae increased its range in the Pliocene, whereas the range of the Echimyidae shrank. The Dasyproctidae and Erethizontidae became less common in the Pliocene than they had formerly been, and the Chinchillidae became less diverse. On the other hand, the Pliocene was a time when some caviomorphs developed spectacular size. *Telicomys*, of the family Dinomyidae, was nearly the size of a rhino, and the hydrochoerid *Protohydrochoerus* must have weighed over 200 kg.

From South America, the hutias (Capromyidae) reached the Lesser Antilles in the mid-Tertiary, probably by rafting, but have had varying degrees of success there (see page 174). By contrast, beginning in the Late Pliocene, the New World porcupines have been extremely successful in moving northward from their ancestral home in South America, and one species is now widely distributed in North American coniferous forests.

FAMILY OCTODONTIDAE. Octodontids are roughly rat-sized rodents that inhabit sparsely vegetated terrain, rocky hills and, locally, cultivated land, in roughly the southern half of South America. Among New World caviomorphs, only in this family are fossorial adaptations strongly developed. Octodontids occur from southern Peru and Mato Grosso, Brazil, southward, and inhabit mountains to elevations of at least 5000 m. The family includes six Recent genera with a total of approximately 33 species, and is recorded from the Oligocene to the Recent in South America.

These rodents are unique among hystricomorphs in having simplified cheek teeth in which re-entrant folds form occlusal surfaces that are roughly the shape of a figure eight (hence the name octodont). In *Ctenomys* the re-entrant folds are shallow and the figure eight is obliterated. The auditory bullae are enlarged in octodontids, as in many mammals that occupy arid or semi-arid regions. The digits bear sharp, curved claws. In some genera, the claws and feet are greatly enlarged and the limbs are powerfully built.

The species of the fossorial octodontid *Ctenomys* (commonly called the tuco-tuco) are remarkable for their resemblance to the pocket gophers (Geomyidae). Tuco-tucos are ecological and, in many ways, structural counterparts of the pocket gophers. Both groups of rodents are highly fossorial. In both tuco-tucos and pocket gophers, the head is large and broad and the stout incisors protrude permanently from the lips; the eyes and ears are small; the neck is short and powerfully built; the forelimbs are powerful; the manus has long claws; and the tail is short and stout. In contrast to pocket gophers, tuco-tucos have greatly enlarged hind feet with powerful claws, and they lack external cheek pouches. Fringes of hair on the toes of fore and hind feet in tuco-tucos are presumably an aid to the animals when they are moving soil.

Octodontids are herbivorous, and some species eat such underground parts of plants as roots, tubers, and rhizomes. Three of the six genera of octodontids are not fossorially adapted, take shelter among rocks, in brush or in burrows, and forage largely on the surface of the ground; but the other three genera are more or less fossorial. All members of the genus *Ctenomys* dig extensive burrow systems in open, often barren areas, and live in colonies composed of many solitary individuals, each with its burrow systems spaced widely apart from those of its neighbors. An animal typically occupies a given burrow system permanently, but periodically seeks adjacent foraging areas by digging new burrows. Leaves, stems, and roots are eaten, and short forays are made from open burrows to gather food (Pearson, 1959).

A different mode of life is typical of *Spalacopus*. In Chile, *Spalacopus cyanus* occupies sandy coastal areas where it occurs in colonies, all members of which occupy a common burrow system. The animals feed entirely below ground, and the tubers and underground stems of huilli, a species of lily (*Leucoryne ixiodes*), form the bulk of the diet. *Spalacopus cyanus* is nomadic, an exceptional mode of life for a rodent. When a colony exhausts the supply of huilli roots at one place, the animals abandon this foraging site and move to a nearby undisturbed area (Reig, 1970). Both *Ctenomys* and *Spalacopus* are unusually vocal for rodents, and give distinctive calls at their burrow openings. Both use their forelimbs and teeth to loosen soil, and use their large hind feet to throw dirt from the mouth of the burrow.

FAMILY ECHIMYIDAE. Members of this important Neotropical family, which includes a variety of roughly rat-sized rodents, are called spiny rats. Most of the 14 Recent genera have flattened, spinelike hairs with sharp points and slender basal portions. Approximately 43 Recent species are recognized. Spiny rats are widely distributed in the Neotropics, occurring from Nicaragua in Central America southward through the northern half of South America to Paraguay and southeastern Brazil.

Echimyids are normally proportioned rodents with prominent eyes and ears. The tail, which in some genera is longer than the head and body, is lost readily, a feature perhaps of value in aiding escape from a predator. The point of weakness is at the centrum of the fifth caudal vertebra. Among 637 *Proechimys* taken in Panama,

18 per cent were tailless (Fleming, 1970:486). The cheek teeth are rooted, and the occlusal surfaces in most species are marked by transverse re-entrant folds. Aside from the reduction of the pollex, the feet are not highly specialized in most genera. In several arboreal genera (members of the subfamily Dactylomyinae), however, the digits are elongate and partially syndactylous. When an animal is climbing, the first two digits grasp one side of a branch in opposition to the remaining digits, which grasp the other side.

Echimyids are an old group, appearing first in the early Oligocene of South America. Two extinct genera of echimyids are known from skeletal material found in Indian kitchen middens in Cuba and Haiti. These genera seemingly became extinct fairly recently. In the case of the genus from Haiti (Brotomys), extinction may have resulted from the introduction of predators by European man.

As far as is known, spiny rats are completely herbivorous. In Panama, fruit was the primary food found in the stomachs of many Proechimys semispinosus (Fleming, 1970:486). Many species apparently seek food by arboreal foraging. Spiny rats are common in many tropical habitats. They frequently occur in heavily vegetated areas near water, and show no tolerance of dry conditions.

FAMILY ABROCOMIDAE. Members of this family, the chinchilla rats, occur in bleak, frequently cold, and mountainous parts of west central South America. Their range includes southern Peru, Bolivia, and northwestern Argentina and Chile. The family is represented today by one genus with two species. This family appears first in the South American Miocene. There is no general agreement about the systematic position of these rodents, and some authors regard them as a subfamily of the Octodontidae.

Abrocoma looks roughly like a large woodrat (Neotoma), and reaches over 400 mm in total length. The pelage is long and dense, somewhat resembling that of the chinchilla. The skull has a long narrow rostrum, and the bullae are enlarged. The cheek teeth are ever-growing; the upper teeth have an internal and an external enamel fold, while the lowers differ in having two internal folds. The limbs appear short, with short, weak nails. The pollex is absent.

These herbivorous rodents are poorly known. They are seemingly colonial, are scansorial (able to climb), and usually seek shelter beneath or among rocks.

FAMILY CAPROMYIDAE. This family is composed of the hutias and the nutria. These rodents have been placed by some in separate families, but are treated here as members of a single family. Both hutias and the nutria have five toes on forefeet and hind feet, have a compact form and a moderately long tail, and have a dental formula of 1/1, 0/0, 1/1, 3/3 = 20.

The hutias (Capromys, Geocapromys, Plagiodontia) are restricted to the West Indies, where living species occupy the Bahama Islands, Cuba, Isle of Pines, Hispaniola, Puerto Rico, and Jamaica. These herbivorous rodents weigh up to 5 kg and look like unusually large rats (Fig. 10–6). They are of little importance today except as interesting and, to some people, alarming examples of a group seemingly on its way toward extinction because of the influence of man. The hutias, adapted to the conditions on the islands of the West Indies before the coming of European man, were unable to cope with predation by the introduced mongoose (Herpestes) or by man and his dogs. Of the 11 known species of capromyids, four are extinct. One only recently became extinct, and some of the surviving species are apparently in danger of extinction. One living species, Capromys nana, was first described from bones found in cave deposits in Cuba, but was subsequently found alive.

The story of the nutria (Myocastor coypu) is quite different. This animal is familiar to many people in North America, Europe, and Asia, because it has been introduced widely and has thrived in certain areas. In some places, it has become a serious pest because of its destruction of aquatic vegetation and crops and its

Figure 10–6. Bahamian hutias (*Geocapromys ingrahami*, Capromyidae). (Photograph by Garrett C. Clough.)

disruption of irrigation systems. In some areas, nutrias have caused a deterioration of waterfowl habitat. Costly Federal study and local control of the nutria have become necessary in some parts of the United States. This animal is native to southern South America, from Paraguay and southern Brazil southward. The family is also represented by nine extinct genera ranging from the Miocene to the Recent in South America.

The nutria is large, up to roughly 8 kg, and looks like a rat-tailed beaver (*Castor*). The skull is heavily ridged and has a deep rostrum. In association with the reduction of the temporal muscles, the coronoid process of the dentary has nearly disappeared, and is represented by a small knob (Fig. 10–1). The hypsodont cheek teeth well illustrate changes in crown pattern that occur with increasing age and wear: the enamel folds become islands under advanced wear (Fig. 10–7). The feet have heavy claws, and a web joins all but the fifth toe of the pes.

Nutrias resemble beavers in some of their habits. They dig burrows in banks, use cleared trails through vegetation, are extremely destructive to plants near their dens, and are skillful swimmers and divers. Nutrias have dense, fine underfur, and have been raised in some fur farms in the United States. Regrettably, animals that have escaped from fur farms have given rise to wild nutria populations in many areas. Great numbers of nutrias are trapped for their fur in the southeastern United States each winter, and they are thus of considerable economic importance. Nonetheless, most biologists strongly oppose the indiscriminate introductions of such animals as the nutria. The activities of nonnative species occasionally result in the alteration of the vegetation, with the resultant disappearance of native species and the destruction, perhaps irretrievable, of the original biotic community.

FAMILY CHINCHILLIDAE. One member of this family, the chinchilla (*Chinchilla*), is somewhat familiar because of

Figure 10–7. First and second upper right molars of the nutria (*Myocastor coypu*, Capromyidae); note the tremendous changes in the crown pattern due to wear. A, Lightly worn molars; B, heavily worn molars. Stippled areas on the occlusal surfaces surrounded by enamel (unshaded) are dentine.

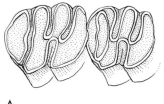 A

 B

the publicity given to chinchilla fur farming. The family also includes the viscachas (*Lagidium* and *Lagostomus*). Three genera with eight Recent species represent the family, which occurs in roughly the southern half of South America in the high country of Peru and Bolivia, and throughout much of Argentina to near its southern tip. The fossil record of this group is entirely South American, and extends from the Oligocene to the Recent.

Chinchillids are densely furred and of moderately large size (1 to 9 kg), with long, well furred tails. Mountain viscachas (*Lagidium*) and chinchillas have fairly large ears and a somewhat rabbit-like appearance, whereas the plains viscacha (*Lagostomus*) has short ears. In all species, the cheek teeth are ever-growing and the occlusal surfaces are formed by transverse enamel laminae with intervening cement. The incisors are narrow. There are some cursorial adaptations, but the clavicle is retained. The forelimbs are fairly short and are tetradactyl; the hind limbs, however, are long, and the elongate feet have four (in *Chinchilla* and *Lagidium*) or three (in *Lagostomus*) toes.

Chinchillids are herbivorous, and occupy a variety of situations, including open plains (pampas), brushlands, and barren, rocky slopes at elevations ranging from approximately 800 to 6000 m. The mountain viscachas and chinchillas are diurnal and seek shelter in burrows or crevices among rocks. Although adept at moving rapidly over rocks and broken terrain, they seem not to depend on speed in the open to escape enemies. The plains viscacha, in contrast, occurs in open pampas areas with little cover, where colonies live in extensive burrow systems marked by low mounds of earth and accumulations of such debris as bones, droppings of livestock, and plant fragments. The habit of collecting items is displayed even by captive animals. Colonies may occupy large areas; one such area measured 20 by 300 m, and this colony was known to have been in existence for at least 70 years (Weir, 1974). Cursorial ability is highly developed: these animals are able to run rapidly with long leaps and to evade a pursuer by abrupt turns. They have considerable endurance, and can run at speeds up to at least 40 km per hour. Plains viscachas occupy a largely rabbitless area, and in locomotor ability and foraging habits strikingly resemble some of the larger rabbits (leporids).

FAMILY DASYPROCTIDAE. Members of this family, the agoutis, occur in the Neotropics from southern Mexico through most of the northern half of South America. Two genera and some nine species are included in this family. The earliest records are from the Early Oligocene of South America.

These rodents are fairly large, up to 2 kg in weight, and are described by some as rodents with rabbit-like heads and piglike bodies (Fig. 10–8). The tail is short. The skull is robust, the incisors are fairly thin, and the crowns of the hypsodont cheek teeth often bear isolated, transversely oriented islands surrounded by enamel. The dental formula is 1/1, 0/0, 1/1, 3/3 = 20. Although agoutis are compactly built, the limbs are slim and have many cursorial adaptations. The forefeet are tetradactyl, and the plane of symmetry passes between digits three and four (as in the Artiodactyla). The hind feet have three toes, and the plane of symmetry passes through digit three (Fig. 10–9C), as in the Peris-

Figure 10–8. An agouti (*Dasyprocta* sp., Dasyproctidae). (Photograph by Lloyd G. Ingles.)

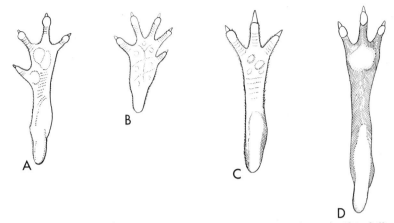

Figure 10–9. Ventral views of the left hind feet of some South American rodents. A, Chinchilla (*Chinchilla* sp., Chinchillidae); B, degu (*Octodon* sp., Octodontidae); C, agouti (*Dasyprocta* sp., Dasyproctidae); D, Patagonian cavy (*Dolichotis* sp., Caviidae), (After Howell, 1944.)

sodactyla. The clavicle is either reduced or absent, and the claws are thick, blunt, and, in some species, nearly hooflike.

Agoutis are herbivorous and typically inhabit tropical forests, where they are largely diurnal and take refuge in burrows that they dig in the banks of arroyos, beneath roots, or among boulders. They are rapid and agile runners, and usually travel along well-worn trails that lead from their burrows. They have a rabbit-like habit of remaining still when approached by a predator and then bursting from cover and running away. Unlike the usually silent rabbits, however, agoutis make high screams as they run. The elongate rump hairs are erected when the animal becomes excited and are frequently erect when the animals run from predators or from other agoutis.

FAMILY CUNICULIDAE. This family includes the pacas, of which there are two species of a single genus (*Cuniculus*). Pacas occur in tropical forests from central Mexico to southern Brazil. Their fossil record is from the Recent only, but pacas presumably arose from agouti ancestry some time in the Tertiary.

Pacas are large, weighing up to about 10 kg, are nearly tailless, and have a conspicuous pattern of white spots and stripes on the body. They have an exceptionally ungraceful form, with short legs and a blunt head (Fig. 10–10). There are four digits on the forefeet and five digits on the hind feet. The cheek teeth are high crowned and the dental formula is 1/1, 0/0, 1/1, 3/3 = 20. Resonating chambers are formed by concavities in the maxillaries and by greatly broadened zygomatic arches (Fig. 10–11A); air is forced through associated pouches, producing a resonant, rumbling sound. The massively enlarged zygoma of the pacas are unique to these animals.

Figure 10–10. A paca (*Cuniculus paca*, Dasyproctidae). (Photograph by Lloyd G. Ingles.)

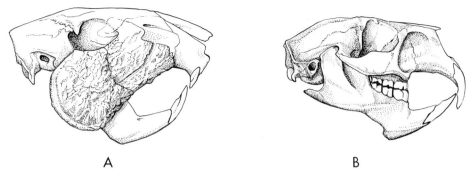

Figure 10–11. Skulls of two hystricomorph rodents. A, Paca (*Cuniculus paca*, Dasyproctidae); length of skull 150 mm. (After Hall, E. R., and Kelson, K. R.: *Mammals of North America*, © 1959 by The Ronald Press Company, New York). Note the great enlargement of the zygomatic arch. B, Porcupine (*Erethizon dorsatum*, Erethizontidae); length of skull 115 mm.

Pacas resemble agoutis in living in burrows and in being forest dwellers. The diet consists of a variety of plant material, including fallen fruit. Unlike agoutis, pacas are nocturnal and are not particularly swift on land; they are good swimmers, however, and often escape enemies in the water. Mortality of young must be low, for pacas have but two pairs of mammae and only rarely bear more than one young.

FAMILY HEPTAXODONTIDAE. Several kinds of West Indian mammals became extinct in Recent or sub-Recent times and are known from fragmentary skeletal material from caves or kitchen middens. Two genera of heptaxodontids, each with a single species, are among this group of recently extinct forms. Heptaxodontids are recorded from Puerto Rico and Hispaniola. These were large rodents; the length of the skull of the largest genus (*Elasmodontomys*) indicates an animal roughly the size of a beaver. The skull is robust, with strongly developed ridges for muscle attachments, and the cheek teeth have four to seven laminae oblique to the long axis of the anteriorly converging tooth row. These rodents were probably terrestrial and herbivorous, and were eaten by man. Little else is known of their biology.

FAMILY DINOMYIDAE. This family includes a single South American species, *Dinomys branickii*, the pacarana. This seemingly rare animal inhabits the foothills of the Andes and adjacent remote valleys in Peru, Colombia, Ecuador, and Bolivia.

The pacarana is a large rodent, up to 15 kg in weight, with a fairly long tail; the dark brown pelage is marked by longitudinal white stripes and spots. Pacaranas lack the cursorial adaptations of the Caviidae and Hydrochoeridae. Instead, the broad tetradactyl feet of pacaranas have long stout claws seemingly adapted to digging, and the foot posture is plantigrade. The clavicle is complete, another departure from the conventional cursorial plan. The unusually hypsodont cheek teeth consist of a series of transverse plates.

These unusual rodents feed on a variety of plant material, and are slow moving and docile in captivity. Although they appear to be near extinction today, they may have been more successful and diverse in the past. Fields (1957:359, 360) assigned eight fossil genera to the Dinomyidae, the oldest of which appears in Miocene beds in South America.

FAMILY CAVIIDAE. This fairly small family, including just 5 genera and 12 species, contains the familiar guinea pig (*Cavia*), as well as several similar types, and Patagonian "hare" (*Dolichotis*), an animal remarkable in having many cursorial adaptations similar to those of rabbits. Caviids occur nearly throughout South America, except in Chile and parts of eastern Brazil.

The guinea pig-like caviids (Caviinae) are chunky and moderately short limbed, and weigh from roughly 400 to 700 g. *Dolichotis* (Dolichotinae), in contrast, has a rabbit-like form, with long slender legs and feet, and weighs up to approximately 16 kg. All caviids have ever-growing cheek teeth with occlusal patterns consisting basically of two prisms. The skull has a deep rostrum, and the dentary has a conspicuous lateral groove into which insert the temporal muscle and the anterior part of the masseter medialis. Although only *Dolichotis* is strongly cursorial, all caviids have certain features typical of cursorial mammals: the clavicle is vestigial, the tibia and fibula are partly fused, and the digits are reduced to four on the manus and three on the pes. Members of the subfamily Caviinae, despite these cursorial adaptations, have a plantigrade foot posture and scuttle about in mouselike fashion. Locomotion in *Dolichotis*, however, is of the bounding style typical of jackrabbits. The foot posture of *Dolichotis* during running is digitigrade, and specialized pads beneath the digits (Fig. 10–9D) cushion the impact when the feet strike the ground. In *Dolichotis*, the resemblance to rabbits is furthered by a deep, somewhat laterally compressed skull and large ears.

Caviids are herbivorous and social, and occupy habitats ranging from grassland and open pampas to brushy and rocky areas and forest edges. Most are nocturnal or crepuscular, and often live in large colonies, the locations of which are marked by conspicuous series of burrows. *Dolichotis*, an inhabitant of open, arid regions, is diurnal, and large groups have been observed together on occasion. In some ways, *Dolichotis* is a remarkably close ecological counterpart of the cursorial jackrabbit (*Lepus*). *Dolichotis* occurs in an area where *Lepus* is absent.

FAMILY HYDROCHOERIDAE. This family contains the largest living rodent, the capybara (*Hydrochoerus*). The two species of capybara occupy Panama and roughly the northern half of South America east of the Andes. Capybaras are known from the Pleistocene of North America.

Capybaras are large (up to 50 kg in weight), robust, rather short-limbed rodents with coarse pelage. The head is large and is made especially unhandsome by a deep rostrum and truncate snout. The skull and dentary are similar to those of members of the Caviidae, but the paroccipital processes are unusually long. The teeth are ever-growing. Both upper and lower third molars are much larger than any other cheek tooth in their respective tooth rows, and these highly specialized teeth are formed by transverse lamellae united by cement (Fig. 10–12B). The tail is vestigial, and the same cursorial features listed for the caviids occur. The digits are partly webbed and are unusually strongly built, an adaptation that probably allows the support of their considerable weight.

Capybaras occur along the borders of marshes or the banks of streams, and forage on succulent herbage. They are largely crepuscular, and although they can run fairly rapidly, usually seek shelter in the water. Capybaras swim and dive well, and in an extremity remain submerged beneath water plants with only the nostrils above water.

Hydrochoerids first appear in Pliocene deposits of South America. Because caviids and hydrochoerids share a series of morphological characters, such as similar cursorial modifications of the limbs, some authors have regarded these groups as having a common ancestry and have placed the capybaras as a subfamily of the Cavii-

Figure 10–12. Crowns of hystricomorph molars. A, Upper right molars one and two of the porcupine (*Erethizon dorsatum*, Erethizontidae); B, third lower left molar of the capybara (*Hydrochoerus hydrochoeris*, Hydrochoeridae). The cross-hatched areas on the porcupine teeth are dentine; the stippled areas on the capybara tooth are cement. (B after Ellerman, 1940.)

A   B

dae (see Ellerman, 1940:237–249; and Landry, 1957:57).

FAMILY ERETHIZONTIDAE. To this small family belong the New World porcupines, a group including four Recent genera with eight living species. These animals are widely distributed in forested areas, and occur from the Arctic Ocean south through much of the forested part of the United States into Sonora, Mexico in the case of *Erethizon*, and from southern Mexico through much of the northern half of South America in the case of the other genera. Porcupines are of interest to most people because of their remarkable coat of quills and because the animals often have little fear of man and can be observed easily.

New World porcupines are large, heavily built rodents, weighing up to 16 kg; all species have quills on at least part of the body. The stiff quills are usually conspicuously marked by dark- and light-colored bands, and the sharp tips have small, proximally directed barbs. These barbs make the quills difficult to remove from flesh and aid in their penetration, which may be at the rate of 1 mm or more per hour. The skull is robust, the rostrum is deep, and the greatly enlarged infraorbital foramen is nearly circular in some species (Fig. 10–11B) and accommodates the highly developed masseter medialis. The dental formula is 1/1, 0/0, 1/1, 3/3 = 20; the rooted cheek teeth have occlusal patterns dominated by re-entrant enamel folds (Fig. 10–12A). New World porcupines have some arboreal adaptations that are lacking in their more terrestrial Old World counterparts. The feet of erethizontids have broad soles that are marked by a pattern of tubercles that increases traction (Fig. 10–13B); in some species, the hallux is replaced by a large, movable pad. The toes bear long curved claws, and the limbs are functionally four-toed. In *Coendou*, a Neotropical genus, the long tail is prehensile, and curls dorsally to grasp a branch.

Erethizontids appeared in the Oligocene of South America, but are not known from deposits earlier than the Pliocene in North America. As suggested by this evidence,

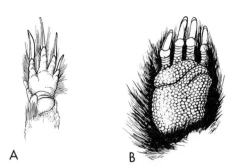

Figure 10–13. Ventral views of the forefeet of two rodents. A, Left manus of the pocket gopher (*Thomomys bottae*, Geomyidae); note the fringes of hairs on the toes. B, Right manus of the porcupine (*Erethizon dorsatum*, Erethizontidae); note the pattern of tubercles on the pads that increase traction.

this family underwent its early evolution in South America and became established in North America only after the emergence of the previously inundated Isthmus of Panama in the Pliocene.

New World porcupines utilize a great variety of plant material. *Erethizon* feeds extensively on the cambium layer of conifers and, in many timberline areas, trees missing large sections of bark and cambium give evidence of long-term occupancy by porcupines. Cambium is a staple winter food in the Rocky Mountains, but in summer a variety of plants is taken. Porcupines have been observed "grazing" at this time on sedges (*Carex*) along the borders of meadows. Most species are able climbers, and *Coendou* spends most of its life in the trees. *Erethizon* is inoffensive and at times almost oblivious of humans, but when in danger the animal directs its long dorsal hairs forward, exposing the quills; it erects the quills and humps its back. The tail is flailed against an attacker as a last resort. Surprisingly, *Erethizon* is killed by a variety of carnivores. Some mountain lions learn to flip porcupines on their backs and kill them by attacking the unprotected belly. Occasionally, however, dead or dying carnivores are found with masses of quills penetrating the mouth and face, indicating that learning to prey on porcupines may be a dangerous undertaking. Erethizontids characteristically take shelter in rock piles, beneath overhanging rocks, or in hollow logs, but do not dig burrows as do Old World porcupines.

SUBORDER MYOMORPHA

This suborder includes over 70 per cent of the Recent species of rodents, and is nearly world-wide in distribution. Within the suborder are rodents adapted to environments ranging from high arctic tundra to desert sand dunes. These are the most important rodents almost throughout the world, and occupy fossorial, terrestrial, semiaquatic, and arboreal niches. Some species within several families (Cricetidae, Muridae, Heteromyidae, Dipodidae) have evolved remarkable behavioral and physiological specializations that allow them to live in arid regions on dry diets.

Undoubted myomorphs first appear in the fossil record in the Early Oligocene of Europe and North America.

FAMILY CRICETIDAE. This family contains a great variety of mouselike creatures, such as deer mice, woodrats, voles, lemmings, muskrats, and gerbils. A total of roughly 567 Recent species of some 97 genera, approximately one third of the kinds of living rodents, belong to this family. Cricetids are nearly ubiquitous, being absent only from some islands, Antarctica, and the Australian and Malayan areas. Habitats ranging from arctic tundras to tropical rain forests are occupied; in many terrestrial communities, cricetid rodents are the most important small mammals in terms of both their effect on the environment and their importance as a staple food item for many predators.

Most cricetids retain a "standard" mouselike form, with a long tail and a generalized limb structure. Cricetids vary in size from roughly 10 g in weight and 100 mm in total length, as in the pygmy mouse (*Baiomys*), to approximately 1500 g and 600 mm, as in the muskrat (*Ondatra*). The skull is quite variable in shape, but always has a somewhat enlarged infraorbital foramen that transmits part of the masseter medialis and a branch of the trigeminal nerve (Fig. 10–14). The masseter lateralis originates partly from the enlarged zygomatic plate. The dental formula is 1/1, 0/0, 0/0, 3/3 = 16, and the molars vary from low-crowned and rooted to high-crowned and ever-growing. The occlusal surface of the cheek teeth is variable, but is based on a pattern of five crests formed by re-entrant enamel folds (Fig. 10–15A–C).

As might be expected from the large number of species within the Cricetidae, a variety of modes of life and structural patterns is represented. By considering three of the subfamilies of cricetids, some of this diversity will be illustrated.

Most species of the subfamily Cricetinae are adapted to a generalized terrestrial or scansorial (climbing) mode of life, but arboreal, semiaquatic, and fossorial habits also occur. Foods include a wide variety of plant and animal material. Three semiaquatic Neotropical genera (*Ichthyomys*, *Daptomys*, and *Anotomys*) are fish eaters, and the Asian mole rats (*Myospalax*) are fossorial and feed partly on below-ground parts of plants. The skulls of cricetines are, in general, not highly specialized and their limbs are not strongly modified, except for such features as fringes of hair on the hind feet or webs between the toes in the case of aquatic forms, and strong and clawed forefeet in the case of the mole rat.

The subfamily Microtinae includes the

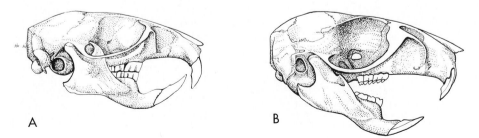

A   B

Figure 10–14. Rodent skulls. A, Stephen's woodrat (*Neotoma stephensi*, Cricetidae); length of skull 42 mm. B, Gerbil (*Tatera humpatensis*, Cricetidae); length of skull 38 mm. (B after Hill and Carter, 1941.)

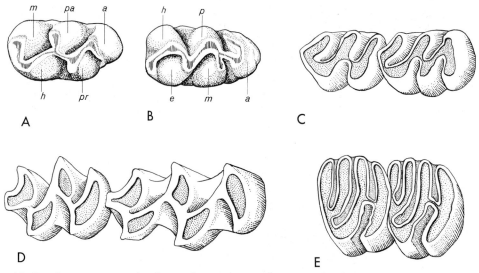

Figure 10–15. Crown patterns of rodent molars. Crowns of the first right upper molar (A) and first left lower molar (B) of a harvest mouse (*Reithrodontomys megalotis*, Cricetidae), showing the cusps (after Hooper, 1952). Abbreviations for A: *a*, anterocone; *h*, hypocone; *m*, metacone; *pa*, paracone; *pr*, protocone. Abbreviations for B: *a*, anteroconid; *e*, entoconid; *h*, hypoconid; *m*, metaconid; *p*, protoconid. First two right upper molars of (C) a cotton rat (*Sigmodon alleni*, Cricetidae); (D) a Mexican vole (*Microtus mexicanus*, Cricetidae); and (E) a beaver (*Castor canadensis*, Castoridae). Unshaded areas on the occlusal surfaces are enamel; stippled areas within enamel folds are dentine.

voles, lemmings, (Fig. 10–16), and musk-rats, a group of rodents that is distributed throughout the Northern Hemisphere. These rodents frequently have short tails, ear openings that are partially guarded by fur, and a chunky, short-legged appear-ance. The cheek teeth often feature complex crown patterns (Fig. 10–15D) adapted to masticating forbs and grasses. The voles and lemmings undergo remark-able population fluctuations in some areas. (Population cycles of rodents are discussed

Figure 10–16. Norwegian lemmings (*Lemmus lemmus*, Cricetidae). The animal on the right is in a threatening posture; the one on the left is in a submissive posture. (Photograph by Garrett C. Clough.)

in Chapter 15.) Because of the reliance on microtines for food by many northern carnivores, the densities and even the distribution of these predators are partially controlled by cycles of microtine abundance (see Table 15–45).

The subfamily Gerbillinae includes the gerbils, a group of rodents that resembles jerboas (Dipodidae) and kangaroo rats (Heteromyidae) in being semifossorial and more or less saltatorial, and in inhabiting mainly desert regions. Gerbils now occur in arid parts of Asia, in the Near East, and in Africa. The hind limbs are large, the central three digits are larger than the lateral ones, and the tail is often long and functions as a balancing organ. The skull does not depart strongly from the general cricetid plan (Fig. 10–14B). In their ability to hop and in their choice of habitats, gerbils resemble heteromyid rodents. Gerbils maintain water balance in the face of hot and arid conditions partly by eating food with a high water content and by concentrating urine, as do heteromyids.

FAMILY MURIDAE.   This is the second largest family of rodents. In this group, including the Old World rats and mice, are roughly 98 genera and 457 species. Some murids live in close association with man in situations ranging from isolated farms to the world's largest cities. As a result of introductions by man, these animals have become nearly cosmopolitan in distribution, and are probably the rodents most familiar to man. Murids that are not commensal with man occur in much of Southeast Asia, Eurasia, Australia, Tasmania and Micronesia, and Africa. Tropical and subtropical areas are centers of murid abundance, but these animals have occupied a wide variety of habitats and some genera are highly adapted to specialized modes of life.

Murids range in size from that of a small mouse to that of a large rat. Some Philippine climbing rats (*Phloeomys*) are roughly 800 mm in length and weigh over 1 kg. Although the tail is usually more or less naked and scaly, it is occasionally heavily furred and bushy. The molars are rooted or ever-growing, and usually have crowns with cusps or laminae (Fig. 10–17); great simplification of the crown pattern occasionally occurs. The dental formula is usually 1/1, 0/0, 0/0, 3/3 = 16. In some murids, the reduction of the cheek teeth has become extreme. The greatest reduction occurs in *Mayermys*, a rare mouse from New Guinea, in which only one molar is retained on each side of each jaw. The zygomasseteric structure is of a myomorph type: the infraorbital foramen is not greatly enlarged, but through it passes some of the anterior part of the medial masseter. The origin of the lateral masseter is partly from the zygoma, and the superficial masseter takes origin from the side of the rostrum. The feet retain all of the digits, but the pollex is rudimentary.

Murids appear fairly late in the fossil record (Late Miocene), but the family has been remarkably plastic from an evolutionary point of view. Among living murids, amphibious, terrestrial, semifossorial, arboreal, and saltatorial types are known. The water rats (*Crossomys*) have greatly reduced ears; large, webbed hind feet; and nearly waterproof fur. These animals live along waterways in New Guinea. At the other extreme is the hopping mouse

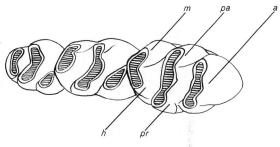

Figure 10–17. Diagram of the occlusal surfaces of the right upper molar of a murid rodent (*Rattus*). With wear, the cross lophs become lakes of dentine (crosshatched areas) rimmed with enamel. Abbreviations are the same as those in Figure 10–15A.

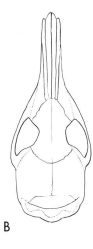

A   B

Figure 10–18. Extremes in skull shape in rodents of the family Muridae. A, Rock mouse (*Delanymys brooksi*), a rock-dwelling omnivore; B, shrew-like rat (*Rhynchomys soricoides*), a *rare species* that apparently feeds on invertebrates. (After Walker, 1968.)

(*Notomys alexis*), a saltatorial inhabitant of extremely arid Australian deserts. This rodent needs no drinking water, and has the greatest ability of any animal in which water metabolism has been studied to concentrate urine as a means of conserving water (MacMillen and Lee, 1969).

Murids feed on a variety of plant material and on invertebrate and vertebrate animals. In association with the great diversity of feeding habits within the murids, the skull form varies widely within the family (Fig. 10–18), with a shrewlike elongation of the rostrum occurring in some genera.

Extremely high population densities have been recorded for feral populations of some murids often commensal with man. A 35-acre area near Berkeley, California, which had only an occasional house mouse for a number of years after population studies began in 1948, supported 7,000 *Mus* in June of 1961 (Pearson, 1964). Among other factors, the high reproductive rate of *Mus* contributes to its ability to reach high densities quickly. *Mus* and *Rattus* (Fig. 10–19) are not restricted to living with or near man, but live away from man over broad areas in such regions as Australia and South America.

Murids that live with man are of great economic importance. Not only do they spread such serious diseases as bubonic plague and typhus, but also the damage they do to stored grains and other foods is tremendous. In some countries *Rattus* and *Mus* compete effectively and devastatingly with man for food.

Figure 10–19. The Australian bush rat (*Rattus fuscipes*, Muridae). (Photograph by Jeffrey Hudson.)

FAMILY GEOMYIDAE. Members of this family, the pocket gophers, are the most highly fossorial North American rodents. Pocket gophers are distributed from Saskatchewan in Canada to northern Colombia in South America. The family includes roughly 40 Recent species in 8 genera. Pocket gophers appeared first in the early Miocene of North America. Although they are not restricted to semiarid habitats today, many of their most characteristic specializations probably evolved in response to the soil conditions and floral assemblages of the semiarid and plains environments that developed in the Miocene.

The most obvious structural characteristics of pocket gophers were developed in response to fossorial life. These animals are moderately small, weighing from roughly 100 to 900 g. They have small pinnae, small eyes, and short tails. The head is large and broad, and the body is stout. External fur-lined cheek pouches are used for carrying food. The dorsal profile of the geomyid skull is usually nearly straight, the zygomas flare widely, and in the larger species the skull is angular and features prominent ridges for muscle attachment. The rostrum is broad and robust, and is marked laterally by depressions from which the masseter lateralis muscles take origin. The large incisors often protrude forward, in some species beyond the anteriormost parts of the nasals and premaxillae; the lips close behind the incisors, which are therefore outside the mouth. The dental formula is 1/1, 0/0, 1/1, 3/3 = 20. The cheek teeth are ever-growing, and have a highly simplified crown pattern. There is no loss of digits. The forelimbs are powerfully built and bear large, curved claws; the toes of the forepaw have fringes of hairs that presumably increase the effectiveness of this foot during digging (Fig. 10–13A).

Pocket gophers occupy friable soils in environments ranging from tropical to boreal. These rodents are entirely herbivorous, and eat a variety of above- and underground parts of forbs, grasses, shrubs, and trees. The extensive burrow system of pocket gophers provides retreats for many vertebrates. In Colorado, burrows occupied by pocket gophers are frequently found also to harbor tiger salamanders, *Ambystoma tigrinum* (Vaughan, 1961). Burrows abandoned by pocket gophers are used by a variety of reptiles and mammals, and occasionally by burrowing owls *(Speotyto cunicularia).* Because pocket gophers keep the entrances to their burrow systems tightly plugged, predators and animals seeking refuge are usually excluded from occupied burrow systems. These burrows provide channels allowing fairly deep penetration of water during periods of snowmelt in mountainous areas, and in some areas they apparently reduce erosion of topsoil. The disturbance of the soil and the mounds of soil thrown up by pocket gophers strongly influence vegetation; often sites of intense pocket gopher activity can be recognized at a distance by the striking dominance of pioneer plants. In some areas of abundant pocket gophers, roughly 20 per cent of the surface of the ground is covered with mounds. Because their digging affects plant composition in rangeland and because their preference for alfalfa and other cultivated plants results in great crop damage, pocket gophers are of considerable economic importance. Large amounts of money have been spent by farmers and by Federal agencies to control pocket gophers on cultivated land in the western United States of America.

FAMILY HETEROMYIDAE. Most members of the family Heteromyidae are adapted to arid or semiarid conditions; kangaroo rats *(Dipodomys;* Fig. 10–20) and pocket mice *(Perognathus)* are, in fact, the most characteristic desert rodents in North America. This family contains 75 Recent species of 5 genera. Heteromyids are restricted to the New World, where they range from southern Canada through the western United States to Ecuador, Colombia, and Venezuela; they occupy tropical, subtropical, arid, and semiarid regions.

Heteromyids first appeared in the Oligocene of North America. The kangaroo rats (subfamily Dipodomyinae) are known from the Pliocene, when the deserts and

Figure 10–20. A chisel-toothed kangaroo rat in a saltbush (*Atriplex confertifolia*). This kangaroo rat is unusual both in its ability to climb and in its use of leaves for food. It uses its chisel-like lower incisors to scrape off the hypersaline peripheral tissues of the leaves of saltbush and eats the starch-rich inner parts (Kenagy, 1972). (Photograph by G. J. Kenagy.)

western North America. Certain diagnostic characteristics of kangaroo rats, such as the greatly enlarged auditory bullae (Fig. 10–21) and features of the hind limbs that favor saltation, probably evolved in association with desert or semidesert conditions.

The bodies of most heteromyids are specialized for jumping. Such adaptations are most strongly developed in the kangaroo rats and kangaroo mice (*Microdipodops*). In these genera, the forelimbs are small, the neck is short, and the tail is long and serves as a balancing organ. The hind limbs are elongate and the thigh musculature is powerful. The hind foot is elongate, but except for the almost complete loss of the first digit in some kangaroo rats (Fig. 10–22E), there is no loss of digits. The cervical vertebrae are largely fused in *Microdipodops*, and in *Dipodomys* they are strongly compressed and partly fused (Fig. 10–22C), producing a short, rigid

semiarid brushlands that heteromyids now frequently inhabit were widespread in

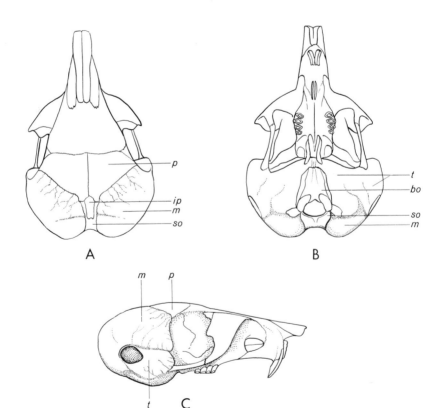

Figure 10–21. Dorsal (A), ventral (B), and lateral (C) views of the skull of Merriam's kangaroo rat (*Dipodomys merriami*, Heteromyidae); note the great enlargement of the auditory bulla, the chamber surrounding the middle ear. Length of skull 45 mm. Abbreviations: *bo*, basioccipital; *ip*, interparietal; *m*, mastoid part of the bulla; *p*, parietal; *so*, supraoccipital; *t*, tympanic part of the bulla (petrosal and tympanic bones). (After Grinnell, 1922.)

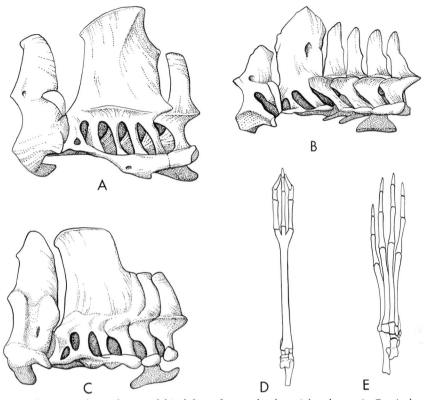

Figure 10–22. The cervical vertebrae and hind feet of several saltatorial rodents. A, Cervical vertebrae of a jerboa (*Jaculus* sp.; Dipodidae); B, of the springhaas (*Pedetes capensis*, Pedetidae); C, of Heermann's kangaroo rat (*Dipodomys heermanni*, Heteromyidae). (After Hatt, 1932.) D, Dorsal view of right hind foot of a jerboa (*Scirtopoda* sp., Dipodidae); note the reduction of digits and the cannon bone formed by metatarsals two, three, and four (after Howell, 1944). E, left hind foot of the desert kangaroo rat (*Dipodomys deserti*, Heteromyidae); note the near loss of the first digit and the elongation of the foot (after Grinnell, 1922).

neck. These species are mostly bipedal when moving rapidly, and when frightened they move by a series of erratic hops.

As in the pocket gophers, fur-lined cheek pouches are present. The skull is delicately built, with thin, semitransparent bones; the zygomatic arch is slender. The auditory bullae are usually large, and in some genera are enormous, formed largely by the mastoid and tympanic bones (Fig. 10–21). The enlargement of the bullae in heteromyids (and in jerboas, Dipodidae) greatly increases auditory sensitivity (see p. 467). The nasals are slender and usually extend well forward of the slender upper incisors. The dental formula is 1/1, 0/0, 1/1, 3/3 = 20. The ever-growing cheek teeth have a strongly simplified crown pattern (Fig. 10–5B) resembling that of pocket gophers, to which heteromyids are seemingly closely related.

Most heteromyids live in areas with strongly seasonal patterns of precipitation, and many must cope with annual cycles of weather and plant growth of the sort that occur in deserts. In such areas, brief periods of precipitation and long periods of drought are typical. Many small annual plants in deserts are able to make the most of irregular moisture by germinating, growing, and flowering rapidly, and by producing abundant seeds that remain dormant until the next rains. This enormously abundant seed crop is the major food source of heteromyids. Many heteromyid features, such as saltatorial ability and seed gathering habits, evolved in response to the demands of desert or semi-desert climates. Perhaps the most remarkable heteromyid adaptation is the ability to survive for long periods on a diet of dry seeds with no free water. This capability

probably does not occur in all heteromyids, nor is it developed to the same
degree in all species adapted to dry climates. In the species of kangaroo rats and
pocket mice (*Perognathus*) that occupy the
desert, however, this ability is well developed (see p. 445).

FAMILY DIPODIDAE. This family includes the jerboas, a group remarkable for
their extreme adaptations for saltation and
for life in arid environments. These specialized rodents occur in arid and semiarid
areas in northern Africa, Arabia, and Asia
Minor, and in southern Russia eastward to
Mongolia and northeastern China. Dipodids are represented today by 10 genera
and 27 species. The family first appeared
in the Pliocene of Asia.

Jerboas have compact bodies, large
heads, reduced forelimbs, and elongate
hind limbs—features associated with saltatorial locomotion. The tail is long and
usually tufted and, as in the New World
kangaroo rats (Heteromyidae), the tuft is
frequently conspicuously black and white.
The posterior part of the skull is broad
(Fig. 10–23A), owing mostly to the enlargement of the auditory bullae, which
are huge in some species. The rostrum is
usually short, the orbits are large, and
through the enlarged infraorbital canal
passes most of the anterior part of the medial masseter, which originates largely on
the side of the rostrum. The zygomatic
plate is narrow and is below the infraorbital canal. The dental formula is 1/1, 0/0, 0-
1/0, 3/3 = 16 or 18; the cheek teeth are

hypsodont, and the crown pattern usually
involves re-entrant enamel folds.

The hind limbs are elongate in all genera, but varying stages of specialization for
saltation are represented. In members of
the subfamily Cardiocraninae, the toes
vary in number from three to five and the
metatarsals are not fused. At the other extreme are such genera as *Dipus* and
Jaculus (subfamily Dipodinae), which
represent the greatest degree of specialization of the hind limbs for saltation that
occurs in rodents. In these genera only
three toes (digits two, three, and four)
remain, and the elongate metatarsals are
fused into a cannon bone (Fig. 10–22D).
An additional specialization that occurs in
some species is a brush of stiff hairs on the
ventral surface of the phalanges. The ears
of jerboas vary from short and rounded to
long and rabbit-like.

Jerboas lead lives that resemble in some
ways those of members of the New World
family Heteromyidae. They live in burrows that are frequently kept plugged during the day, a habit that favors water conservation by keeping the humidity in the
burrow as high as possible. They are nocturnal, and many species sift seeds from
sand or loose soil with the forefeet, although some species depend largely on insects for food. Unlike kangaroo rats, jerboas hibernate during the winter in fairly
deep burrows. Locomotion in jerboas is
chiefly bipedal, but when they are moving
slowly, the forefeet may be used to some
extent. When frightened, jerboas move rap-

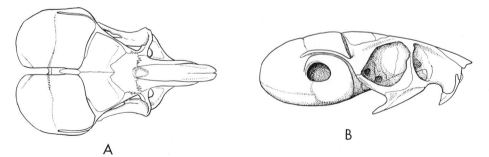

A

B

Figure 10–23.   Dorsal (A) and lateral (B) views of the skull of a jerboa (*Salpingotus kozlovi*, Dipodidae). Length
of skull 27 mm. Note the greatly enlarged auditory bullae and the general resemblance between this skull and that
of the kangaroo rat (Fig. 10–21). (After Allen, 1940.)

idly in a series of long leaps, each of which may cover three meters. Such a rapid and, more important, erratic mode of escape from predation is especially effective in the barren terrain jerboas occupy.

Of the two well known saltatorial rodent families, the Heteromyidae and Dipodidae, the latter is far more highly specialized for saltation. This apparently is not due to the dipodids being the older group and therefore having had more time available for evolutionary change, for heteromyids are known from the Oligocene and the dipodids do not appear until the Pliocene. Instead, perhaps, the different degrees of adaptation are related to the habitats occupied by the two groups. Jerboas have probably been restricted throughout their history, as they are now, to deserts or dry plains, where adaptations leading to greater perfection of saltation have been advantageous. Such common heteromyid genera as *Dipodomys* and *Perognathus*, by contrast, occupy habitats ranging from arid deserts to dense chaparral and in such broadly adapted animals extreme saltatorial modifications would seemingly be disadvantageous.

FAMILY ZAPODIDAE. Members of this small family, which contains only four Recent genera with four species, are called jumping mice (Zapodinae) or birch mice (Sicistinae), and are typically boreal rodents. These mice occur widely in North America as far south as approximately the southern part of the Rocky Mountains. (In the western United States they occur almost exclusively in the mountains.) In the Old World they inhabit Germany, Norway, Russia, Mongolia, and China. Zapodids appeared in Europe in the Oligocene, and possibly as early as Late Eocene in North America.

Zapodids are small, graceful mice, from roughly 10 to 25 g in weight, with long tails and, in all genera but *Sicista*, elongate hind limbs. The coloration in most species is striking: the belly is white and the dorsum is bright yellowish or reddish brown. Much of the anterior part of the medial masseter muscle originates on the side of the rostrum and passes through the enlarged infraorbital foramen (Klingener, 1964:11). The dental formula is 1/1, 0/0, 0-1/0, 3/3 = 16 or 18; the check teeth are brachyodont or "semihypsodont," and have quadritubercular crown patterns with re-entrant enamel folds. The hind limbs in members of the subfamily Zapodinae are elongate and are somewhat adapted for hopping, but unlike more specialized saltatorial rodents, all digits are retained. As an additional contrast with specialized saltators, the cervical vertebrae of zapodids are unfused.

Jumping mice usually inhabit boreal forests. Some species occur typically in coniferous forests, while others appear in birch stands or in mixed deciduous forests. Usually, jumping mice favor moist situations and *Zapus princeps* of the western United States is most abundant in many areas of dense cover adjacent to streams or in wet meadows. These mice hibernate in the winter and emerge during or after snowmelt. Food consists of a variety of seeds and other vegetable material, but insect larvae and other animal material made up approximately half of the food of *Z. hudsonius* in New York (Whitaker, 1963), and roughly one third of the diet of *Z. princeps* in Colorado (Weil, 1968).

The development of saltation in a boreal mammal that inhabits dense vegetation is rather unusual, for this style of locomotion has almost without exception evolved in mammals frequenting open, sparsely vegetated situations offering little concealment, where erratic evasion of predators is advantageous. *Zapus* seems to use saltation for brief intervals; a few jumps are usually sufficient to enable it to reach concealment.

FAMILY SPALACIDAE. This family contains animals usually called mole rats. There are three Recent species of one genus (*Spalax*). These rodents occur in the eastern Mediterranean region and southeastern Europe, and are strongly fossorial. The eyes are small, and the eye muscles and optic nerve are reduced or absent. The ears are small and the tail is vestigial. Unlike many fossorial rodents, *Spalax* digs primarily with its large incisors and moves

Figure 10–24. A root rat (*Tachyoryctes splendens,* Rhizomyidae) from Kenya, East Africa. (Photograph by Jennifer U. M. Jarvis.)

soil with its blunt head. As adaptations to this style of digging, the neck and jaw musculature are powerful, the incisors are robust, and the nose is protected by a broad, horny pad. The feet, surprisingly, are not unusually large; the claws have been described as blunt, round nubbins. These nocturnal rodents burrow in both alluvial and stony soils, and eat both below- and aboveground parts of plants.

FAMILY RHIZOMYIDAE. Members of this group, called root rats, occur in Southeastern Asia and in East Africa. The family includes 3 genera represented by 18 species. These rodents range from a total length of roughly 200 to 500 mm and have short, robust limbs and compact bodies

(Fig. 10–24). In two genera the incisors are procumbent (Fig. 10–25).

Root rats live primarily in areas with at least 500 mm of annual precipitation, and occupy habitats ranging from dense bamboo thickets in Asia to subalpine slopes at 4000 m on Mount Kenya in East Africa. The digging behavior of the African root rat *Tachyoryctes* has been described by Jarvis and Sale (1971). This animal burrows by slicing away the soil with powerful upward sweeps of the protruding lower incisors. The dislodged soil is moved behind the animal by synchronous thrusts with the hind limbs. When the burrow becomes blocked with freshly dug soil, *Tachyoryctes* turns and pushes the load to the surface with the side of its head and one forefoot. The conspicuous mounds, up to 6 m in diameter, that are associated with the activities of *Tachyoryctes* on Mount Kenya resemble the Mima mounds that occur in some parts of the western United States and are thought to be formed by pocket gophers (see Fig. 15–35). *Tachyoryctes* is solitary and aggressive, and eats a variety of below- and aboveground parts of plants.

FAMILY GLIRIDAE. This family includes the dormice, a group (9 genera with 25 Recent species) of rather squirrel-like,

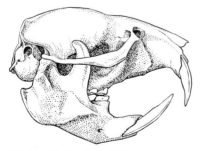

Figure 10–25. Skull of an African root rat (*Tachyoryctes splendens*). Note the procumbent incisors that are used for digging. Length of skull 41 mm.

Old World rodents known first from the Eocene of France. Included here within the Gliridae are the genera *Platacanthomys* and *Typhlomys*; these genera were assigned to the family Platacanthomyidae by Simpson (1945:92), and have variously been included in the families Cricetidae, Muridae, and Gliridae by other authors. Dormice are entirely Old World in distribution, occurring in much of Africa south of the Sahara, England, Europe from southern Scandinavia southward, Asia Minor, southwestern Russia, southern India, southern China, and Japan.

Dormice are small (up to 325 mm in length) and most genera have bushy or well-furred tails. The skull has a smooth, rounded braincase, a short rostrum, and large orbits. The dental formula is 1/1, 0/0, 0-1/0-1, 3/3 = 16-20. The crowns of the brachyodont molars have parallel cross ridges of enamel, or in some cases the ridges are reduced and the crowns have basins. The infraorbital foramen is somewhat enlarged and transmits part of the masseter muscle. The limbs and digits are fairly short, and the sharp claws are used in climbing. The manus has four toes and the pes has five.

Glirids are typically climbers that occupy trees and shrubs, rock piles, or rock outcrops. These rodents are omnivorous, but are able little predators, capable of killing small birds and large insects. In temperate areas the animals are active and breed in the spring and summer, but hibernate in the winter. A unique feature of glirids is their ability to lose and regenerate their tails (Mohr, 1941:63).

FAMILY SELEVINIIDAE. This family is represented by one species, which bears the common name of dzhalman, and is restricted to the Betpak-dala Desert of Russia. This small rodent is insectivorous and, as is typical of many desert mammals, has greatly enlarged auditory bullae. The dental formula is 1/1, 0/0, 2/0, 3/3 = 20; the cheek teeth are small and short crowned, and the much-simplified crown pattern features smooth, concave surfaces. This unusual desert rodent may have evolved from dormouse-like (gliroid) ancestors.

LINEAGES NOT ASSIGNED TO SUBORDERS

The validity of each of the following families as distinct taxonomic units is clear, but their relationships to other families remain unclear. Wood (1937, 1974, 1975) has repeatedly stressed the widespread occurrence of parallelism within the Rodentia, of which some of the following families are excellent examples. The Ctenodactylidae parallel caviomorph rodents in the arrangement of the masseter muscles, but probably had a separate origin from Eocene paramyids (Lavocat, 1973). The Anomaluridae and the Pedetidae are also troublesome. Although each family possesses features typical of other groups of rodents, neither family seems closely related to any other group of rodents.

Two of the families discussed below are perhaps closely related. The Thryonomyidae and the Petromyidae were probably derived from the family Phiomyidae, a group of hystricognathous rodents that first appeared in the highly fossiliferous Fayum beds of the Oligocene of Egypt. In the absence of effective competition, phiomyids underwent rapid radiation in Africa and attained considerable diversity by the Miocene (Lavocat, 1962). But in the Miocene, a major invasion of Africa by Eurasian rodents provided overwhelming competition for some phiomyids, and this group declined. The families Thryonomyidae and Petromyidae are probably relict descendants of the extinct Phiomyidae.

FAMILY SCIURIDAE. This successful and widespread family includes some 261 Recent species representing 51 genera. Squirrels, chipmunks, marmots (Fig. 10–26), and prairie dogs (Fig. 10–27) belong to this family. Sciurids appeared first in the middle Oligocene of North America. Ground squirrels, tree squirrels, and flying squirrels appear in the fossil record in the Miocene. Sciurids remain widespread today, absent only from the Australian region, Madagascar, the polar regions, southern South America, and certain Old World desert areas.

Sciurids are fairly distinctive structur-

Figure 10–26. Yellow-bellied marmot (*Marmota flaviventris*, Sciuridae). (Photograph by O. D. Markham.)

ally. The skull is usually arched in profile, and the front of the zygomatic arch is flattened (forming the so-called zygomatic plate) where the anterior part of the masseter lateralis rests against it. The dental formula is 1/1, 0/0, 1-2/1, 3/3 = 20-22. The cheek teeth are rooted and usually have a crown pattern that features transverse ridges. Sciurids have relatively unspecialized bodies: a long tail is usually retained, and the limbs seldom have a loss of digits or reduction of freedom of movement at the elbow, wrist, and ankle joints. Several semifossorial types, including ground squirrels (*Spermophilus*), prairie dogs (*Cynomys*), and marmots (*Marmota*), have variously departed from this plan in the direction of greater power in the forelimbs and, in some cases, reduction of the tail.

Sciurids are basically diurnal herbivores, but a great variety of food is utilized. Tree squirrels occasionally eat young birds and eggs; chipmunks (*Eutamias*) and the antelope ground squirrels (*Ammospermophilus*) are seasonally partly insectivorous in some areas. Sciurids are tolerant of a great range of environmental conditions. Some, such as marmots, prairie dogs, chipmunks, and some ground squirrels, hibernate during cold parts of the year. Red squirrels (*Tamiasciurus*) in boreal coniferous forests of North America remain active throughout the winter, when temperatures frequently stay well below 0°F for days at a time. The antelope ground squirrel is adapted to live under the tremendously different conditions typical of Sonoran deserts (see page 429).

Styles of locomotion vary among sciurids; the most specialized style occurs in the flying squirrels. This group of 13

Figure 10–27. Black-tailed prairie dogs (*Cynomys ludovicianus*, Sciuridae) (Photograph by O. J. Reichman.)

genera, comprising the subfamily Petauristinae, is characterized by gliding surfaces formed by broad folds of skin between the forefoot and hindfoot. These animals are able to glide fairly long distances between trees. The giant flying squirrel (*Petaurista*) of southeast Asia, for instance, can glide up to 450 m, and can turn in midair (Walker, 1968:716).

FAMILY CASTORIDAE. To this family belong the beavers. The family is represented today by but two species, *Castor canadensis* of the United States, Canada, and Alaska, and *Castor fiber*, of northern Europe and northern Asia. Beavers had a pronounced effect on the history of the United States. Much of the early exploration of some of the major river systems in the western United States was done by trappers in quest of the valuable beaver pelts.

The fossil record of beavers begins in the Oligocene, and several lines of descent developed in the Tertiary. One line developed fossorial adaptations; the fossil remains of the Miocene beaver *Paleocastor* have been found in the spectacular corkscrew-shaped burrows apparently dug by these animals. Another evolutionary line led to the bear-sized giant beaver (*Castoroides*) of the North American Pleistocene. Throughout their history castorids have been restricted to the Northern Hemisphere.

Living beavers are semiaquatic, and some of their distinctive structural features are adaptations to this mode of life. The animals are large, reaching over 30 kg in weight. Their large size is associated with a mass to surface ratio that is more advantageous in terms of heat conservation than that of smaller rodents. In addition, the body is insulated by fine underfur protected by long guard hairs. These are important adaptations in animals that frequently swim and dive for long periods in icy water. The large hind feet are webbed, the small eyes have nictating membranes, and the nostrils and ear openings are valvular and can be closed during submersion.

Because of two structural specializations, beavers can open their mouths when gnawing under water; while they are

swimming, they can carry branches in the submerged open mouth without danger of taking water into the lungs (Cole, 1970). The epiglottis is internarial (it lies above the soft palate); it allows efficient transfer of air from the nasal passages to the trachea, but does not allow mouth breathing or panting. Also, the mid-dorsal surface of the back part of the tongue is elevated and fits tightly against the palate and, except when the animal is swallowing, blocks the passage to the pharynx (Cole, 1970).

The tail is broad, flat, and largely hairless. The skull is robust. The zygomasseteric structure is of a specialized type, in which the rostrum is marked by a conspicuous lateral depression (Fig. 10–28B) from which a large part of the masseter lateralis muscle originates. The jugal is broad dorsoventrally, and the external auditory meatus is long and surrounded by a tubular extension of the auditory bulla. The dental formula is 1/1, 0/0, 1/1, 3/3 = 20. The premolars are molariform, and the complex crown pattern features transverse enamel folds (Fig. 10–15E).

Beavers are always found along waterways. Although they are most typical of regions supporting coniferous or deciduous forests, they also live in some hot desert regions, as for example along the lower Colorado River of Arizona and California. In the southwestern and middle-western United States, beavers dig burrows in the river banks, but in northern and mountainous regions they build lodges of sticks and mud in ponds formed behind their dams. They remain active beneath the ice throughout the winter, feeding on the cambium of aspen and willow branches that they have stuck in the mud bottoms of the ponds.

Beavers are remarkable in their ability to modify their environment be felling trees and by building dams. Many high mountain valleys in the Rocky Mountains have been transformed by the dams of beavers from a series of meadows through which a narrow, willow-lined stream meandered, to a terraced series of broad ponds bordered by willow thickets and soil saturated with water. A valley suitable for

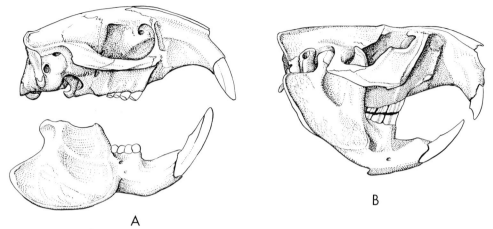

Figure 10–28. Two rodent skulls. A, A mole rat (*Cryptomys mechowii*, Bathyergidae); note the large, procumbent incisors that are used for digging (after Hill and Carter, 1941). Length of skull 57 mm. B, A beaver (*Castor canadensis*, Castoridae); note the depression in the side of the rostrum from which the anterior part of the masseter lateralis originates. Length of skull 139 mm.

grazing by cattle before occupancy by beavers may be more suitable afterward for trout and waterfowl.

FAMILY CTENODACTYLIDAE. Members of this family, commonly called gundis, inhabit arid parts of northern Africa from Senegal, French West Africa, on the west, to Somaliland on the east. There are four Recent genera and eight Recent species. The fossil record of this family begins in the Oligocene of Asia.

These are small, compact, short-tailed rodents with long, soft pelages. The ears are round and short, and are protected from wind-blown debris is some species by a fringe of hair around the inner margin of the pinnae. The infraorbital canal is enlarged, and through it passes part of the medial masseter muscle. The skull is flattened and the auditory bullae and external auditory meatus are enlarged. The cheek teeth are ever-growing and the crown pattern is simple. The dental formula is 1/1, 0/0, 1/1, or 2/2, 3/3 = 20 or 24. The limbs are short; the manus and pes each have four digits.

These herbivorous rodents occur in arid and semiarid areas, where they are restricted to rocky situations. They are diurnal and crepuscular, and scurry into jumbles of rock or fissures in rock when threatened.

FAMILY ANOMALURIDAE. This family, composed of some 12 Recent species of 4 genera, includes the scaly-tailed squirrels. These animals occupy forested, tropical parts of western and central Africa, where they are locally common. They resemble the flying squirrel (Sciuridae) in some structural features and in gliding ability.

Increased surface area for gliding is provided in anomalurids by a fold of skin that extends between the wrist and the hind foot, and is supported and extended during gliding by a long cartilaginous rod, roughly the length of the forearm, that originates on the posterior part of the elbow. (In "flying" members of the Sciuridae, in contrast, a short cartilaginous brace arises from the wrist.) In anomalurids, folds of skin similar to the uropatagium of a bat occur between the ankles and the tail a short distance distal to its base. The fur over most of the membranes is fine and soft, but a tract of stiff hairs occurs along the outer edge of the membrane behind the cartilaginous elbow strut. This tract may possibly improve the efficiency of gliding by controlling the flow of the boundary layer of air sweeping over the membrane during gliding. Compared to those of most rodents, the limbs of anomalurids are unusually long and lightly built; they provide for a wide spreading of the flight membranes.

One genus (*Zenkerella*) does not have a gliding membrane. The anomalurid tail is usually tufted and has a bare ventral area near its base that has two rows of keeled scales. These scales seemingly serve to keep the animals from losing traction when they cling to the trunks of trees. The feet are strong and bear sharp, recurved claws.

In the anomalurid skull, the infraorbital canal is enlarged and transmits part of the medial masseter muscle. The dental formula is 1/1, 0/0, 1/1, 3/3 = 20, and the cheek teeth are rooted.

Anomalurids are handsome animals that are beautifully adapted to an entirely arboreal life. They are apparently largely vegetarians. Derby's anomalure (*Anomalurus derbianus*) feeds primarily on bark of a number of kinds (Rahm, 1970) and fruit, but is known to also eat flowers and some insects. This species is a graceful and highly maneuverable glider, capable of glides of over 100 meters and of mid-air turns. The dwarf anomalure (*Idiurus zenkeri*) has been described by Durrell (1954) as gliding "with all the assurance and skill of hawking swallows." Kingdon (1974b: 454) has observed that animals launching themselves do not immediately spread their membranes but gain speed by leaping powerfully out from the trunk of a tree and dropping at least a meter before beginning to glide.

Groups of anomalurids often take shelter during the day in cavities in trees. Group size varies from 6 or 8 animals to colonies of over 100 anomalurids of several species. Rosevear (1969) reports four species occupying the same hole in a tree. On occasion these rodents share a hollow tree with dormice (*Graphiurus*) and with several species of bats. The propensity of anomalurids to sleep during the day in densely packed clusters and their use of sunbathing suggest that they may at times utilize adaptive hypothermia for conserving energy.

FAMILY PEDETIDAE. This family is represented by one distinctive species,

Pedetes capensis, the spring hare. Its distribution includes East Africa and southern Africa, where it inhabits sandy soils in semiarid regions. Sparsely vegetated areas or places where the vegetation has been heavily grazed by ungulates are preferred.

The spring hare is saltatorial and is roughly the size of a large rabbit, weighing up to about 4 kg. The eyes are extremely large, suggesting perhaps a reliance on vision for detection of predators. The forelimbs are short but robust and bear long claws that are used in digging. The hind limbs are long and powerfully built, the fibula is reduced and is fused distally to the tibia, and the feet have only four toes. The long tail is heavily furred throughout its length. Through the enormous infraorbital foramen passes the large anterior division of the medial masseter. As in a number of saltatorial rodents, the cervical vertebrae are partly fused (Fig. 10–22B). The dental formula is 1/1, 0/0, 1/1, 3/3 = 20, and the cheek teeth are ever-growing with a simplified crown pattern. A tragus fits against the ear opening and excludes sand and debris when the animal digs.

Spring hares dig fairly elaborate burrows. Because of their restriction to friable soils, these animals are not evenly distributed and appear to occur in colonies. When frightened, spring hares can make tremendous bipedal leaps of over six meters, but when foraging and moving slowly, the animals are quadrupedal. A variety of plant material is eaten, including bulbs, seeds, and leaves. Water balance may be maintained during some seasons by eating succulent vegetation or insects. For a rodent, the spring hare has an unusually low reproductive rate. There are but two pectoral mammae, and typically the female bears only one young. Newborn young are large, roughly one third the size of the adult (Hediger, 1950; Cos, 1967), are well developed, and remain in the maternal burrows until they weigh at least half as much as the adults.

FAMILY HYSTRICIDAE. To this family belong the Old World porcupines, a widely distributed group of rodents (4 genera with 15 Recent species) that resemble the New World porcupines (Erethizontidae) in having quills for protection. Hystricids occur throughout most of Africa, in southern and central Italy, in southern Asia and South China, and in Borneo, southern Celebes, Flores, and the Philippines. Hystricids appear first in the Oligocene of Europe.

These large rodents weigh up to 27 kg and have a stocky build. The occipital region of the skull is unusually strongly built and provides attachment for powerful neck muscles. The nasoturbinal, lacrimal, and frontal bones are highly pneumatic in some species (Fig. 10–29). The dental formula is 1/1, 0/0, 1/1, 3/3 = 20. The hypsodont cheek teeth have re-entrant enamel folds that, with wear, become islands on the occlusal surfaces. Some of the hairs are stiff, sharp spines that reach at least 40 cm in length in some species; in some species, open-ended, hollow spines make a noise when rattled that appears to have a warning function. One genus (Trichys, of Borneo, the Malay Peninsula, and Sumatra) lacks stiff spines. The large, plantigrade feet are five-toed, and the soles are smooth.

Hystricids eat a wide variety of plant material and locally may damage crops. In contrast to New World porcupines, hystricids are terrestrial rather than partly arboreal, and dig fairly extensive burrows that are used as dens. The quills are often conspicuously marked with black and white bands and this visual signal, together with the rattling of the quills, deters some predators. When threatened, hystricids erect the quills and often rush an attacker. Aside from man, the larger cats are the main predators of hystricids.

FAMILY THRYONOMYIDAE. One genus with six species comprises this small family, the members of which are known as cane rats. These animals are broadly distributed in Africa south of the Sahara Desert. The earliest records of cane rats are from the Miocene of Africa. Fossil cane rats, from the Pliocene of Asia and Europe, and the occurrence of an extinct species of the Recent genus *Thryonomys*, from the central Sahara Desert, indicate that the range of cane rats was once far greater than it is now.

Cane rats are large rodents, from 4 to 6 kg in weight, with a coarse, grizzled pelage. The snout is blunt and the ears and tail are short. The robust skull has prominent ridges and a heavily built occipital region. The cheek teeth are hypsodont, and the large upper incisors are marked by three longitudinal grooves. The dental formula is 1/1, 0/0, 1/1, 3/3 = 20. The fifth digit of the forepaw is small, and the claws are strong and adapted to digging.

Cane rats are capable swimmers and divers, and are largely restricted to the vicinity of water, where they take shelter in matted vegetation or in burrows. Males indulge in ritualized snout-to-snout pushing contests, and the odd, blunt shape of the snout seems to enable the animals to avoid damage during these bouts. They are herbivorous and do considerable local damage to crops, particularly sugar cane. Cane rats are prized for food in many parts

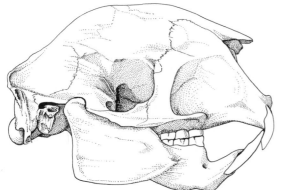

Figure 10–29. Skull of the African crested porcupine (*Hystrix cristata*, Hystricidae). Note the greatly inflated rostrum and frontal part of the skull.

of Africa. The animals are often taken during organized drives using dogs, or are driven from their hiding places and captured when natives set fire to reeds.

FAMILY PETROMYIDAE. This relict family includes but a single species, *Petromus typicus*, the dassie rat. This animal is restricted to parts of southwestern Africa and is not represented by fossil material.

The dassie rat is a small rodent with a squirrel-like appearance. The rooted, hypsodont cheek teeth have a simplified crown pattern; the dental formula is 1/1, 0/0, 1/1, 3/3 = 20. Structurally, these animals are most remarkable for specializations enabling them to seek shelter in narrow crevices. Such specializations include a strongly flattened skull, flexible ribs that allow the body to be dorsoventrally flattened without injury, and mammae situated laterally, at the level of the scapulae, where the young can suckle while the female is wedged in a rock crevice.

Dassie rats are diurnal and feed largely on leaves. They are restricted to rocky sections of foothills and mountains, where immediately available shelter in the form of crevices in rocks is available.

FAMILY BATHYERGIDAE. This family contains the African mole rats, a group of highly specialized fossorial rodents. The family includes 5 genera with approximately 20 Recent species. These mole rats occupy much of Africa from Ghana, Sudan, Ethiopia, and Somaliland southward. The earliest fossil records of bathyergids are from the Miocene of Africa.

Bathyergids are from roughly 120 to 330 mm in total length, and they possess a number of unique structural features associated with their fossorial life. The eyes are small in all species and vision is apparently poorly developed, but the unusually thick cornea is sensitive to air currents and seemingly allows the animals to detect an opening in the burrow system (Eloff, 1958). The ears lack or nearly lack pinnae. The skull is robust, the powerful incisors are procumbent in all species (Fig. 10–28A), and the roots of the upper incisors extend above or behind the molars. The lips close tightly behind the incisors so that dirt does not enter the mouth when the animal is burrowing. The cheek teeth are hypsodont but rooted, and typically have a simplified crown pattern. The dental formula is variable (1/1, 0/0, 2/2 or 3/3, 0/0 to 3/3 = 12-28). In *Heliophobius*, there are six cheek teeth but not all are functional simultaneously. The zygomasseteric structure is distinctive. The infraorbital foramen transmits little or no

Figure 10–30. The naked mole rat (*Heterocephalus glaber*, Bathyergidae) of East Africa. The lips close behind the incisors and a fold of skin guards the nostrils; these adaptations seemingly avoid the ingestion and inhalation of soil when the incisors are involved in digging. (Photograph by Jennifer U. M. Jarvis.)

muscle. The masseter muscles, however, are highly specialized: the large anterior part of the masseter medialis originates from the upper part of the medial wall of the orbit and the superficial part of the masseter lateralis originates partly on the anterior face of the zygoma. The mandibular fossa and angular part of the dentary are greatly enlarged (Fig. 10–28A) and provide an extensive area for the insertion of the masseter muscles.

The limbs are fairly robust in all species, but are apparently used for digging only in *Bathyergus*. The hind feet are broad, and the animals back up against a load of soil and push it from the burrow with the hind feet. The tail is short and is used as a tactile organ. The pelage is normal in most species, but in *Heterocephalus glaber*, the skin is nearly naked with only a sparse sprinkling of long hairs (Fig. 10–30).

The African mole rats are basically herbivorous and utilize largely bulbs, roots, and rhizomes that are reached by burrowing. They seldom appear above ground, and typically occupy soft loamy or sandy soils in desert and savanna areas. The huge incisors are the major digging tools in all species of bathyergids except *Bathyergus suillus*, which seems to use its feet also. *Heterocephalus* and *Cryptomys* are colonial. Jarvis and Sale (1971) found that members of a *Heterocephalus* colony form a "digging chain," which enables the colony to work together during excavation of a burrow. The remarkable division of labor associated with this communal digging is described on page 360. The burrow systems of the colonial species are far more extensive than are those of the solitary types.

CETACEANS: WHALES, PORPOISES, DOLPHINS

11

Cetaceans are notable for being the mammals most perfectly adapted to aquatic life. Of further interest, however, is their frequent position at the top of the food chain in marine environments generally thought to be "dominated" by fish. The large baleen whales are the largest living or fossil animals known, and cetaceans are among the fastest creatures in the sea. Remarkable swimming ability, the capability to echolocate, considerable intelligence, and well-developed social behavior have all contributed to the success of cetaceans.

MORPHOLOGY

All cetaceans are completely aquatic, and their structure reflects this mode of life. The body is fusiform (cigar-shaped), lacks sebaceous glands, is nearly hairless, and is insulated by thick blubber. Regional differentiation of vertebrae is not pronounced and most vertebrae have high neural spines (Fig. 11–1). The clavicle is absent, the forelimbs (flippers) are paddle-shaped, and no external digits or claws are present. The joints distal to the shoulder

Figure 11–2. Dorsal view of the right forelimb of the bottle-nosed dolphin (*Tursiops truncatus*, Delphinidae).

allow no movement. The proximal segments of the forelimb are short, whereas the digits are frequently unusually long because of the development of more phalanges per digit than the basic eutherian number (Fig. 11–2). The hind limbs are vestigial, do not attach to the axial skeleton, and are not visible externally. The flukes (tail fins) are horizontally oriented. The skull is typically highly modified as a result of the migration of the external nares to the top of the skull. The premaxillaries and maxillaries form most of the roof, and the occipitals form the back of the skull. The nasals, frontals, and parietals are telescoped between these bones and form only a minor part of the skull roof (Figs. 11–3 and 11–4). The tympanoperiotic bone (the bone that houses the

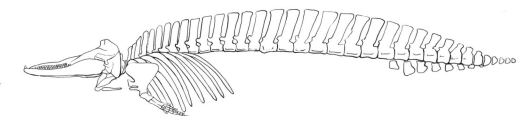

Figure 11–1. The skeleton of the Tasmanian beaked whale (*Tasmacetus shepherdi*, Ziphiidae).

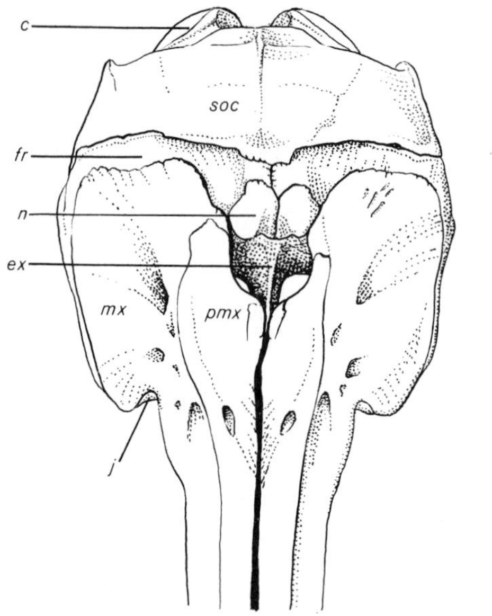

Figure 11–3. Dorsal view of part of the skull of a bottle-nosed dolphin (*Tursiops truncatus*, Delphinidae). Note the asymmetry of the bones surrounding the external nares (*ex*). The function of this remarkable asymmetry is not known. Abbreviations: *c*, occipital condyle; *fr*, frontal; *j*, jugal; *mx*, maxillary; *n*, nasal; *pmx*, premaxillary; *soc*, supraoccipital.

"pinhole camera" effect of the cetacean's pupil. Mechanisms for fine adjustment of lens shape or displacement are seemingly absent in cetaceans, but in bright light the central part of the pupil closes completely, leaving two tiny apertures, which yield great depth of field, as does a pinhole camera. This allows the dolphin to receive a sharp image from distant objects when its head is above water.

PALEONTOLOGY

The cetacean fossil record is not consistently good, and the progenitors of cetaceans are unknown. Presumably, they diverged early from primitive eutherian stock, and the evidence suggests independent lines of descent for each of the three orders. The most primitive cetaceans comprise the order Archaeoceti, which is known from the middle Eocene to the Miocene. Although these animals were well adapted to aquatic life (the hind limbs are vestigial and the body is elongate), the skull is primitive and the migration of the external nares to the top of the skull so typical of the advanced orders is not evident (Fig. 11–6). The dentition is heterodont, and the primitive eutherian number of teeth (44) is not exceeded. Some archaeocetes attained large size. The skull of *Basilosaurus*, an Eocene type, is 1.5 m long, and the slim body is roughly 17 m long.

middle and inner ear) is not braced against adjacent bones of the skull, and is partly insulated from the rest of the cranium by surrounding air sinuses (Fig. 11–5).

Experimental work on the bottle-nosed dolphin *(Tursiops truncatus)* by Herman et al. (1975) showed that visual acuity was similar above and below the water. The above-water acuity may be due to the

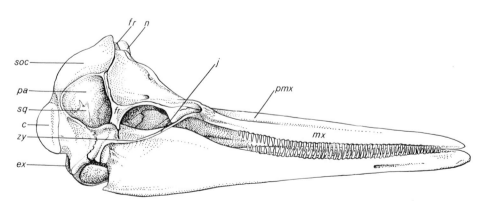

Figure 11–4. The skull of a dolphin (*Delphinus bairdii*, Delphinidae); length of skull 475 mm. Note the highly telescoped skull with the maxillary and frontal bones roofing the small temporal fossa. The frontal is barely exposed on the skull roof. Abbreviations: *c*, occipital condyle; *ex*, exoccipital; *fr*, frontal; *j*, jugal; *mx*, maxillary; *n*, nasal; *pa*, parietal; *pmx*, premaxillary; *sq*, squamosal; *soc*, supraoccipital; *zy*, zygomatic arch.

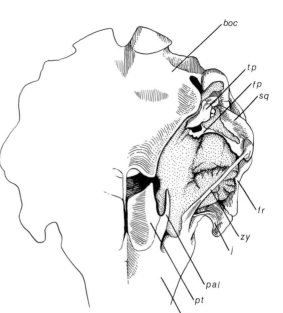

Figure 11–5. Ventral view of part of the skull of the grampus (*Grampus griseus*, Delphinidae), showing the large air sinuses (stippled) anterior to the tympanoperiotic bone (*tp*) and partly surrounding it. Abbreviations: *boc*, basioccipital; *fp*, falciform process of the squamosal; *fr*, frontal; *j*, jugal; *mx*, maxillary; *pal*, palatine; *pt*, pterygoid; *sq*, squamosal; *zy*, zygomatic arch. (After Purves, P. E., in Norris, K. S.: *Whales, Dolphins, and Porpoises*, originally published by the University of California Press; redrawn by permission of The Regents of the University of California.)

The earliest records of the order Mysticeti, the baleen whales, are from the middle Oligocene of Europe. The origin, assumed to be from primitive toothed forms, is not known. The earliest known family, Cetotheriidae, had a partially telescoped skull and, as in modern Mysticeti, lacked teeth. The gray whale (*Eschrichtius gibbosus*) is the only surviving member of the family Eschrichtiidae, a group closely related to the extinct Cetotheriidae. The order Odontoceti, the toothed whales, appears in the late Eocene of North America.

By the Miocene, cetaceans had undergone considerable radiation; 99 of the 173 known cetacean genera, fossil and Recent, are known from this epoch. Advanced odontocetes with highly telescoped skulls, homodont dentitions, and with many more teeth than the primitive eutherian complement, are known from the Miocene (Fig. 11–7).

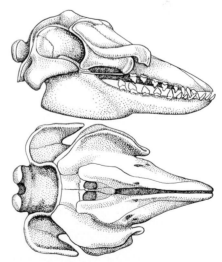

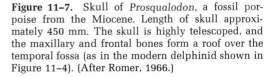

Figure 11–6. Skull of *Prozeuglodon*, a fossil archaeocete from the Eocene. Length of skull approximately 600 mm. Note the lack of telescoping of the skull and the heterodont dentition. (After Romer, 1966.)

Figure 11–7. Skull of *Prosqualodon*, a fossil porpoise from the Miocene. Length of skull approximately 450 mm. The skull is highly telescoped, and the maxillary and frontal bones form a roof over the temporal fossa (as in the modern delphinid shown in Figure 11–4). (After Romer, 1966.)

The two Recent orders of cetaceans can be distinguished on the basis of a number of morphological characters. Members of the order Mysticeti lack teeth, but have baleen plates of horny material that grow from the upper jaw and function as a sieve (Fig. 11–8). The ascending processes of the maxillae are narrow, interlock with the frontals, and do not spread laterally over the supraorbital processes. The respiratory and alimentary canals are not permanently separated. Mysticetes are not known to use echolocation.

Odontocetes, in contrast, have simple, peglike teeth, frequently far exceeding the typical eutherian maximum number, or secondarily reduced in extreme cases to 0/1. The ascending processes of the maxillae are large, do not interlock with the frontals, and spread laterally over the supraorbital processes (Figs. 11–3, 11–4). The respiratory and alimentary tracts are permanently separated by specializations of the glottis and the laryngeal apparatus. The bones surrounding the blowhole in odontocetes, but not in mysticetes, depart from the usual mammalian pattern of bilateral symmetry (Fig. 11–3). Probably all odontocetes use echolocation.

SWIMMING ADAPTATIONS

The Achilles' heel of cetaceans is their need to breathe air but, unlike most other mammals, many cetaceans are able to alternate between periods of eupnea (normal breathing) and long periods of apnea (cessation of breathing). Some whales remain submerged for roughly two hours, but many small delphinids surface to breathe several times a minute. The ability of cetaceans to remain active during periods of apnea probably depends on many adaptations. Rapid gaseous exchange is enhanced by two layers of capillaries in the interalveolar septa. During expiration, most of the air can be exhausted from the lungs, and up to 12 per cent of the oxygen from inhaled air is utilized (the corresponding figure for terrestrial mammals is only 4 per cent). Compared to terrestrial mammals, cetaceans have up to twice as many erythrocytes per given volume of blood, and about two to nine times the myoglobin (a molecule able to store oxygen and to release it to tissue) in the muscles. During deep dives, the heart rate drops to roughly half the "surface rate," and vascular specializations allow blood to bypass certain muscle masses. According to Rice (1967:293), the most important physiological adaptations to prolonged submersion include (1) anaerobic glycolysis, (2) tolerance to high levels of lactic acid, and (3) a relative insensitivity to carbon dioxide. Good discussions of cetacean adaptations to deep diving are given by Elsner (1969), Irving (1966), Kooyman and Andersen (1969), and Lenfant (1969).

Most small odontocetes are seemingly shallow divers, but some large odontocetes and some mysticetes can perform deep dives (Table 11–1). During deep dives, cetaceans are subjected to tremendous pressures, for with every 10 m increase in depth an additional atmosphere of pressure is exerted on their bodies. A cetacean swimming at a depth of only 200 m, proba-

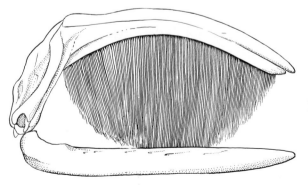

Figure 11–8. Skull of the Atlantic right whale (*Balaena glacialis*, Balaenidae); length of skull roughly 4 m. Note the baleen plates attached to the maxilla.

Table 11–1. Depths at Which Cetaceans Have Been Recorded and Methods by Which Observations Have Been Made

Species	Depth (m)	Method of Observation
Balaenopteridae		
Fin whale, *Balaenoptera physalus*	500	Harpooned and collided with bottom
	355	Depth manometer on harpoon
Physeteridae		
Sperm whale, *Physeter catodon*	900	Entangled in deep-sea cable
	1134	Entangled in deep-sea cable
	520	Echo-sounder
Stenidae		
Rough-toothed dolphin, *Steno bredanensis*	30	Attached to depth recorder
Delphinidae		
North Pacific pilot whale,		
Globicephala scammoni	366	Inferred from feeding behavior
Bottle-nosed dolphin, *Tursiops truncatus*	92	Visual observations from underwater craft
	185	Vocalizations near underwater craft
	170	Trained to activate buzzer

Data from Kooyman and Andersen (1969:72), who cite the sources of the observations.

bly a commonplace depth for many species, is subjected to 20 atmospheres of pressure, or 294 lbs per square inch. Because of their lungs and air sinuses, cetaceans have semihollow bodies, which tend to collapse during deep dives. Any gases that remain within body cavities are therefore subjected to great pressures, resulting in a decrease in their volumes and an increase in the amounts of gases that go into solution in the body solvents, such as blood. One serious result in man is "bends," or decompression sickness. When man uses equipment that allows him to breathe under water and undergoes prolonged exposure to high pressures during diving, greater than normal amounts of gases in the lungs are dissolved in the tissues and the blood. If decompression is too rapid, these gases cannot be carried to the lungs rapidly enough to be removed from the body; instead, they quickly leave solution and appear as bubbles in the tissues. Intravascular bubbles may occlude capillaries and result in injury to tissues or even in death.

Cetaceans are not known to have these problems, although they stay for considerable periods at great depths and ascend to the surface rapidly. This is partly due to their ceasing breathing while beneath the water, but is also a result of structural modifications. The following anatomical specializations are seemingly adaptations to deep diving: (1) a large proportion of the ribs lack sternal attachment or lack attachment either to the sternum or to other ribs; (2) the lungs are dorsally situated above the oblique diaphragm; (3) in deep divers the lungs are small and the volume of the air passages is relatively large; (4) the trachea is short and often of large diameter, and the cartilaginous rings bracing the trachea are nearly complete, have small intermittent breaks, or are fused (Slijper, 1962); (5) the bronchioles are reduced in length and the entire system of bronchioles, to the very origin of the alveolar ducts, is braced by cartilaginous rings; (6) the lungs, especially the walls of the alveolar ducts and the septa, contain unusually large concentrations of elastic fibers; (7) in some of the small odontocetes, a series of myoelastic sphincters occur in the terminal sections of the bronchioles.

The adaptive importance of all these features is not yet completely understood. Clearly, the specializations of the ribs and the placement of the lungs relative to the diaphragm permit the collapse of the lungs, and bracing of the respiratory passages may aid in the regulation of air in the lungs during diving. Probably the alveoli fully collapse during deep dives, forcing air into the respiratory passages and the air sinuses that surround the tympanoperiotic bone (Fig. 11–5). Experimental work with a trained bottle-nosed dolphin by Ridgeway et al. (1969) indicated that alveolar collapse is complete at about 100 m beneath the surface, and that little pulmonary respiration occurs at greater depths. Fraser and Purves (1955) hypothesize that this air, plus mucus and fat droplets, forms a foam that absorbs nitrogen (nitrogen dissolves roughly six times as rapidly in some oils as it does in water) and other gases; this prevents the gases from going into solution in the blood stream and leading to decompression problems. This foam is expelled in the "blow" of a cetacean. (The possible function of this foam in connection with echolocation is discussed on p. 204).

Cetaceans are remarkably fast swimmers. Powerful dorsoventral movements of the tail provide propulsion, and the flippers are used for steering. Dolphins have been observed to maintain speeds of about 32 km per hour for up to 25 minutes, and for short periods to reach 38 km per hour (Johannessen and Harder, 1960). These authors report that a killer whale (Orcinus orca) approached a ship at some 56 km per hour and cruised around the ship for 20 minutes at speeds in excess of the ship's 38 km per hour. A blue whale (which weighs some 200,000 pounds) has been observed traveling for 10 minutes at 37 km per hour (Gawn, 1948), and a pilot whale swam in a tight circle around a tank at the same speed (Norris and Prescott, 1961:343).

This remarkable swimming performance of cetaceans has proved difficult to explain. Recent studies have demonstrated that their speed is not due to muscles that are vastly more powerful than those of other mammals, but to specializations that greatly reduce resistance (drag) as the animals swim. The amount of resistance depends on the type of flow of water over the surface of the body. If the flow is smooth—parallel to the surface—it is said to be laminar. When such smooth flow is interrupted by movements of the water

that are not consistently parallel to the surface of the body, however, turbulent flow occurs. All other conditions being equal, laminar flow creates much less resistance than does turbulent flow. If the bodies of small dolphins were subjected to turbulent flow, swimming at 24 mph would require their muscles to be five times as powerful as those of man (Lang, 1966:426). But assuming flow to be nearly laminar, this speed is approximately what would be expected if their power output were that of a well-trained human athlete. Scientists for many years have attempted without success to design bodies shaped so that flow of air or water over their surfaces is laminar or nearly so. What is the cetacean solution to this problem?

Several factors seemingly contribute importantly to furthering the reduction of resistance as a porpoise swims rapidly (these are discussed by Hertell, 1969:33–49). The body is hairless, and no obstructions except the streamlined appendages break the extremely smooth surface. In addition, the body form of the dolphin is approximately parabolic; this form creates even less resistance than the rounded (elliptical) head end and tapered body of such a rapid swimmer as a trout. Perhaps most remarkable is the structure of the skin (described by Kramer, 1960). The skin consists basically of two layers. The soft outer layer is 80 per cent water and has narrow canals filled with a spongy material; the stiff inner layer has stringy connective tissue. Pressure differences caused by turbulent flow of water are transmitted to the soft outer skin, which yields in areas of increased pressure and expands where pressure is reduced. The liquid in the skin tends to move parallel to the surface in response to distortion of the skin, but this movement is inhibited by the channels and their spongy filling. These effects result in a strong damping of vibrations caused by turbulent flow and a more complete laminar flow over the body. According to Hertel (1969:40), further reduction of resistance may be associated with movements of the flukes and body during swimming. The story, however, is still incom plete; more must be learned in order for us to fully account for the remarkable high-speed swimming ability of cetaceans.

ORDER MYSTICETI

In this order are the huge baleen whales, which inhabit all oceans. There are only ten Recent species, grouped in five genera and three families. In abundance and numbers of species, this order is less important than the Odontoceti.

All mysticetes are filter-feeders. Plankton-rich water is taken into the large mouth and forced out through the plates of baleen by the enormous tongue; the plankton strained from the water by the baleen is swallowed. Under some conditions these whales "skim" plankton from the surface with the head partly exposed, but they may also feed down to several hundred meters below the surface. One species is known to feed on the floor of shallow seas.

FAMILY BALAENIDAE. These, the right whales, are creatures that were killed in such great numbers during the height of the whaling activities that they are rare today and are protected by international treaty. The family includes three Recent species of two genera, and inhabits most marine waters except tropical and south polar seas. Balaenids are known as fossils from the Miocene to the Recent.

These are large, robust whales that reach about 18 m in length and over 67,000 kg in weight. The head and tongue are huge—the head amounts to nearly one third of the total length. The flippers are short and rounded, and the dorsal fin is usually absent. There are more than 350 long baleen plates on each side of the upper jaw; these plates fold on the floor of the mouth when the jaws are closed. No furrows are present in the skin of the throat or chest. The cervical vertebrae are fused, and the skull is telescoped to the extent that the nasals are small and the frontals are barely exposed on the top of the skull.

Right whales feed largely on planktonic crustaceans and molluscs, and are most common near coastlines or near pack ice. These whales are not highly migratory, and some species remain in far northern waters throughout the year. According to Rice (1967:299), Eskimos claim that the bowhead whale (Balaena mysticetus) can break through ice nearly one meter thick to reach air.

FAMILY ESCHRICHTIIDAE. This family is represented today only by the gray whale (Eschrichtius gibbosus), which occupies parts of the North Pacific. There is no fossil record of this family, but there are subfossil records from the North Atlantic (Rice and Wolman, 1971:20).

The gray whale is fairly large, weighing up to about 31,500 kg and measuring some 15 m in length, and has a slender body with no dorsal fin. The baleen plates are short, and the telescoping of the skull is not extreme; the nasals are large and the frontals are broadly visible on the roof of the skull. The throat usually has two longitudinal furrows in the skin.

Gray whales perform the longest known mammalian migration. They occupy parts of the North Pacific (the Bering, Chukchi, and Okhotsk seas) in the summer. Here they feed largely on bottom-dwelling crustaceans (amphipods), which they take by stirring up the sediments with their snouts. In late autumn they migrate southward along the coastlines. The western Pacific population winters along the coast of Korea, and the eastern Pacific gray whales winter along the coast of Baja California. Young are born in shallow coastal lagoons in the wintering areas. The round trip distance of the migration is from about 10,000 to 22,000 km (some 6,500 to 14,500 miles!). Many people each year watch migrating gray whales from a vantage point at Cabrillo National Monument, near San Diego, California.

The future of the gray whale today seems reasonably bright. Driven near extinction by whaling activities between 1850 and 1925, they are now protected by the International Convention for the Regulation of Whaling, and increased greatly in recent years. Unfortunately, however, the gray whale is being molested to some extent by sightseers in its wintering and breeding areas in bays of Baja California.

FAMILY BALAENOPTERIDAE. This group of whales, frequently known as rorquals, includes six Recent species of two genera. The distribution includes all oceans.

These whales vary in size from fairly small (for whales) to extremely large. The huge blue whale *(Balaenoptera musculus)*, a giant even among whales, reaches a length of about 31 m and a weight estimated at some 160,000 kg. In some species of rorquals the body is slender and streamlined, but it is chunky in others. The baleen plates are short and broad, and the skin of the throat and chest is marked by numerous longitudinal furrows. The nasals are small and the frontals are either not exposed or only barely exposed on the skull roof.

These whales feed in cold waters, often near the edges of the ice where upwelling water results in great growths of plankton in summer. Planktonic crustaceans and small schooling fish are eaten. During the northern winter, the Northern Hemisphere populations move southward toward equatorial areas, and during the southern winter, southern populations move northward. Wintering adults do not feed, but live off stored blubber. Breeding occurs in the wintering areas, but because the southern and northern winters are six months out of phase, no interbreeding between populations occurs. The humpback whale *(Megaptera novaeangliae)*, an animal given to spectacular leaps, makes remarkably melodious and varied underwater sounds that have been beautifully recorded by Payne (1970) and discussed by Payne and McVay (1971).

Excessive commercial exploitation has resulted in a tremendous decline in the populations of fin whales *(Balaenoptera physalus)*, humpbacks, and blue whales. Over broad areas, blue and humpback whales are so scarce as to be "commercially extinct" (Rice, 1967:304), but intense hunting of whales continues over broad areas and populations continue to decline at alarming rates. A bleak prospect seems imminent: within this century man could destroy the blue whale, the largest and one of the most remarkable species of animals that has ever lived.

ORDER ODONTOCETI

The toothed whales, porpoises, and dolphins comprise the most important group of cetaceans. Not only are odontocetes vastly more numerous than mysticetes, both with respect to abundance and diversity, but they are more widely distributed. The order Odontoceti includes about 74 Recent species within 33 genera and seven families, and occurs in all oceans and seas connected to oceans. Members of five families also inhabit some rivers and lakes in North America, South America, Asia, and Africa. Odontocetes are readily observed; they frequently forage close to shore, often make spectacular leaps, and roll repeatedly out of the water; some are prone to ride the bow waves of ships much as man rides shore waves.

The classification of the odontocetes is still uncertain; the arrangement used here is that of Fraser and Purves (1960b).

FAMILY ZIPHIIDAE. The beaked whales are widely distributed, occupying all oceans, but are rather poorly known; some species have never been seen alive. Eighteen Recent species of five genera are recognized. The earliest fossil record of ziphiids is from the early Miocene.

These are medium-sized cetaceans with fairly slender bodies. The length varies from 4 to over 12 m, and the weight reaches some 11,500 kg. The snout is usually long and narrow, and in some species the forehead bulges prominently. Only one species *(Tasmacetus shepherdi)* has a large number of teeth; in the others the dentition is strongly reduced. Only two lower teeth on each side occur in the two species of *Berardius*; in all of the remaining ziphiids there is only a single functional lower tooth on each side (Fig. 11-9). In some species the lower jaw is "undershot" and the teeth are outside the mouth. Two to seven cervical vertebrae are

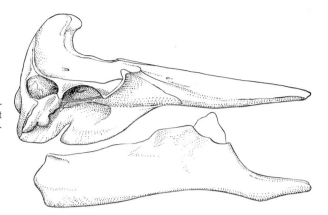

Figure 11–9. Skull of a beaked whale (*Meso-plodon* sp., Ziphiidae); length of skull about 590 mm. Note the single large tooth in the dentary.

fused. The stomach is divided into from 4 to 14 chambers.

Beaked whales are deep divers that are able to remain submerged for long periods. The North Atlantic bottle-nosed whale (*Hyperoodon ampullatus*) may dive for periods of up to two hours. Some species forage in the open ocean well away from coasts. Whereas some species are solitary, others are highly social and travel in schools in which all members surface and dive in synchrony. The primary food is squid, but deep sea fishes are also taken. The North Atlantic bottle-nosed whale is known to make annual migrations, and other species are probably also migratory.

FAMILY MONODONTIDAE. This family contains two species, the narwhal (*Monodon monoceros*), remarkable for its long, straight, forward-directed tusk, and the beluga (*Delphinapterus leucas*), also called the white whale. They occur in the Arctic Ocean and the Bering and Okhotsk seas, in Hudson Bay, in the St. Lawrence River in Canada, and in some large rivers in Siberia and Alaska. Fossil monodontids are known from the Pleistocene and Recent of arctic areas of North America and Eurasia.

These are medium-sized cetaceans; belugas reach about 6 m in length and 2000 kg in weight, and narwhals, without the tusk, are a similar length. The facial depression in the skull is large, and the maxillary and frontal bones roof over the reduced temporal fossa (refer to Figs. 11–3 and 11–4); the zygomatic process of the squamosal is strongly reduced; and the cervical vertebrae are not fused. The beluga has 9/9 teeth, and the narwhal has 1/0. One tooth of the male narwhal (usually the left) forms a straight, spirally grooved tusk up to 2.7 m long; the corresponding tooth in the other jaw is normally rudimentary.

These gregarious cetaceans are characteristic of northern seas, where in winter they assemble in areas of open water. In summer belugas move far up large rivers. Belugas feed largely on fish, both benthic (bottom-dwelling) kinds and those that live at intermediate depths, and squid. Narwhals are seemingly largely pelagic (open-sea dwellers). The function of the narwhal's tusk is not known, but this tusk, discovered separately from the skull by man, probably gave rise to the myth of the unicorn. Both species are quite vocal, and the trilling sounds made under water by belugas account for their common name of "sea canary."

FAMILY PHYSETERIDAE. The sperm whales occur in all oceans, and the giant sperm whale (*Physeter catodon*), of Moby Dick fame, has long been an important species to the whaling industry. Fossil sperm whales are known from the early Miocene.

Physeter is large, attaining a length of over 18 m and weight in excess of 53,000 kg; the pigmy sperm whales (*Kogia*) are small, reaching about 4 m in length and some 320 kg in weight. The head is huge in *Physeter*, accounting for over one third of the total length. In both genera the rostrum is truncate, broad, and flat. The facial depression of *Physeter* contains a spermaceti organ, which contains great quantities of oil. The blowhole is toward the end of the left side of the snout. The left nasal passage serves in respiration, but the right one is specialized as a sound-producing organ. The upper jaw lacks functional teeth; the lower jaw has some 25 functional teeth on each side in *Physeter*, and from 8 to 16 in *Kogia*. All of the cervicals are fused in *Kogia*, and all but the atlas are fused in *Physeter*.

The habits of *Kogia* are not well known, but those of *Physeter* are better understood, probably because man has persistently hunted this animal for many years. *Physeter* is social and assembles in groups with occasionally as many as 1000 individuals. Schools of females with their calves, together with male and female subadults, are overseen by one or more large adult males, whereas younger males congregate in "bachelor schools." Some adult males are solitary. Sperm whales generally forage in the open sea at depths where little or no light penetrates (the use of echolocation by *Physeter* is discussed on page 465). Dives to depths of roughly 1000 m are probably usual, and dives of 1130 m have been recorded (Heezen, 1957). *Physeter* feeds largely on deepwater squids, including giant squids, and a variety of bony fishes, sharks, and skates. Males commonly migrate far north to the edge of the pack ice in summer, but females remain in temperate and tropical waters. *Kogia* is either solitary or travels in small schools, and feeds largely on cephalopods such as squid and cuttlefish.

FAMILY PLATANISTIDAE. This group, the long-snouted river dolphins, is remarkable because its members live largely in rivers. The distribution includes some large river systems in India, the Amazon and Orinoco river systems of South America, coastal waters along the east coast of South America, and Tungt'ing Lake in China. There are four Recent genera, each with a single species, and fossil members are recorded back to early Miocene.

These are small cetaceans, from 1.5 to 2.9 m in length and from about 40 to 125 kg in weight. The jaws are unusually long and narrow and bear numerous teeth (from about 26/26 to 55/55); the forehead rises abruptly and is rounded, giving the head an almost birdlike aspect. In *Platanista* there are large maxillary crests (discussed on page 465). The large temporal fossa is not roofed by the maxillary and frontal bones. None of the cervical vertebrae are fused. The eyes of all members of the family are reduced, and presumably food and obstacles are detected largely by echolocation. The Ganges dolphin (*Platanista*), which usually swims on its side (Fig. 11–10), lacks eye lenses and can perhaps only detect light and dark; the eyes of the white-flag dolphin (*Lipotes*) are greatly reduced and vision is presumably poor. The eyes of the other river dolphins are small but are presumably functional.

These strange cetaceans often inhabit rivers that are made nearly opaque by suspended sediment, and under these conditons echolocation may completely supplant vision. A variety of fishes and crustaceans are eaten, some of which are captured by probing muddy river bottoms. The Amazon dolphins (*Inia*) feed entirely on fish, and during the rainy season may move deep into flooded tropical forests (Humboldt and Bonpland, 1852). River dolphins are seemingly not as social as are many other cetaceans. Only 14 per cent of Layne's (1958:18) observations of Amazon dolphins were of groups with more than four individuals, and these animals typically did not form closely knit groups. Layne (1958:16) made observations that suggest fairly acute vision above water. Individuals approaching a narrow channel used their eyes above water, presumably to scan the banks for danger.

FAMILY STENIDAE. This family includes eight Recent species representing three genera. They inhabit much of the east coast of South America and the Amazon and Orinoco river systems (*Sotalia*), tropical and warm temperate waters of all oceans (*Steno*), and coastal waters, river mouths, and estuaries of southern Asia and some of the coasts of Africa.

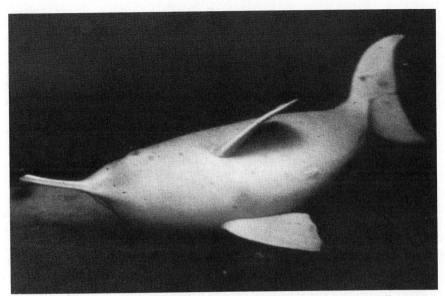

Figure 11–10. The Ganges dolphin (*Platanista gangetica*, Platanistidae), which normally swims on its side. (From Herald, E. S., et al.: Blind river dolphin: first side-swimming cetacean. Science, 166:1408–1410, 1969. Copyright 1969 by the American Association for the Advancement of Science.)

The members of this family resemble the Delphinidae except for certain details of the air-sinus system. The snout is slender and has from approximately 24/24 to 32/32 teeth. These are fairly small cetaceans, weighing from about 50 to 70 kg.

The habits of these animals are very poorly known. Scattered information indicates primarily a fish and cephalopod diet. *Sotalia*, which occurs together with the platanistid *Inia* in some large rivers in South America, may be more prone than the latter to leave the river channels and penetrate inundated jungles (Layne, 1958:4).

FAMILY PHOCOENIDAE. The members of this family are generally called porpoises. Seven Recent species of three genera are recognized. They occur widely in coastal waters of all oceans and connected seas of the Northern Hemisphere, as well as in some coastal waters of South America and some rivers in southeastern Asia. The earliest fossil record of phocoenids is from the late Miocene.

Phocoenids are small, from about 1.5 to 2.1 m in length and from roughly 90 to 118 kg in weight, and have fairly short jaws and no beak. The dorsal fin is either low or absent. The skull resembles that of the Delphinidae, but has conspicuous prominences anterior to the nares. The teeth of phocoenids are distinctive in being laterally compressed and spadelike; the crowns have two or three weakly developed cusps. The number of teeth varies from 15/15 to 30/30. From three to seven cervical vertebrae are fused.

Some phocoenids (*Phocoena* and *Neophocaena*) inhabit inshore waters, such as bays and estuaries, whereas the swift white-flanked porpoises (*Phocoenoides*) generally inhabit deeper water. Small schools of at least 100 phocoenids may assemble, and crescentic formations associated with feeding have been noted (Fink, 1959). A variety of food is taken, including such cephalopods as cuttlefish and squid, crustaceans, and fish.

FAMILY DELPHINIDAE (Figs. 11–11, 11–12). This is by far the largest and most diverse group of cetaceans. Because some species come close to shore and roll and jump conspicuously, they are the most frequently observed cetaceans. About 32 Recent species representing 14 genera are known. Delphinids inhabit all oceans and some large rivers in southeastern Asia. Fossil delphinids appear first in the early Miocene.

Figure 11–11. Dorsal view of the head of the Pacific pilot whale (*Globicephala scammoni*, Delphinidae), showing the "blowhole" (the opening of the external nares). (Photograph courtesy of Marineland of the Pacific.)

Figure 11–12. Bottle-nosed dolphin (*Tursiops truncatus*, Delphinidae) giving birth. (Photograph courtesy of Marineland of the Pacific.)

210

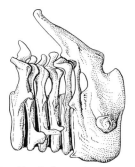

Figure 11–13. Cervical vertebrae of a dolphin (*Delphinus bairdii*, Delphinidae). Only the axis and atlas are fused, whereas most of the series are fused in some cetaceans.

Small delphinids are roughly 1.5 m in length and some 100 kg in weight, but the killer whale (*Orcinus orca*) reaches 9.5 m in length and at least 7000 kg in weight. The facial depression of the skull is large, and the frontal and maxillary bones roof over the reduced temporal fossa (Fig. 11–4). The "melon," a lens-shaped fatty deposit that lies in the facial depression, is well developed and gives many delphinids a forehead that bulges prominently behind a beaklike snout. Some delphinids, such as the killer whale, lack a beak and have a rounded profile. The number of teeth varies from 65/58 to 0/2. From two to six cervical vertebrae are fused (Fig. 11–13). Males are typically larger than females, and in some species there is considerable sexual dimorphism in the shapes of the flippers and dorsal fin. Coloration is varied: some species are uniformly black or gray, some have beautiful contrasting patterns of black and white, and still others have colored stripes or spots.

Delphinids characteristically feed by making shallow dives and surfacing several times a minute. They are rapid swimmers, and some species regularly leap from the water during feeding and traveling. In the Gulf of California I observed a bottle-nosed dolphin (*Tursiops truncatus*) leaping completely out of the water and catching mullet in midair, much as a trout catches a fly. Pacific striped dolphins (*Lagenorhynchus obliquidens*) have been trained at Marineland of the Pacific to leap over a wire 4.8 m (over 15 feet!) above the water. Most small delphinids eat fish and squid, but the killer whale is known to take a great variety of items, including large bony fish, sharks, sea birds, sea otters, seals and sea lions, porpoises, dolphins, and whales.

Most delphinids are highly gregarious, and assemblages of approximately 100,000 individuals have been observed. Some groups of delphinids kept in large tanks establish a dominance hierarchy, with an adult male having the highest position (Bateson and Gilbert, 1966). From an underwater vehicle, Evans (Evans and Bastian, 1969:448) observed that narrow-snouted dolphins (*Stenella attenuata*) had three major types of groups. The first had five to nine adult females and juveniles; the second consisted of a lone male, occasionally accompanied by a female; and the third included four to eight subadult males. Individuals of the adult female and juvenile group were spaced one above the other, with the topmost individual near the surface and the others occupying successively deeper levels. Members of the subadult male group, in contrast, were spread out horizontally. Similar spatial arrangement of groups has been observed in other species by Evans. Many recent studies have indicated that cetaceans are remarkably intelligent and inventive (see, for example, Tavolga, 1966). Their behavioral adaptability is demonstrated by the observations of Hoese (1971), who watched two bottle-nosed dolphins cooperatively pushing waves onto a muddy shore and stranding small fish. The dolphins rushed up the bank and snatched the fish from the mud before sliding back into the water.

CARNIVORES

Predation in mammals is an ancient and profitable, if not entirely honorable, occupation. Primitive carnivorous mammals (creodonts) appear in the early Paleocene, before the appearance of most of the Recent mammalian orders. Mammalian carnivores probably evolved in response to the food source offered by an expanding array of terrestrial herbivores, and underwent adaptive radiation as herbivores diversified.

The classification of carnivorous mammals used in this discussion is that of Romer (1966; see also Romer, 1968:189, 190), and recognizes the order Creodonta (primitive carnivorous mammals) and the order Carnivora. Included in the Carnivora are the suborders Fissipedia (terrestrial carnivores) and Pinnipedia (aquatic carnivores: seals, sea lions, and walruses). Other schemes of classification have been proposed. Simpson (1945), for example, regarded the Creodonta, Fissipedia, and Pinnipedia as suborders of the order Carnivora.

Most Recent fissiped carnivores are predaceous and have a remarkable sense of smell. Cursorial ability may be limited, as in the Ursidae and Procyonidae, or may be strongly developed, as in the cheetah and some canids. The braincase is large; the orbit is usually confluent with the temporal fossa; the turbinal bones are usually large, and their complex form provides a large surface area for olfactory epithelium. There are usually 3/3 incisors (3/2 in the sea otter, *Enhydra lutris*), and the canines are large and usually conical; the cheek teeth vary from 4/4 premolars and 2/3 molars in long-faced carnivores, such as the Canidae and Ursidae, to 2/2 premolars and

1/1 molars, as in some cats. The fourth upper premolar and the first lower molar are carnassials (specialized shearing blades). The teeth are rooted. The condyle of the dentary and the glenoid fossa of the squamosal are transversely elongate and allow no rotary jaw action and only limited transverse movement.

Cursorial adaptations evident in the carpus include the fusion of the scaphoid and lunar bones and the loss of the centrale (Fig. 12–1). The foot posture is plantigrade, as in ursids and procyonids, or digitigrade, as in canids, hyaenids, and felids. Little reduction of digits has occurred; the greatest reduction occurs in the hyenas and in the African hunting dog (*Lycaon pictus*), in which the manus and pes have four toes.

ORDER CREODONTA

The oldest carnivorous mammals, order Creodonta, appeared in the early Paleocene, were the typical carnivores of the Paleocene and Eocene, and persisted in

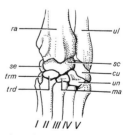

Figure 12–1. Anterior view of the left carpus of the grey fox (*Urocyon cinereoargenteus*). Abbreviations: *cu*, cuneiform; *ma*, magnum; *ra*, radius; *se*, sesamoid; *sc*, scapholunar; *trd*, trapezoid; *trm*, trapezium; *ul*, ulna; *un*, unciform.

Old World tropical refugia into the early Pliocene. Some creodonts appear to have retained the insectivorous food habits of their ancestors, whereas some were carnivorous and some were omnivorous. The creodont skull (Fig. 12–2) differs from that of the Carnivora in lacking an ossified auditory bulla, and in either having no carnassial pair or in having M^1 and M_2 or M^2 and M_3 forming the carnassials. (Abbreviations of teeth utilize capital letters for the type of tooth and a number either at the top or the bottom of the letter to designate the number of the tooth. For example, upper premolar four is PM^4, and lower molar number one is M_1.)

The creodont braincase was small, and intelligence was presumably low. The limbs were primitive. The feet were usually five-toed and plantigrade, and the limbs were often short; the scaphoid and lunar of the carpus were not fused and the centrale was present; the distal phalanges were fissured, and in some species bore flattened, rather than clawlike, nails. Creodonts diversified in the Paleocene and Eocene, and the Late Eocene genus *Andrewsarchus* is the largest known land carnivore. Its skull is roughly a meter in length.

ORDER CARNIVORA

SUBORDER FISSIPEDIA

Modern terrestrial carnivores, suborder Fissipedia, are advanced, basically predatory types. Fissipeds largely replaced the smaller brained and less cursorial creodonts early in the Tertiary. The basal fissiped family is the Miacidae, which first appears in middle Paleocene beds of North America and is not known after the Eocene. Resemblances between early miacids and *Cimolestes*, a Late Cretaceous insectivore of the extinct family Palaeoryctidae, suggest that the order Carnivora evolved from an ancestral stock within the order Insectivora. Miacids were small and perhaps mostly arboreal carnivores. Morphologically transitional between the Creodonta and Carnivora in some features, the miacids had the modern carnassial arrangement, but lacked ossified bullae and had separate scaphoid and lunar bones. In contrast to the creodonts, the brain was fairly large, and the distal phalanges were not fissured. Except for the Miacidae, all of the families of the suborder Fissipedia survive today.

FAMILY CANIDAE (Figs. 12–3, 12–4). Fifteen genera with about forty-one Recent species comprise this familiar family. The

Figure 12–3. The handsome face of an Alaskan wolf. This photograph is from the classic study *The Wolves of Mount McKinley,* by Adolf Murie (1944).

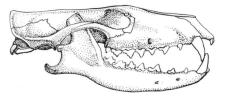

Figure 12–2. The skull of *Sinopa,* an Eocene creodont in which the carnassials are M^2 and M_3; length of skull approximately 150 mm. (After Romer, 1966.)

Figure 12–4. A young kit fox (*Vulpes macrotis*) at the entrance to its burrow. This fox occurs in the deserts of the southwestern United States. (Photograph by O. J. Reichman.)

crushing surfaces, indicating a more flexible diet than that of the more strictly carnivorous cat family (Felidae). The limbs in most species are long, and rotation at the joints distal to the shoulder and hip joints is reduced in the interest of cursorial ability. The clavicle is absent. The feet are digitigrade, and the well-developed but blunt claws are nonretractile. The forepaw usually has five toes, and the hind paw has four. The weight of canids ranges from roughly 1 to 75 kg.

As a family, the Canidae are the most cursorial carnivores. Many species forage tirelessly over large areas, and lengthy pursuit is frequently part of the hunting technique. (The canid hunting style is discussed in Chapter 17, p. 356.) The coyote (*Canis latrans*), probably one of the swiftest canids, can run at speeds of up to about 65 km per hour. The fact that coyotes in many areas depend partly on jackrabbits for food is an impressive testimonial to this carnivore's speed. Canids often hunt in open country, and wolves (*Canis lupus*) and the African hunting dog (*Lycaon pictus*) seem to rely more on endurance than on speed when hunting. These canids, and the east Asian dholes (*Cuon alpinus*), habitually hunt in packs and kill larger prey than could be overcome by a solitary hunter. The gray fox (*Urocyon cinereoargenteus*) does not generally forage in open areas, but is amazingly agile and can run rapidly through the maze of stems beneath a canopy of chaparral. The food of canids includes vertebrates, arthropods, mollusks, carrion, and many types of plant material. Black-backed jackals have recently become a

canids—foxes, wolves, dogs, and jackals—occupy a great array of environments from the arctic to the tropics. Prior to their dispersal with man, canids occurred nearly worldwide except on most oceanic islands. The dingo (*Canis dingo*) was probably brought to Australia by early man. The canids appeared in the late Eocene in Europe and North America and have occupied these areas continuously to the Recent.

Canids are broadly adapted carnivores, a feature reflected in their morphology. The canid skull typically has a long rostrum (Fig. 12–5A) that houses a large nasal chamber with complex turbinal bones, a feature associated with a remarkable sense of smell. Most canids have a nearly complete placental complement of teeth (3/3, 1/1, 4/4, 2/3 = 42); the canines are generally long and strongly built, and the carnassials retain the shearing blades (Fig. 12–6A, B). The postcarnassial teeth have

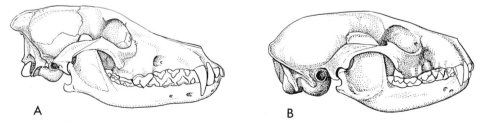

Figure 12–5. Skulls of two carnivores. A, The coyote (*Canis latrans*; length of skull 202 mm). B, Raccoon (*Procyon lotor*; length of skull 115 mm).

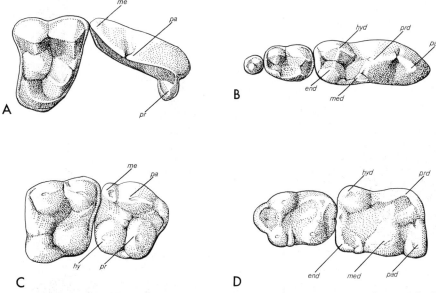

Figure 12-6. Occlusal view of some cheek teeth of two carnivores; the right upper teeth are on the left, and the left lower teeth are on the right. The cusps of the carnassials (P^4 and M_1) are labeled. A and B, The coyote (*Canis latrans*); the cheek teeth serve both shearing and crushing functions. C and D, A raccoon (*Procyon lotor*); note that the upper carnassial has a hypocone and that all of the teeth are adapted to crushing. Abbreviations: *end*, entoconid; *hy*, hypocone; *hyd*, hypoconid; *me*, metacone; *med*, metaconid; *pa*, paracone; *pad*, paraconid; *pas*, parastyle; *pr*, protocone; *prd*, protoconid.

problem in parts of South Africa because of their extensive feeding on pineapples (Ewer, 1968:30), and coyotes in parts of the western United States feed heavily on cultivated crops such as melons and non-cultivated plant material such as juniper berries and prickly-pear cactus fruit. The average canid is clearly an opportunist; this may in large part account for the great success of this family.

FAMILY URSIDAE. The bears are notable for their large size and their departure from a strictly carnivorous mode of life. This family contains six genera and eight species. Morphological and physiological evidence suggests that the giant panda *Ailuropoda melanoleuca*, is a bear, and it is included here in the Ursidae. The distribution includes most of North America and Eurasia, the Malay Peninsula, the South American Andes, and the Atlas Mountains of extreme northwestern Africa. Bears inhabit diverse habitats, from drifting ice in the arctic to the tropics, but are most important in boreal and temperate areas.

The bears are an offshoot from the canid

evolutionary line, and first appeared in Europe in the middle Miocene. Ursids apparently did not enter North America until late Pliocene. They probably reached South America and northwest Africa in the Pleistocene and Recent, respectively.

The bear skull retains the long rostrum typical of the canids, but the orbits are generally smaller and the dentition is very different (Fig. 12-7). The postcarnassial teeth are greatly enlarged, and the occlusal surfaces are "wrinkled" and adapted to crushing (Fig. 12-8B). On the other hand,

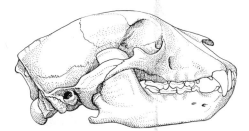

Figure 12-7. Skull of a black bear (*Ursus americanus*); length of skull 289 mm. Note the diastema behind the anteriormost cheek tooth and the small upper carnassial (the third tooth in back of the canine).

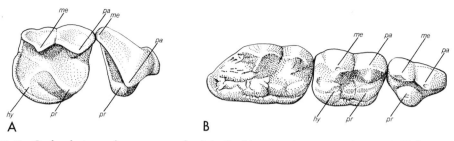

Figure 12–8. Occlusal views of some upper cheek teeth of two omnivorous carnivores. *A*, Right upper carnassial (P⁴) and first molar of the hog-nosed skunk *(Conepatus mesoleucus)*; note the blade on the carnassial and the broadened crushing surface on M¹. *B*, Right upper carnassial and two molars of the black bear *(Ursus americanus)*; note the small, nontrenchant carnassial and the greatly lengthened molars. Abbreviations are as in Figure 12–6.

the first three premolars are usually rudimentary or they may be lost, and a diastema usually occurs between premolars. The upper carnassial is roughly triangular because of the posterior migration of the protocone, and is much smaller than the neighboring molars (Fig. 12–8B); both upper and lower carnassials no longer have a shearing function. The dental formula is usually 3/3, 1/1, 4/4, 2/3 = 42, but premolars may be lost with advancing age. The limbs, especially the forelimbs, are strongly built; the plantigrade feet have long, nonretractile claws. There are five toes on each foot. The ears are small and the tail is extremely short. In size, bears range from that of a large dog to over 760 kg (1650 lbs).

The abandonment of cursorial ability in favor of power of the limbs, and the loss of the shearing function of the cheek teeth in favor of a crushing battery, has accompanied the adoption of omnivorous feeding habits. The strong forelimbs can aid in the search for food by rolling stones or tearing apart logs, and the crushing surfaces of the molars can cope with many kinds of food, from insects and small vertebrates to berries, grass, and pine nuts. Carrion is also avidly sought. The polar bear *(Ursus maritimus)* has a more restricted diet, consisting largely of seals, and the giant panda eats mostly bamboo shoots. In areas with cold winters, bears sleep for much of the winter in caves or other retreats protected from drastic temperature fluctuations. This is not a true hibernation, however, because the temperature and metabolic rate do not drop precipitously.

FAMILY PROCYONIDAE. This family, which includes the familiar raccoon *(Procyon)* and its relatives, probably had a common ancestry with the canids. As with the bears, in procyonids omnivorous feeding habits have become predominant. Some 7 genera and 18 species are known. The taxonomic position of the lesser panda *(Ailurus)* is controversial; it is here regarded as a procyonid. Procyonids occupy much of the temperate and tropical parts of the New World, from southern Canada through much of South America. The lesser panda occurs in south central China, northern Burma, Sikkim, and Nepal. Procyonids chiefly inhabit forested areas, but the range of one species of ring-tail *(Bassariscus astutus)* includes arid desert mountains and foothills. Procyonids are known from the late Oligocene to the Recent in North America, from the late Miocene to the late Pliocene in Europe, and from the Pliocene to the Recent in Asia. They reached South America from North America in the Pliocene.

The structural and functional departure of procyonids from the carnivorous norm has included adaptations favoring both omnivorous feeding habits and climbing ability. Associated with the omnivorous trend has been a specialization of the cheek teeth. The premolars are not reduced, as in the bears, but the shearing action of the carnassials is nearly lost. Instead, the carnassials are high-cusped crushing teeth; a hypocone was added to the upper, and in the lower the talonid was enlarged and broadened (Fig. 12–6C, D) In contrast to the elongate upper molars of bears, those of procyonids are broader than they are long. The dental formula is usually 3/3, 1/1, 4/4, 2/2 = 40. (A procyonid skull is shown in Fig. 12–5B). There are five toes on each foot; the foot posture is usually plantigrade, and the claws are nonretractile or semiretractile. The limbs are fairly long. The toes are separate, and the forepaw has considerable dexterity in some species and is used in food handling. Tracks left by the manlike hand of the raccoon are familiar to many. The tail is long, is generally marked by dark rings, and is prehensile in the arboreal kinkajou (*Potos flavus*). Procyonids are of modest size, weighing from less than a kilogram to about 20 kg.

The familiar raccoon often takes advantage of man's crops. Corn is a staple food item for Midwest raccoons, and they eat grapes, figs, and melons in parts of California (Grinnell et al., 1937:159,160). In addition, they prey on a variety of small vertebrates, and some invertebrates. Some tropical procyonids are largely vegetarians. Hall and Dalquest (1963) report that in Veracruz the coati (*Nasua narica*) eats corn, bananas, and the fruit of the coyol palm, and that kinkajous eat mostly fruit. The ringtail, on the other hand, is known to feed mostly on small rodents in some areas. Procyonids reach their greatest diversity and greatest densities in the Neotropics, where the animals are largely arboreal; in some tropical forests several species may occur together. In such areas the nocturnal, quavering cries of kinkajous can be heard regularly.

FAMILY MUSTELIDAE. This large family, with 25 genera and some 70 Recent species, includes the weasels, badgers, skunks, and otters. Mustelids occupy virtually every type of terrestrial habitat, from arctic tundra to tropical rain forests, and they occur in rivers, lakes, and the sea. The distribution is nearly cosmopolitan, but they do not inhabit Madagascar, Australia, or oceanic islands. Mustelids appear

in the fossil records of North America and Eurasia in the early Oligocene, but they did not reach South America and Africa until the Pliocene.

These are typically fairly small, long-bodied carnivores with short limbs and "pushed in" faces (Fig. 12–9). The skull generally has a long braincase and a short rostrum (Figs. 12–10, 12–11), and the postglenoid process partially encloses the glenoid fossa so that in some species the condyle of the dentary is difficult to disengage from the fossa. Obviously, little lateral and no rotary jaw action is possible. The dentition is quite variable, but is generally 3/3, 1/1, 3/3, 1/2 = 34. The carnassials are trenchant in many species (Fig. 12–12C, D), but have been modified into crushing teeth in others; in the sea otter (*Enhydra lutris*), for example, none of the cheek teeth are trenchant, the carnassials have rounded cusps adapted to crushing, and the postcarnassial teeth, M^1 and M_2, are broader than they are long (Fig. 12–12E, F). The first upper molar is frequently hourglass-shaped in occlusal view (Fig. 12–12C), or may be expanded into a large crushing tooth, as in skunks (Fig. 12–8A). The limbs are usually short, the five-toed feet are either plantigrade or digitigrade, and the claws are never com-

Figure 12–9. A pair of river otters (*Lontra canadensis*); one animal is grooming the fur of the other.

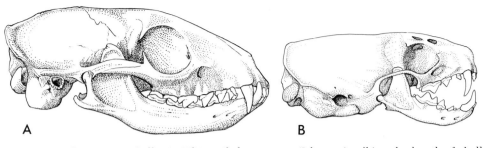

Figure 12–10. Small carnivore skulls. A, White-tailed mongoose (*Ichneumia albicauda,* length of skull 106 mm); B, least weasel (*Mustela nivalis;* length of skull 31 mm).

Figure 12–11. The skull of the sea otter *(Enhydra lutris);* length of skull 152 mm. The heavy cheek teeth are adapted to crushing marine invertebrates (see also Fig. 12–12).

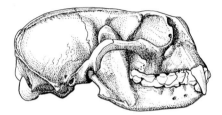

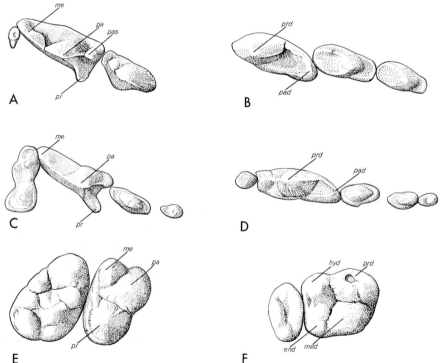

Figure 12–12. Occlusal views of the cheek teeth of three carnivores; the right upper teeth are on the left, and the left lower teeth are on the right. The cusps of the carnassials (P^4 and M_1) are labeled, and abbreviations are as in Figure 12–6. In A the parastyle *(pas)* is also shown. A and B, and entire sets of cheek teeth of the bobcat *(Lynx rufus).* Note the lack of crushing teeth. Only sectorial teeth are present; the parastyle *(pas)* of P^4 increases the length of its shearing blade, and the loss of the talonid of M_1 makes this tooth entirely bladelike. C and D, the entire sets of cheek teeth of the least weasel *(Mustela nivalis);* the crushing teeth, M^1, and M_2 are reduced, and the shearing function of the cheek teeth is of major importance. E, P^4 and M^1, and F, M_1 and M_2, of the sea otter *(Enhydra lutris).* Note that all of the teeth have rounded cusps and are adapted to crushing; the "carnassials" retain no shearing function.

pletely retractile. Anal scent glands are usually well developed; they are extraordinarily large in skunks and are used for defense. The tail is generally long, and the pelage may be conspicuously marked, as in skunks and badgers. Some mustelids have beautiful, glossy fur that has considerable value in the fur trade. In size, mustelids range from the smallest member of the order Carnivora, a circumboreal weasel (Mustela nivalis) which weighs some 35 to 50 gm, to the fairly large sea otter (about 35 kg).

Mustelids, although basically carnivorous, pursue many styles of feeding. Most mustelids aggressively search for prey in burrows, crevices, or dense cover. Many mustelids are able killers. In Colorado, I observed a male long-tailed weasel (M. frenata) killing a young cottontail rabbit (Sylvilagus audubonii) roughly twice its own weight. The weasel killed by biting the back of the rabbit's skull repeatedly. According to Errington (1967:24), a mink (Mustela vison) "hugs its victim with the forelegs while it scratches violently with the hind legs and bites vital parts, especially about head and neck." Some mustelids, such as the beautiful and graceful marten (Martes americana), are swift and agile climbers and feed partly on arboreal squirrels. Otters (subfamily Lutrinae) are semiaquatic, or almost completely aquatic in the case of the sea otter, and feed on a wide array of vertebrates and invertebrates. The skunks, with no claim to remarkable agility or killing ability, seem to feed on whatever animal material is most readily available, which during the summer is generally insects.

FAMILY VIVERRIDAE. This is the largest family in the order Carnivora with regard to numbers of species, and includes the civets, genets, and mongooses. Thirty-six genera and roughly seventy-five Recent species are recognized, but the taxonomy of this family is still uncertain. Viverrids inhabit much of the Old World; however, the center of their distribution is in tropical and southern temperate areas, and they are absent from northern Europe and all but southern Asia, as well as from New Guinea and Australia. This is an old group: it appeared in the late Eocene of Europe, but did not reach Africa until the Pleistocene, and only reached Madagascar in the Recent.

Viverrids are usually small, fairly short-legged and long-tailed carnivores (Figs. 12–13, 12–14). Like mustelids, some viverrids have well-developed scent glands. The viverrid skull frequently has a moderately long rostrum (Fig. 12–10A). The premolars are large and the carnassials are usually trenchant. The upper molars are tritubercular and are wider than they are long; the lower molars have well-developed talonids. The dental formula is generally 3/3, 1/1, 3-4/3-4, 2/2 = 36-40. The five toes on each foot include a much

Figure 12–13. A, African civet (Viverra civetta), a large and rather dog-like viverrid; B, large spotted genets (Genetta tigrina), adroit climbers that often feed on birds. (Photographs taken in Kenya, East Africa; civet by Richard G. Bowker, genets, Thomas R. Huels.)

Figure 12–14. Two African viverrids. A, White-tailed mongoose *(Ichneumia albicauda)*, a solitary, nocturnal species; B, the dwarf mongoose *(Helogale parvula)*, a highly social, diurnal species. (Photographs taken in Kenya, East Africa, by Richard G. Bowker.)

reduced pollex or hallux. The foot posture is plantigrade or digitigrade, and the claws are partly retractile. The ears are generally small and rounded. Some species are banded, others are spotted, and still others are striped. The smallest viverrid weighs less than a kilogram, and the largest weighs 14 kg. Because of the great morphological variation within this family, it is difficult to frame a definitive structural diagnosis applicable to all species.

Viverrids make their livings in a variety of ways. Most are carnivorous, and eat small vertebrates or insects. Some feed on the ground, and some in trees. In some parts of East Africa, viverrids are remarkably common and diverse for carnivores, with half a dozen or more species occurring in the same area. Some diurnal spe-

cies are highly social and are easily seen (Fig. 12–14B). Foraging parties of the banded mongoose *(Mungos mungo)* feed heavily on the dung beetles attracted to the feces of large ungulates. The palm civets (subfamily Paradoxurinae) are omnivorous and often feed primarily on fruit or other plant material. Some viverrids are semiaquatic and feed largely on aquatic animals, and some mongooses *(Herpestes)* kill and eat snakes, including certain highly venomous species. The habits of several viverrids are almost unknown; the Congo water civet *(Osbornictis piscivora)*, for example, is represented in museums by only a few specimens and has never been observed in the wild by a biologist. Probably no family of carnivores is so poorly known as the viverrids; clearly, much field study could profitably be concentrated on this group.

FAMILY HYAENIDAE. Many carnivores will eat carrion if the opportunity arises, but most members of the family Hyaenidae have become specialized for carrion feeding. This is a small family, with but three genera and four Recent species. The distribution includes Africa, southwestern Asia, and parts of India. The Hyaenidae, probably derived from viverrid stock, appeared in Eurasia in the late Miocene and, except for *Chasmaporthetes* (which probably crossed the Bering Strait land bridge and is known from the Pleistocene of North America), has been an entirely Old World family.

Leaving the unusual aardwolf *(Proteles cristatus)* aside temporarily, hyenas are characterized by rather heavy builds, forelimbs longer than the hind limbs, strongly built skulls, and powerful dentitions (Fig. 12–15A). The carnassials are well developed, and all of the cheek teeth have heavily built crowns adapted to bone crushing. The dental formula is 3/3, 1/1, 4/3, 1/1 = 34. The feet are digitigrade, and both the forepaws and the hind paws have four toes that bear blunt, nonretractile claws. The pelage is either spotted *(Crocuta;* Fig. 12–16) or variously striped *(Hyaena).* Hyenas weigh up to 80 kg.

The aardwolf (subfamily Protelinae), in comparison with the hyenas (Hyaeninae), is lightly built and has a delicate skull and smaller teeth (Fig. 12–15B). All teeth except the canines are small, and the cheek teeth are simple and conical. The dental formula is generally 3/3, 1/1, 3/2-1, 1/1-2 = 28-32, but frequently some of these teeth are lost

(as in the skull shown in Fig. 12–15B). The forefeet have five toes and the hind feet have four. The animal is striped and has a mane of long hair from neck to rump. The tail is quite bushy. The hair of the mane and tail is erected but the mouth remains closed when the animal is threatened and adopts a defensive posture. *Proteles* has abandoned the open-mouthed threat used by most carnivores, which in its case would merely advertise the weakness of the dentition, in favor of extensive erection of the long hair. The aardwolf also releases fluid from the well-developed anal glands when attacked.

Hyenas in some areas specialize in scavenging on the kills of lions and other large carnivores, and are able to drive cheetahs *(Acinonyx)* from their kills. They may also forage around in villages at night for edible refuse. In their ability to crush large bones they are unsurpassed. But recent studies by Kruuk (1966) and others have shown that spotted hyenas are also powerful predators. Often hunting in packs of up to 30 animals, these nocturnal hunters can bring down even zebras. Indeed, in the Ngorongoro Crater in Tanzania, a reversal of the usual pattern of interactions between lions and hyenas has occurred: spotted hyenas are better able than the lions to make regular kills, and lions live by driving hyenas from their kills and eating the carrion (Ewer, 1968:103).

The strange aardwolf eats mostly termites. This animal is directed to termites largely by the sounds they make (Kruuk and Sands, 1972), and its unusually large auditory bullae may be associated with an

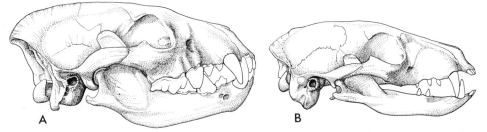

Figure 12–15. Hyaenid skulls. A, Spotted hyena *(Crocuta crocuta;* length of skull 248 mm); aardwolf *(Proteles cristatus;* length of skull 148 mm).

enhancement of the sense of hearing. In contrast to many termite feeders, the aardwolf does not dig for termites, but laps them from the surface of the ground.

FAMILY FELIDAE (Figs. 12–17, 12–18). Within this family are 4 or 5 Recent genera (depending on whether or not the genus *Lynx* is regarded as separate from *Felis*) with some 37 species. (No general agreement has been reached on felid taxonomy.) The cats are quite a uniform group; all cats, from the pampered house cat (*Felis catus*) to the tiger (*Panthera tigris*), bear a strong family resemblance. This family

Figure 12-17. Two small, New World cats that occur largely in tropical and subtropical areas: A, ocelot (*Felis pardalis*); B, jaguarundi (*Felis yagouaroundi*).

Figure 12–18. The cheetah (*Acinonyx jubatus*), the fastest cursorial mammal. Note the nonretractile claws. This animal is in a semicrouch and is stalking some Thompson's gazelles. (Photograph taken in Nairobi National Park, Kenya, East Africa, by Matthew P. Vaughan.)

occurs worldwide, with the exceptions of Antarctica, Australia, Madagascar, and some isolated islands. Of all of the carnivores, the cats are the most proficient killers; some species regularly kill prey as large as or considerably larger than themselves.

The evolution of the cats, the most specialized carnivores, was rapid, and the evolutionary pattern was unusual. There are differences of opinion concerning the interpretation of the fossil record of felids, but the following broad features of their evolution are agreed upon by many. (The taxonomic scheme used here for the felids is that of Simpson, 1945.) Two lines of descent, represented by the subfamilies Nimravinae and Machairodontinae, were established in the early Oligocene, and members of these groups were probably widely sympatric (occupying the same area at the same time) in North America and Europe from the Oligocene through the Pleistocene. The subfamily Nimravinae includes species with somewhat enlarged upper canines and small lower canines. A progressive reduction of the upper canines and development of "true cat" features are evident in Miocene and Pliocene nimravines, and the Recent cats (Felinae) perhaps descended from nimravine ancestry in the late Miocene or early Pliocene. Although the structure of members of the felid ancestral stock is not documented by the fossil record, the progenitors of felids probably lacked enlarged upper canines. In the Nimravinae, then, an early trend

toward the sabertooth specialization was seemingly reversed; the upper canines became smaller, and certain specializations of the skull and lower jaw were lost.

Forms representing the second line of descent, the Machairodontinae (sabertoothed cats), are first known from the Oligocene, when already they had larger canines than did nimravines (Fig. 12–19). In contrast to the nimravines, however, the sabertooth specializations became progressively more pronounced in machairodonts. This felid line, together with many of the large ungulates, became extinct at the end of the Pleistocene. The most extreme, almost grotesque, degree of sabertooth adaptation occurred in the Late Pliocene machairodont *Barbourofelis fricki* of North America (Schultz et al., 1970). This animal was roughly the size of an African lion and had enormous sabers, an orbit protected by a heavy postorbital bar, and an oddly shortened cranium that housed a relatively small brain (Fig. 12–20). Probably the most impressive and fearsome of all cats, this animal became extinct at the end of the Pliocene, perhaps as a result of the extinction at this time in North America of the large Pliocene rhinos and mastodonts.

Many machairodont specializations are adaptations to the unique mode of attack these animals employed. It is generally believed that the sabers were plunged into large prey by powerful downward and forward thrusts of the head, while the cat clung to the prey with the heavily muscled forelimbs. The greatly enlarged mastoid

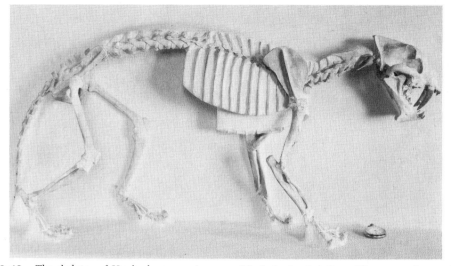

Figure 12–19. The skeleton of *Hoplophoneus primaevus* (Machairodontinae), an Oligocene saber-toothed cat. (Courtesy of Robert W. Wilson; South Dakota School of Mines, No. 2528.)

process is one of many adaptations to the stabbing action. This enlarged process provided a large surface area and unusual mechanical advantage for the origin of the sternomastoideus, the muscle that pulled the snout downward. The receding nasal bones probably allowed the nose pad to be withdrawn so that the animal could breathe while its sabers were deeply imbedded in its prey. The upper carnassial had a prostyle (in addition to the usual parastyle that occurs in Recent felids; see Fig. 12–12A) that increased the length of the shearing blade formed by this tooth.

The felid rostrum is short, an adaptation furthering a powerful bite, and the orbits

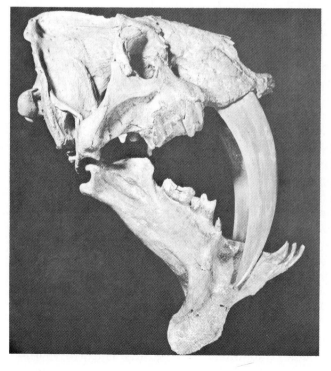

Figure 12–20. Skull of *Barbourofelis fricki*, an extinct Pliocene felid from Nebraska that represents the extreme in sabertooth specialization. Length of skull about 290 mm. (Photograph from Schultz et al., 1970.)

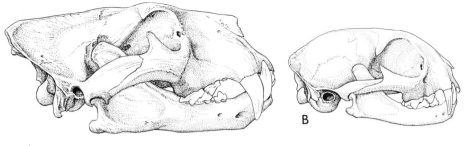

Figure 12–21. Felid skulls. A, African lion (*Panthera leo*; length of skull 366 mm); B, bobcat (*Lynx rufus*; length of skull 120 mm).

in most species are large (Fig. 12–21). The number of teeth is reduced. The typical dental formula is 3/3, 1/1, 3/2, 1/1 = 30, and the anteriormost upper premolar is strongly reduced or lost (as in *Lynx*). The carnassials are well developed and have specializations that enhance their shearing ability (Fig. 12–12A, B). The foot posture is digitigrade. The forelimbs are strongly built, and the manus can be supinated; the claws are sharp and recurved and are completely retractile, except in the cheetah (*Acinonyx*), in which they are partly retractile. These features of the forelimbs allow cats to clutch and grapple with prey with the forelimbs. Some species are spotted or striped; these color patterns enable the animals to conceal themselves effectively (Fig. 12–22). The weight of cats varies from about that of a domestic cat (3 kg) to 275 kg, in the tiger.

As described in Chapter 17 (p. 357), cats usually catch prey by a stealthy stalk followed by a brief burst of speed. They are typically sight hunters, and some species spend considerable time watching for prey and waiting for it to move into striking distance. Many kinds of animals are eaten,

Figure 12–22. A leopard (*Panthera pardus*) well concealed against a background of mottled light and shadow. (Photograph by W. Leslie Robinette.)

from fish, mollusks, and small rodents to ungulates as large as buffalo. *Felis planiceps* of southeastern Asia seems to be the only cat with noncarnivorous tendencies; this species prefers fruit (Goodwin, 1954:570).

SUBORDER PINNIPEDIA

Many distinctive morphological features of pinnipeds are adaptations to marine life. Compared to land carnivores, pinnipeds are large (approximately 91 to 3600 kg). Large size improves metabolic economy in cold environments because of the favorable mass to surface ratio of large animals (as discussed on page 418). According to Scheffer (1958:8), large size in pinnipeds is primarily an adaptation to a cold environment. The body is insulated by thick layers of blubber. The pinnae (external ears) are either small or absent, the external genitalia and mammary nipples are withdrawn beneath the body surface, the tail is rudimentary, and only the parts of the limbs distal to the elbow and knee protrude from the body surface. As a result, the torpedo-shaped body has smooth contours and creates little drag during swimming. The nostrils are slitlike (Fig. 12–23), are normally closed, and are opened by voluntary effort.

The skull is partially telescoped, with the supraoccipital partially overlapping the parietals; the rostrum is usually shortened and the orbits are usually large and encroach on the narrow interorbital area. Either one or two pairs of lower incisors are present. The canines are conical; the cheek teeth are homodont (none are modified as carnassials), two-rooted, and usually simple and conical (Fig. 12–24); they vary in total number from 12 to 24. In some pinnipeds, cheek teeth are characteristically lost with advancing age. The limbs and girdles are highly specialized. The clavicle is absent, and the humerus, radius, and ulna are short and heavily built; the pollex is the longest and most robust of the five digits, and forms the leading edge of the winglike fore flipper. The pelvic girdle is small and is nearly parallel to the vertebral column. The femur is broad and flattened. The first and fifth are the longest digits of the pes, and both the manus and pes are fully webbed.

The reduction of the vertebral zygapo-

Figure 12–23. The face of a California sea lion (*Zalophus californianus*). Note the valvular nostrils and the external ear. (Photograph courtesy of the United States Navy.)

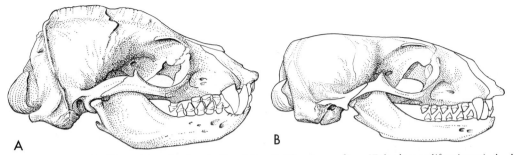

Figure 12–24. Sexual dimorphism in the skulls of the California sea lion (*Zalophus californianus*). A, An adult male, length of skull 290 mm; B, an adult female, length of skull 214 mm.

physes and the absence of the clavicle allow the vertebral column and the fore-limbs considerable flexibility and freedom of movement; these features may favor rapid maneuvers during the pursuit of prey. Although terrestrial locomotion is characteristically slow and laborious in most species, the importance of terrestrial locomotion when the animals are hauled out on rocks or ice, or are on breeding grounds, has probably limited the extent to which the limbs of pinnipeds have become specialized for swimming.

Although not so completely adapted to aquatic life as are cetaceans, pinnipeds make impressively deep dives. Scattered records of seals with depth recording instruments and of individuals caught on the hooks of deepset fishing lines indicate that some seals reach depths of at least 600 m (Kooyman and Andersen, 1969:72). Dives lasting 20 minutes have been recorded for several species of seals (Degerbøl and Freuchen, 1935:210; Scheffer and Slipp, 1944; Backhouse, 1954). The following responses to diving occur in pinnipeds: the heart rate slows to as low as one tenth normal; the metabolic rate and body temperature drop; peripheral vasoconstriction occurs, but normal blood flow to vital organs continues. In addition, their tolerance of CO_2 in the tissues is unusually high, and large amounts of myoglobin in the muscles aid in the storage of oxygen.

FAMILY OTARIIDAE (Figs. 12–23, 12–25). This family, containing 6 genera and 12 Recent species, includes the eared seals and the sea lions. These animals inhabit

many of the coastlines of the Pacific Ocean and parts of the South Atlantic and Indian Oceans. They are common along the Pacific coast of North America. The earliest otariids are known from the middle Miocene, and may have originated in the food-rich kelp reefs of the North Pacific (Scheffer, 1958:33).

Otariids differ from other pinnipeds in being less highly modified for aquatic life and better able to move on land. The hind flippers can be brought beneath the body and used in terrestrial locomotion, and well developed nails occur on the three middle digits. A small external ear is present (Fig. 12–23). Males are much larger than females; in the northern fur seal (*Callorhinus ursinus*), males weigh four and a half times as much as females. Considerable sexual dimorphism in the shape of the skull occurs in some species (Fig. 12–24), and in males the skull becomes larger and more heavily ridged with advancing age. The dental formula is 3/2, 1/1, 4/4, 1-3/1 = 34-38. The body is covered with fur which is uniformly dark. Weights of otariids range from roughly 60 to 1000 kg.

These seals are generally highly vocal, and utter a great variety of sounds. They tend to be gregarious all year round and are social during the breeding season, when they assemble in large breeding rookeries (see page 383). Propulsion in the water is accomplished by powerful downward and backward strokes of the forelimbs; speeds up to 27 km per hour have been recorded (Scheffer, 1958:13). Otariids

Figure 12–25. Steller sea lions (*Eumetopias jubata*) on Año Nuevo Island, off the coast of California. Above, a bull with cows and young. Below, cows with young. (Photographs by Robert T. Orr.)

eat mostly squid and small fish that occur in schools, and maneuver rapidly in pursuit of prey. In the Gulf of California, sea lions frequently assemble around commercial fishing boats, signal their presence by barking, and eat discarded fish.

FAMILY ODOBENIDAE. This is a monotypic family; that is, it contains only one species, *Odobenus rosmarus*, the walrus. This species occurs near shorelines in arctic waters of the Atlantic and Pacific Oceans, but may stray southward to some extent along the coastlines. Odobenids first appear in the late Miocene and, like the otariids, may have had a North Pacific origin.

The walrus is a large pinniped (up to 1270 kg), with a robust build, a nearly hairless skin, and no external ears. The hind flippers can be brought beneath the body and are used for terrestrial locomotion, which is ponderous and slow. In both sexes the upper canines are modified into long tusks (Fig. 12–26), which in the adult lack enamel. There are no lower incisors in adults, and 12 cheek teeth are usually present. The dental formula is 1-2/0, 1/1, 3-4/3-4, 0/0 = 18-24. On the huge mastoid

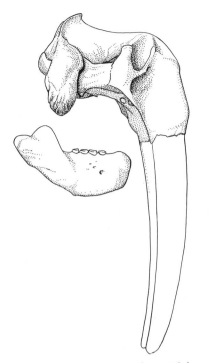

Figure 12–26. The skull of a walrus *(Odobenus ros-marus)*; length of skull approximately 355 mm. The tusks are enlarged upper canines.

processes attach the powerful neck muscles that pull the head downward.

Walruses feed on mollusks, which they rake from the sea floor by means of the lips and the huge tusks. They have been observed to eat cetaceans (Scheffer, 1958:15,

16), but this must be an extremely unusual food item. Walruses are gregarious and polygynous, and frequently assemble in large groups of more than 1000 individuals. They are migratory to some extent, moving southward in winter. Walruses make a variety of loud noises when out of water, and make a "church bell" sound and rasps and clicks underwater (Schevill et al., 1963; Schevill et al., 1966). The fact that the rasps and clicks are made during swimming suggests their use in echolocation.

FAMILY PHOCIDAE (Fig. 12–27, 12–28, 12–29). There are 18 species and 13 genera of earless seals, the most abundant pinnipeds. They occur along most northern (above 30° north latitude) and most southern (below 50° south latitude) coastlines and in some intermediate areas. They appear in the middle Miocene, and presumably originated in the Northern Hemisphere.

The earless seals are more highly specialized for aquatic life than are other pinnipeds. As the vernacular name implies, there is no external ear. The hind flippers are useless on land, but, as a result of lateral undulatory movements of the body, are the primary propulsive organs in the water. The fore flippers are short and well furred. The structure of the cheek teeth is

Figure 12–27. Northern elephant seal *(Mirounga angustirostris)* on Año Nuevo Island, off the coast of California. (Photograph by Robert T. Orr.)

Figure 12–28. Weddell seals (*Leptonychotes weddelli*) near McMurdo Station, Antarctica. (Photograph courtesy of the United States Navy.)

highly variable, but is usually fairly simple. In the crabeater seal *(Lobodon)*, however, the cheek teeth have complex cusps (Fig. 12–30). The pelage of most phocids is spotted, banded, or mottled. These seals frequently have extremely heavy layers of subcutaneous blubber that give the bodies smooth contours and, in some cases, a nearly perfect fusiform shape. Most species weigh from about 80 to 450 kg, but

Figure 12–29. Harbor seals *(Phoca vitulina)* on Año Nuevo Island, off the coast of California. (Photograph by Robert T. Orr.)

Figure 12–30. Medial view of two right lower cheek teeth of the crab-eater seal (*Lobodon carcinophagus*). These complex teeth enable this animal to depend on filter feeding. (After Walker, 1968.)

male elephant seals occasionally weigh as much as 3600 kg (7920 lbs).

Many phocids are monogamous and form small, loose groups in which no social hierarchy is evident; but some, such as the elephant seal, are gregarious and polygynous, and have a dominance hierarchy. The monogamous species are quiet, whereas the polygynous species are highly vocal (Evans and Bastian, 1969:437). The sole function of the proboscis of the male elephant seal (Fig. 12–27) is the production of vocal threats (Bartholomew and Collias, 1962).

The usual food of phocids is fish, cephalopods, and other mollusks. Large prey may also be taken. Sterling (1969) reports a fish weighing 29.5 kg removed from the stomach of a Weddell seal (*Leptonychotes weddelli*, Fig. 12–28), and the powerful leopard seal (*Hydrurga leptonyx*) eats penguins and small seals. Two species of phocids are filter feeders, and use the complex cheek teeth (Fig. 12–30) to filter crustaceans and other plankton from the water. The Weddell seal is well adapted to life in Antarctic waters and remains beneath the ice all winter; it uses anticlinal, air-filled ice domes as places to breath, and rests on "interior ice shelves" (Perkins, 1945:278–279). Mortality in some seals is due to worn teeth and the resulting inability to keep breathing holes open through the ice. Some phocids that rest on ice are able to leap onto ice seven feet above the water.

SUBUNGULATES: ELEPHANTS; HYRAXES; SIRENIANS

If general appearance were used as the single criterion for evaluating relationships, elephants, the largest land mammals hyraxes, rodent-like creatures, and sirenians, ungainly aquatic mammals, would be judged to be but distantly related. In this case, appearances are deceptive, for the fossil record suggests that these groups evolved in Africa from a common ancestral stock related to ungulates. Because of this presumed ungulate ancestry these groups have long been referred to as subungulates.

ELEPHANTS: ORDER PROBOSCIDEA

Through much of the Cenozoic, some of the largest and most spectacular herbivores were proboscideans and, in the late Tertiary, a varied array of these animals occurred widely in North America, Europe, and Africa. The diversity of proboscideans was reduced in the Pleistocene, and today only two species represent this remarkable group. Because elephants now often threaten the interests of man, and because of the great value of their tusks, they are being extirpated over wide areas as human populations increase. Elephants occur today only in Africa south of the Sahara Desert (Loxodonta), and in parts of southeastern Asia (Elephas). Regrettably, we may be witnessing the final stages in the history of one of the most interesting mammalian orders.

The fossil record of proboscideans begins in the late Eocene of Egypt with Moeritherium. This tapir-sized animal had a moderately primitive complement of teeth (3/3, 1/0, 3/3, 3/3), but the second incisors above and below were enlarged into short tusks. Proboscideans apparently reached North America in the late Miocene. Late Tertiary proboscideans had brachyodont teeth with few ridges; most or all of the cheek teeth were in place at one time, and both an upper and a lower set of tusks were usually present (Fig. 13–1).

FAMILY ELEPHANTIDAE. This family, to which both living species belong, is represented first by the Pliocene and Pleistocene genus Stegodon. Although the skull was not as short as in more advanced elephants and the teeth were brachyodont, the occlusal surfaces of the cheek teeth had laminae and no more than 2/2 cheek teeth were functional at one time. The lower tusks were vestigial, whereas the uppers were long and curved. The Pleistocene woolly mammoth (Mammuthus primigenius) was in some ways more specialized than the living elephants. It had a

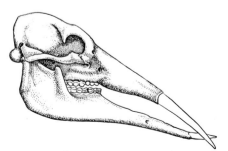

Figure 13–1. The skull of Gomphotherium, a Miocene proboscidean. Length of skull and tusks roughly 1 m. (After Romer, 1966.)

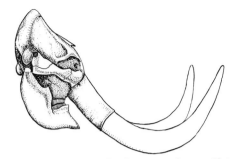

Figure 13–2. The skull of *Mammuthus*, a Pleistocene elephantid. Length of skull and tusks roughly 2.8 m. (After Romer, 1966.)

remarkably short, high skull (Fig. 13–2) and long tusks that occasionally crossed; the last molar had up to 30 laminae, more than occur in the living elephants. Entire frozen woolly mammoths have been found in Siberia and Alaska, and many graceful drawings made by Paleolithic man on the walls of caves depict these animals.

The two living proboscideans—the African elephant, *Loxodonta africana* (Fig. 13–3), and the Indian elephant, *Elephas maximus*—are the largest land mammals, reaching weights of 5900 kg. They have a long dextrous proboscis (trunk) with one or two finger-like structures at its tip, large ears, and graviportal limbs. The limb bones are heavy and the proximal segments of the limbs are relatively long; the ulna and tibia are unreduced, and the bones of the five-toed manus and pes are short, robust, and have an unusual, spreading, digitigrade posture (Fig. 13–4). A heel pad of dense connective tissue braces the toes and largely supports the weight of the animal. As an adaptation allowing the efficient support of great weight, the long axis of the pelvic girdle is nearly at right angles to the vertebral column, and the acetabulum faces ventrally. In addition, when the weight of the body is supported by the limbs, there is little angulation between limb segments; that is to say, each segment is roughly in line with other segments. The gait is unusual. As described by Howell (1944:53), an elephant "relies exclusively upon the walk or its more speedy equivalent, the running walk, which permits it to keep at least two feet always upon the ground. Not only does the weight make it advisable that this be distributed among each of the four feet when the animal is in motion, but the bulk doubtless requires that the equilibrial stresses be shifted as gradually as possible to each foot, rather than more abruptly as in the trot or gallop."

The skull is unusually short and high, perhaps in response to a need for great mechanical advantage for the muscles that attach to the lambdoidal crest and raise the

Figure 13–3. An African elephant *(Loxodonta africana)* immediately after wallowing in the mud. (Photograph taken in Tsavo West National Park, Kenya, East Africa, by Richard G. Bowker.)

Figure 13–4. The right hind foot of *Mastodon*, a late Tertiary and Pleistocene proboscidean. (After Romer, 1966.)

front of the head and the tusks. The skull contains numerous large air cells, particularly in the cranial roof.

The highly specialized dentition consists of the tusks (each a second upper incisor), and six cheek teeth in each half of each jaw. The pattern of cheek tooth replacement is remarkable. The cheek teeth erupt in sequence from front to rear, but only a single tooth, or one tooth and a fragment of another, is functional in each half of each jaw at one time. As a tooth becomes seriously worn it is replaced by the next posterior tooth. The usage intervals for the cheek teeth in the African elephant are: 1, birth to 2nd year; 2, 1.5 to 5th year; 3, 2nd to 11th year; 4, 5th to 19th year; 5, 15th to about 60th year; 6, 23rd to 60+ years (Krumrey and Buss, 1968). The hypsodont cheek teeth are formed of thin laminae (cross ridges), each consisting of an enamel band surrounding dentine, with cement filling the spaces between the ridges (Fig. 13–5). The last molar, the tooth that must serve for much of the animal's adult life, has the greatest number of laminae; the premolars are considerably

smaller, simpler, and less durable than the molars. In old elephants some of the anterior laminae of the third molar may be lost while the remainder of the tooth is still functional. The unique pattern of tooth replacement and the complex occlusal surface provide for an enduring dentition and a long life. The life span of elephants is thought by Perry (1954) to be about 70 years.

Elephants occupy forests, semiopen or dense scrub, and savanna, and are restricted to areas near water. They feed on a variety of trees, shrubs, grasses, and aquatic plants, and characteristically strongly influence their environment (see page 282). Each individual eats up to 410 kg of forage daily. Their great size and strength enable them to "ride down" fairly large trees in order to feed on the leaves. Female elephants are highly social. They live largely in kinship groups from which adult males are excluded; these groups are held together by close social ties between adult females and between mothers and their young.

Locally, elephants do great damage to crops. They have been killed for this reason and for sport, and illegal killing of elephants for their ivory is a significant cause of mortality. The large African game preserves, where hunting is prohibited, offer some hope for the survival of wild elephants.

HYRAXES: ORDER HYRACOIDEA

Members of this unusual order are small, rather rodent-like creatures commonly called hyraxes or dassies; their external appearance (Fig. 19–10) gives little indication of their relationships to the ungulates. This is a small order, with a single Recent family, Procaviidae, and two fossil families of doubtful validity. Recent members include 3 genera with 11 species. Hyracoids occupy nearly all of Africa except the arid northwestern part. Hyraxes appear first in lower Oligocene beds in Egypt, from which members of an extinct family and of the living family are known.

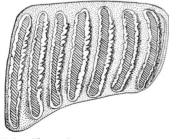

Figure 13–5. The occlusal surface of a molar of the Asiatic elephant *(Elephas maximus)*. The ridges of the lamellae are enamel; the crosshatched areas are dentine, and the stippled areas are cement.

The relationships of the hyracoids are uncertain, but they perhaps descended from an early "ungulate" stock that was also ancestral to the elephants and sirenians. Some early members of an extinct hyracoid family (Geniohyidae) reached the size of a tapir. The distribution of the structurally more conservative surviving family extended north of its present limits during the Pliocene, when a giant procaviid occurred in western Europe as far north as France.

The roughly rabbit-sized procaviids of today have short skulls with deep lower jaws (Fig. 13–6). The dental formula is 1/2, 0/0. 4/4, 3/3. The incisors are specialized: the pointed, ever-growing uppers are broadly separated and triangular in cross section, and the flattened posterior surfaces lack enamel; the lowers are chisel-shaped and are generally tricuspid. Behind the incisors is a broad diastema, and the cheek teeth are either brachyodont or hypsodont. The molars resemble those of a rhinoceros: the uppers have an ectoloph and two cross lophs, and the lowers have a pair of V-shaped lophs. The body is fairly compact and the tail is tiny. The forefoot has four toes and the hind foot has three, and the feet are mesaxonic (the plane of symmetry goes through the third digit). The digits are united to the bases of the last phalanges (Fig. 13–7), and except for the clawed second digit of the pes, all digits bear flattened nails. The plantigrade feet have specialized elastic pads on the soles that are kept moist by abundant skin glands; in addition, the soles may be "cupped" by specialized muscles and, by

Figure 13–7. The sole of the right hind foot of a rock hyrax (*Dendrohyrax dorsalis*).

functioning as suction cups, may provide for remarkable traction. Although the clavicle is absent, as in cursorial mammals, the centrale of the carpus is present, a feature decidedly not characteristic of runners. The stomach is simple, but digestion is aided by microbiota in the pair of caecae of the colon and in the single iliocolic caecum .

Hyraxes are herbivorous and to some extent insectivorous, and are nimble climbers and jumpers. They occur in a variety of habitats, from forests and scrub country to grassland and lava beds, and up to elevations of over 4000 m. The rock hyraxes live in cliffs, ledges, and talus, and are capable of running rapidly over steep rock faces. These diurnal and crepuscular animals are colonial, and on occasion occur in groups of up to 50 individuals.

The body temperature of one species of rock hyrax (*Heterohyrax brucei*) was found by Bartholomew and Rainy (1971) to be quite variable. According to these authors, basking in the sun, a conspicuous behavior among rock hyraxes, is probably important because it reduces the metabolic cost of elevating the temperature in the morning from its slightly depressed nocturnal level. Evaporative cooling, aided by profuse sweating of the feet, dissipates metabolic heat at high ambient temperatures. These animals huddle together at low ambient temperatures (Fig. 19–10), a behavior that reduces the rate of heat loss from their bodies. Tree hyraxes are less gregarious, and are largely nocturnal. They climb rapidly over the nearly vertical trunks of trees and can leap from branch to branch. At night they are highly vocal, producing an impressive array of croaks and screams that reverberate through the forest galleries.

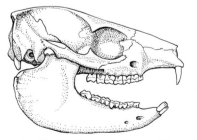

Figure 13–6. Skull of a rock hyrax (*Heterohyrax sp.*); length of skull 98 mm. (After Hatt, 1936.)

SIRENIANS: ORDER SIRENIA

The sirenians—the dugongs, sea cows, and manatees—are the only completely aquatic mammals that are herbivorous, and comprise one of the most anomalous mammalian orders. There are four living species of two genera *(Dugong,* Dugongidae; *Trichechus,* Trichechidae). According to Jones and Johnson (1967:367), sirenians occur in "coastal waters from eastern Africa to Riu Kiu Islands, Indo-Australian Archipelago, western Pacific and Indian oceans; tropical western Africa; coastlines of Western Hemisphere from 30°N to 20°S, Caribbean region, and Amazon and Orinoco drainages in South America; formerly also in Bering Sea." Sirenians probably shared a common ancestry with the proboscideans, and are known from Eocene deposits at such scattered points as Europe, Africa, and the West Indies. This group was once far more diverse and widely distributed than it is today.

Sirenians are large, reaching weights in excess of 600 kg. They are nearly hairless except for bristles on the snout, and have thick, rough or wrinkled skin. The nostrils are valvular, the nasal opening extends posterior to the anterior borders of the orbits, and the nasals are either reduced or absent. The skull is highly specialized and the dentary is deep (Fig. 13–8); the tympanic bone is semicircular and the external auditory meatus is small. The skeleton is dense and heavy—perhaps an increase in specific density is adaptive. Postcranially, sirenians somewhat resemble cetaceans. The five-toed manus is enclosed by skin and forms a flipper-like structure, the pelvis is vestigial, and the tail is a horizontal fluke. There is no clavicle and, unlike cetaceans, the scapula is narrow and blade-like.

The teeth are unusual. Functional teeth are present in *Dugong,* but they are large and columnar, lack enamel, and are cement covered. They have open roots, and the occlusal surfaces are wrinkled and bunodont. *Trichechus,* in contrast, has an indefinite, large number of teeth; the teeth are enamel covered and lack cement, and each has two cross ridges and closed roots. As teeth at the front of the tooth row wear out, they are replaced by the posterior teeth pushing forward. Five or six teeth in each side of each jaw are functional at one time. This unusual style of replacement is basically similar to that of the proboscideans. Horny plates cover the front of the palate and the adjacent surface of the mandible in all genera. The skull of *Trichechus* is curiously modified by elongation of the nasal cavity, and this animal, alone among mammals, has only six cervical vertebrae. Some differences between the two families of sirenians are shown in Table 13–1.

Sirenians are heavy-bodied, slow-moving animals that inhabit coastal seas, large rivers, and lakes, and graze while submerged on aquatic plants. They remain submerged for periods of up to about 15 minutes. Some individuals inhabiting the coasts of Florida move into rivers and to springs in the winter, perhaps in an effort to avoid cold water (Layne, 1965:168). Manatees are known to make sounds underwater, but these are probably used for communication rather than for echolocation (Schevill and Watkins, 1965). Man has been responsible for a great restriction of the range of sirenians, and succeeded in exterminating the Steller's sea cow *(Hydrodamalis)* in about 1769, only some 27 years after the animal's discovery (Walker, 1968:1334). This giant sirenian may have reached 4000 kg in weight, and

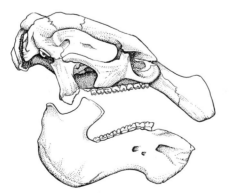

Figure 13–8. The skull of a manatee *(Trichechus manatus).* Length of skull 360 mm. (After Hall, E. R., and Kelson, K. R.: *Mammals of North America,* © 1959 by the Ronald Press Company, New York.)

Table 13–1. Comparison of Sirenian Characteristics

Dugongidae	Trichechidae
Functional dentition 1/0, 0/0, 0/0, 2-3/2-3	No functional incisors; numerous cheek teeth
Cheek teeth columnar; no enamel, cement covered; roots single	Teeth with cross ridges, covered with enamel; cement absent; roots double
Premaxillaries large; nasals absent; nasal cavity short	Premaxillaries small; nasals present; nasal cavity long
Slender neural spines and ribs	Robust neural spines and ribs
Flippers lacking nails	Flippers with nails in one species
Tail notched, like that of whales	Tail not notched but spoon-shaped

inhabited parts of the Bering Sea. Present serious declines in the populations of dugongs (Dugong) in some areas are due to persistent hunting by man.

14

The ungulates are the hoofed mammals, members of the orders Perissodactyla (horses, rhinos, and tapirs) and Artiodactyla (pigs, camels, deer, antelope, cattle, and their kin). The term *ungulate* has no taxonomic status, but refers to this broad group of herbivorous mammals that are more or less specialized for cursorial locomotion.

CURSORIAL SPECIALIZATION

Exceptional running ability has developed independently in a number of mammalian groups. It has provided a means of escaping predators (as in some rodents, rabbits, and ungulates) or of capturing prey (as in carnivores). The refinement of cursorial adaptations in ungulates, the most cursorial mammals, was favored by their occupation of the expanding grasslands in the Miocene. For ungulates living in such open country there was probably a high adaptive premium on running ability. Speed became the primary means of avoiding predation, and seasonal movements to seek water or appropriate food probably became an important part of the ungulate mode of life.

Running speed is determined basically by two factors: the length of the stride and the rate of the stride (the number of strides per unit of time). Most important cursorial specializations serve to lengthen the stride or increase its rate. Perhaps the most universal cursorial adaptation that lengthens the stride is lengthening of the limbs. In generalized mammals, or in many powerful diggers, the limbs are fairly short and the segments are all roughly the same length (Fig. 14–1). But in cursorial species the limbs are long and, in the most specialized runners, the metacarpals and meta-

tarsals have become greatly elongate, and the manus and pes are the longest segments (Fig. 14–1). The loss or reduction of the clavicle contributes further to the length of the stride. This occurs in carnivores, leporids, and ungulates. With the loss of the clavicle, the scapula and shoulder joint are freed from a bony connection with the sternum, and the scapula can change position to some extent and can rotate about a pivot point roughly at its center. Because the scapula is not anchored to the axial skeleton, as the forelimb reaches forward during the stride the shoulder joint pivots upward and forward, and when the forelimb moves back the shoulder joint swings downward and backward. Hildebrand (1960) estimated that such movements of the scapula added roughly 115 mm to the length of the stride of the running cheetah. As an additional advantage, when the forefeet strike the ground at the end of a forward bound the impact is cushioned by the muscles that bind the scapula to the body, rather than the shock being transferred directly from the shoulder joint to the axial skeleton via the clavicle.

Substantial lengthening of the stride also results from an inchworm-like flexion and extension of the spine (Fig. 14–2). In small or moderate-sized runners, the flexors and extensors of the vertebral column are powerfully developed; the vertebral column extends as the forelimbs reach forward and the hindlimbs are driving against the ground, and it flexes when the front feet move backward while braced against the ground as the hind limbs swing forward. Hildebrand estimated that such movements of the vertebral column could propel the cheetah at nearly 10 km per hour if the animal had no legs!

The speed of limb movements and thus

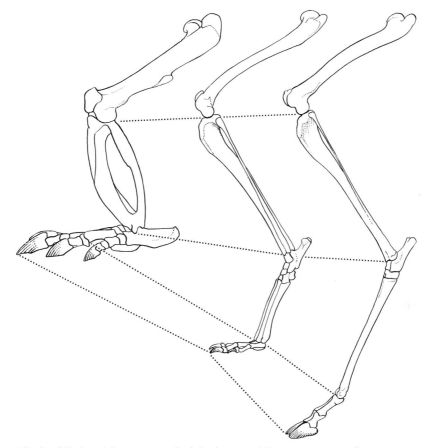

Figure 14–1. The hind limbs of three mammals: left, the armadillo (*Dasypus novemcinctus*), a powerful digger with plantigrade feet; middle, the coyote *(Canis latrans)*, a good runner with digitigrade feet; right, the pronghorn (*Antilocapra americana*), an extremely speedy runner with unguligrade feet. Note the lengthening of the shank and foot of the coyote and of the pronghorn especially; the metatarsals have undergone the greatest lengthening. The limbs are not drawn to scale, but the femur is the same length in each drawing.

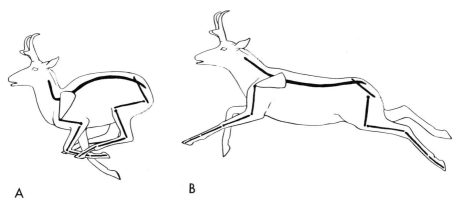

A   B

Figure 14–2. Two positions of a running pronghorn, showing the flexion and extension of the vertebral column and the changing position of the scapula. A, The forelimbs have just left the ground; the hind limbs are reaching forward and will contact the ground as the forelimbs swing forward. B, The animal is bounding ahead after the limbs have driven against the ground; the forelimbs are reaching forward.

the rate of the stride are similarly increased by a combination of structural modifications. The total speed of the foot, which drives against the ground and propels the animal, depends on the speed of movement at each joint of the limb. If another movable joint is added to a limb, the speed of the limb will be increased by the speed of movement at the new joint. The greater the number of joints that move in the same direction simultaneously, the greater the speed of the limbs. For this reason nearly all cursorial mammals have abandoned a plantigrade foot posture in favor of a digitigrade or unguligrade stance. This lifting of the heel from the ground allows another limb joint, that between the metapodials and the phalanges, to contribute to the speed of the limb. In addition, the movable scapula and vertebral column, which help to increase the length of the stride, also contribute their motion to the total speed of the feet.

Specializations of the musculature also add importantly to limb speed, and hence to running speed. The trend in many cursorial mammals is toward a lengthening of the tendons of some limb muscles (in association with the elongation of the distal segments of the limbs), and in some cases there has been a proximal migration of the points of insertion of these muscles. Generally, the nearer the insertion of a muscle approaches the joint that it spans and at which it causes motion, the greater advantage for speed it possesses; such specialized muscles are primarily geared for speed (but not for power). In these animals a complex division of labor among muscles probably allows other muscles, having attachments that confer considerable mechanical advantage for power, to control certain relatively slow movements that get the animal in motion; the less powerful "high-gear" muscles are brought into play during high-speed running.

Speedy limb movements are further facilitated by a reduction of the weight of the distal parts of the limbs and the resultant reduction of the kinetic energy to be overcome at the end of one limb movement and the start of another. Inasmuch as the distal part of the limb moves more rapidly than the proximal part during a stride, reduction of weight of the distal parts is especially advantageous. Several specializations commonly serve this end. The most obvious is the loss of digits. In the most extreme mammalian cases—the front and hind limbs of the horse, and the hind limbs of the pig-footed bandicoot (Fig. 5–22, page 62) and of the red kangaroo (Fig. 5–33)—only one digit that functions during running is retained. The strengthening of the distal joints of the limbs by modifications of bones and ligaments, and the limiting of movements at these joints to a fore and aft plane, obviate the need for muscular bracing and for muscles that produce rotary motions, further reducing distal weight. Also, the heaviest muscles are mostly in the proximal segment of the limb, thus keeping the center of gravity of the limb near the body. The combined effect of these modifications that reduce and redistribute weight is to favor rapid limb movement and to reduce the outlay of energy associated with that movement. (For excellent discussions of cursorial adaptations in mammals, see Hildebrand, 1959, 1960, 1965, 1974:487–515).

The graceful legs of an antelope, with slim distal segments and the largest muscles bunched near the body (Fig. 14–3), display beautifully many of the cursorial modifications discussed here.

Living ungulates typically have many of the cursorial specializations just discussed. The feet are modified by the loss of toes, by the alteration of the foot posture so that only the tips of the toes touch the ground, and by the development of hoofs. The limbs are usually slender, and the tendency in the most rapid runners is toward great elongation of the second segments (the forearm of the forelimb and the shank of the hind limb) and distalmost segment (the manus and pes). The distal parts of the ulna and the fibula are strongly reduced in advanced ungulates, and the joints distal to the shoulder and hip joints tend to limit movement to the anteroposterior plane.

In the ungulate ankle joint, the calcan-

Figure 14–3. The pronghorn antelope (*Antilocapra americana*), one of the fastest cursorial mammals. The long slender limbs are typical of the more cursorial ungulates. (Photograph by F. D. Schmidt; San Diego Zoo photo.)

eum appears to be pushed aside, so to speak. In mammals in which no drastic reduction of digits has occurred, the distal surface of the astragalus articulates with the navicular and the calcaneum articulates with the cuboid (Fig. 2–27), and the weight of the body is transferred through the digits, the distal carpals, both the astragalus and calcaneum, and the tibia and fibula. In ungulates, a different arrangement occurs in association with the reduction of digits: the astragalus rests more or less directly on the distal tarsal bones, which may be highly modified by fusion and loss of elements (Figs. 14–4, 14–5), and the weight of the body is borne by the central digits (or digit, in the case of the horse), the distal tarsals, and the astragalus. The astragalus thus becomes the main weight-bearing bone of the two proximal tarsals. The calcaneum remains important as a point of insertion for extensors of the foot, but it no longer is a major weight-bearing bone of the tarsus. A similar bypassing of the calcaneum occurs in the cursorial peramelid marsupials (see Fig. 5–22, page 62).

Two distinctive ungulate specializations involve connective tissue. The nuchal ligament is a heavy band of elastin (an elastic protein found in vertebrates) that is anchored posteriorly to the tops of the neural spines of some of the anteriormost thoracic vertebrae and attaches anteriorly high on

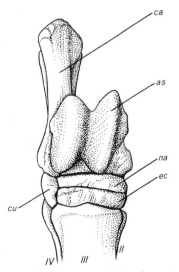

Figure 14–4. The tarsus of the domestic horse (*Equus caballus*). Abbreviations: *as*, astragalus; *ca*, calcaneum; *cu*, cuboid; *ec*, ectocuneiform; *na*, navicular. The metatarsals are numbered.

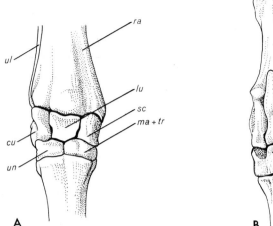

Figure 14–5. The ankle and wrist joints of artiodactyls: A, the right carpus of the mule deer (*Odocoileus hemionus*); B, the right tarsus of the pronghorn (*Antilocapra americana*.). Abbreviations: *as*, astragalus; *ca*, calcaneum; *can*, cannon bone (fused third and fourth metatarsals); *cu*, cuneiform of the carpus; *cu + na*, fused cuboid and navicular; *ec + me*, fused ectocuneiform and middle cuneiform; *ent*, internal cuneiform; *lu*, lunar; *ma + tr*, fused magnum and trapezoid; *ra*, radius; *sc*, scaphoid; *ul*, ulna; *un*, unciform.

the occipital part of the skull (Fig. 14–6). This ligament, especially robust in large, heavy-headed ungulates such as the horse and moose, helps support the head, so that the burden on the muscles that lift the head is greatly lightened. The elasticity of the ligament allows the head to be lowered during eating or drinking. A second specialized ligament, the springing ligament, occurs in the front and hind feet of ungulates, and evolved from muscles that flexed the digits (Camp and Smith, 1942). In the hind foot of the pronghorn (*Antilocapra*), for example, the springing ligament arises from the proximal third of the back of the cannon bone and inserts distally on the sides of the first phalanges of digits three and four (Fig. 14–7). When the

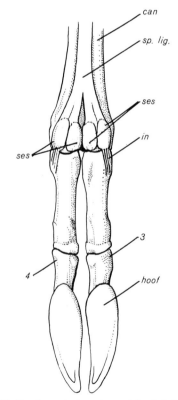

Figure 14–7. A posterior view of the left hind foot of an artiodactyl (*Antilocapra americana*), showing the position of the springing ligament. Abbreviations: *can*, cannon bone; *in*, insertion of the springing ligament; *ses*, sesamoid bone; *sp lig*, springing ligament; *3*, third digit (second phalanx); *4*, fourth digit (second phalanx).

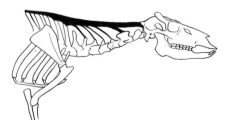

Figure 14–6. A schematic drawing of the nuchal ligament (in black) of an ungulate.

foot supports the weight of the body, the phalanges are extended, thereby stretching the springing ligament. But as the foot begins to be relieved of the weight of the body toward the end of the propulsion stroke of the stride, the elastic ligament begins to rebound, and when the foot is leaving the ground the phalanges snap toward the flexed position (Fig. 14–8C). The familiar backward flip of the horse's foot just as it leaves the ground is controlled by the springing ligament. This flip gives a final increase in speed and thrust to the stride, and serves to increase the ungulate's speed afoot without the use of muscular effort.

FEEDING SPECIALIZATION

The herbivorous diet characteristic of most ungulates has favored the development of cheek teeth with large and complex occlusal surfaces that function to finely section plant material as an aid to digestion. Premolars tend to become molariform, and thus to increase the extent of the grinding battery, and the anterior dentition becomes variously modified. In advanced types, there is a diastema between the anterior dentition and the cheek teeth.

Many ungulates have become large; indeed, the largest members of most mammalian faunas are ungulates. Large size reduces the number of predators to which an animal is vulnerable and is advantageous in terms of temperature regulation and energy requirements. Probably these and other factors have influenced the size of ungulates.

Unusual demands are put on the digestive systems of ungulates by their diet. Vegetation is far less concentrated food than is meat, and is more difficult to digest. In addition, plant material is frequently low in protein. An herbivore must break down the cell wall, a fairly rigid structure formed largely of cellulose, not so much for the energy it yields but to gain access to the proteins within the cell. This breakdown is difficult, however, for mammals lack enzymes that digest cellulose. All herbivores, therefore, must have alimentary canals that are specialized to cope with cellulose by other means than direct enzymatic action. Both perissodactyls (horses, rhinos, and tapirs) and ruminant artiodactyls (camels, deer, antelope, sheep, goats, and cattle) utilize a fermentation process that involves the breakdown of cellulose by the cellulolytic enzymes of microorganisms that live in the alimentary canal.

In perissodactyls, if we can assume that their digestive systems are all similar to the system of the domestic horse, microorganism-aided fermentation takes

Figure 14–8. The left hind foot of the pronghorn (*Antilocapra americana*): A, anterior view of the distal end of the cannon bone; B, posterior view of this bone; C, position of the phalanges when the foot is supporting the weight of the body (right) and the springing ligament (shown in black) is stretched, and the position when the foot leaves the ground and the springing ligament flexes the phalanges (left). Abbreviations are the same as in Figure 14–7.

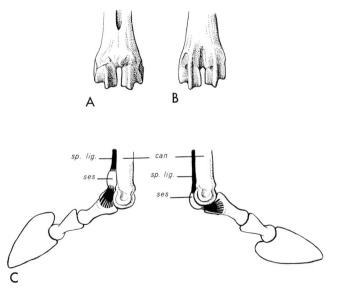

place in the large intestine and enlarged colon. Protein is digested and absorbed in the relatively small and simple stomach. In ruminants, in contrast, the enlarged and several-chambered stomach harbors microorganisms that break down cellulose. Large food particles float on top of the fluid in the rumen, are passed to the diverticulum (see Fig. 2–14, p. 14), and are then regurgitated and remasticated. In the vernacular, the animals chew their cud. This cycle is repeated until the chemical and mechanical breakdown of the food has reached a point at which the particles sink in the fluid and pass on to the intestine.

As a further refinement, in the stomach a complex system of recycling and reconstitution of food constituents ensures that after protein is extracted from food it is used to greatest advantage. When the ruminant's food is high in cellulose and lignin (a substance similar to cellulose that contributes the hard, woody characteristics to plant stalks and roots), it is digested slowly, and the rate of passage of food through the gut is low. The horse's system, however, has no regulation of the rate at which food is processed. Food takes some 30 to 45 hours to pass through the gut of a horse, as compared to 70 to 100 hours to make the corresponding journey in the cow. The central difference, then, is that the digestive system of the horse is less efficient than that of the ruminant, but in compensation the horse eats greater quantities of food; the emphasis in ruminants is on highly efficient digestion and on more selective feeding, but not on high rates of food intake. Therefore, if food is in short supply, the ruminant will probably survive after the horse has died.

An extremely interesting African grazing succession, strongly influenced by differences between digestive efficiencies and food requirements, has been described by Bell (1971). He studied primarily the most abundant ungulates: the zebra (*Equus burchelli*), a nonruminant, and the wildebeest (*Connochaetes taurinus*) and Thompson's gazelle (*Gazella thompsonii*), both ruminants. He found that in the Serengeti Plains the zebra was the first of these

ungulates to be forced by food shortages to move from the preferred shortgrass area down into the longer, coarser grasses of the lowlands. After the zebras' feeding and trampling activities in the lowlands had removed the coarse upper parts of the grass and had made the lower, more nutritious plant parts more readily available, the wildebeest, a more selective feeder, moved in. By this time the zebras were becoming less able to get sufficient quantities of forage and were moving to new tallgrass pastures. A similar replacement of wildebeest by Thompson's gazelles occurred after the wildebeest had removed still more grass and had made available to the small, highly selective gazelles the fruits and leaves of low-growing forbs. Not only was competition between these abundant ungulates minimized by this grazing pattern, but the activities of the early members of the grazing succession were highly advantageous to the later, more selective members. Bell's study clearly illustrates that differences in the digestive systems of ungulates have pronounced effects on food preferences, migratory patterns, and, in fact, many basic interactions of a grazing ecosystem.

ORDER PERISSODACTYLA

Since Eocene times some of the most specialized cursorial mammals have been perissodactyls. Throughout the early Tertiary these were the most abundant ungulates, but their diversity was reduced in the Oligocene, and with the diversification and "modernization" of the artiodactyls in the Miocene the fortunes of perissodactyls began to decline. The surviving perissodactylan fauna (consisting of 6 genera and 16 species) is but an insignificant remnant of this once important group, and is vastly overshadowed by an impressive Recent artiodactylan assemblage (consisting of 171 species). Perissodactyls occur today largely in southern areas—Africa, parts of central and southern Asia, and tropical parts of southern North America and northern South America.

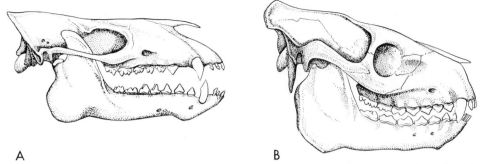

Figure 14–9. Skulls of extinct ungulates. A, *Phenacodus*, a Lower Eocene primitive ungulate (Condylarthra); length of skull roughly 230 mm. B, *Oreodon*, a primitive Oligocene ruminant artiodactyl; length of skull roughly 125 mm. (After Romer, 1966.)

Perissodactyls evolved from herbivorous condylarths of the family Phenacodontidae (Fig. 14–9A). The order Condylarthra includes a diverse group of ancient ungulates that occurred from the Cretaceous to the Oligocene, and was probably the basal stock for some 18 mammalian orders (Szalay, 1969), including the Recent orders Artiodactyla, Perissodactyla, Hyracoidea, Proboscidea, and Sirenia. Perissodactyls appeared in the late Paleocene in North America and underwent rapid diversification. Eleven of the 12 families appeared in the Eocene, but in addition to the living families Tapiridae, Rhinocerotidae, and Equidae, only the anomalous extinct family Chalicotheriidae survived into the Pleistocene.

The features of several important perissodactylan families illustrate the considerable structural and functional diversity within the group. The dentition and cranial morphology of perissodactyls developed in response to herbivorous feeding habits. Living perissodactyls have elongate skulls, owing to an enlargement of the facial region to accommodate a full series of large cheek teeth (often hyposodont), and some have a complete complement of 44 teeth. The teeth are usually lophodont and are either hypsodont in grazing types (all Equidae, and *Ceratotherium* of the Rhinocerotidae) or brachyodont in browsers (all Tapiridae, and *Rhinoceros* and *Didermocerus* of the Rhinocerotidae). Many postcranial specializations further cursorial ability. The clavicle is absent, and usually the manus has three or four digits and the pes, three digits; but in the equids only one functional digit is retained on each foot (Fig. 14–10C). The feet are mesaxonic; that is, the plane of symmetry of the foot passes through the third digit, whereas this plane passes between digits three and four in the paraxonic foot of artiodactyls.

FAMILY EQUIDAE (Fig. 14–11). Horses, the most highly cursorial and graceful perissodactyls, now occur wild only in Africa, Arabia, and parts of western and central Asia. There is but one genus with seven Recent species.

Wild horses are in general not as large as domestic breeds. The average weight of a female zebra (*Equus burchelli*) is given by Bell (1971) as 219 kg, but some domestic breeds weigh over 1000 kg. The skull has a

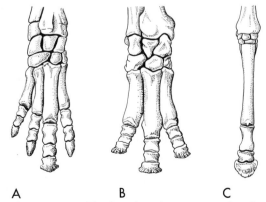

Figure 14–10. The front feet of some perissodactyls. A, A tapir (*Tapirus*); B, a rhinoceros (*Rhinoceros*); C, a horse (*Equus*). (After Howell, 1944.)

Figure 14-11. Two species of equids. A, Burchell's zebra (*Equus burchelli*); B, Grevy's zebra (*Equus grevyi*). Note the striking differences in the patterns of stripes between these species. (Photographs taken in Kenya, East Africa.)

fairly level profile, and the rostrum is long and deep (Fig. 14–12); the dental formula is 3/3, 0-1/0-1, 3-4/3, 3/3 = 36-42. The cheek teeth are hypsodont and have complex patterns on the occlusal surfaces (Fig. 14–13). The limbs are of a highly cursorial type: only the third digit is functional, all but the proximal joints largely restrict

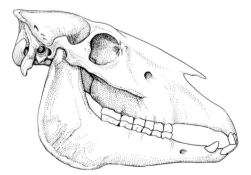

Figure 14-12. The skull of the domestic horse (*Equus caballus*). Length of skull 530 mm.

movement to one plane, and the foot is greatly elongate. In the tarsus the main weight-bearing bones are the ectocuneiform, navicular, and astragalus; the calcaneum is mostly posterior to the astragalus (Fig. 14–4).

The evolution of horses is well documented by an excellent and largely New World fossil record, and is discussed by Simpson (1951). Equids are first represented by *Hyracotherium* from the late Paleocene of Wyoming (Jepsen and Woodburne, 1969). This primitive type had a generalized skull with 44 teeth (Fig. 14–14A). The upper and lower molars were brachyodont and basically four cusped. The upper molars bore a protoconule and a metaconule, and the paraconid of the lower molars was reduced (Fig. 14–13A). The premolars were not molariform. The limb structure reflected considerable running ability: the limbs were fairly long and

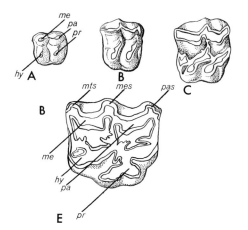

Figure 14-13. Right upper molars of equids. A, *Hyracotherium*; B, *Mesohippus*; C, *Parahippus*; D, *Pliohippus*; E, *Equus*. These teeth illustrate stages in the evolution of the equid molars. Abbreviations: *hy*, hypocone; *me*, metacone; *mes*, mesostyle; *mts*, metastyle; *pa*, paracone; *pas*, parastyle; *pr*, protocone. (After Romer, 1966.)

slender; the front foot had four toes and the hind foot had three toes, but the animal was functionally tridactyl. Each digit terminated in a small hoof, and the foot posture was unguligrade. *Hyracotherium* was the size of a small dog, and presumably browsed on low-growing vegetation in forested or semiforested areas.

Side branches from the main stem of equid evolution developed at various times, but the main evolutionary line can be traced through such intermediate genera as *Merychippus* and *Pliohippus* to the Pleistocene and Recent *Equus*. *Merychippus*, a pony-sized Miocene type, was functionally tridactyl, but retained short lateral digits. The dentary bone was deep, the face was long, and the orbit was fully enclosed. The cheek teeth were high crowned, were covered with cement, and had an occlusal pattern similar to that of

Equus (Fig. 14–13E). *Pliohippus* occurred in the Pliocene; it had the skull features of its progenitor, *Merychippus*, but was more progressive in having higher-crowned teeth and lateral digits reduced to splintlike vestiges. *Equus*, as well as an extinct evolutionary side branch of short-legged South American horses (typified by *Hippidium*), evolved from *Pliohippus*. *Equus*, the genus to which all living horses belong, differs from *Pliohippus* in greater size and in a more complex crown pattern of the cheek teeth. Major evolutionary trends of the Equidae listed by Colbert (1969:396) include: increase in size; lengthening of legs and feet; reduction of lateral toes and emphasis on the middle toe; molarification of the premolars; increase in height of the crowns of the cheek teeth; lengthening of the facial part of the skull to accommodate the large cheek teeth; and

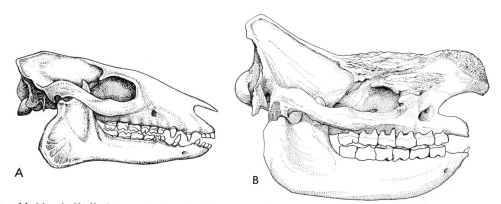

Figure 14-14. A, Skull of *Hyracotherium*, first known equid; length of skull 134 mm. B, Skull of black rhinoceros (*Diceros bicornis*); length of skull 692 mm. (*Hyracotherium* after Romer, 1966.)

deepening of the maxillary and dentary to accommodate the high-crowned teeth. In addition, the profile of the angular border of the dentary swept progressively farther forward, and the origin of the masseter muscles migrated forward. These adaptations increased the force the masseter muscles could exert on the dentary.

Cenozoic changes in climate and in the flora of North America may have had a critical influence on the evolution of horses. Especially important was the Miocene development of grasslands over much rolling or nearly level land within the present Great Plains, the Great Basin, and the southwestern deserts of the U.S.A. Many of the most progressive equid skull and dental features probably arose in response to the shift to a diet high in grasses. Grass, at least at certain times of the year, has low nutritional value, and must be eaten in large quantities to sustain life. High-crowned, persistently growing teeth were necessary to cope with large amounts of grasses made highly abrasive by silica in the leaves and by particles of soil deposited on leaves by wind and the splash effect of rain. Also of great adaptive value were the highly cursorial limbs with single-toed feet that facilitated rapid and efficient locomotion on the firm, level footing of the grasslands. Cursorial ability was perhaps as advantageous for traveling between widely scattered concentrations of food and distant water holes in semiarid regions as for escaping from predators.

For some unknown reason, horses disappeared from the New World, seemingly their place of origin and primary center of their evolution, before historic times. Although wild horses now occupy only Africa and parts of Asia, within historic times they occurred throughout much of Eurasia. Wild equids inhabit grasslands in areas ranging from tropical to subarctic in climate, and feral horses and burros thrive in western North America.

FAMILY TAPIRIDAE. Tapirs occupy tropical parts of the New World and the Malayan area. The family includes one living genus and four species. Structurally, tapirs are notably primitive, and according to Romer (1966:269) are still very close in many respects to the common ancestors of all perissodactyls. "True" tapirs are known first from the Oligocene, but possible ancestral types occurred in the Eocene.

Tapirs have stocky builds and weigh up to about 300 kg. The limbs are short, and both the ulna and fibula are large and separate from the radius and tibia respectively; the front feet have four toes (Fig. 14–10A) and vestiges of the fifth (the pollex), and the hind feet have three toes. Tapirs retain a full placental complement of 44 teeth. Three premolars are molariform, and the brachyodont cheek teeth retain a simple pattern of cross lophs. The short proboscis (Fig. 14–15) and reduced nasals are among the few specializations of tapirs.

These animals have probably always occupied moist forests, where their primitive feet serve well on the soft soil, and the teeth are adequate for masticating plant material that is not highly abrasive. Tapirs today inhabit mostly tropical areas, and usually are found near water. They are rapid swimmers, and often take refuge from predators in the water. Tapirs are solitary and nocturnal, and their presence is frequently made known chiefly by their systems of well-worn trails between feeding areas, resting places, and water. Their food is largely succulent plant material, including fruit.

FAMILY RHINOCEROTIDAE. This family is represented today by four genera and five species, and is restricted to parts of the tropical and subtropical sections of Africa and southeastern Asia. These ponderous creatures—the armored tanks of the mammal world—are surviving members of the spectacular late Tertiary and Pleistocene ungulate fauna. Although apparently a declining group, rhinos have an illustrious past.

The fossil record of the rhinos and their relatives (superfamily Rhinocerotoidea) is remarkably complex, and parallels that of the horses in documenting the early Tertiary and mid-Tertiary success and the late Tertiary decline of the group. Two genera that illustrate well the diversity of early

Figure 14–15. South American tapir *(Tapirus terrestris).* (Photograph by Ron Garrison; San Diego Zoo photo.)

Tertiary rhinocerotoids are *Hyracodon* and *Baluchitherium*. *Hyracodon* (Hyracodontidae), a small North American Oligocene "running rhinoceros," had slender legs and tridactyl feet, and was probably similar in cursorial ability to Oligocene horses. Perhaps as a result of competition with horses, hyracodonts became extinct in late Oligocene. A contemporary of *Hyracodon* in the Oligocene was the Asian form *Baluchitherium* (Rhinocerotidae), the largest known land mammal. This giant was 18 feet high at the shoulder, and the skull (small in proportion to the great size of the rest of the animal) was four feet long. The neck was long, and *Baluchitherium* perhaps browsed on high vegetation in giraffe-like fashion. The limbs were long and graviportal, but the tridactyl feet were unique in that the central digit was greatly enlarged and terminated in a broad hoof, whereas the lateral digits were more strongly reduced than in any other rhinocerotoid. Rhinos died out in the New World in the Pliocene, but remained common and diverse in Eurasia through the Pleistocene. The Pleistocene woolly rhinoceros *(Coelodonta)* was apparently adapted to cold climates. Entire preserved specimens of this rhino have been found in an oil seep in Poland.

All Recent rhinos are large, stout-bodied herbivores with fairly short, graviportal limbs (Fig. 14–16). Weights range up to about 2800 kg. The front foot has three or four toes (Fig. 14–10B), and the hind foot is tridactyl. The nasal bones are thickened and enlarged, often extend beyond the premaxillaries, and support a horn. Where there are two horns, the posterior one is on the frontals. The horns are of dermal origin and lack a bony core. The occipital part of the skull is unusually high (Fig. 14–14B) and yields good mechanical advantage for neck muscles that insert on the lambdoidal crest and raise the heavy head. The incisors and canines are absent in some rhinos and are reduced in number in others; the dental formula is 0-2/0-1, 0/0-1, 3-4/3-4, 3/3 = 24-34. The cheek teeth have a pattern of cross lophs far simpler than that of equids (Fig. 14–17).

Rhinos inhabit grasslands, savannas, brushlands, forests, and marshes in tropical and subtropical areas. Some species are solitary *(Diceros)*, whereas others occur in family groups *(Ceratotherium)* or even in assemblages including up to 24 animals (Heppes, 1958). Rhinos practice scent marking by establishing dunghills along well-worn trails. A variety of plant material is taken; some species are browsers

Figure 14–16. The black rhinoceros (*Diceros bicornis*) of East Africa. (Photograph taken in Amboseli National Park, Kenya, East Africa.)

and some are grazers. Adults are nearly invulnerable to predation, except by man, but young rhinos may be attacked by carnivores as small as spotted hyenas (Kruuk and Van Lawick, 1968:53). The Asian rhinos (*Rhinoceros* and *Didermocerus*) are probably facing extinction. Because of the supposed medicinal properties of the horn and other parts, Asian rhinos have been hunted persistently for at least 1000 years. Remnants of the formerly more widespread populations of the one-horned rhino (*Rhinoceros*) are restricted to India, Nepal, and Java. Regrettably, the survival of the Asian rhinos seems unlikely.

ORDER ARTIODACTYLA

The order Artiodactyla—containing the pigs, peccaries, hippopotami, camels, deer, cows, sheep, goats, and antelope—is by far

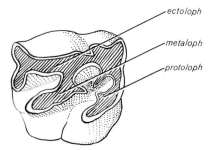

Figure 14–17. The right upper molar of a rhinoceros (*Rhinoceros*).

the most important ungulate group today. Whereas perissodactyls were abundant and reached their greatest diversity in late Eocene, artiodactyls underwent their most important evolution much later, in the Miocene. Since this epoch the perissodactyls have steadily declined, but the artiodactyls have remained diverse and successful. One is tempted to relate the decline of the perissodactyls to the rise of the artiodactyls, and to regard the latter as the more effective competitors. However, although many structural differences between perissodactyls and artiodactyls are apparent, the functional advantages conferred by many features are not easily recognized.

In the order Artiodactyla, the structure of the foot is especially diagnostic. The foot is paraxonic; that is, the plane of symmetry passes between digits three and four (Fig. 14–18). The weight of the animal is borne primarily by these digits: the first digit is always absent in living members, and the lateral digits (digits two and five) are always more or less reduced. Four complete and functional digits occur in the families Suidae, Hippopotamidae, and Tragulidae, and in the forelimb of the Tayassuidae (the hind limb has the medial digit suppressed). Two complete toes, with the lateral digits absent (Camelidae, some Bovidae, Antilocapridae, and Giraffidae), or with incomplete remnants of the lateral digits (Cervidae and some Bovidae), occur

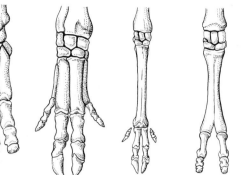

Figure 14–18. The right front feet of some artiodactyls. A, Hippopotamus (*Hippopotamus*); B, swine (*Sus*); C, elk (*Cervus*); D, camel (*Camelus*). (After Howell, 1944.)

in the more cursorial families. The cannon bone (a single bone formed by the fusion of the third and fourth metapodials) is present in the families Camelidae, Cervidae, Giraffidae, Antilocapridae, and Bovidae. Typically, the terminal phalanges are encased in pointed hoofs. The astragalus has a "double pulley" arrangement of articular surfaces (Fig. 14–5B) that completely restricts lateral movement. The proximal articulation (with the tibia) and the distal articulation (with the navicular and cuboid, which are fused in many advanced types) of the astragalus are of critical importance in allowing great latitude of flexion and extension of the foot and digits, as is the extension of the articular surfaces and keels on the distal ends of the cannon bones to the anterior surfaces (Fig. 14–8A, B). The distinctive artiodactyl astragalus is regarded by some as a primary key to the success of the group. Perhaps more important, however, is the remarkable efficiency of the artiodactyl (ruminant) digestive system (described on pages 14 and 244).

The limbs of artiodactyls, especially the distal segments, are usually elongate and fairly slim. The femur lacks a third trochanter. Whereas this prominence serves as a point of insertion of the gluteal muscles in perissodactyls, in artiodactyls these muscles insert more distally, on the tibia. The distal parts of the ulna and the fibula are usually reduced and may fuse with the radius and tibia respectively; this fusion is associated with the restriction of movement of the limb to one plane. The clavicle is seldom present. The intrinsic muscles of the feet (those that both originate and insert on the feet) are usually absent, being replaced by specialized tendons and ligaments.

The skull usually has a long preorbital section, and a postorbital bar or process is always present. Horns, always of bone or with a bony core, are most often borne on the frontals, which are enlarged at the expense of the parietals. The teeth are brachyodont or hypsodont and vary from 30 to 44 in number. The crown pattern is bunodont or, more often, selenodont. The premolars are not fully molariform, in contrast to the perissodactyl situation, and considerable specialization of the anterior dentition occurs in advanced types.

Although the classification of the Artiodactyla has occupied the attention of a number of the leading paleontologists and mammalogists, it is still not fully resolved (Romer, 1968:209). The system used here is that of Romer (1966), and recognizes the extinct suborder Palaeodonta and the living suborders Suina and Ruminantia, all of which are first known from the Eocene. The Palaeodonta contains the most primitive known artiodactyls from the early Tertiary. Members of this group are small, and some are so primitive in structure as to be nearly unrecognizable as artiodactyls if it were not for the distinctive astragalus.

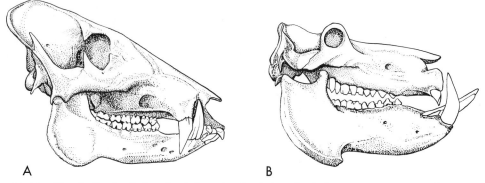

Figure 14–19. The skulls of artiodactyls of the suborder Suina: A, javelina (*Tayassu tajacu*); length of skull 225 mm. B, Hippo (*Hippopotamus*); length of skull 600 mm. (B after Romer, 1966.)

SUBORDER SUINA

In members of the suborder Suina, which includes, pigs, peccaries, and hippopotami, the molars are bunodont, the canines (and in the case of the hippopotami, the incisors also) are tusklike (Fig. 14–19), and the feet nearly always retain four toes with complete and separate digits. The skull contrasts with those of other artiodactyls in having a posterior extension of the squamosal that meets the exoccipital and conceals the mastoid bone. The stomach is of a nonruminant type (the animals do not chew their cud) but may have as many as three chambers.

FAMILY SUIDAE. Swine are omnivorous, and lack many structural modifications typical of more specialized artiodactyls. The Suidae is an Old World family; the present distribution includes much of Eurasia and the Oriental region, and Africa south of the Sahara. There are eight Recent species of five genera. Suids appeared in the Oligocene. The entelodonts (Entelo-

dontidae), an early branch from the "swine" evolutionary line, were huge pig-like creatures with skulls up to three feet in length.

Most suids resemble the domestic swine (*Sus*). Adults may weigh as much as 275 kg, and typically have thick, sparsely haired skin. The skull is long and low, and usually has a high occipital area and a concave or flat profile (Fig. 14–20). The large canines are ever-growing, and the upper canines form conspicuous tusks that protrude from the lips and curve upward. In *Babyrousa*, an Indonesian member of the Suidae, the upper canines protrude from the top of the snout (Fig. 14–21). Suid molars are bunodont, and the last molars are often elongate, with many cusps and a complexly wrinkled crown surface (Fig. 14–22A). The dental formula is variable even within a species; the total number of teeth ranges from 34 to 44. The limbs are usually fairly short, and the four-toed feet never have cannon bones (Fig. 14–18B).

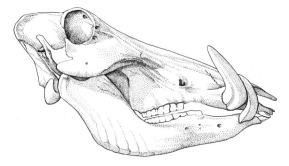

Figure 14–20. Skull of the wart hog (*Phacochoerus aethiopicus*) of Africa. (Length of skull 376 mm.)

Figure 14–21. The babirussa (*Babyrousa babyrussa*), a suid that occurs in the Celebes and several nearby small islands in Indonesia. Note that the tusklike upper canines emerge from the top of the snout. (San Diego Zoo photo.)

Swine inhabit chiefly forested or brushy areas, but the wart hog (*Phacochoerus*) favors savanna or open grassland and is entirely herbivorous. Most suids are gregarious, and some assemble in groups of up to 50 individuals. Most species eat a broad array of plant food and carrion, and given the opportunity will kill and eat such animals as small rodents and snakes. By comparison with ruminant artiodactyls, cursorial ability in suids is modest. Wart hogs are fairly swift, however, and a family moving away with their unique animated-cartoon trot and their tails straight in the air is one of the most engaging sights of East Africa (Fig. 14–23).

FAMILY TAYASSUIDAE. These animals, usually called javelinas or peccaries, are restricted to the New World, where they occur from southwestern United States to central Argentina. There are but three Recent species of two genera (*Dicotyles, Catagonus*). *Catagonus* was first known from Pleistocene fossils and was for many years thought to be extinct, but Wetzel et al. (1975) reported a surviving population in the biologically poorly known Gran Chaco area of Paraguay, South America. The fossil record of tayassuids begins in the Oligocene. Presumably javelinas evolved from primitive Old World pigs, but javelinas are not known from the Old World after the Miocene. Javelinas are more progressive in limb structure than are suids, and are less carnivorous.

Javelinas are much smaller than suids; weights range up to only about 30 kg. The

Figure 14–22. The molars of artiodactyls. A, The second and third right upper molars of the swine (*Sus scrofa*); B, the comparable teeth of the elk (*Cervus elaphus*), with the enamel ridges unshaded, the enamel-lined depressions stippled, and the dentine crosshatched. The molars of *Sus* are bunodont; those of *Cervus* are selenodont.

A

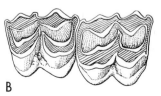

B

Figure 14-23. A female wart hog and her young running in usual tail-in-air fashion. (Photograph taken in Tsavo East National Park, Kenya, East Africa.)

skull has a nearly straight dorsal profile, and the zygomatic arches are unusually robust (Fig. 14–19A). The canines are long and are directed slightly outward, but never turn upward, and have sharp medial and lateral edges. These teeth slide against one another, and the anterior surface of the upper and the posterior surface of the lower are planed flat by this contact. The molars are roughly square and have four cusps; they lack the complex wrinkled and multicusped pattern typical of suids. The dental formula is 2/3, 1/1, 3/3, 3/3 = 38. The feet are slender and appear delicate, and the side toes are small compared to those of suids and usually do not reach the ground. In tayassuids, the front feet have four toes and the hind feet have three, and a cannon bone in the hind foot is partially formed by the proximal fusion of the medial metatarsals. Surprisingly, modern javelinas are not as advanced in foot structure as is *Mylohyus,* an extinct Pleistocene javelina, in which the side toes of the forefoot were very strongly reduced and the didactyl hind foot had a fully developed cannon bone.

Javelinas occupy diverse habitats, from deserts and oak-covered foothills in Arizona to dense tropical forests and thorn scrub in southern Mexico, Central America, and South America. They are gregarious, and on occasion form groups including several dozen individuals. Javelinas are omnivorous, but seem to rely more heavily on plant material than do suids. The presence of javelinas is often indicated by shallow excavations where roots have been exposed beneath bushes or patches of prickly-pear cactus. Despite their rather chunky build, javelinas are rapid and extremely agile runners in the broken terrain they often inhabit.

FAMILY HIPPOPOTAMIDAE. This family is represented today by the genera *Hippopotamus* and *Choeropsis* (the pigmy hippo), each with one species. The group first appeared in the upper Pliocene in Africa and Asia and occurred widely in the southern parts of the Old World in the Pleistocene. Hippos now occur only in Africa; in North Africa they are restricted to the Nile River drainage, but they occur widely in the southern two thirds of the continent.

Hippos are bulky ungraceful creatures with huge heads and short limbs (Fig. 14–24). They are large, weighing up to roughly 3600 kg in *Hippopotamus* and about 180 kg in *Choeropsis.* Some of the distinctive features of hippos probably evolved in association with their amphibious mode of life. Specialized skin glands secrete a pink, oily substance that protects the sparsely haired body. The highly specialized skull has elevated orbits and enlarged and tusk-like canines and incisors (Fig. 14–19B). The bunodont molars are basically four cusped; the dental formula is 2-3/1-3, 1/1, 4/4, 3/3 = 38-40. The limbs are robust, and the feet are four toed (Fig. 14–18A). The foot posture is semidigitigrade; only the

Figure 14-24. Hippopotami (*Hippopotamus amphibius*) in East Africa. (Photograph by W. Leslie Robinette.)

distal phalanx of each toe touches the ground. The broad foot is braced by a sturdy "heel" pad of connective tissue, and the central digits are nearly horizontal.

Hippos are restricted to the vicinity of water. *Hippopotamus* is gregarious, and groups spend much of the day in the water. When bodies of water are at a premium during the dry season, hippos often concentrate in stagnant ponds that the animals churn into muddy morasses. They are good swimmers and divers, and when submerged are able to walk on the bottoms of rivers or lakes by using a strange slow-motion type of gait. At night hippos may move far inland to feed on vegetation. *Choeropsis* is solitary or occurs in pairs, and inhabits forested areas. Instead of seeking shelter in the water when disturbed, as is characteristic of *Hippopotamus*, *Choeropsis* seeks refuge in dense vegetation.

SUBORDER RUMINANTIA

This suborder includes camels, giraffes, deer, antelope, sheep, goats, and cattle. Members of this most advanced artiodactylan suborder have been in the past, and remain, the dominant artiodactyls. In general, these animals are committed strictly to a herbivorous diet and to highly cursorial locomotion. Ruminants chew their cud; the stomach has three or four chambers (Fig. 2-14, page 14) and supports microorganisms that have cellulolytic enzymes. Ruminants have selenodont molars (Fig. 14-22B), and the anterior dentition is variously specialized by loss or reduction of the upper incisors, by the development of incisiform lower canines, and commonly by the loss of upper canines. The skull differs from those of members of the suborder Suina in the exposure of the mastoid bone between the squamosal and the exoccipital. Antlers or horns, often large and complex structures, are present in the most progressive families. In the limbs there is a pronounced trend toward the elongation of the distal segments, the fusion of the carpals and tarsals, and the perfection of the two-toed foot. The ruminants may conveniently be separated into two divisions (infraorders): Tylopoda, the camels and llamas and their extinct relatives; and Pecora, which includes all of the remaining, more progressive ruminants. A diagnostic feature of the pecorans is the fusion of the navicular and cuboid, over which the astragalus is nearly centered (Fig. 14-5B).

An early tylopod family (now extinct), the Merycoidodontidae, usually called oreodonts, deserves mention because they are by far the most abundant mammals in some Oligocene and early Miocene strata

in North America. Oreodonts were typically piglike in general build, with short
limbs, digitigrade and four-toed feet, and a
continuous tooth row with no loss of teeth
(Fig. 14–9B). In contrast to the Suina, however, the cheek teeth were selenodont and
became fairly high crowned in advanced
types, suggesting the acquisition of grazing
habits. Probably as a result of competition
with more advanced artiodactyls, oreodonts declined in the upper Miocene, and
disappeared in the Pliocene.

FAMILY CAMELIDAE. These primitive
ruminants are restricted to arid and semiarid regions. *Camelus*, with two species,
occupies the Old World, and wild populations persist in the Gobi Desert of Asia;
Lama (Fig. 14–25), with two species,
occurs in South America from Peru
through Bolivia, Chile, Argentina, and
Tierra del Fuego. Camels probably arose in
the Old World and migrated in the late
Eocene to North America, where their fossil record begins. Of special interest, as an
example of a reversal of a well established
evolutionary trend, is the development of
the camelid foot. By the Oligocene, camels
were already highly specialized in foot
structure; they were nearly unguligrade in
foot posture and were didactyl; the distalmost phalanges probably bore hoofs. The

distinctive distal divergence of the metapodials (Fig. 14–18D), however, was already recognizable. In the Miocene the
central metapodials fused to form a cannon bone, but at this same time a retro-
grade trend toward the secondary development of a digitigrade foot posture began,
and from the Pliocene onward camels
were digitigrade. Camels are the only living, fully digitigrade ungulates. Because
semiarid conditions developed and became widespread in the Miocene, one is
tempted to relate the changes in the camelid foot posture to changing soil conditions. In any case, the peculiar camelid
foot clearly provides effective support on
soft, sandy soil, into which the feet of
"conventional" unguligrade artiodactyls
sink deeply. Taking advantage in the Pleistocene of land bridges between North
America and Asia, and between North and
South America, camels spread to the Old
World and to South America.

Although highly specialized in foot
structure, camelids are the most primitive
living ruminants. These are fairly large
mammals, ranging in weight from about 45
to 500 kg, and have long necks and long
limbs. The dentition has advanced less
toward herbivorous specialization than
has that of the Pecora. In camelids, only

Figure 14–25.   The vicuña (*Lama vicugna*), a camelid that inhabits the central Andes of South America. (Photograph by Carl B. Koford.)

the lateral upper incisor is present, but it is caniniform; the lower canines are retained and are little modified (Fig. 14–26) The lower incisors are inclined forward and occlude with a hardened section of the gums on the premaxillaries. A broad diastema is present, and the premolars are reduced in number (to 3/1-2 in *Camelus* and 2/1 in *Lama*). As in other ruminants, the limbs are long and the ulna and fibula are reduced; in the carpus, the trapezium is absent, and in the tarsus, the mesocuneiform and ectocuneiform are fused. The digitigrade feet are didactyl, but the cannon bone is distinctive in that the distal ends of the metapodials remain separate and flare outward (Fig. 14–18D). The toes are separate, and each is supported by a broad cutaneous pad, which largely encases the second phalanx and serves to increase the surface area of the foot greatly. The short ungual phalanges do not bear hoofs, but have nails on the dorsal surfaces.

Camels are remarkably well adapted to arid areas, and their ability to go for long periods in hot weather without water is remarkable (see p. 443). They are grazers, and can survive in regions with only sparse vegetation. The guanaco (*Lama glama*) of South America is gregarious and usually occurs in small groups led by an adult male. Guanacos are fairly speedy runners, but are especially skillful at moving rapidly over extremely rough terrain. Simpson (1965a:189) observed that in Argentina guanacos spent considerable time running up and down cliffs and were able to leap delicately up a nearly vertical trail that men could only climb laboriously. He also noted that guanacos are highly vocal. They make a strange yammering noise, and Simpson comments that "there is something distinctly indecent about the noise as it issues from the beast's protrusile and derisive lips."

FAMILY TRAGULIDAE. This family, containing the chevrotains or mouse deer, has only two living genera with four species, but is of interest because these animals probably resemble in many ways the ancestors of the more advanced pecorans. Chevrotains are small, delicate creatures, weighing from 2.3 to 4.6 kg, that occur in tropical forests in central Africa (*Hyemoschus*) and in parts of southeast Asia (*Tragulus*; Fig. 14–27). The tragulid fossil record begins in the Miocene.

Although apparently related to higher pecorans, chevrotains have many primitive features not characteristic of other pecorans. The tragulid skull never bears antlers, but, seemingly in compensation, the upper canines are unusually large, defensive weapons. Otherwise, the dentition resembles that of higher pecorans: the upper incisors are lost, the lower canine is incisiform, and the cheek teeth are selenodont. The limb structure is an unusual mosaic of primitive and advanced features. Although the limbs are long and slender, and the carpus is highly specialized in having the navicular, cuboid, and ectocuneiform fused, the lateral digits of chevrotains are complete, a condition never present in higher pecorans. In addition, although a cannon bone occurs in the hind limb, the metacarpals of the central digits are separate in the African tragulid and are partly fused in the Asian form, whereas the cannon bone is represented by fully fused metapodials in all other pecorans. A further contrast is perhaps associated with differences in bounding ability. In tragulids the articular surfaces at the distal end of the cannon bone do not extend onto its dorsal surface, while in other pecorans these articular surfaces are extended.

Tragulids are secretive, nocturnal creatures that inhabit forests and underbrush

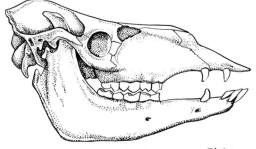

Figure 14–26. The skull of an extinct Pleistocene New World camel (*Camelops*); length of skull 565 mm. (After Romer, 1966.)

Figure 14–27. A chevrotain (*Tragulus napu*, Tragulidae). Note the large upper canine and the lack of antlers or horns. (Photograph by Ron Garrison; San Diego Zoo photo.)

and thick growth along water courses. They escape predators by darting along diminutive trails into dense vegetation. Their food consists of grass, leaves of shrubs and forbs, and some fruit.

Advanced Pecorans. The remaining pecorans (the Cervidae, Giraffidae, Antilocapridae, and Bovidae) are advanced artiodactyls and share a series of progressive features. The upper incisors are absent, the upper canine is usually absent, the lower canine is incisiform, and the cheek teeth are selenodont. The dental formula is typically 0/3, 0/1, 3/3, 3/3 = 32. The cannon bone is present in fore and hind limbs, its distal articular surfaces are extensive, and the lateral digits are always incomplete (Fig. 14–18C) and are often lacking. Movement of the foot is strongly limited to a single plane by the tongue-in-groove contacts between the astragalus and the bones with which it articulates, and by the specialized articular surfaces of the joint between the cannon bone and the first phalanges (Fig. 14–8A, B). Some fusion of

carpal elements always occurs and serves further to restrict movement to one plane. The navicular and cuboid are always fused (Fig. 14–5B), and a variety of patterns of fusion of the other elements occurs. The four-chambered stomach is of a ruminant type. Although all pecorans but the tragulids share this basic structural plan, each family has distinctive features usually related to diet and degree of cursorial ability.

FAMILY CERVIDAE. Members of this family, which includes the deer, elk, caribou, and moose, occur throughout most of the New World, and in the Old World in Europe, Asia, and northwest Africa; they have been introduced widely elsewhere. Living members include some 16 genera and 37 species. Cervids appeared in the early Oligocene in Asia, and reached North America in the early Miocene.

Antlers are the most widely recognized characteristic of members of the family Cervidae (Fig. 14–28). Antlers attain spectacularly large size in some species and vary widely between species. Their com-

Figure 14-28. A mule deer (*Odocoileus hemionus*, Cervidae) in northern Colorado. This species inhabits much of the western United States. (Photograph by O. D. Markham.)

plexity and symmetry have long interested both biologists and outdoorsmen. All but 2 of the some 37 species of cervids have antlers, and they occur only in males, except in caribou (*Rangifer*). In some antlered cervids, the upper canines are retained but reduced (as in the elk, *Cervus elaphus*). Two cervids with short antlers have enlarged canines (*Cervulus* and *Elaphodus*), and in two deer that have no

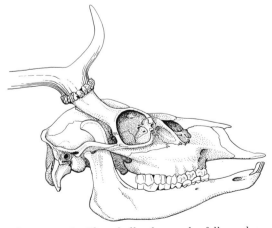

Figure 14-29. The skull of a male fallow deer (*Dama dama*, Cervidae). The bony antlers are shed yearly. Length of skull 265 mm.

antlers (*Hydropotes* and *Moschus*) the canines are large sabers. Antlers usually arise from a short base on the frontals (the pedicel, Fig. 14-29) and are entirely bony. Of particular interest is the annual cycle involving rapid growth of the antlers, their use during the breeding season in ritualized social interactions, and their subsequent loss.

The annual cycle of growth and loss of antlers has been carefully studied in the white-tailed deer (*Odocoileus virginianus*) of North America (Wislocki, 1942, 1943; Wislocki et al., 1947). The cycle is primarily under the control of the testicular and pituitary hormones. Secretions from the pituitary, activated by increasing daylength in the spring, initiate antler growth in April or May, and somewhat later pituitary gonadotropin stimulates growth of the testes. The growing antlers are covered by "velvet." This is fur-covered skin that carries blood vessels and nerves to the growing antlers. In the fall, androgen from the enlarging testes inhibits the action of the pituitary antler-growth hormone, leading to the drying up and loss of the velvet. At this time the animals rub and thrash their antlers against vegetation, and as the velvet is removed the antlers are stained by resins and take on a brown, polished look. Androgen in the fall and early winter maintains the connection between the dead bone of the antlers and the live frontal bones. In winter, pituitary stimulation of the testes declines as daylength is reduced and androgen secretion diminishes. This results in decalcification in the pedicel, weakness at the point of connection between the antler and the pedicel, and shedding of the antlers. For several months in late winter, before reinitiation of antler growth, the males are antlerless.

During the spring and summer the rate of growth of the developing antlers is remarkable. In such large cervids as elk (*Cervus elaphus*) and caribou, the antlers attain a length of some 150 cm in 90 days or less. The growth rate of the antler and its attending nervous and vascular tissue is therefore some 1.7 cm per day. Touch and pain receptors of the velvet protect the growing antler from injury, and damage to nerves supplying a growing antler may be followed by enough injury to the antler to cause abnormal development.

The cheek teeth of cervids are brachyodont, reflecting the browsing habit of these animals. These animals have a wide size range: the musk deer (*Moschus*) weighs but 10 kg, whereas the moose (*Alces*) weighs up to roughly 800 kg. The feet are always four toed, but the lateral toes are often greatly reduced. Distal to the astragalus and calcaneum, the tarsus is usually composed of three bones: the fused navicular and cuboid, the fused ectocuneiform and mesocuneiform, and the internal cuneiform (Fig. 14–5B).

Cervids occur from the arctic to the tropics. Many cervids are well adapted to boreal regions, and occupy mountainous or subarctic areas with severely cold winters. Effective insulation is provided in many cervids by the long, hollow hairs of the pelage. Some species are gregarious for much of the year, and may assemble in large herds during the winter and during migratory movements. In the western United States, mule deer (*Odocoileus hemionus*; Fig. 14–28) and, in the eastern United States, white-tailed deer (*O. virginianus*) are common over wide areas, and are heavily hunted in many states. Some states employ many biologists to study and manage deer; the expense of this work is generally covered by revenue from the sale of hunting licenses.

FAMILY GIRAFFIDAE. This group is represented today by but two genera, *Giraffa* and *Okapia*; each has a single species. The family occurs in much of Africa south of the Sahara Desert.

The robust cheek teeth are brachyodont and are marked with rugosities. Short, bony horns, covered with furred skin, are borne on the front part of the parietals, and a medial thickening of the bone in the area where the nasals and the frontals join is conspicuous (Fig. 14–30); in some populations from north of the equator in East Africa this thickening produces a median horn. Horns occur in both sexes and are never shed. The lateral digits of the elongate limbs are entirely gone, and the tarsus is highly specialized. Distal to the astragalus and calcaneum, only two tarsal bones are present. One is formed by the fusion of the navicular and cuboid, and the other by the fusion of the three cuneiform bones. The okapi lacks the extreme elongation of the neck and legs that is typical of giraffes, but has an even more specialized tarsus in which all bones distal to the astragalus and calcaneum are fused. The fossil record of the giraffids begins in the Miocene. The okapi, not known to zoologists until 1900, is remarkable in its close resemblance to primitive upper Miocene and lower Pliocene giraffids long known to paleontologists.

Giraffes occurs in savanna and lightly wooded areas, where their exceptional height (Fig. 14–31) enables them to browse on branches of leguminous trees up to some 6 m above the ground. Some of the species of *Acacia* that are fed upon are

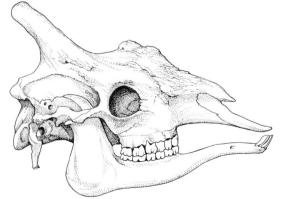

Figure 14–30. Skull of a male giraffe (*Giraffe camelopardalis*; length of skull 708 mm). The heavy deposits of bone on the frontals and nasals seemingly protect the skull when the head is used as a weapon in fights between males.

Figure 14–31. A giraffe among umbrella trees (*Acacia tortilis*) in Tsavo West National Park, Kenya, East Africa. The remarkably beautiful pattern of markings on the body enables this huge animal to blend inconspicuously into savanna vegetation.

FAMILY BOVIDAE. The family Bovidae, including the African and Asian antelope, the American pronghorn, the bison, sheep, goats, and cattle, is the most important and most diverse living group of ungulates. The family includes some 45 genera and 111 species, and wild species occur throughout Africa, in much of Europe and Asia, and in most of North America. (A classification of living bovids is given in Table 14–1). The domestication of bovids began in Asia roughly 8000 years ago (Darlington, 1957:405), and domestic bovids are nearly as cosmopolitan in distribution as is man.

Bovids were derived from traguloid ancestry in the Old World, and first appeared in the early Miocene of Europe. Judging from the many kinds of bovids known from the Pliocene, this group underwent a rapid radiation. More Pleistocene than Recent genera of bovids are known (100 versus 45), but toward the end of the Pleistocene most bovids were driven from Europe by the southward advance of cold climate. Ancestors of the American pronghorn (*Antilocapra americana*) appear first in the middle Miocene of North America, and a few bovids reached the New World in the Pleistocene via the Bering Strait land bridge. Because this avenue of dispersal was under the influence of severe boreal climatic conditions at this time, it functioned as a "filter bridge" (Simpson, 1965b:88), and only animals adapted to these conditions dispersed across the bridge.

As a consequence, the New World received from Asia such bovids as bighorn sheep *(Ovis)*, the mountain "goat" (*Oreamnos*; Fig. 14–32), the musk ox (*Ovibos*), and the bison (*Bison*), all animals able to withstand cold. Bovids less able to withstand boreal conditions—the Old World antelopes and the gazelles are prime examples—were forced from the northern parts of Europe and Asia in the Pleistocene to their present strongholds in Africa and Asia, and hence did not disperse across the Bering bridge to North America. An exception is the saiga antelope *(Saiga),* which now inhabits arid parts

armed with wicked thorns, but giraffes adroitly use the long tongue and prehensile upper lip to gather leaves from even the most thorny acacias. Despite their considerable weight (to nearly 1820 kg in males), giraffes can gallop for short distances at over 55 km per hour (White, 1948). Relative to lighter, shorter-limbed artiodactyls, the limbs of giraffes are flexed little during each stride, producing a curiously stiff-legged gait. During walking or galloping, the center of gravity of the animal is partly controlled by fore-and-aft movements of the head and neck (Dagg, 1962). Males fight by delivering powerful blows with the head against the opponent's head, neck, and body. The okapi lives in dense tropical forests and eats leaves and fruit.

Table 14–1. A Classification of Bovidae and Distribution of Recent Genera*

Subfamily Antilocaprinae
  Tribe Antilocaprini
    Antilocapra†   Pronghorn (North America)

Subfamily Bovinae
  Tribe Bovini
    Bison   Bison (Europe, North America)
    Bos   Cattle (Worldwide)
    Bubalus   Asiatic buffalo (Asia)
    Synceros   African buffalo (Africa)
  Tribe Boselaphini
    Boselaphus   Nilgai (Asia)
    Tetracerus   Four-horned antelope (Asia)
  Tribe Tragelaphini
    Tragelaphus   Bushbuck, nyala, kudu, bongo, etc. (Africa)
    Taurotragus   Eland (Africa)

Subfamily Cephalophinae
  Tribe Cephalophini
    Cephalophus   Duiker (Africa)
    Sylvicapra   Bush duiker (Africa)

Subfamily Reduncinae
  Tribe Reduncini
    Redunca   Reedbuck (Africa)
    Kobus   Waterbuck, kob, lechwe (Africa)

Subfamily Hippotraginae
  Tribe Hippotragini
    Hippotragus   Roan, sable (Africa)
    Oryx   Oryx (Africa)
    Addax   Addax (Africa)

Subfamily Alcelaphinae
  Tribe Connochaetini
    Connochaetes   Wildebeest (Africa)
  Tribe Alcelaphini
    Alcelaphus   Hartebeest (Africa)
    Damaliscus   Hartebeest, topi, blesbok (Africa)

Subfamily Aepycerotinae
  Tribe Aepycerotini
    Aepyceros   Impala (Africa)

Subfamily Antilopinae
  Tribe Antilopini
    Antilope   Blackbuck (Asia)
    Gazella   Gazelles (Africa)
    Antidorcas   Springbok (Africa)
    Litocranius   Gerenuk (Africa)
    Procapra   Black-tailed gazelle (Asia)
  Tribe Ammodorcadini
    Ammodorcas   Dibatag (Africa)
  Tribe Neotragini
    Oreotragus   Klipspringer (Africa)
    Madoqua   Dik-dik (Africa)
    Dorcatragus   Beira antelope (Africa)
    Ourebia   Oribi (Africa)
    Rhaphicerus   Steinbuck, grysbuck (Africa)
    Neotragus   Royal, pigmy, and suni antelopes (Africa)

Subfamily Peleinae
  Tribe Peleini
    Pelea   Rhebuck (Africa)

Subfamily Caprinae
  Tribe Saigini
    Pantholops   Chiru (Asia)
    Saiga   Saiga (Asia, Europe)

Table 14-1.   A Classification of Bovidae and Distribution of Recent Genera (*Continued*)

Tribe Rupicaprini	
Naemorhedus	Goral (Asia)
Capricornis	Serow (Asia)
Oreamnos	Rocky mountain "goat" (North America)
Rupicapra	Chamois (Southwest Asia)
Tribe Ovibovini	
Budorcas	Takin (Asia)
Ovibos	Musk ox (North America, Greenland)
Tribe Caprini	
Hemitragus	Tahr (Asia)
Capra	Ibex, goat (Asia, Europe, North America)
Pseudois	Nahur (Asia)
Ammotragus	Barbary sheep (North Africa)
Ovis	Mouflon, argali, bighorn sheep, sheep (Asia, Europe, North America, North Africa)

*(Partly from Ansell, 1971, and Simpson, 1945.)
†Regarded as a bovid following O'Gara and Matson, 1975.

of Asia but which occurred in Alaska in the Pleistocene. Bison were seemingly extremely abundant members of grassland faunas in the Pleistocene and Recent in North America, where they occurred as far south as El Salvador. Some structural divergence occurred in Pleistocene bison, and in some areas several species may have occupied common ground. Some Pleistocene bison were considerably larger than the present *Bison bison*. Specimens of the Pleistocene species *Bison antiquus* from California indicate that this animal was over 2 m high at the shoulder and had horns that in larger individuals spanned more than 2 m.

Bovids characteristically inhabit grasslands, and the advanced dentition and limbs of bovids probably developed in association with grazing habits. The cheek teeth are high crowned and the upper canine is reduced or absent. Preorbital vacuities in the skull are present in some bovids and absent in others (Figs. 14–33, 14–34). The lateral digits are reduced or totally absent, the ulna is reduced distally and is fused with the radius, and only a distal nodule remains as a vestige of the fibula. Horns, formed of a bony core and a keratinized sheath, are present in males of all wild species, and females usually also bear horns. The entire horns (including both sheath and core) are never shed and in some species grow throughout the life of the animal. With the single exception of *Antilocapra*, bovid horns are never branched, but are often spectacularly long and form graceful curves or spirals (Fig. 14–35). Males of the Indian four-horned

Figure 14-32. A mountain goat (*Oreamnos americanus*) in the mountains of Idaho. (Photograph by Stewart M. Brandborg.)

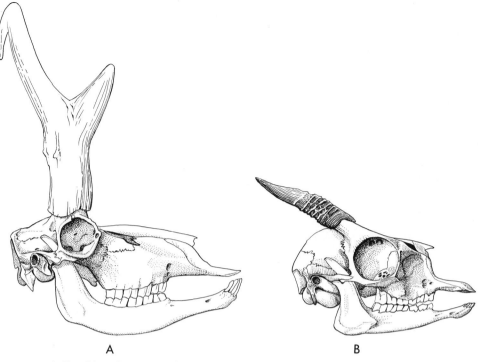

Figure 14–33. Skulls of bovids. A, Pronghorn antelope *(Antilocapra americana;* length of skull 292 mm); *B,* Kirk's dik-dik *(Madoqua kirki;* length of skull 108 mm). The receding nasal bones of the dik-dik are an adaptation allowing mobility of the short proboscis.

antelope *(Tetracerus quadricornis)* are unique in having four short, dagger-like horns. The horns are frequently used in fights between males during the breeding season, but in the Grant's gazelle *(Gazella granti),* and in many other species, styl-

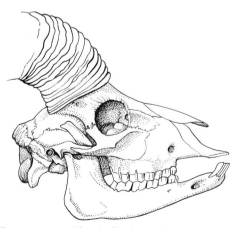

Figure 14–34. The skull of a male mouflon sheep *(Ovis musimon).* The horns are never shed, but continue to grow throughout the animal's life. Length of skull 242 mm.

ized contests of strength are often substituted for the use of the dangerous horns in encounters between rival males (see p. 390). Some bodids, such as the sable antelope *(Hippotragus niger)* and oryx *(Oryx),* can use their horns as awesome defensive weapons, respected even by lions.

The last great strongholds of bovids are the grasslands and savannas of East Africa. Here a diverse bovid fauna occurs (Fig. 14–35), and seemingly every conceivable niche that could be filled by bovids has been occupied. Some antelope, such as the Bohor reedbuck *(Redunca redunca)* and the lechwe *(Kobus leche),* inhabit river borders and swampy ground whereas, at the other extreme, the oryx *(Oryx)* and addax *(Addax)* live in arid plains and desert wastes, where they may seldom have access to drinking water. The protection afforded game in some parts of Africa will allow, it is hoped, the survival of many species of this remarkable group for some time to come.

Figure 14–35. A few members of the diverse bovid fauna of East Africa. A, Klipspringer (*Oreotragus oreotragus*), a small, rock-dwelling antelope that walks on the blunt tips of its hoofs; B, Grant's gazelle (*Gazella granti*), often an inhabitant of dry areas; C, wildebeest (*Connochaetes taurinus*), famous for spectacular migrations; D, African buffalo (*Syncerus caffer*), which live in large, closely knit herds. (Photographs taken in Kenya, East Africa.)

(*Illustration continued on following page*)

Because it is the only New World "antelope" and is unique in several ways, the pronghorn (subfamily Antilocaprinae) merits comment. For many years the pronghorn was regarded as the sole living member of the family Antilocapridae, but O'Gara and Matson (1975) present convincing evidence in favor of the pronghorn's being included in the Bovidae.

The fossil record of pronghorns, which begins in the Miocene, is entirely North American, and today the pronghorn occupies open country from central Canada to north central Mexico. The pronghorn fauna was at one time more diverse than it is now. From the middle Pleistocene Tacubaya Formation in central Mexico, which contains numerous fossils of mammals that lived in a small area at one time, Mooser and Dalquest (1975) list four species of extinct pronghorns ranging in size from a tiny species to one at least as large as the living species.

The pronghorn is unique in being the

Figure 14–35. *Continued*

only mammal that sheds the horn sheaths annually. The sheaths are formed of a specialized growth of skin. The old sheath, beneath which a new sheath is beginning to develop, is shed annually in early winter, and the new sheath is fully grown by July (O'Gara et al., 1971). Whereas the mature sheath is forked, the bony core is a single, laterally compressed blade. Both sexes have horns, but those of the females are small and inconspicuous.

Perhaps as an adaptation allowing the pronghorn to keep watch for danger while its head is down close to the ground, the orbits are unusually far back in the skull (Fig. 14–33). The legs are long and slender, and all vestiges of lateral digits are gone. The tarsus distal to the astragalus and calcaneum consists of but three bones (Fig. 14–5B).

Pronghorns inhabit open prairies and deserts that support at least fair densities of low grasses, shrubs, and forbs. The numbers of antelope were seriously reduced during the pioneering period of the western United States, but, according to Einarsen (1948:7), "At no time has the antelope been driven from its original range, being continuously represented by a few individuals in widely scattered sections of the plains country." Today pronghorns are common in a number of the western states, and even the most unobservant tourist, when driving across Wyoming, cannot fail to notice these animals. Pronghorns are among the fastest of cursorial mammals. At full speed on level footing, they can attain a speed of about 95 km per hour for short distances (McLean, 1944; Einarsen, 1948). Their endurance is remarkable. On the shortgrass prairie of north central Colorado, I observed pronghorns that had run more than two miles put on a burst of speed and run away from a closely pursuing light plane that was traveling at 72 km per hour.

15

ECOLOGY

A modern inclusive definition used by Kendeigh (1961:1) and others describes ecology as "a study of animals and plants in their relations to each other and to their environment." Man's interest in ecology is not new. Paleolithic man understood some of the relationships between the major game animals and their environments. He could predict where certain species could be found, and he used this knowledge to increase his hunting success, hence his survival. Many thousands of years later, we "civilized" modern people are finally understanding that our very survival may depend on a knowledge of basic ecological relationships.

The value of an ecological approach to the study of mammalogy is difficult to overemphasize. Knowledge of the biology of any species of mammal is clearly incomplete if the relations of the animal to its environment are unknown. An adequate knowledge of the ecology of mammals has been long in emerging, however, not because mammalogists lack interest in ecology but because an understanding of the ecology of any single species demands detailed knowledge of a great many aspects of that species' biology. This problem is even more acute in considering complex interactions among many species in natural communities.

Because the scope of ecology is extremely broad, only an incomplete and selective coverage is given here. I will concentrate on ecological relationships or principles well or best illustrated by mammals.

The environments of animals can be described in terms of physical and biotic factors. Physical factors include temperature, humidity, climatic patterns, precipi-

tation, and soil types; biotic factors are those associated with interactions between organisms.

PHYSICAL FACTORS OF THE ENVIRONMENT AND THE DISTRIBUTION OF MAMMALS

TEMPERATURE AND CLIMATE

Solar radiation provides the energy, in the form of heat and light, on which living organisms depend. The intensity of solar radiation at the earth's surface is influenced largely by the directness with which the sun's rays strike the earth. The angle of these rays decreases, and the climates become progressively cooler, the farther north or south of the equator areas are situated.

Warm air holds more moisture than does cool air, and equatorial areas, especially areas near 25° N and 25° S latitude, receive relatively heavy precipitation (Fig. 15–1). In addition, major global patterns of air circulation are set up as the warm air that rises from equatorial regions moves north and south. In a belt centered some 30° north and south of the equator, the equatorial air masses reach a stage of cooling at which they tend to sink; as they sink they are warmed and their ability to carry moisture is increased. Some of the major deserts of the world, such as those in southwestern United States and northern Mexico, are roughly 30° north of the equator and are under the influence of this system of descending air. (See Ricklefs, 1973, Chapter 10, for a more detailed description of global climatic patterns.)

Even in many tropical areas, rainfall is

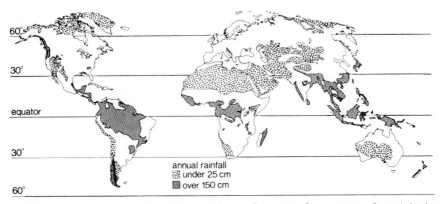

Figure 15–1. The important desert regions (those with less than 10 inches, 25 cm., of precipitation annually) and the major wet regions (those with more than 80 inches, 125 cm., of precipitation annually). (After Espenshade, 1971.)

seasonal (Fig. 15–2), and periodic droughts may strongly affect mammalian survival. In Tsavo National Parks, in Kenya, East Africa, the death of over 5000 elephants in 1971–1972 was attributed to the devastating drought of this period. In many tropical areas, mammals must cope with annual or biannual shortages of food or water. Migrations of some tropical bats coincide with seasonal shifts in the abundance of insects, fruits, or flowers, and the dramatic migrations of wildebeest in East Africa are in response to seasonal changes in the availability of nutritious forage. In the tropics, some small mammals, like the hedgehog, become dormant during prolonged dry periods.

Superimposed on global weather patterns are a myriad of local variations and complexities. Many factors determine local differences in climate. In western United States, for example, in winter moist air masses that sweep inland from the west

or northwest are forced upward as they pass over high, north-south oriented mountain ranges. As the air flows up the western slopes of the mountains, it is cooled and tends to lose moisture; as it descends the east faces of the mountains, it becomes warmer and its ability to hold moisture increases. Typically, then, the western slopes receive high precipitation, the eastern slopes receive lower precipitation, and the basins at the eastern bases of the mountains are often deserts. These abrupt local differences in precipitation are attended by striking changes in the fauna and flora. Pine forests supporting boreal animals are often visible from desert flats supporting scattered creosote bushes and kangaroo rats.

Even minor topographic features may have pronounced local effects on precipitation and on plant communities. The Santa Ana Mountains of southern California are often enshrouded by fog sweeping

Figure 15–2. Mean monthly rainfall at Bushwhacker's Safari Camp, Kenya, East Africa. Note the sharply bimodal annual pattern of precipitation. (From Vaughan, 1976.)

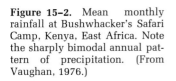

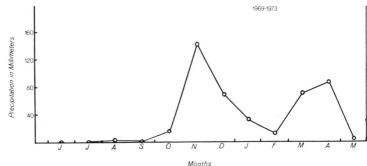

inland from the Pacific Ocean. The fog funnels through the major passes in the mountains, condenses on vegetation, and drops of water fall to the ground. The Coulter pine (*Pinus coulteri*) is restricted locally to the vicinity of these passes (Pequegnat, 1951:9). In the spring and early summer, when fogs are most frequent, condensation drip from the needles of the knob-cone pine *(Pinus attenuata)* in these mountains can total a remarkable 10.2 cm of precipitation per month, and growth can thus be extended into the dry summer period (Vogl, 1973). The influence of these pockets of relatively heavy precipitation on the distributions of small mammals remains to be studied, but may well be important.

In most animals, temperature is a critically important aspect of the physical environment, for optimal regulation of vital metabolic processes in an organism occurs only within a narrow temperature range. Although most mammals can maintain a constant body temperature under varying environmental conditions by regulating the rate of production or dissipation of heat, many species are adapted to such a limited temperature range that this single climatic factor is of prime importance in controlling their distribution.

As mentioned earlier, the angle at which the sun's rays strike the ground strongly affects global temperature patterns; of additional importance is daylength. Daylength varies seasonally little in regions near the equator, but progressively greater extremes in daylength are encountered toward the poles. During their long winters, antarctic and arctic areas remain essentially dark for long periods; the intense cold is thus unrelieved by solar radiation. The cold air from these areas tends to flow toward the equator, affecting temperatures and weather in temperate regions. In winter, in arctic and cold temperate areas, there is no plant growth, and few species of terrestrial vertebrates are able to remain active.

In sharp contrast, in midsummer in the high arctic there is nearly continuous daylight, and such plant growth as does occur is telescoped into the brief period of warmth and light. Here both plants and animals must be able to cope with low annual solar energy budgets. The adaptations of some plants to short growing seasons are remarkable: the corollas of two high arctic flowers maximize their intake of heat by tracking the sun and reflecting solar radiation on their reproductive parts (Kevan, 1975). Herbivorous mammals of areas with brief growing seasons must gear their behavior to making the most of the short summer.

Belding ground squirrels (*Spermophilus beldingi*) live at fairly high elevations in the western United States and are active for only four or five summer months. During the remainder of the year they live on fat reserves. They must literally eat enough in the short summer to last them throughout the year. A careful study of these animals by Morhardt and Gates (1974) showed that they are able to enhance their ability to store fat by their stingy use of energy for thermoregulation. In summer foraging, the squirrels' body temperatures fluctuate some 3° to 4° C to a high of 40°C, thus avoiding the energy cost of maintaining a more constant body temperature and permitting the animals to keep feeding longer while exposed to solar heating. By retreating briefly into a cool burrow, or orienting their bodies to reduce or increase absorption of radiant energy from the sun, the animals adjust their body temperatures at little metabolic cost. Almost constant breezes allow the animals to lose heat by convection in the hottest parts of the day. These physiological and behavioral adaptations enable the ground squirrels to feed almost constantly through the long summer days and to expend a minimum of energy on thermoregulation. Adult squirrels are able to replenish as fast as possible the fat stores depleted the previous winter, and young squirrels can undergo rapid growth and storage of fat.

It should be mentioned parenthetically that the very different temperature regimes and weather patterns in tropical as compared to temperate or boreal regions are obviously not only associated with very different seasonal patterns of productivity, but this productivity is limited by different factors. In tropical and subtropical areas, temperatures are always high enough to permit plant growth; the critical factor here is precipitation. In some cold areas, on the other hand, water in sufficient amounts for plant growth is available for much of the year, but the limiting factor is temperature. These same factors influence distributions of animals and plants in mountains; lower distributional limits are often set by lack of water or heat or both, whereas upper limits may be set by lack of heat.

Vertical temperature gradients are encountered as one ascends mountain ranges. The lowering of the temperature with increased elevation is of the magnitude of approximately 1°C for every 150 meters. This effect, coupled with increased precipitation at higher elevations, shorter growing seasons for plants, and drastic diurnal-nocturnal fluctuations in temperature, is associated with a distinct separation of climatic zones in high mountains throughout the world. The distributions of some mammals clearly reflect this zonation. In northern Arizona, the Abert's

squirrel (*Sciurus aberti*) occupies the ponderosa pine belt, but is abruptly replaced by the red squirrel (*Tamiasciurus hudsonicus*) where spruce and fir forests appear at higher elevations. In some areas in western United States, an assemblage of "desert" mammals, resembling those typical of arid lands as far south as central Mexico, may occupy the arid or semiarid land at the foot of a mountain range, while the crests of the mountains, a few airline miles away, may support boreal genera that occur as far north as northern Canada or Alaska (Fig. 15–3).

Local conditions may also strongly affect the amount of heat the surface of the earth receives. In many mountainous sections of western United States, steep slopes and precipitous canyon walls are common. Because the main axes of most mountain ranges lie north and south, the drainage systems are oriented more or less east and west and the canyon walls face roughly north or south. In northern latitudes, no matter what the time of year, the sun's rays strike a south-facing slope more directly than a north-facing slope. South-facing slopes are consequently drier and warmer than are nearby north-facing

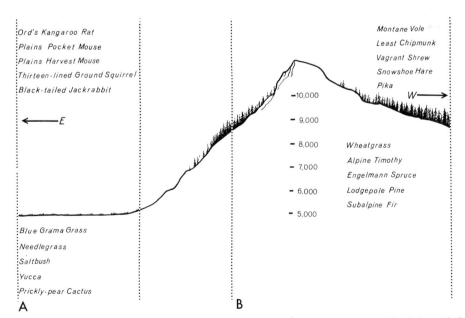

Figure 15–3. Assemblages of plants and mammals inhabiting short-grass prairie (A) and subalpine habitats (B) in northern Colorado (Larimer County).

slopes. The effects of slope (steepness of incline) and exposure (the direction the slope faces) are strongly reflected by the flora, and the compositions of small mammal faunas are frequently as conspicuously different on adjacent opposing slopes as are the assemblages of plants. In the precipitous chaparral-covered mountains of southern California, for example, contrasting biotas occupy adjacent north- and south-facing slopes (Vaughan, 1954). Some species of mammals that occur on one slope are absent from the opposing slope (Fig. 15–4).

Temperature even imposes rigid constraints on the structure and behavior of whales, the largest mammals. The enormous body size of many whales allows them to conserve heat more effectively than can small mammals (see page 418); but because heat is dissipated from a warm body many times faster in water than in air, temperature has exerted strong selective pressures on marine mammals.

Brodie (1975) discussed some of the thermoregulatory and energetic problems faced by the fin whale (*Balaenoptera physalis*). This whale is a plankton feeder of great size: its length is about 20 m and its weight is up to 48 metric tons (roughly 100,000 lbs). Fin whales feed in summer in the cold (near 0°C), plankton-rich waters of Antarctica and the far north, spending an average of 120 days and 183 days respectively in these regions. The whales migrate slowly to subtropical waters in the winter, where they eat little or nothing, deriving energy from the layer of blubber deposited in the summer. In the case of the Antarctic fin whales, the period of fasting lasts some 245 days. This migration allows the animals to spend their fasting period in warm waters (25° to 30°C), where energy can be conserved because of the greatly reduced cost of thermoregulation. Brodie argues that the demands of the thermal environment of this whale have influenced the evolution of body size. The optimum body size and surface area in these whales are those allowing sufficient subcutaneous fat to be deposited to maintain the whales through the long fasting period in warm waters poor in food.

Degrees of tolerance for large and rapid temperature changes differ widely between species of mammals. Most tropical mammals are adapted to the fairly narrow seasonal and diurnal nocturnal temperature fluctuations characteristic of the tropics, where freezing temperatures never

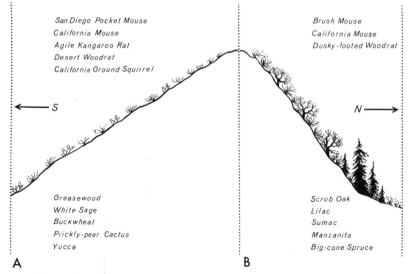

Figure 15–4. Assemblages of plants and mammals inhabiting a south-facing slope (A) and a north-facing slope (B) in lower San Antonio Canyon, San Gabriel Mountains, Los Angeles County, California. (Data from Vaughan, 1954.)

occur. Some tropical mammals, such as the tree sloths (Bradypodidae), have limited thermal adaptability and are under cold stress at ambient temperatures as high as 22°C (Scholander, 1955), and some species, such as the vampire bat (Desmodus rotundus), are sensitive to ambient temperatures above roughly 35°C (Lyman and Wimsatt, 1966). Mammals occupying cold temperate or boreal areas, on the other hand, must have considerable thermal flexibility, and must adapt structurally, physiologically, and behaviorally to the stresses imposed by great seasonal and diel shifts in temperature.

Because the cover provided by such features as vegetation or rock alters the environment locally, the environment at a terrestrial locality is not uniform, but consists of a complex mosaic of microenvironments. As a general rule, few terrestrial mammals can withstand the most extreme temperatures (or other conditions) that occur in the habitats they occupy, but are able to select microenvironments in which temperature extremes are moderated or eliminated. A notable example of such a microenvironment is shown in Figure 15–5. Although occupying a temperate region, a pocket gopher may live for much of the year in a "tropical" microenvironment that features even, fairly high temperatures and high humidities. A group of beavers oc-cupying a beaver lodge in the winter is by no means subjected to the extreme air temperatures outside the lodge (Fig. 15–6). Similarly, most species of shrews forage beneath litter, under logs or rocks, or beneath dense foliage; not only is their food abundant in such places, but temperature and humidity are moderated by such cover. These animals are actually not adapted to the general climatic conditions of the regions they occupy, but are instead adapted to a limited set of conditions which occur in a chosen microenvironment.

WATER

In mammals, as in other animals, water forms an essential part of protoplasm and body fluids, and the maintenance of water balance is basic to life. The availability of free water and the amount of water in the air affect the habitat selection of mammals.

Fossorial mammals occupy microenvironments generally characterized by high humidities. Under such conditions, pulmocutaneous water loss (water loss from breathing and loss through the skin) is minimal, and fossorial mammals that eat moist food can maintain water balance without drinking water. Careful studies of the microenvironmental conditions of burrows of the plains pocket gopher (Geomys

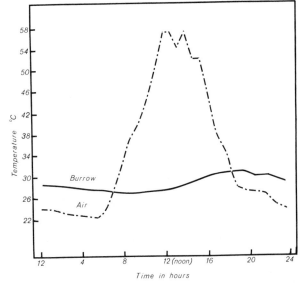

Figure 15–5. Temperature fluctuations of the air within a pocket gopher burrow (solid line) and just above the surface of the ground (dashed line). Temperatures were recorded on 24 June 1961, in McLennan County, Texas. (After Kennerly, 1964.)

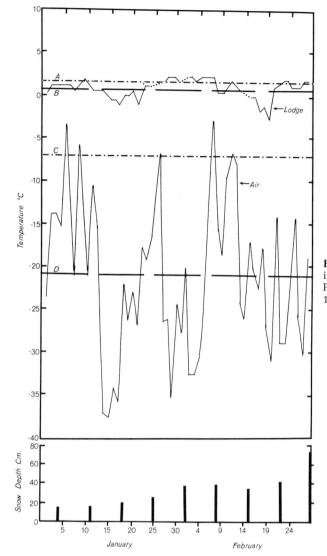

Figure 15-6. Daily minimum temperatures inside and outside a beaver lodge in Algonquin Park, Ontario, Canada. (After Stephenson, 1969.)

bursarius) in Texas by Kennerly (1964) showed that relative humidities within burrows weré usually between 86 and 95 per cent. Humidities within the sealed burrows may be high despite low soil moisture: Kennerly recorded a relative humidity of 95 per cent in a burrow in July when soil from the floor of the burrow contained but 1 per cent water. Although the microenvironment within burrows is such that stresses resulting from high temperatures and low humidities are avoided, other stresses may occur. Concentrations of carbon dioxide from 10 to 60 times that of atmospheric air were recorded from pocket gopher burrows by Kennerly!

Flooding or high soil moisture may cause seasonal changes in mammal distributions. The mole rat (*Cryptomys hottentotus*) in Southern Rhodesia centers its activity during the rainy season in the bases of large termite mounds that rise a few feet above the surrounding grassland and provide relatively dry islands in a sea of waterlogged terrain (Genelly, 1965).

Snow is an environmental feature of great importance in some boreal areas. The continuous snow cover that persists through the winter in many areas is a severe hardship for some large mammals. To most North American artiodactyls, including deer, elk, bighorn sheep, and moose, even moderately deep snow imposes a burden by covering some food and making locomotion difficult. In mountainous areas, deer and elk avoid deep snow by abandoning summer ranges and moving to lower elevations. South-facing slopes and windswept ridges, where snow is shallow or periodically absent, are preferred at these times. In areas with fairly level terrain, deer and moose respond to deep snow by restricting their activities to small areas, called "yards," where they have established trails through the snow. Prolonged winters with deep snow commonly cause high mortality among deer and elk. On Vancouver Island, British Columbia, in the winter of 1948–1949, snow covered the ground even at low elevations to depths of about one meter; deer were denied access to their accustomed winter foods and mortality from starvation was high. Forestry crews in the summer of 1949 counted 18 deer carcasses per square mile in some areas (Cowan, 1956:560).

Rather than being a source of winter hardship for small mammals, snow is actually a blessing. It forms an insulating mantle that provides a microenvironment at the surface of the ground where activity, including breeding in some species, continues through the winter. To these small mammals, such as shrews (Sorex), pocket gophers (Thomomys), voles (Microtus, Clethrionomys, Phenacomys) and lemmings (Lemmus, Dicrostonyx), the most stressful periods are those in the fall, when intense cold descends but snow has not yet moderated temperatures at the surface of the ground (Formozov, 1946), and in the spring, when rapid melting of a deep snowpack often results in local flooding of much of the ground (Jenkins, 1948; Ingles, 1949: Vaughan, 1969).

A deep snowpack also ensures a food supply to some small mammals through the winter. In cold areas that remain snow-free, the soil freezes to considerable depths in the winter and virtually no succulent above-ground vegetation is available, while in areas with deep snowpack, there is often only a thin layer of frozen soil and green vegetation remains available throughout the winter. Beneath the snow some perennial forbs develop basal rosettes of leaves in preparation for a period of rapid growth after spring snowmelt, and some plants retain green leaves from the previous summer.

Even in the summer, snow may be important to some mammals. Alpine or northern snowfields often persist through much of the summer on northern exposures and provide a cool microenvironment unfavorable for insects. Caribou (Rangifer tarandus) and bighorn sheep (Ovis spp.) congregate at times in these places to seek relief from warble flies.

SUBSTRATES

Many small mammals seek diurnal refuge in burrows, and many terrestrial mammals of a variety of sizes have specialized modes of locomotion that are effective on reasonably smooth surfaces. To these mammals, the type or texture of the soil or substrate is a critical environmental feature. Burrowing species may be narrowly restricted to a particular type of soil; for example, some heteromyid rodents that are weak diggers occur only where the soil is sandy. Some scansorial (climbing) species occur only where there are large rocks or cliffs.

Perhaps the most striking examples of mammalian preferences for specific types of substrate are to be found among desert rodents (see Grinnell, 1914a, 1933; Hardy, 1945). In most desert localities no single species of rodent occurs on all types of substrate, and some species are tightly restricted to a single type of soil. In Nevada the four species of pocket mice (Perognathus) have largely complementary soil preferences (Hall, 1946:358, 364, 371, 376). One (P. longimembris) lives "on the firmer soils of the slightly sloping margins of the valleys." A second (P. formosus) is "closely confined to slopes where stones from the size of walnuts up to those 8 inches or even more in diameter are scattered over and partly imbedded in the ground...." A third (P. penicillatus) occupies "the fine, silty soil of the bottom land...." The most broadly adapted (*P. parvus*) "occurs in a wide variety of habitats as regards soil...."

Lines of snap traps set from the desert floor into rocky desert hills will frequently reveal that one assemblage of rodents is associated with the sandy desert floor, another inhabits the gravelly lower slopes of the hills, and another occupies areas marked by boulders and rock outcrops (Fig. 15–7). Although these preferences may in some cases reflect relative digging ability, they appear often to be a reflection of locomotor style and foraging technique. Merriam's kangaroo rat (*Dipodomys merriami*) escapes danger by rapid and erratic hops. It will inhabit several types of soil, but favors fairly open terrain with fine-grained soils where the animal's distinctive type of locomotion can be used effectively. The cactus mouse (*Peromyscus eremicus*), in contrast, is not a speedy runner, but is a capable climber and scrambler. It always occurs where rocks, or in some places cactus or brush, offer immediately accessible retreats from danger.

Large rocks or cliffs are essential environmental features for some mammals. As a few examples, pikas (*Ochotona*) seldom occur away from talus or extensive boulder piles, some wood rats (*Neotoma*) build their houses only in cliffs or steep rock outcrops, and the dwarf shrew (*Sorex nanus*) is apparently restricted to rocky areas in alpine or subalpine situations (Brown, 1967; Marshall and Weisenberger, 1971). To many species of bats, crevices in outcrops or cliffs provide retreats in areas otherwise lacking suitable roosting places. Even some saltatorial species have adopted a rock-dwelling mode of life. The Australian rock wallabies (*Petrogale* and *Peradorcas*) have granular patterns on the soles of the hind feet that increase traction; these animals leap adroitly among rocks and are restricted to rocky areas.

The extreme importance of a suitable substrate to some mammals is well illustrated by the habits of African rock hyraxes (*Heterohyrax, Procavia*; Fig. 19–10). The entire life style of these mammals is built around their occupancy of rock piles and cliffs. Most of their food consists of plants growing among or immediately adjacent to the rocks and, in connection with the maintenance of their social systems, the rocks are elaborately scent-marked with urine and feces. The rocks also provide vantage points where watch is kept for predators, and where there are immediately available refuges in the form of crevices. In addition, rocks provide an economical means of conserving energy (see page 430).

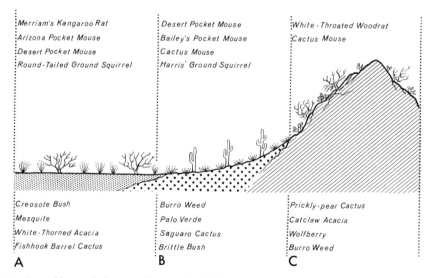

Figure 15–7. Assemblages of plants and mammals inhabiting contrasting types of substrate: A, sandy and silty soil; B, gravelly soil; C, rock.

BIOTIC FACTORS IN THE ENVIRONMENT

VEGETATION

Not only are plants important as food for many mammals, but the cover, escape routes, and retreats they provide, as well as the degree to which the plants facilitate or obstruct rapid locomotion, are important aspects of the environments of many terrestrial mammals. Therefore, plants that are never used as food by a mammal are often as essential a part of this animal's environment as are staple food plants.

A species of mammal is seldom evenly distributed, even within an area of seemingly homogeneous vegetation. On the contrary, the actual distributional patterns of most mammals are discontinuous, indicating that all requirements for the species are not met over broad areas. Close observation of even a limited area usually indicates that there are local changes in the relative densities and the spacing of the plants; further, a given species of mammals is usually restricted to a habitat characterized by plants of a certain life form. The size, shape, foliage density, and pattern of branching of a plant determines its life form. Analyses of the environmental requirements of a mammal, therefore, must include considerations of not only the species of plants with which the animal is associated, but also (and frequently equally important) the life forms of these plants and the "aspect" they give to the habitat.

Consider, for example, two species of East African ungulates. Kirk's dik-dik (*Madoqua kirki*) is a small, delicately built bovid (Fig. 17-7). When disturbed, it dashes into the nearest patch of brush, where it "freezes" and stares at the source of danger. This animal is restricted to brushy areas, and although when in cover it may be approached fairly closely, one's view of the dik-dik is often through a formidable screen of thorny branches. At the other extreme is the Grant's gazelle (*Gazella granti*). This extremely swift animal seeks to escape its predators by outrunning them, at times over long distances. Although this gazelle occupies a variety of settings, from semideserts to grasslands, it is restricted to areas where escape can be sought in the open, where shrubs are scattered or absent, and where the grasses are not tall enough to limit high speed running.

FOOD

Food is necessary to animals as a source of energy and for building and maintaining protoplasm, and is therefore one of the most important biotic factors in the environment of a mammal. Mammalian adaptive radiation has involved a progressively broader exploitation of food sources, but although mammals as a group utilize many types of food, a single species usually eats a fairly limited array of foods that it is structurally, physiologically, and behaviorally capable of utilizing efficiently. Much of mammalian evolution has been "guided" by the necessity of achieving the most favorable possible balance between energy and time expended in securing and metabolizing food on the one hand and energy gained from the food on the other. The relationships of mammals to their environments must ideally be considered against a background of knowledge of feeding biology.

The most abundant and omnipresent foods for terrestrial mammals are plants and insects. It is not surprising that the three most important mammalian orders in terms of numbers of species—Insectivora, Chiroptera, and Rodentia—depend primarily on these major food sources. In addition, members of the orders Edentata, Pholidota, and Tubulidentata are primarily insect eaters, whereas some members of the Marsupialia, Primates, and Rodentia, as well as the Lagomorpha, Proboscidea, Hyracoidea, Sirenia, and Perissodactyla, and some artiodactyls are herbivores.

Although many herbivores are selective in their feeding, a fairly wide variety of food is generally utilized, and seasonal variations in feeding habits are typical of temperate zone species. Western wheatgrass (*Agropyron smithii*) forms about 35

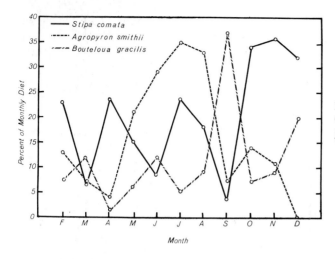

Figure 15–8. Seasonal differences in the diet of the plains pocket gopher *(Geomys bursarius)* in eastern Colorado. (After Myers and Vaughan, 1964.)

per cent of the diet of the plains pocket gopher (*Geomys bursarius*) in July but is not eaten in December (Fig. 15–8). A number of studies have shown that, given a wide variety of plants to choose from, most herbivorous mammals are selective foragers (see, for example, Ward and Keith, 1962; Yoakum, 1958; Zimmerman, 1965); accordingly, a herbivore may show a great preference for one of the least abundant plants in its habitat (Fig. 15–9).

The nutrient content of plants may determine in part their choice as food by herbivores. Lindlöf et al. (1974) found that in February mountain hares (*Lepus timidus*) in Sweden preferred plants high in crude protein and phosphorus; in addition, in habitats preferred by hares the concentrations of these nutrients in the major food plants were unusually high. The nutrient content of various deciduous browse plants utilized by mule deer (*Odocoileus*

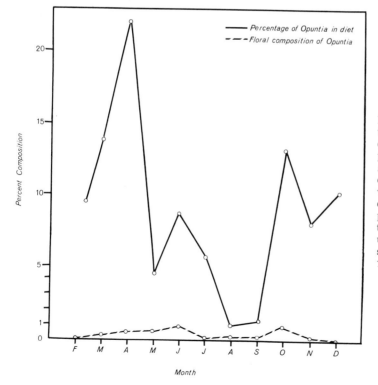

Figure 15–9. Seasonal changes in the utilization of prickly-pear cactus *(Opuntia humifusa)* by the plains pocket gopher *(Geomys bursarius)*. The per cent composition of prickly-pear cactus in the diet varies markedly, whereas the floral composition of prickly-pear (an expression of the percentage of the total coverage of vegetation contributed by *Opuntia*) is nearly constant. (After Myers and Vaughan, 1964.)

hemionus) was shown by Short et al. (1966) to vary seasonally, whereas nutrient levels of evergreen plants were less variable. The deer in many areas eat a great variety of plants, but may select those from which nutrients can be extracted most easily by digestion, rather than those plants in which levels of certain critical nutrients are highest. In rodents, food preferences of different sympatric species (species that live in the same area) are often highly specific and may differ to the extent that there is remarkably little interspecific dietary overlap (Fig. 15–10).

The choice of specific plants or parts of plants by herbivorous animals is influenced by a variety of factors, including the occurrence of secondary compounds that are toxic. Glander (1977) found that in Costa Rica leaf-eating howler monkeys (Alouatta palliata) were forced to be highly selective in their feeding by toxic compounds present in the leaves of trees. The leaves of only certain individuals of some species of trees were eaten, and the monkeys generally ate the petioles (leaf stalks), which have lower concentrations of toxins than do the leaves. The price of carelessness is high: three of six dead howler monkeys that Glander examined had been eating the leaves of either of two trees known to have toxic leaves, as had a female that was observed to go into convulsions and fall from a tree.

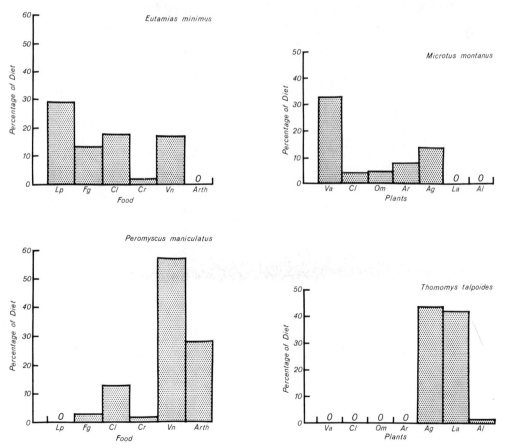

Figure 15–10. The diets of four partly or completely herbivorous rodents in part of the summer of 1965 in a subalpine area in Routt County, Colorado. The rodents are the least chipmunk (Eutamias minimus), the montane vole (Microtus montanus), the deer mouse (Peromyscus maniculatus), and the northern pocket gopher (Thomomys talpoides.). Note the lack of competition among the species for food. Abbreviations: Ag, Agoseris glauca; Al, Achillea lanulosa; Ar, Arnica cordifola; Arth, arthropods; Cl, Collomia linearis; Cr, Carex sp.; Fg, fungus; La, Lupinus argenteus; Lp, Lewisia pygmaea; Om, Oenothera micrantha; Va, Vicia americana; Vn, Viola nuttali. (Based on unpublished data.)

Insects have probably been a major food source of mammals for over 180 million years, and have had a marked influence on the patterns of mammalian evolution. The dentitions of some late Triassic mammals were well adapted to sectioning insects, and the radiation of the Lepidoptera (moths and butterflies) and the termites (Isoptera) in the Cretaceous broadened the food base for insectivorous mammals. The early Tertiary adaptive radiation of the highly successful microchiropteran bats probably occurred primarily because of the great nocturnal abundance of moths. In tropical areas, termites are of tremendous importance to mammals today. At least some members of 10 of the 19 orders of mammals commonly eat termites, and the spiny anteaters (Tachyglossidae), the South American anteaters (Myrmeco-phagidae), the pangolins (Manidae), and the aardvark (Orycteropodidae) specialize on termites, as do some members of the order Insectivora. Members of a community of insectivorous bats choose prey not only on the basis of size but on the basis of prey type. Some are "moth strategists" and some are "beetle strategists" (Table 15–1).

Protective devices, such as thick, chit-inous exoskeletons, toxic sprays, and po-tent stings, protect some insects from pre-dation, but some mammals have evolved countermeasures. In the Serengeti Plain of Africa the primary food of the aardwolf is the termite *Trinervitermes bettonianus*. Aardwolves feed on foraging parties of this termite for periods averaging only 22 sec-onds, probably in an effort to avoid the soldier caste. As a protective device, the soldiers spray distasteful terpenoids from a pointed projection of the head capsule. When the column of workers is disturbed, the soldiers stream from the nest and soon mix in large numbers with the retreating workers. Apparently the aardwolf adjusts its foraging time accordingly and moves on to undisturbed termites before this occurs (Kruuk and Sands, 1972). The striped skunk (*Mephitis mephitis*) handles "stink bugs" in a unique way. Some large beetles of the family Tenebrionidae discourage predators by spraying toxic quinones from abdominal glands. The striped skunk takes such a beetle between his front feet and rolls it roughly in the soil; when the beetle has exhausted its spray and the spray has been absorbed by the soil, the skunk eats the beetle (Slobodchikoff, 1977).

Many mammals prey on higher verte-brates, including reptiles, birds, and mam-mals. Such carnivores occur in the orders Marsupialia, Insectivora, Chiroptera, Ceta-cea, and Carnivora. The size of a predator obviously determines the range of size of its prey. Many mammalian faunas consist of a wide variety of prey species fed upon by various carnivores; in such faunas each carnivore differs from the others in size

Table 15–1. Density of Moth Scales and Per Cent Frequency of Beetles and Moths in Feces of Bats of a North Temperate Bat Community

	Species	Sample Size	Mean Scales/g	Per Cent Frequency Moths	Per Cent Frequency Beetles
M	*L. noctivagans*	19	145,591	100	0
M	*L. cinereus*	39	10,943	100	5
M	*P. hesperus*	7	5,016	100	0
M	*I. phyllotis*	3	2,397	100	0
B	*E. fuscus*	165	679	61	84
B	*A. pallidus*	12	32	17	92
?	*M. californicus-leibii*	16	19,929	94	69
M	*M. volans*	29	4,031	96	17
M	*M. auriculus*	10	1,868	90	20
?	*M. yumanensis*	16	913	53	24
B	*M. evotis*	13	98	62	92
B	*M. thysanodes*	11	48	36	73

M, moth strategist; B, beetle strategist. (From Black, 1974.)

and takes a different sized class of prey. Many carnivores are highly specialized structurally for killing and eating their prey, and most have behavioral specializations that further their predatory ability. The learning of efficient hunting and killing methods is critical to the survival of carnivores. Prey must be captured without an excessive outlay of energy. Of equal importance, particularly in the larger carnivores that kill prey as large as or larger than themselves, hunting and killing behavior must be geared to avoiding serious injury to the predator. The high incidence of skeletal damage in the fossil remains of saber-tooth cats (Smilodon) suggests that in the Pleistocene, as today, preying on large game was a dangerous undertaking.

Still other mammals are omnivorous and are opportunistic feeders; such mammals eat a wide variety of plant and animal material. Omnivores occur among the orders Marsupialia, Insectivora, Chiroptera, Primates, Rodentia, Carnivora, and Artiodactyla. Omnivores are typically less specialized in structure than are mammals adapted to narrower diets, and within some orders the omnivorous mode of life has been highly successful. Among rodents, for example, the nearly ubiquitous North American genus Peromyscus includes species seemingly adapted to a wide variety of plant and animal foods (Fig. 15–11). Some terrestrial sciurids are also omnivorous, and in the order Carnivora, such widespread and successful families as Canidae, Ursidae, and Procyonidae have many omnivorous members.

A variety of additional foods is eaten by mammals. Planktonic crustaceans are the major food of the filter-feeding baleen whales and some phocid seals; most odontocete cetaceans eat fish, squid, and, less commonly, a variety of other invertebrates, but the killer whale (Orcinus) preys on porpoises, whales, diving birds, and a variety of large sharks and bony fish. Fish, squid, and mollusks are taken by pinnipeds; fish are also eaten by some members of the Carnivora and even by some bats. The leopard seal (Hydrurga) of the Atlantic preys on a variety of marine vertebrates and is an important enemy of penguins. A number of marine invertebrates, including sea urchins (Strongylocentrotus), provide food for sea otters (Enhydra).

A few mammals do considerable scavenging; hyaenas (Crocuta and Hyaena) and jackals (Canis spp.) frequently feed on the leftovers of kills of larger carnivores. Nectar and pollen feeding is common among pteropodid and phyllostomatid bats, and one phalangerid marsupial (Tarsipes) is highly specialized for a diet that consists partly of nectar. Vampire bats, which have one of the most specialized mammalian feeding techniques, feed entirely on blood. Some carnivores—the coyote is a good example—are highly opportunistic and, under the pressure of hunger, may take almost any vulnerable

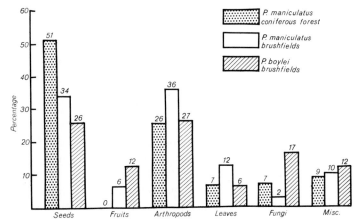

Figure 15–11. Foods of deer mice (Peromyscus) in different habitats in the northern Sierra Nevadas of California. (After Jameson, 1952.)

animal, vertebrate or invertebrate, as well as plant material.

An unusual example of opportunism is described by Huey (1969). Part of the range of the Peruvian desert fox *(Dusicyon sechurae)* includes the extremely barren Sechura Desert of northwestern Peru, where the only other mammal is the rodent *Phyllotis gerbillinis*. On an extremely sterile area away from the Pacific Coast this fox subsists almost exclusively on the seeds of shrubs. All of the fecal material from this area contained seeds, and 77 per cent contained seeds alone; the only other food of any importance was tenebrionid beetles, which occurred in small amounts in 23 per cent of the stomachs. Near the coast this fox had access to animals washed up on the beach; here, although seeds still occurred in most feces, invertebrates such as crabs and beetles, and vertebrates such as gulls, mice, and lizards were important foods.

ENVIRONMENTAL IMPACT OF MAMMALS

No animals have had, nor now have, a greater impact on terrestrial environments than mammals. The ability of man to modify, or in some cases to devastate, his environment is well known, and this ability is of ever-increasing importance as human populations continue to expand. But other mammals also strongly effect their environments, and in "natural" areas, where the hand of man has lain lightly on the land, the activities of wild mammals often drastically alter the character of the vegetation, the availability of water, the patterns of erosion, and the diversity and nature of the vertebrate and invertebrate fauna. If the importance of a group of animals is equated with their effect on the environment, mammals are clearly the most important terrestrial animals.

The impact of mammals results from a variety of activities, including feeding, patterns of migration or daily movement, the quest for water, and the construction of shelters or refuges. These activities are typically interrelated, and an understanding of the impact of mammals can be best approached by a consideration of daily or annual cycles of activity. Even the capture of a moth larva by a shrew or the hoarding of a seed by a white-footed mouse has an effect on the environment; but, as might be expected, large mammals can modify their environments most drastically and conspicuously.

In some of the National Parks and game preserves of Africa, there are high elephant populations and, because of the encroachment of man and his agriculture, the elephants are no longer free to range widely when pressed by local or seasonal shortages of food or water. Studies on elephants in various parts of East Africa document the impact elephants can have on the landscape (Laws, 1970) under these conditions. The diet of elephants seems ideally to consist of a mixture of browse from trees and shrubs and grass (Laws and Parker, 1968), and preferred habitat is thus forest edge, woodland, or bush-grass mosaic. During the dry seasons in the drier areas, such as Tsavo National Parks in Kenya, sources of water are not generally distributed; because elephants at these times need water daily, they concentrate within a radius of some 20 to 30 km of water. A very large group of elephants may have a daily food requirement of 50,000 kg or more, and when high densities of elephants occupy restricted areas near water rapid destruction of trees and shrubs results.

In Tsavo National Parks, elephants are effecting the transformation of the original bushland to grassland (Fig. 15–12). Aerial photographic transects studied by Watson (1968) indicated that in a period of some five years elephants killed from 26 to 28 per cent of the trees above 65 cm in crown diameter. To the south, in Lake Manyara National Park of Tanzania, Douglas-Hamilton observed similar destruction: in one area of especially acute damage, elephants killed 8 per cent of the umbrella trees *(Acacia tortilis)* in one year. Particularly notable in Tsavo is the destruction of baobab trees (Fig. 15–12). These huge and picturesque trees live to be several hundred

Figure 15-12. Elephants destroying a baobab tree in Tsavo West National Park, Kenya, East Africa.

years old, and their reproductive rate is low. At the present rate of destruction by elephants, the baobabs will be eliminated from Tsavo National Parks within a few years.

In Tsavo, as elsewhere, elephants have not affected the vegetation alone. There have also been striking changes in the fauna. Ungulates such as zebras (*Equus burchelli*), Grant's gazelles (*Gazella granti*), and oryx (*Oryx gazella*) have been favored by the shift toward grassland and, as pointed out by Sheldrick (1972), the visitor to Tsavo is now able to observe a much greater diversity of game animals than could be seen when the area was bushland. Other effects cited by Sheldrick include enlargement of wallows to form water holes in the wet season and local

compaction of riverbeds resulting in the bringing to the surface of previously subsurface water. These processes are clearly dynamic, and the rates and direction of change are controlled by a great many factors, including fluctuating patterns of rainfall and changes in elephant densities. Seemingly, the grassland cannot continue for long to support high densities of elephants, and Sheldrick (1972, 1973) suggests that we may be observing but one phase of a cycle that has recurred over and over again.

Feeding activities of a variety of mammals have pronounced effects on vegetation. Peterson (1955:162, 163) noted that moose (*Alces alces*) were responsible for the local decimation of ground hemlock, quaking aspen, and balsam on Isle Royale,

Figure 15-13. When two species overlap broadly in some niche characteristic (above), such as choice of prey size, there is reduced survival of the most strongly competing members of each species. An evolutionary trend toward niche divergence typically results (below). (After Whittaker, R. H.: *Communities and Ecosystems*, The Macmillan Company, 1970. © by R. H. Whittaker, 1970.)

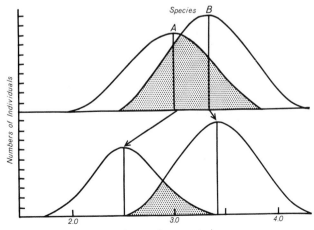

Michigan, and these animals have caused considerable damage to vegetation in Finland (Kangas, 1949). In Rocky Mountain National Park in Colorado, heavy browsing by deer and elk in some areas caused the death of 85 per cent of the sagebrush and 35 per cent of the bitterbrush plants in a five-year period (Ratcliff, 1941). In California, a tract of bitterbrush was killed by overbrowsing by deer within roughly five years (Fischer et al., 1944). Small mammals may also have a marked effect on vegetation. Batzli and Pitelka (1971) found that, during cyclic changes in density, California voles (*Microtus californicus*) had significant effects on preferred food plants. During high mouse densities (160/acre), the mouse's major food plants contributed 85 per cent less volume to the vegetation outside experimental plots from which the mice were excluded than to the vegetation inside the plots. In addition, the fall of seeds of preferred grasses was reduced by 70 per cent on grazed areas.

PATTERNS OF ECOLOGICAL DISTRIBUTION

COMPETITION AND THE ECOLOGICAL NICHE

Just as no two species of animals are structurally identical, no two are functionally identical nor have exactly the same environmental requirements. The very morphological, physiological, and behavioral characters that determine the distinctness of species also determine the distinctness of habitat requirements. Each species requires a specific environment—a particular combination of physical and biotic factors—and each is functionally unique, pursuing a particular mode of life within its environment. This specific environmental setting that a species occupies and the functional role it plays in this habitat constitute the animal's ecological niche.

The fundamental niche of Hutchinson (1957) is an abstract formalization of the usual concept of an ecological niche. The fundamental niche is an "n-dimensional hypervolume" defined by all of the values limiting the survival of a species and within which every point "corresponds to a state of the environment which would permit the species S to exist indefinitely" (Hutchinson, 1957:417).

Niche segregation among animals tends to reduce the loss of time and energy resulting from competition between species. Competition occurs when two or more individuals occupying the same habitat at the same time are utilizing some environmental resource in short supply. Competition can be between members of the same species (intraspecific competition) or between members of different species (interspecific competition). Competition may also be direct or indirect (Miller, 1969). Individuals competing indirectly may never come in contact—a chipmunk may eat so many cutworms during the day that it becomes unprofitable for a white-footed mouse to search for this preferred food item at night. This sort of mutual use of the same resource is the most important type of competition. Individuals competing directly, on the other hand, are in direct confrontation, as when a pair of lions takes over a freshly killed wildebeest from a group of spotted hyenas. Direct competition often involves the defense of a space, an area within which an individual (or a social group) seeks exclusive access to food, den sites, and shelter from predators. This is typically intraspecific competition. Not only is direct competition the most conspicuous type, but it can result in important losses of time and energy.

There is a high adaptive premium both on behavioral patterns that minimize the outlay of energy associated with intraspecific competition for space and on the use of different resources by two sympatric species. When two species of animals are in strong competition for a limited resource, such as food, there will be one of the following outcomes: one species will become extinct and will be replaced by the other (a process probably repeated countless times during the three or four billion years of the history of life on earth): one species will emigrate (move to another area): or one species will change with regard to its use of the disputed resource.

Interspecific competition may take the form of a variety of aggressive behaviors, including fights, chases, or threats. The evolutionary trend in such behaviors has generally been in the direction of maintaining their effectiveness but reducing their energetic cost and avoiding physical damage. Aggressive behavior that involves high expenditures of energy is maladaptive in a number of ways: it may lower vigor, leaving an animal susceptible to disease or predation; it may force an animal to take refuge in marginal habitats; or it may render an animal conspicuous and thus more vulnerable to predation. Of greatest importance, these effects may seriously reduce the amount of energy available for reproduction and may thus lower the reproductive output.

Many animals live through certain critical times of food scarcity on tight time and energy budgets; during these periods, energy-costly interspecific interactions may jeopardize survival. Equally critical in these times may be the use of time and energy in shifting from a familiar but depleted food to a more abundant but unfamiliar (and perhaps less nutritious) food. Niche segregation is one result of natural selection favoring structural, physiological, or behavioral modifications that decrease the probability of interspecific encounters and reduce the energy expended in interspecific competition (Fig. 15–13). These modifications decrease the energetic cost of such vital activities as feeding, reproducing, and escaping from predators, but may also restrict the environmental sphere within which the animal functions. These adaptations only survive, however, if they are associated with no loss of, or an increase in, reproductive success. Interspecific competition, then, is brought to limits that sympatric species can bear by niche segregation.

A full description of an animal's ecological niche must include a consideration of the animal's behavior and biotic interactions, that is, its functional role within its environment. A primary consideration is how an animal seeks its food. Does it burrow to reach a root, as does the pocket gopher, or does it dig roots from above, as does the javelina; does it kill its prey or does it consume prey killed by other animals? The reproductive cycle of the animal is equally important, for the timing of breeding and the seasonal demands put on the habitat due to increased energy requirements during the breeding season, the choice of sites for rearing young, as well as yearly population fluctuations due to reproduction, relate directly to an animal's use of, and its impact on, the environment. The role an animal plays in the story of predator-prey relationships is also of vital importance in understanding the animal's niche, as is its mortality rate. In a terrestrial community, the role of an abundant mouse that consumes vegetation is very different from that of a rare carnivore that preys upon this mouse. (Food chains, food webs, and trophic levels are discussed on page 309.)

Similar species may avoid competition in a number of ways. Each can utilize a different microenvironment, they can have different diets, or the species can utilize the same foraging area but with their activity cycles out of phase.

In many desert areas of western United States and northern Mexico, several species of heteromyid rodents occupy common ground and are seemingly competing for seeds. In the Sonoran desert of southern Arizona, Merriam's kangaroo rat (*Dipodomys merriami*) and the Arizona pocket mouse (*Perognathus amplus*) live together in the desert flats, while the rock pocket mouse (*P. intermedius*) and Bailey's pocket mouse (*P. baileyi*) occupy the rocky hills. The flatland dwellers avoid competition in several ways. The diets differ to some extent (Reichman, 1975): *D. merriami* eats large numbers of insects but relatively few seeds of the creosote bush (*Larrea divaricata*), whereas *P. amplus* eats few insects but large numbers of *Larrea* seeds. In addition, the seasonal activity cycles of these rodents are out of phase. During the hottest time of the summer, *D. merriami* reduces its aboveground activity, but it remains active through the winter, whereas *P. amplus* remains active all summer but is below ground most of the winter (Reichman and Van De Graaff, 1973). These species are further segregated spatially: *D. merriami* tends to forage in the open, while *P. amplus* forages mostly near the cover of shrubs. Regarding the hill-dwellers, both seem to remain active throughout the year, but their diets are in part complementary: *P. baileyi* eats mostly seeds of forbs (nongrasslike herbs); *P. intermedius* eats twice as many grass seeds as does *P. baileyi* and eats the seeds of shrubs and trees more often.

There is clearly an allocation of food resources between these pairs of desert rodents, and of equal importance may be differences between species in the use of space and time. Asynchronous (out of phase) breeding of sympatric ungulates was found to occur in Southern Rhodesia (Dasmann and Mossman, 1962); this asynchrony offsets to some extent the times at which females of the different species have increased food requirements, and may therefore lessen interspecific competition for food (Table 15–2). In East Africa, a yearly pattern of grazing succession by some of the large ungulates results

in reduced interspecific competition for forage (see page 244). Food resource allocation may effectively segregate sympatric herbivores in certain situations. As an example, the foraging areas of several small mammals of a subalpine area in Colorado were found to overlap broadly, but the diets of no two species were the same (Fig. 15–10).

Erlinge (1972) described an interesting competitive situation resulting from the introduction in the late 1920s of mink (*Mustela vison*) from the United States into the range of the European otter (*Lutra lutra*) in Sweden. The establishment of mink seemed to cause a decline in otter populations, but the two carnivores continue to coexist despite their heavy mutual use in winter of such foods as fish and crayfish. Erlinge concluded that the mink is able to exploit a wide array of foods, whereas the otter specializes on aquatic vertebrates. Also, otters may directly interfere with mink and are even known to kill them occasionally (Egorov, 1966); this may reduce the use of otter habitat and preferred otter foods by mink. The specialized otter thus seems to occupy a relatively narrow niche within the broader niche of the less specialized mink (Fig. 15–14), a situa-

Table 15–2. Reproductive Seasons of Some Ungulates in Southern Rhodesia, Africa

Species	Wet Season			Dry Season							Wet Season	
	JAN.	FEB.	MAR.	APR.	MAY	JUN.	JUL.	AUG.	SEPT.	OCT.	NOV.	DEC.
Burchell zebra	Y	Y	Y	Y	Y	Y	YP	L	L	Y	Y	YP
Warthog	Y						P	P				B
Giraffe					B	Y	Y	Y	Y			
African buffalo		YP			Y				LY			
Blue wildebeest			Y						P		P	
Waterbuck		YB	YB	Y	L							
Eland								Y				
Kudu		P	BY	Y				LP	L		LY??	
Bushbuck						Y						
Impala	Y	L	L	L	R	R	LP	P	P	P	P	B
Common duiker		Y			Y	Y		YP			PY	
Steenbok		Y	Y		Y	Y	Y	YP			YP	
Klipspringer								Y				

Key: Y = Young, estimated under 1 month of age, observed
$\quad$ Y = Maximum number of young observed during period
$\quad$ B = Most births occur during this period
$\quad$ L = Lactating females observed or collected
$\quad$ R = Rutting season behavior observed
$\quad$ P = Pregnant females collected
(After Dasmann and Mossman, 1962.)

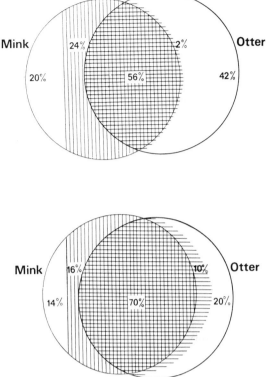

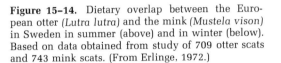

Figure 15–14. Dietary overlap between the European otter *(Lutra lutra)* and the mink *(Mustela vison)* in Sweden in summer (above) and in winter (below). Based on data obtained from study of 709 otter scats and 743 mink scats. (From Erlinge, 1972.)

tion discussed in connection with other animals by Hutchinson (1957) and Miller (1964, 1967).

Sexual dimorphism reduces intraspecific competition in some species of mammals. The great size difference between the sexes of some small predators may reduce competition for food between the sexes by enabling males and females to exploit different sizes of prey. The males of two species of weasel (*Mustela erminea* and *M. frenata*) are roughly twice as heavy as the females, and the skulls of the males are more robust (Hall, 1951:26–28). This situation seems to parallel that described for some birds (Storer, 1955; Selander, 1966), in which pronounced sexual dimorphism allows the sexes to utilize different feeding niches. Where the species of weasel mentioned above occur together, they differ considerably in size. This difference, and the considerable sexual variation in size within each species, not only may reduce interspecific and intraspecific competition for food, but also may increase the effi-

ciency of utilization of the food by forcing each predator to specialize on a specific size of prey. Lockie (1966) and Erlinge (1974) have noted unusually high numbers of males in declining populations of the weasel *Mustela nivalis* in Europe. Under conditions of low prey populations, males may have a survival advantage over females because the relatively large size of the males allows them to exploit a wider size range of prey (Fig. 15–15; Erlinge, 1975) and may make them superior aggressors.

Although the niches of species within a community tend to be complementary, some partial niche overlap occurs, and there is occasionally some interspecific competition between sympatric mammals. Obvious indications of competition in nature are relatively rare, however. Reduced populations or reproductive rates or both may occur in the most severely affected species. When competition occurs, its severity is often difficult to assess. As shown in Figure 15–10, two sympatric herbivores

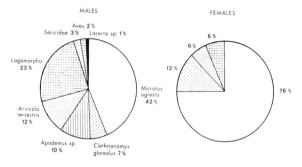

Figure 15–15. Diets of male and female weasels (*Mustela nivalis*) in Sweden, based on analyses of 94 male scats and 15 female scats. (From Erlinge, 1975.)

may utilize some of the same foods. Where they occurred in the same habitat, both the northern pocket gopher and the montane vole ate *Agoseris* (mountain dandelion), but this plant was not equally important to both species. Although it was a major food of the pocket gopher, it was of relatively minor importance in the diet of the montane vole. In addition, *Agoseris* was so abundant where these animals occurred together that they were seemingly not utilizing a resource in short supply.

Suitable roosts are of prime survival value to bats and seem to determine the distribution of colonial species (Humphrey, 1975); local interspecific competition for roosting sites would therefore be expected to be intense. My observations in East Africa illustrate a case of such competition. Hollow baobab trees are favorite roosting sites for some bats in this area, and a single hollow may harbor several species. In areas where the large, slightly carnivorous *Cardioderma cor* (Fig. 6–44A) is common, however, these hollows are occupied virtually exclusively by this bat. The competitive dominance of *Cardioderma* may thus strongly affect the local diversity of bats.

BIOTIC COMMUNITIES

Man has long recognized that animals and plants with similar environmental requirements form recognizable communities. A perceptive description of a community was given by Mobius (1877), who considered an oyster bed as "a community of living beings, a collection of species and a massing of individuals, which find here everything necessary for their growth and continuance...." A community, however, is characterized not only by its unique plant and animal assemblage but by complex interactions between organisms and by the effects the physical environment has on the biota. The term community has been used to designate plant-animal assemblages of differing size and importance (Odum, 1971:140). This term can be appropriately used in reference to the biota of a woodrat nest or to that of the extensive deciduous forests of eastern United States.

As described by Ricklefs (1973:603), ". . . community structure and dynamics are the sum of the fates of the organisms in the community." The structure of a community and its dynamics are determined by a web of interactions between physical and historic factors and the biota. The history of the community is of importance in two ways. First, the kinds of animals and plants that compose its biota depend on the geographic position of the community and the land masses to which it has been connected or near which it lies. Consider a community in southern Mexico. It may have some species of South American ancestry and some of North American origins; avenues of dispersal for land-dwelling mammals have allowed "North American" and "South American" species to contribute to the biota of the community. Secondly, the antiquity of the community is of importance. Because the land bridge between North and South America was established some two million years ago, the association between the mammals in our Mexican community is probably of long standing and coadaptation between species would be expected. This community has probably reached an equilibrium lacking in areas where species of different geographic origins were newly thrown together.

In some communities, both in tropical and temperature areas, the faunal composition undergoes radical seasonal changes; here residents must adapt to a seasonal influx of migrants. Dramatic wet season rises in insect populations are in some tropical areas associated with increased densities and diversities of insect predators, and a parallel situation occurs in the summer in temperate communities. Even in semiarid tropical areas, seasonal productivity of vegetation is high: nineteen percent of the total terrestrial primary productivity of the earth comes from the semiarid tropics (Rodin et al., 1975).

The highest number of species are encountered in tropical communities, with decreasing diversity in areas progressively farther toward the poles. As an example of this pattern in mammals, an area of some 40 square km near Point Barrow, Alaska supports only about 16 species of land mammals, whereas comparable figures for eastern Kansas and an area near Panama City, Panama, are 55 and 140 respectively (Hall and Kelson, 1959:xxiv). A number of factors seem to influence this trend. Mammals near Point Barrow can live only on the surface of the ground or in a shallow stratum of soil above the permafrost, which extends locally to depths of about 300 m. Large bushes and trees are absent, and bodies of water are frozen throughout the winter. There are, therefore, neither arboreal mammals nor those that depend on open water. For only two months of the year are insects available; understandably, there are no bats. Primary productivity is contributed entirely by low-growing layers of vegetation.

The contrast between this community and that of a tropical rain forest is extreme. In the rain forest an abundance of evergreen vegetation extends from ground level to over 30 m above the ground, and temperatures are moderate year long. Throughout the year, a variety of foods is available that never occurs even briefly in the arctic. Large nectar-bearing flowers and fruit are present at all seasons and support nectar-feeding and fruit-eating bats, rodents, and primates; green leaves are perennially available to primates, rodents, ungulates, and (in the Neotropics) tree sloths. Of tremendous importance are kinds of insects absent or rare in the far north. Immense numbers of moths and beetles support a great variety of insectivorous bats; termites support anteaters (in some Neotropical forests) and elephant shrews (in some African forests) and contribute to the diets of many small mammals; and dung beetles are eaten by everything from bats to small carnivores. In addition, many kinds of shelter from beneath the ground to high in the trees are provided. For mammals, and for animals in general, the arctic offers a restricted assortment of modes of life, whereas the tropics offer an extraordinarily rich variety of possibilities.

The stability of a community seems to depend in part on its complexity. The greater the species diversity among both plants and animals and the greater the number of energy pathways, the more resistant the community to such changes as strong shifts in the densities of common species. Although little definitive research bears directly on the matter, some ecologists accept the view that "stability increases as the number of links increases" (MacArthur, 1955).

Under some circumstances, as in precipitous mountainous areas where the combined effects of slope, exposure, edaphic (soil) factors, and local patterns of air flow produce remarkably complex distributions of organisms, transitions between communities are abrupt. At the interface where one community adjoins another, as where a grassy meadow meets a coniferous forest, an "edge effect" is produced. At this edge where the communities adjoin, a greater diversity and density of animals may occur than within either adjacent community, owing to the increased diversity of vegetation and types of shelter. Carnivores frequently concentrate their efforts on these edge situations, and habitat manipulation by game managers often includes making maximum use of the edge effect.

Often there are no sharp dividing lines between adjacent communities. A sub-alpine forest, for example, may become progressively more extensively interrupted by open "parks" or alpine meadows until the treeless alpine tundra predominates. The zone of intergradation between communities, whether broad or narrow, is called an ecotone. Ecotonal belts between communities are broad in some regions, as in western Mexico, where one may travel southward through the region of transition between the Sonoran desert community and the tropical thorn forest community for many miles. Under such conditions there is a continuous gradient from one major community to another. Despite the difficulty in assigning georgraphic limits to some communities, the major terrestrial communities are usually readily recog-

nized. Some terrestrial communities and their distributions are shown in Table 15–3, which is based on the community and plant-formation types of Whittaker (1970: 52–64). Some of these communities are illustrated in Figures 15–16 to 15–30.

Over broad areas, this pattern of gradual clinal changes in the biota in response to gradual climatic and edaphic changes is more typical than is a sudden shift from one community to another. Because the classification of units or areas within such a continuum involves arbitrary choices as to the limits of the units and takes little account of variation within the units, systems of classification of broad environmental units have not been completely satisfactory. Nevertheless, some of them have been widely used. Such a classification is that involving the recognition of terrestrial biomes. According to Odum (1971:378), "The biome is the largest land

Table 15–3. Some Major Plant Communities and Their Distributions

Community	Distribution
Tropical rain forest (Fig. 15–16)	South and Central America; Africa; S. E. Asia; East Indies; N. E. Australia
Tropical deciduous forest (Fig. 15–17)	Mexico; Central and South America; Africa; S. E. Asia
Temperate rain forest (Fig. 15–18)	Pacific Coast from northern California to northern Washington; parts of Australia, New Zealand, and Chile
Temperate deciduous forest (Fig. 15–19)	Eastern U.S.A.; parts of Europe and Eastern Asia
Subarctic-subalpine coniferous forest (Fig. 15–20)	Northern North America; Eurasia; high mountains of Europe and North America
Thorn scrub forest (Fig. 15–21)	Parts of Mexico; Central and South America; Africa; S. E. Asia
Temperate woodlands (Fig. 15–22)	Parts of western and southwestern U.S.A. and Mexico; Mediterranean area; parts of Southern Hemisphere
Temperate shrublands (Fig. 15–23)	California; Mediterranean area; South Africa; parts of Chile; West and South Australia
Savanna (tropical grasslands) (Figs. 15–24, 15–25)	Parts of Africa, Australia, southern Asia, South America
Temperate grasslands (Fig. 15–26)	Plains of North America; steppes of Eurasia; parts of Africa and South America
Arctic and alpine (Fig. 15–27)	Tundras north of treeline in North America and Eurasia; some areas in Southern Hemisphere
Deserts (Figs. 15–28, 15–29)	On all continents; widespread in North America, North Africa, and Australia

Figure 15–16. A, Tropical rain forest near Catemaco, southern Veracruz, Mexico. Common mammals in this area include the Mexican mouse-opossum *(Marmosa mexicana)*, opossums *(Didelphis marsupialis* and *Philander opossum)*, many kinds of leaf-nosed bats (Phyllostomatidae), the howler monkey *(Alouatta villosa)*, agouti *(Dasyprocta mexicana)*, and coati *(Nasua narica)*. B, Tropical rain forest in the Chyulu Range, Tsavo West National Park, Kenya, East Africa. In addition to a diverse group of smaller mammals, this forest supports African elephants *(Loxodonta africana)* and African buffalo *(Syncerus caffer)*.

community unit which it is convenient to recognize." The biome is defined in terms of climate, biota, and substrate. The following biomes are recognized by Odum (1971: 379): tundra; northern coniferous forest; temperate deciduous and rain forest; temperate grassland; chaparral; desert; tropical rain forest; tropical deciduous forest;

Figure 15–17. Tropical deciduous forest near Kibwezi, southern Kenya, East Africa. This forest, shown in the dry season, is dominated by several species of *Commiphora*. Common mammals in the area when the photograph was taken were elephant shrews *(Elephantulus rufescens)*, vervet monkeys *(Cercopithecus aethiops)*, baboons *(Papio anubis)*, bushbabies *(Galago senegalensis)*, genets *(Genetta tigrina)*, African civets *(Viverra civetta)*, a variety of bovid ungulates including dik-diks *(Madoqua kirki)* and bushbuck *(Tragelaphus scriptus)*, and bats of the families Pteropodidae, Emballonuridae, Nycteridae, Megadermatidae, Rhinolophidae, Vespertilionidae, and Molossidae.

Figure 15–18. Temperate evergreen rain forest in Olympic National Park, Washington. Mammals typical of this area are the shrew mole *(Neurotrichus gibbsii)*, Townsend's mole *(Scapanus townsendii)*, mountain beaver *(Aplodontia rufa)*, western red-backed mouse *(Clethrionomys occidentalis)*, marten *(Martes americana)*, black bear *(Ursus americana)*, and elk *(Cervus elaphus)*. (Photograph by Ray Atkeson.)

Figure 15–19. Temperate deciduous forest in Indiana in summer. Mammals typical of this type of community are the short-tailed shrew *(Blarina brevicauda)*, eastern chipmunk *(Tamias striatus)*, gray squirrel *(Sciurus carolinensis)*, flying squirrel *(Glaucomys volans)*, white-footed mouse *(Peromyscus leucopus)*, gray fox *(Urocyon cinereoargenteus)*, and white-tailed deer *(Odocoileus virginianus)*. (Photograph by U.S. Forest Service.)

Figure 15–20. Subalpine coniferous forest near Rabbit Ears Pass, Routt County, Colorado. The following mammals are common in this community: shrews (*Sorex vagrans* and *S. cinereus*), the red squirrel (*Tamiasciurus hudsonicus*), least chipmunk (*Eutamias minimus*), pocket gopher (*Thomomys talpoides*), montane vole (*Microtus montanus*), red-backed vole (*Clethrionomys gapperi*), beaver (*Castor canadensis*), porcupine (*Erethizon dorsatum*), red fox (*Vulpes vulpes*), mule deer (*Odocoileus hemionus*), and elk (*Cervus elaphus*).

tropical scrub forest; tropical grassland and savanna; and mountains, with complex zonation of plants and animals.

In North America, the "life zone" has been used widely in descriptions of the distributions of vertebrates. Life zones were originally described as temperature zones by C. Hart Merriam (1894), but later were used as community zones characterized by assemblages of plants and animals.

Figure 15–21. Thorn scrub near San Carlos Bay, Sonora, Mexico. Typical mammals are bats (including several members of the Neotropical families Phyllostomatidae and Mormoopidae), the antelope jackrabbit (*Lepus alleni*), Merriam's kangaroo rat *(Dipodomys merriami)*, hispid cotton rat *(Sigmodon hispidus)*, white-throated woodrat (*Neotoma albigula*), and javelina (*Tayassu tajacu*).

Figure 15–22. Temperate woodlands. A, Piñon-juniper woodland near Flagstaff, Coconino County, Arizona. Most of the trees are piñons; the tree at the extreme left is a juniper. Common mammals include several bats of the genus *Myotis*, the desert cottontail *(Sylvilagus audubonii)*, piñon mouse *(Peromyscus truei)*, northern grass-hopper mouse *(Onychomys leucogaster)*, Stephen's woodrat *(Neotoma stephensi)*, gray fox *(Urocyon cin-ereoargenteus)*, bobcat *(Lynx rufus)*, and mule deer *(Odocoileus hemionus)*. B, Oak woodland near Claremont, Los Angeles County, California. A number of vespertilionid bats, the gray squirrel *(Sciurus griseus)*, brush mouse *(Peromyscus boylii)*, dusky-footed woodrat *(Neotoma fuscipes)*, and raccoon *(Procyon lotor)* are common in this habitat.

Figure 15–23. Temperate shrubland (chaparral) near San Antonio Canyon, Los Angeles County, California. A, Steep slopes covered by dense brush; B, interlacing stems and branches beneath the chaparral in the right foreground of A. Typical mammals in this area are the western pipistrelle *(Pipistrellus hesperus)*, Merriam's chipmunk *(Eutamias merriami)*, brush mouse *(Peromyscus boylii)*, California mouse *(P. californicus)*, dusky-footed woodrat *(Neotoma fuscipes)*, gray fox *(Urocyon cinereoargenteus)*, bobcat *(Lynx rufus)*, and mule deer *(Odocoileus hemionus)*.

Figure 15–24. Savanna in the Serengeti Plain, Tanzania, Africa. This area supports large numbers of ungulates. Some of the typical kinds are zebra *(Equus burchelli)*, buffalo *(Syncerus caffer)*, wildebeest *(Connochaetes taurinus)*, and Thompson's gazelle *(Gazella thompsonii)*. Lions *(Panthera leo)*, hyenas *(Crocuta crocuta)*, and African hunting dogs *(Lycaon pictus)* prey on these ungulates. The African savanna supports a richer ungulate fauna than does any other area. (Photograph by W. Leslie Robinette.)

Figure 15–25. Savanna (mallee scrub) in the southern part of New South Wales, Australia. The trees are eucalyptus. In this area occur a number of marsupials, including marsupial "mice" *(Sminthopsis crassicaudata* and *Antechinus flavipes)*, the vulpine phalanger *(Trichosurus vulpecula)*, the "mallee gray" kangaroo *(Macropus giganteus)*, and the red kangaroo *(Megaleia rufa)*. (Photograph by Diana Harrison.)

Figure 15–26. Temperate grassland: shortgrass prairie near Nunn, Weld County, Colorado. Among the common mammals are the white-tailed jackrabbit *(Lepus townsendii)* and black-tailed jackrabbit *(L. californicus)*, the thirteen-lined ground squirrel *(Spermophilus tridecemlineatus)*, prairie dog *(Cynomys ludovicianus)*, Ord's kangaroo rat *(Dipodomys ordii)*, northern grasshopper mouse *(Onychomys leucogaster)*, coyote *(Canis latrans)*, badger *(Taxidea taxus)*, and pronghorn *(Antilocapra americana)*. (Photograph courtesy of Robert E. Bement, Agricultural Research Service.)

Figure 15–27. Arctic tundra on the northern slope of the Brooks Range, northern Alaska. Among the common mammals of this area are the arctic shrew (*Sorex arcticus*), collared lemming (*Dicrostonyx groenlandicus*), brown lemming (*Lemmus trimucronatus*), singing vole (*Microtus miurus*), wolf (*Canis lupus*), grizzly bear (*Ursus arctos*), moose (*Alces alces*), caribou (*Rangifer tarandus*), and Dall sheep (*Ovis dalli*). (Photograph by James W. Bee.)

Merriam's thinking was influenced by his recognition of the sharp elevational zonation of biotas in some mountains of western United States

BIOTIC INTERACTIONS

Considerations of interactions between species of animals and between plants and animals are essential to an understanding of mammalian ecology. A biotic community is a tremendously complex functional unit within which animals live, feed, reproduce and die; it has an evolutionary history and some degree of "dynamic equilibrium." The role of an organism in a community depends on its interactions with other members of the community and with the physical environment; the fabric of the entire community depends on the combined effects of the interlacing threads of interaction.

Figure 15–28. High sand dunes in the Namib Desert near Gobabeb, South West Africa. Several kinds of mammals that occur in this area are gerbils (*Gerbillus sp.*), ground squirrels (*Xerus sp.*), bat-eared foxes (*Otocyon megalotis*), and gemsbok (*Oryx gazella*). (Photograph by C. K. Brain.)

Figure 15–29. Sandy desert near Lake Victoria, southwestern New South Wales, Australia. Mammals of this community include a marsupial "mouse" *(Sminthopsis crassicaudata)*, a kangaroo *(Macropus giganteus)*, and a rodent, the Australian Kangaroo mouse *(Notomys mitchelli).* (Photograph by Diana Harrison.)

Figure 15–30. Pronounced zonation of vegetation on the San Francisco Peaks in northern Arizona. Observations he made in this area influenced C. Hart Merriam's thinking on life zones. In the foreground, at about 1900 m. elevation, is shortgrass prairie; in the near distance are flats supporting piñon and juniper trees. The mesas and slopes in the middle distance are covered with ponderosa pine; the higher slopes in the distance have spruce and fir forests; and on the treeless and barren tops of the peaks, between roughly 3350 and 3660 m. elevation, grow small alpine forbs and grasses.

EFFECTS OF MAMMALS ON ECOSYSTEM AND COMMUNITY STRUCTURE; COEVOLUTION

Some species of mammals impose such critical influences on their ecosystems that they can be viewed as "keystone species." Such species ". . . are the keystone of the community's structure, and the integrity of the community and its unaltered persistence through time, that is, stability, are determined by their activities and abundance" (Paine, 1969).

The wildebeest (*Connochaetes taurinus*) is a keystone species. The annual migration of wildebeest from the Serengeti Plains in Tanzania, East Africa, northward and westward toward the bush country of northern Tanzania and the Maasai-Mara Game Reserve of Kenya, is a justly famous spectacle. The movement begins at the end of the rainy season, in May or June, and the animals return to the shortgrass in November. In May, 1974, some half million wildebeest were concentrated in the shortgrass country along the western border of the Serengeti Plains. On May 22, the great herd began to move northward and, during the four-day period of the passage of the main mass, the animals had a tremendous impact on the dynamics of this shortgrass community (McNaughton, 1976). The green biomass of this *Themeda-Pennisetum* grassland was reduced by grazing in this brief time by 84.9 per cent, and the height of the vegetation by 56 per cent (Table 15–4). This apparent devastation markedly affected the subsequent growth of the grasses, and, indirectly, the dry season distribution of the abundant Thompson's gazelle (*Gazella thompsonii*). This antelope, next to the wildebeest, is the most abundant ungulate in the Serengeti Plains.

During the 28-day period following the main migration, areas grazed by wildebeest had a net productivity of green vegetation of 2.6 g per square meter per day, whereas green biomass *declined* in experimental plots protected from grazing at a rate of 4.9 g per square meter per day. A dense mat of new and nutritious vegetation was produced in the grazed area by abundant tillering (growth of new shoots from nodes of stems or of rhizomes) of the grasses, while in the ungrazed plots the bulk of the biomass was tall and relatively non-nutritious grass stems. One month after the exodus of the wildebeest, the area was occupied by Thompson's gazelles, which selectively grazed the areas of vigorous regrowth, the areas previously grazed heavily by wildebeest. As an indication of the high selectivity of the gazelles, consumption of vegetation in these latter areas averaged 1.05 g per square meter per day, whereas virtually no grazing by gazelles occurred in the plots where wildebeest did not graze.

Table 15–4. Effect on Grassland Vegetation (largely grasses *Themeda* and *Pennisetum*) of the Four-day Passage of Herds of Migratory Wildebeest in the Serengeti Plains of Tanzania, East Africa

	Biomass (g/m²)	Height (cm)	Biomass Concentration (mg/10 cm²)
FENCED VEGETATION, WILDEBEEST EXCLUDED			
Before	501.9	64	7.9
After	449.2	63	7.1
	N.S.	N.S.	N.S.
VEGETATION SUBJECT TO WILDEBEEST GRAZING			
Before	457.2	66	6.9
After	69.0	29	2.4
	P = .005	P = .005	P = .05

Before and After refer to before and after the passage of the wildebeest. P = level of significance; N.S. = not significant.

(After McNaughton, S. J.: Serengeti migratory wildebeest: facilitation of energy flow by grazing. Science, *191*:92–94, 1976. Copyright 1976 by the American Association for the Advancement of Science.)

McNaughton concluded that by their intense grazing the wildebeest transformed a senescent grassland into a productive community on which the Thompson's gazelle depended in the dry season. Further, the coexistence of the wildebeest and the Thompson's gazelle is favored by coevolution resulting in the partitioning of the grassland in the dry season, the period of the year when competition between the two most abundant ungulates would be mutually disadvantageous. Without the wildebeest the Serengeti Plain would certainly not have the community structure that it has today, and this community would presumably lack its present diversity.

The activities of a single mammal, the sea otter (*Enhydra lutris*), seem to play a critical role in structuring nearshore marine communities along the Aleutian Islands of Alaska (Estes and Palmisano, 1974). The Rat Islands support high populations of sea otters, and the nearshore community is characterized by abundant beds of macrophytes, consisting mostly of brown algae (kelp) and red algae; filter feeders, such as barnacles and mussels, are scarce, as are motile herbivores, such as sea urchins and chitons. In the Near Islands, some 400 km to the northwest, sea otters are absent; here macrophytes are scarce below the lower intertidal zone, and barnacles, mussels, sea urchins, and chitons are many times as dense as they are along the Rat Islands (Fig. 15–31). These differences seem related to the activities of

otters. These animals prey heavily on sea urchins (*Strongylocentrotus* spp.), which graze on macrophytes. When high populations of otters drastically reduce the abundance of urchins, kelp beds flourish and the complexion of the nearshore community is conspicuously altered. The abundance of vertebrates is also affected. Probably because of the relative abundance of fish around kelp beds, harbor seals (*Phoca vitulina*) and bald eagles (*Haliaeetus leucocephalus*) are common in the Rat Islands but are rare or absent in the Near Islands. It seems, then, that the character of the nearshore communities depends on the presence or absence of otters, and it may well be that productive kelp beds and a stable nearshore community on the Pacific Coast of the United States are maintained only in the presence of the sea otter. The otter thus seems to be the keystone of the nearshore community.

Mammals often influence community structure by serving as important agents of dispersal for the seeds of many kinds of plants. In Mexico, sprouts from seeds of leguminous trees commonly grow from piles of cow manure deposited far from the plants that produced the seeds. In western United States, the seeds of prickly-pear cactus (*Opuntia*) are distributed in a similar fashion by skunks and foxes, as are the seeds of manzanita (*Arctostaphylos*) by black bears and coyotes. By dispersing seeds of the Washington palm (*Washingtonia filifera*), coyotes may have helped to spread that plant to suitable sites in the

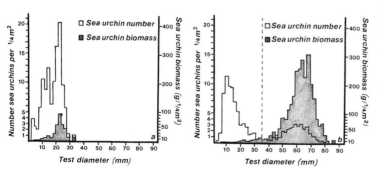

Figure 15–31. Distributions of size classes and biomass contributions of sea urchins. Left, based on data from Amchitka Island, which supports high densities of sea otters; right, based on data from Shemya Island, where sea otters are absent. The dotted line indicates the largest class of sea urchins present at Amchitka Island. (From Estes, J. A., and Palmisano, J. F.: Sea otters: their role in structuring nearshore communities, *Science*, 185:1058–1060, 1974. © 1974 by the American Association for the Advancement of Science.)

deserts of southern California (Jaeger, 1950).

For a particularly interesting example of mutually beneficial interactions between mammals and a plant, we return to Serengeti National Park. Here, and over wide areas of East Africa, the umbrella tree (*Acacia tortilis*) is a conspicuous, picturesque, and important savanna tree. The green, leathery pods of the plant are eaten avidly by a variety of herbivores, ranging from the tiny dik-dik to the elephant, and elephants often feed heavily on the foliage. Lamprey et al. (1974) found that after ingestion and defecation of the acacia seeds by herbivores, the germination rates were strikingly higher than in uningested seeds. In some vertebrate-adapted seeds, the digestive process is known to erode the seed coat and hasten germination, but the high germination rate in ingested acacia seeds is thought by Lamprey and his coworkers not to depend primarily on this effect. Beetles of the family Bruchidae lay their eggs on acacia seed pods and the larvae feed and grow within the seeds; if the larvae damage the embryos of the seeds or destroy a large amount of the cotyledon material, the seed will not germinate. But if the seed pods are eaten by herbivores soon after they fall to the ground, as is typically the case in the Serengeti, the digestive process kills the larval bruchids at an early stage of development before they have killed the seeds. Some 500 seed samples were collected from the ground and were stored for a year; over 95 per cent of these seeds had bruchid damage and the germination rate was only 3.0 per cent. The rate of damage and germination in herbivore-ingested seeds, however, ranged from 26 to 28 per cent respectively for the impala (*Aepyceros melampus*) to 45 and 11 per cent for the dik-dik. The interactions seem clearly to be mutually advantageous: from the acacia the herbivores get a seasonally important food; from the mammals, in turn, the acacia gains an effective means of escape from a seed predator and wide dissemination.

Some mutually advantageous plant-animal associations are of long standing, and in some of these cases remarkable mutual adaptations have evolved. Especially noteworthy examples of such coevolution are offered by nectar-feeding bats and the plants on which they feed. In the Old World, a number of bats of the family Pteropodidae feed on nectar; in the Neotropics, nectar-feeding occurs in the family Phyllostomatidae. Chiropterophily (the dependence of a plant largely on bats for pollination) has been discussed by Baker (1961, 1973), Alcorn et al. (1959), and Faegri and Van Der Pijl (1966:111–118), and laboratory studies by Howell (1974) have demonstrated several physiological adaptations of a bat (*Leptonycteris sanborni*) that feeds from flowers.

Rarely is pollination in a plant that depends on animal pollination accomplished by a single species. A common situation is that involving pollination by many agents, including perhaps a number of insects and birds. Plants of this sort are termed polyphilous. Some plants, on the other hand, are pollinated by relatively few agents, and the flowers are open only during the day or night. In these plants, characteristics of the flower that favor one or several pollinators would be expected to be under strong positive selective pressure; the plants would be expected to evolve features making their flowers especially attractive to a small group of pollinators. This seems to have occurred in a number of bat-pollinated plants, but in no case is the bat the sole pollinating agent.

Most chiropterophilous plants occur in the tropics and subtropics. Baker (1961) ascribes this to the fact that the form of tropical trees and their inflorescences encourage nectar-feeding bats, whereas trees of temperate regions are less hospitable. Many tropical trees bear flowers on spreading, temporarily leafless branches; the flowers are often large and hang down from long sturdy stalks; tropical trees frequently have a spreading growth form; and some have extended flowering seasons. All of these features make the flowers attractive and accessible to bats. Chiropterophilous flowers are usually not brightly colored, they open at night, and typically

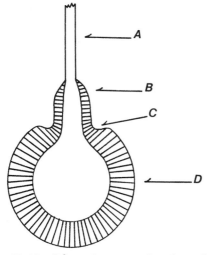

Figure 15–32. Schematic cross section of an inflorescence of *Parkia clappertoniana*. A, Stout peduncle; B, sterile, nectar-producing flowers; C, ring in which nectar collects and where bats feed; D, potentially fertile flowers. (After Baker and Harris, 1957.)

they have a strong (and to man unpleasant) smell that attracts bats.

Two general types of flowers are chiropterophilous. The first type accommodates large bats. An example is the large flower of the African baobab tree. The feeding bat clutches the ball of stamens of this flower while lapping nectar from the pillar-like stamen column. *Eidolon helvum*, a large pteropodid bat with a wingspread of about 760 mm, has been observed visiting these flowers. The heads of the bats become covered with pollen, which is likely to be transferred to the stigma on the spreading style.

Another variation on the theme occurs in an African species of *Parkia* that was studied by Baker and Harris (1957). *Parkia clappertoniana* displays a division of labor between nectar-producing and reproductive stamens (Fig. 15–32); bats of the genera *Epomophorus* and *Nanonycteris* were observed by these authors lapping nectar from the circular depression in the top of the ball of the flowers, which was grasped by the bat's feet (Fig. 15–33). Most of the flowers are in a staminate condition one night and a pistillate condition the next, thus insuring cross fertilization. Evidence of the importance of bats as agents of pollination in *P. clappertoniana* is provided by the position of the fruits; these tend to form only on the side of the inflorescence that faces the periphery of the tree, the very side from which bats characteristically approach.

The second group of flowers includes those adapted to smaller bats, and these flowers often have corolla tubes into which the bat pushes its head. A number of Neotropical flowers are of this type. Although these flowers may be approached in a variety of ways, during most approaches the bat's back is liberally dusted with pollen and the pollen may be transported to another flower.

Sanborn's long-nosed bat (*Leptonycteris sanborni*) is a New World bat of the family Phyllostomatidae that feeds from flowers; this bat has been reported by some observers as feeding entirely on nectar. Nectar is rich in carbohydrates, but has no

Figure 15–33. An African pteropodid bat (*Epomophorus gambianus*) feeding at an inflorescence of *Parkia clappertoniana*. (Photograph by H. G. Baker and B. J. Harris.)

more than trace amounts of protein. Vertebrate animals are known to need protein in their diets. Has *Leptonycteris* developed the remarkable and unique ability to subsist on a pure carbohydrate diet? Or does this bat take from flowers other food in addition to nectar? Howell (1974) showed that *Leptonycteris* must indeed have protein in its diet: experimental bats fed on artificial or real nectar fortified with vitamins and NaCl, but containing no proteins, lost weight rapidly, and most died within eight days, whereas bats given nectar and a protein source remained healthy.

Under natural conditions, the source of protein for these bats (and presumably for most nectar-feeding bats) is protein from pollen, which these bats ingest in large quantities. Some pollen may be taken directly from flowers, but probably the bats ingest pollen largely when grooming their fur. Nectar-feeding bats have specialized hairs to which pollen adheres extremely readily (Fig. 6–39). Of particular importance is the fact that the pollens of the plants (saguaro, *Carnegeia gigantea*, and agave, *Agave palmeri*) visited by *Leptonycteris* are unusually high in protein when compared to the pollens of closely related plants that are pollinated by a variety of agents (Table 15–5). Of additional interest, the "bat pollens" contain at least 18 amino acids, and are unusually high in the amino acid proline, which comprises over 80 per cent of collagen. Because it composes much of the extensive network of connective tissue that braces the wings

Table 15–5. Amounts of Protein in the Pollen of Plants Pollinated by Bats and in Pollens Dispersed by Other Means

Species	Protein Content of Pollen (%)
"Bat-adapted" pollen	
Agave palmeri	22.9
Carnegeia gigantea	43.7
"Nonbat" pollen	
Opuntia versicolor	8.93
Ferocactus wislizeni	10.17
Agave schottii	8.30
Agave parviflora	15.90

(Data from Howell, 1974.)

and tail membrane, collagen is especially important in bats.

But *Leptonycteris* faces another problem, for pollen grains are notoriously resistant to being broken down mechanically or biologically. In a sugar solution such as that in the stomach of a bat that has fed on nectar, however, the cellular contents of pollen are known to extrude through the pores. Also, HCl solutions extract proteins from some kinds of pollen, and unusual concentrations of HCl- secreting glands are present in the gastrointestinal submucosa of *Leptonycteris*. Finally, the habit that this bat has of ingesting its urine on occasion could possibly be of importance in connection with digestion (Howell, 1974:270). Urea is known to degrade pollen protein.

In the particular case of *Leptonycteris*, then, the "bat flowers" have seemingly evolved a type of pollen that supplements the nectar and provides the bat a balanced diet. The bat, in turn, has specializations of the hair that facilitate rapid accumulation of this pollen, and a digestive system able to utilize it. The plants are effectively pollinated and the bat is well fed.

ASSORTED INTERACTIONS

Interactions between mammals and many other kinds of animals have been observed. Some of these interactions are seemingly of no importance to the mammal involved, and some parasite-mammal interactions are harmful to mammals; on the other hand, mammals derive considerable benefit from other types of interactions. Mammal collectors have frequently noticed that some kinds of pocket mice concentrate their foraging efforts around the prominent mounds of harvester ants, where the rodents presumably find seeds dropped by the insects. In some cases the activities of mammals improve the habitat for other vertebrates, as in the case of the beaver, which by its dam building creates ponds that often support high densities of trout. Mammals that feed to some extent on carrion may be guided to dead animals by concentrations of carrion-feeding birds and by the calls of the birds. Murie (1940) found that such a relationship occurred between coyotes and ravens (*Corvus corax*) in Yellowstone National Park, and African jackals (*Canis mesomelas* and *C. aureus*) are careful vulture watchers.

One of the most remarkable instances of mutually beneficial interactions concerns a bird, the African honey guide (*Indicator indicator*), and the African honey badger (*Mellivora capensis*). The bird attracts the badger's attention by raucous chattering and then leads the way to a bee nest. After the nest is torn apart by the mammal (African tribesmen sometimes perform this service), the honey guide eats fragments of the wax, which it has the ability to digest. Two species of African birds, the ox-peckers (*Buphaga*), eat ticks that they remove from big game mammals. The mammals are oblivious of the birds' attentions, but seem to derive benefit from the loss of ticks and from the alarm calls given by the birds when predators approach.

Some species of mammals profit in various ways from the activities of other mammals. Often one species will use shelter provided by another species, as in the case of certain white-footed mice (*Peromyscus*) and shrews (*Notiosorex*) that use woodrat houses for shelter, or of the wart hog (*Phacochoerus*), which seeks refuge in burrows of aardvarks (*Orycteropus*). The burrows of pocket gophers are used by a variety of vertebrates, including amphibians, reptiles, and other mammals (Vaughan, 1961). Scraps from the kills of large carnivores may be eaten by mammalian scavengers, such as hyenas (*Crocuta* and *Hyaena*), which specialize in this type of feeding.

Of particular interest are the close and seemingly mutually beneficial associations that occur between two species of mammals. A badger and a coyote have been observed hunting together on many occasions in widely scattered areas (Young and Jackson, 1951:95). Herds of impala (*Aepyceros*) were observed staying with baboons (*Papio*) through much of the day by De Vore and Hall (1965:48,49), who judged that the excellent eyesight of the baboons supplemented the acute senses of smell and hearing of the impalas and made the mixed group difficult for a predator to approach undetected. On one occasion these authors observed a large male baboon discouraging three cheetahs that were approaching a mixed group of impalas and baboons.

TERRITORIALITY AND HOME RANGE

Because individuals of the same species usually have identical niches and are therefore potentially competing for the same environmental resources at the same place and time, intraspecific competition might be expected to be intense. Indeed, this seems to be the case, with the result that individuals of the same species customarily occupy separate, or nearly separate, home ranges. Burt (1943) has described the home range of a mammal as "that area traversed by the individual in its normal activities of food gathering, mating, and caring for the young." Home ranges may have irregular shapes and may be partially overlapping (Fig. 15–34). Within the home range of some animals is an area that is actively defended against other members of the species. This area, usually not including the peripheral parts of the home range, is called the territory, and species that apportion space in this fashion are termed territorial. A home range or territory may be occupied by one individual, by a pair, by a family group, or by a social group consisting of a number of families.

To solitary animals, or to members of a group, the occupancy of a home range has several important advantages. Each home range provides all the necessities of life for an individual or group, permitting self-sufficiency within as small an area as possible; the less extensively the animal must range, the less chance there is for encounters with predators. The home range quickly becomes familiar to the individual, who can then find food and shelter with the least possible expenditure of energy and can escape predators more effectively because escape routes and retreats are familiar and

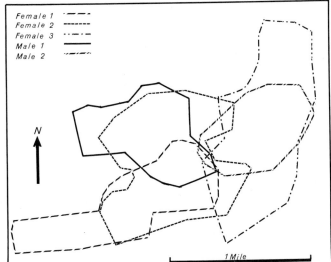

Female 1
Female 2
Female 3
Male 1
Male 2

N

1 Mile

Figure 15-34. Distributions of red fox home ranges at the University of Wisconsin Arboretum. (After Ables, 1969.)

no time or movement is lost in seeking shelter. Some species, such as rabbits and meadow voles (*Microtus*), maintain trails that serve as routes to food and as avenues of escape.

Reproductive success may be increased by an animal's knowledge of areas occupied by animals inhabiting adjoining home ranges (in the case of solitary species) or by familiarity with animals sharing his home range (in the case of social species). During early life, young can develop under parental care largely free from interference by other individuals of their own species. The spacing of home ranges is often such that the individual or the group is assured a food supply largely untouched by "foreign" members of the species; territorial species tend not to exceed the carrying capacity of a habitat (the maximum number of individuals an environment can support).

The sizes of home ranges vary tremendously, from a fraction of an acre for some small rodents and shrews to an area of 100 square miles or more for some carnivores (Table 15-6). Many mammals within the orders Insectivora, Primates, Rodentia, Lagomorpha, Carnivora, Perissodactyla, and Artiodactyla are known to be territorial. The recognition of territorial boundaries in some species depends on scent marking and other means of territorial

marking, and much remarkable behavior is associated with the maintenance of territories (some of this behavior is discussed in Chapter 17). Some territorial species are distributed according to a pattern of home ranges that may persist throughout the lives of many generations.

Hansen (1962b) found such a pattern to be typical of northern pocket gophers (*Thomomys talpoides*) in some areas. Each animal occupies an area of raised ground called a mima mound (Fig. 15-35), which is some 10 m in diameter. The mima mounds are more productive of food than are the relatively narrow intermound areas, which usually have shallow soil. Except in the winter, the intermound areas are used little by pocket gophers, and the chances of survival are slim for an animal that is unable to establish itself in a mima mound. Likewise, woodrat houses may be used over periods of thousands of years (Wells and Jorgensen, 1964), as indicated by the presence in them of plants that no longer occur in the area but lived there thousands of years ago.

FEEDING STRATEGIES

Biologists stress that the success of organisms can be measured in terms of their relative reproductive rates; the organism with the greatest fitness from an evolution-

Table 15–6. Sizes of Home Ranges of Some Mammals

Species	Home Range (acres)	Source
Common shrew (Sorex araneus)	0.7	Buckner, 1969
Varying hare (Lepus americanus)	14.5	O'Farrell, 1965
Mountain beaver (Aplodontia rufa)	0.3	Martin, 1971
Least chipmunk (Eutamias minimus)	2.1–4.7 (summer only)	Martinsen, 1968
Yellow-pine chipmunk (Eutamias amoenus)	3.89 (males); 2.49 (females)	Broadbooks, 1970
White-footed mouse (Peromyscus)	.08–10.66	Redman & Sealander, 1958; Blair, 1951
Red-backed mouse (Clethrionomys gapperi)	.25 (winter only)	Beer, 1961
Prairie vole (Microtus ochrogaster)	.11 (males); .02 (females)	Harvey & Barbour, 1965
Timber wolf (Canis lupus)	23,040 (pack of 2) (36 sq. mi.) 345,600 (pack of 8) (540 sq. mi.)	Stenlund, 1955 Rowan, 1950
Red fox (Vulpes vulpes)	1280 (2 sq. mi.)	Ables, 1969
Grizzly bear (Ursus arctos)	50,240 (1 mother + 3 yearlings) (78.5 sq. mi.)	Murie, 1944
Russian brown bear (Ursus arctos)	6400–8320 (10–13 sq. mi.)	Bourliere, 1956
Raccoon (Procyon lotor)	13.3–83.4	Shirer & Fitch, 1970
Badger (Taxidea taxus)	2100 (3.3 sq. mi.)	Sargant & Warner, 1972
Mountain lion (Felis concolor)	9600–19,200 (males) (15–30 sq. mi.) 3200–16,000 (females) (5–25 sq. mi.)	Hornocker, 1970
Lynx (Lynx canadensis)	3840–5120 (6–8 sq. mi.)	Saunders, 1963
Black-tailed deer (Odocoileus hemionus)	90 (winter); 180 (summer)	Leopold et al., 1951
Mule deer (Odocoileus hemionus)	502–2534 (.78–4 sq. mi.)	Swank, 1958
White-tailed deer (Odocoileus virginianus)	126–282	Ruff, 1938
Pronghorn antelope (Antilocapra americana)	160–480 (.25–.75 sq. mi.)	Bromley, 1969

Figure 15–35. Mima mounds in Mima Prairie, Thurston County, Washington. These mounds, some 10 m. in diameter, are probably formed by the burrowing activities of pocket gophers. (Photograph by Victor B. Scheffer.)

ary point of view is the one that leaves the greatest number of progeny. A morphological, physiological, or behavioral character and the genotype determining it is adaptive if the organisms possessing the feature leave more progeny than do those without it. Evolutionary change thus depends in part on the differential fitness of organisms with different genotypes. The primacy of reproduction in determining the success of a species is virtually axiomatic—but what about energy?

For an individual to survive to reproductive age requires energy, and the reproductive process itself is energetically expensive. Clearly, no aspect of an animal's life more directly affects its survival and reproductive success than its efficiency in gaining energy. The increased reproductive success with age in some species is probably a reflection of the animals' learning to forage more efficiently (Ricklefs, 1973). Adaptive radiation in mammals has involved a progressively more complete exploitation of the possible food sources, and competition between and within species for these sources has forced each species to evolve the most efficient possible feeding strategy.

Feeding strategies are most appropriately viewed in terms of cost and benefit. The optimal strategy assures the greatest yield of energy from the food, in relation to the cost in energy of pursuit, handling, and eating, per unit of time spent in these activities. Put more gracefully by MacArthur and Pianka (1966), natural selection will achieve optimal allocation of time and energy expenditures. Schoener (1971) gives four key aspects of feeding strategies: optimal diet, optimal foraging space, optimal foraging period, and optimal foraging-group size. A great variety of strategies occurs among mammals; following are a few examples.

The pronghorn antelope (*Antilocapra americana*), an herbivore that feeds primarily on forbs and small shrubs, utilizes a strategy common to many ungulates. Most of a pronghorn's time is spent eating or processing food. Pronghorns alternately feed and bed down through the day and night, and ruminating (chewing the cud) occupies 60 to 80 per cent of the bedding time (Kitchen, 1974). The use of open country and group foraging allows the animals to make the most of their sharp vision and speed in avoiding predation. Herbivory is popular among mammals and has many advantages: the biomass of plants vastly exceeds that of animals, plants need not be pursued, and they are equally available throughout the 24-hour cycle. But the digestion of leaves and stems is a demanding process, the energy yield for each item is low, the energy yield per unit of weight is low relative to that of animal material, and a great deal of time is invested in feeding.

Among some herbivores, however, the allocation of time is strikingly different. Seed-eating heteromyid rodents, for example, spend relatively little time seeking and gathering food. As seeds are the most concentrated source of energy that plants offer, these rodents are not locked into the ungulate feeding system. Instead, kangaroo rats and pocket mice forage solitarily. They often search in depressions in the soil and other places where seeds are concentrated, and gather seeds rapidly and rather unselectively into their cheek pouches. These rodents forage at night, when they are least vulnerable to predators. They make a series of foraging trips, returning from each to deposit pouch contents in their burrows. This gathering phase of feeding requires considerable energy, and during gathering the rodents are exposed to predation and perhaps to low temperatures, but this phase takes relatively little time. As little as an hour or two per night is spent outside the burrow. The selection of the preferred seeds from the randomly collected bunches brought to the burrow probably takes place in the safety of the burrow (Reichman, 1975). The exact amount of time allocated to feeding by kangaroo rats is unknown, but is clearly vastly less than that used by the pronghorn and other ungulates.

By utilizing a high energy and clumped food source, the heteromyids increase the yield of energy per unit of time expended in feeding, and remain in the safety of burrows for most of the 24-hour cycle. But here again the advantages are gained at a cost. During years when fall and winter rains fail, there may be no green annuals nor seed production by annuals, and at these times heteromyids do not reproduce and their populations decline (Beatley, 1969). Desert rodents are therefore at the mercy of the vagaries of desert weather.

Predatory animals employ a variety of feeding strategies, but two major types can represent the extremes. The "sit-and-wait" predator (Pianka, 1966) remains quietly at a vantage point and surveys its surroundings; when prey is detected, the predator makes a brief attack and typically avoids lengthy pursuit. The giant left-nosed bat (*Hipposideros commersoni*) of Africa exemplifies this predatory style (Vaughan, 1977). This bat hangs from an acacia branch and uses echolocation to scan for insects. It is discriminating in its choice of prey, and feeds exclusively on large, straight-flying insects. These are captured during brief and precise interception flights, and the bat returns to its perch to consume the insect. Such predators are typically solitary, are territorial, and gear their periods of watchfulness to the times of greatest activity and vulnerability of the prey.

The sit-and-wait strategy is profitable when prey is common and is motile. Although the predator spends considerable time waiting for appropriate prey, it is at rest and probably uses little more energy than it would if simply remaining alert for predators or competitors. The investment in searching time is great, therefore, but the energy outlay for this activity is low. In contrast, the rate of energy expenditure is high during the capture of prey, but the attack and capture occupy little time. This strategy is used by a few bats and by many kinds of birds, but is not widely used by mammalian carnivores.

More common among mammalian predators is the search-and-chase system. Wolves, for example, form social groups capable of dealing with large prey, and search over wide areas. The pursuit and capture of prey is often a lengthy process that taxes the endurance of the predators and clearly involves the outlay of considerable energy. Although the cost is high, however, the benefit is great. Packs of wolves probably average only one large kill every two or three days. But if the prey is an adult moose weighing some 400 kg, as it often is in the case of the wolves studied by Mech (1966) on Isle Royal, they can gorge on the carcass for at least two days. Small carnivores typically hunt solitarily, but here again search is lengthy, although a series of chases of small prey is usual.

Associated with differing predatory strategies are differences in such abilities as locomotion. The sit-and-wait predator must be capable of brief bursts of activity and often employs a highly stereotyped style of attack, but it need not be well adapted to efficient, enduring locomotion. For the search-and-chase predator, however, refinements of locomotor efficiency are highly adaptive, and the animals may be more opportunistic in their methods of capture.

The development of feeding strategies can be considered as a case of evolutionary gamesmanship between resourceful opponents. The success of every evolutionary move of one species is tested against a position or a countermove by the competing opponent species. A species heading for a desirable square on the evolutionary chess board may be forced to occupy another square if the first is already occupied by a competing species. Or mutual vulnerability to devastating countermoves may make the simultaneous occupancy by two opponents of adjacent squares extremely costly.

In areas with high species diversity the game is complex, involving many opponents utilizing numerous complementary feeding strategies and diets. Pulliam (1974) defines the optimal diet as "a set of successive prey choices which maximizes the rate of caloric intake or, alternately, minimizes the time required to find the food ration." In the case of Black's (1974) bat community, for example, each species of bat is making a unique set of prey choices (Table 15-1).

FEEDING INTERACTIONS

Just as energy transfers between parts of an organism are vital to life, complex patterns of energy transfer within a biotic community maintain this "superorganism" of interdependent and interacting species. The organisms involved in the transfer of energy within a community—from photosynthetic plants that utilize solar energy and inorganic materials to produce protoplasm, to animals that eat the plants, and thence to animals that eat animals—constitute the food chain. Typically, the transfer in a food chain goes from photosynthetic plants (primary producers), to herbivores that eat these plants (primary consumers), to first carnivores (secondary consumers) that eat the herbivores, to secondary or perhaps tertiary carnivores in some extended chains (Fig. 15-36).

For the complex pathways of energy transfer that usually occur in nature, often involving predators and primary consumers that figure importantly in more than one food chain, the term food web has been used. (The intricacy of a food web is suggested by Fig. 15-37). Animals that occupy comparable functional positions in the food chain—the position of primary consumers, for example—are at the same trophic level. Green plants occupy the first trophic level and are referred to as autotrophs (self-feeders). Using mammals as examples, the second trophic level is occupied by herbivorous rodents, rabbits, and ungulates. Small carnivores such as weasels occupy the third trophic level, and large carnivores may be in the third or fourth level.

The food chain is often depicted as a pyramid, in an attempt to stress the relationships of biomass (the total weight of organisms of a given type in the community), numbers of organisms, and available energy at the different trophic levels. Food chains typically rest on a broad food base of plant material, but energy available to animals in each successively higher trophic level becomes progressively reduced.

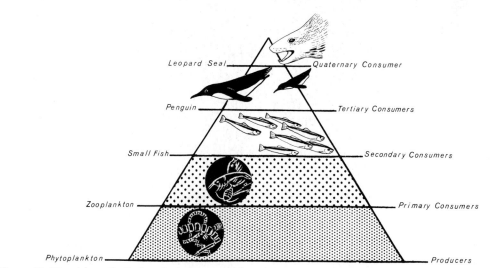

Figure 15–36. A hypothetical ecological pyramid based on an Antarctic food chain. The higher the step in the food chain (the higher the trophic level), the larger the individuals and the lower their numbers. Ultimately, gigantic numbers of tiny planktonic plants and animals are necessary to support, through several intermediate steps, one leopard seal.

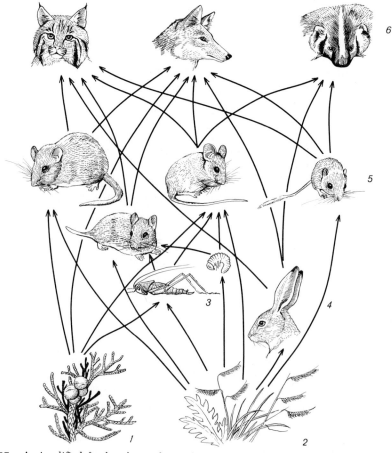

Figure 15–37. A simplified food web involving the mammals of a piñon-juniper community in Coconino County, Arizona. The arrows indicate the foods utilized by the mammals. The plants (1, juniper; 2, grasses and forbs) support arthropods (3), rabbits (4), and rodents (5). The rodents (*Neotoma stephensi, Onychomys leucogaster, Peromyscus truei, Perognathus flavus*) and the rabbits (*Lepus californicus* and *Sylvilagus audubonii*) are preyed upon by bobcats, coyotes, and badgers.

Figure 15–38. Pyramids of numbers, calories, and energy utilization for one acre of annual grassland near Berkeley, California. The pyramid showing calories is also approximately to scale for biomass. (After Pearson, 1964.)

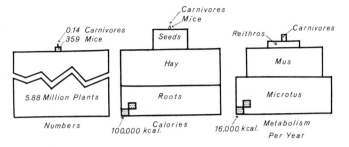

The reduction results from loss of energy by respiration and by way of organisms that die and are utilized by decomposers (organisms that decompose organic material) rather than by animals of the next higher trophic level. Also, energy is lost because of inefficient transfers between levels. Consider, for example, the pyramids of numbers, caloric content, and energy utilization shown in Figure 15–38 and Table 15–7.

The typical relationship of size and abundance of animals in a food chain involves small but numerous primary consumers, larger but much less abundant secondary consumers, and still larger but relatively scarce tertiary consumers. Although the animals at the top of the food chain seem to be in the commanding position of potentially being able to prey on all animals at lower trophic levels without being vulnerable to predation themselves, the top predators occupy precarious positions because they frequently depend on animals from trophic levels with low pro-

ductivity. Under conditions of food stress, therefore, the fate of the species at the top of the food chain will be not death at the hands of a predator but death by starvation as a result of low availability of food.

Because of the great loss of energy accompanying food transfer between successive trophic levels, the total available energy is largest for consumers at the lower levels. Thus, predators feeding on primary consumers have more energy available to them than do the predators feeding on secondary consumers. Considered in this light, the adaptive importance of filter feeding in some marine mammals becomes clear. The baleen whales and certain plankton-feeding seals exploit tremendously larger sources of energy in primary-consumer plankton than they could if they fed on large fish that are secondary or tertiary predators. Only some 10 to 20 per cent of the energy entering a trophic level can be utilized by the next higher level; this factor limits the length of food chains.

Diagrams of food chains or food webs,

Table 15–7. Standing Crop of Plants, Prey, and Predators on One Acre of California Grassland and Rate of Use of Vegetation by Rodents and of Prey by Carnivores at Peak Population Levels

	Standing Crop		Rate of Use Per Year	
	Kg (Dry Wt)	Kcal	Kcal	% of Crop
Roots	2,131	7,269,000		
Hay	2,097	8,141,000		
Seeds	442	1,920,000		
Microtus	1.24	6,402	1,368,750	71
Mus	0.88	4,543	876,000	46
Reithrodontomys	0.084	434	81,650	4
Other prey	0.13	671	27,000	
Carnivores	0.126	650	11,700	97

(After Pearson, 1964; the sources of the figures in this table are not listed here but are given in Pearson's paper.)

although valuable for purposes of illustration, usually of necessity strongly simplify what is really an extremely intricate meshwork of interactions. A broadly adapted carnivore like a coyote, or an omnivore such as the possum *(Didelphis)*, may function in all trophic levels above that of the primary producer. For a coyote, the fruit of the prickly-pear cactus or juniper berries may form one meal, while a jackrabbit or deer fawn may be the next. More frequently, seasonal differences in the position of an animal in the food chain may occur. Johnson (1961, 1964) found that the deer mouse *(Peromyscus maniculatus)* in Colorado and Idaho became strongly insectivorous in the summer, and thus functioned during this season as a secondary consumer, whereas it ate largely plant material during the cooler seasons, functioning as a primary consumer at those times. Occasionally in lean times the ability to utilize various "alternate" foods may be of critical importance to large and powerful species. Even the mountain lion *(Felis concolor)*, generally a predator of deer, will stoop under pressure of hunger to such lowly prey as skunks, raccoons, bobcats, or even porcupines (Grinnell et al., 1937: 574–575).

Recent ecological research has revealed important differences between small mammal energetics in various ecosystems and in the proportions of energy utilized by rodents from different trophic levels. The excellent study by French et al. (1976) compared the small mammal energetics in a series of grassland ecosystems, ranging from tallgrass and midgrass sites in Oklahoma and South Dakota, respectively, to a desert grassland ecosystem in New Mexico. The rodent communities of the ecosystems differed strikingly in composition and in biomass (Fig. 15–39). Of particular interest, foods from different trophic levels were utilized to different extents by the rodents at different sites: microtine herbivores specialized on green primary production dominated (in terms of biomass) at the tallgrass site; sciurid omnivores, utilizing both primary production and primary consumers (insects), dominated in the

northern shortgrass prairie; and heteromyid seed eaters (granivores) dominated the desert site. Broadly speaking, small mammals depended chiefly on herbage in the tallgrass and midgrass sites, and on herbage, seeds, and invertebrates in the desert site.

Although the energy consumed by rodents was highest in the tallgrass prairie (where 172×10^3 kilocalories/hectare supported 935 g live weight of small mammals/hectare), the amount of small mammal biomass relative to consumption by rodents was greatest in the northern shortgrass prairie (where 32×10^3 kilocalories/hectare supported 277 g of small mammals/hectare). This was probably due largely to the high assimilation and digestion rates for granivores and omnivores (Table 15–8). Usually less than 10 per cent of the primary production at the various sites was utilized by small mammals; but, in striking contrast, invertebrates were heavily utilized, in extreme cases to close to 100 per cent (Table 15–9). This information would suggest that microtine herbage eaters underutilize their food resources, whereas seed-eating heteromyids overexploit food reserves, an idea presented by Baker (1971).

Herbage eaters seem in general to be r-selected types; in r-selected animals, selection has favored high rates of reproduction and growth, and with this pattern goes high mortality. Such r-selected species would be expected to have a high population turnover and to be able to respond rapidly to optimal conditions by increased population densities. Seed-eating rodents are seemingly K-selected; selection in K-selected rodents has favored characters furthering survival of the individual and, typically, these mammals have lower reproductive rates than do the herbage eaters. In heteromyids, for example, periodic torpor and a well-developed social system (described by Eisenberg, 1966) tend to preserve the individual and to lead to relatively low rates of mortality and population turnover. Under conditions of the tight food supplies typical of arid regions, K-selected rodents would seem to have the

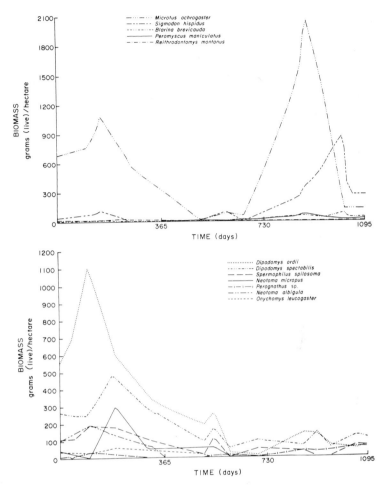

Figure 15–39. A plotting through time of the biomass of small mammals at a tallgrass prairie site (above) and a desert grassland site (below). (From French et al., 1976.)

adaptive edge, whereas the boom-or-bust productivity of some mesic grasslands perhaps favors r-selected microtine species.

French et al. (1976) point out that where mesic grasslands are receding northward,

Table 15–8. Assimilation and Digestibility (Expressed as Organic Matter Consumed or Per Cent of Energy) of Natural Foods in Some Small Mammals

Group	Number of Species	Di-gest-ibility	Assimi-lation
Grazing herbivore (microtines)	14	67	65
Omnivore (sciurids, cricetines)	8	75	73
Granivore (heteromyids)	6	90	88
Insectivore (*Blarina*)	1	—	80

(From French et al., 1976.)

Table 15–9. Fraction of the Available Energy Utilized by Small Mammals, According to Diet

Grassland Sites	Year	Herbage	Seed	Animal
Tallgrass	1970	0.063	—	0.88
	1971	0.010	—	0.38
	1972	0.047	0.080	all?
Midgrass	1970	0.002	—	0.08
	1971	0.001	—	0.08
	1972	0.002	—	0.02
Northern shortgrass	1970	—	0.002	all?
	1971	—	—	0.42
	1972	0.005	—	0.83
Southern shortgrass	1970	—	—	0.96
	1971	0.013	—	0.61
	1972	0.091	—	0.11
Desert	1970	0.032	—	all?
	1971	0.190	—	all?
	1972	0.004	—	0.68

(From French et al., 1976.)

as they are in some parts of the midgrass areas in the Great Plains of the United States, herbage-eating microtines are being replaced by the specialized seed eaters. Sharp between-habitat differences in the relative importance of herbivorous, omnivorous, granivorous, and insectivorous small mammals can also be observed in areas other than grasslands (Fig. 15–40), and reflect differences in the composition of the vegetation.

FACTORS INFLUENCING POPULATION DENSITIES

The abundance of an animal species at a given time and a given locality depends on (1) the carrying capacity of the habitat and (2) the relationship between the rate at which the animals are added to the population (by reproduction or immigration) and the rate at which they are lost from the population (by death or emigration). Many mammalian populations are in a state of "dynamic equilibrium," and tend to be stabilized within certain limits of density by such interacting processes as competition, reproduction, predation, dispersal, and disease. These factors are usually regarded as density dependent; that is, they fluctuate in direct relationship to the density of the species involved. The intensity of predation, for example, is generally dependent on the population density of the prey species. A higher proportion of a prey species is frequently taken during high densities of that prey species. Other

factors, such as space requirements and weather, are generally density independent; that is, they do not change in response to changes in population density.

NATALITY

The number of individuals added to a population through reproduction depends on the reproductive potential of a species, which refers to the greatest number of individuals that a pair of animals or a population can produce in a given span of time. The reproductive potential is a function of age and sex ratios, the age at which a female first bears young, litter size, and the frequency of litters, Even species of the same genus occupying the same area can have markedly different reproductive potentials as indicated by litter size (Table 15–10). Reproductive performance is also responsive to environmental differences, such as temperature or rainfall, for sharp regional shifts in litter size occur within some species (Table 18–4).

The ability of some species to vary reproductive performance in response to environmental conditions or population levels may be of considerable adaptive importance. For example, the reproductive potential of mule deer is lower in poor habitats than in habitats productive of high quality browse; whereas well-nourished does may breed first at 17 months of age, those that occupy poor ranges may not breed first until as late as 41 months of age (Taber and Dasmann, 1957). The litter size of carnivores is also

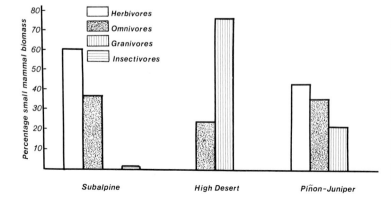

Figure 15–40. Relative biomass of herbivorous, omnivorous, granivorous, and insectivorous small mammals at a subalpine site (Rabbit Ears Pass, Routt County, Colorado), a high desert site, and a piñon-juniper site. The latter two sites are in Wupatki National Monument, Coconino County, Arizona. (Data for the subalpine site from Vaughan, 1974; biomass for the other sites based on unpublished data from the summer of 1976, courtesy G. C. Bateman.)

Table 15–10. Differences Between Reproductive Patterns of Three Species of *Peromyscus* That Are Sympatric in Some Areas

Characteristic	*P. maniculatus*	*P. truei*	*P. californicus*
Number of litters per season	4.00	3.40	3.25
Number of young per litter	5.00	3.43	1.91
Number of offspring per breeding female per season	20.00	11.66	6.21

(From McCabe and Blanchard, 1950.)

affected by food supply; Stevenson-Hamilton (1947) reported that the litter size of the African lion dropped when food was scarce. Batzli and Pitelka (1971) found delays in the start of breeding in the California vole *(Microtus californicus)* following times of peak densities; these delays may have resulted from decreased availability of preferred foods during periods of high densities of voles. These authors and other workers (Hoffmann, 1958; Greenwald, 1957) have observed seasonal changes in litter size that are presumably caused by changes in forage quality.

In Finnish Lapland, Kalela (1957) found that reproductive maturity was delayed in young male voles *(Clethrionomys rufocanus)* in several localities in a year of high populations and, in one area supporting especially high densities, breeding was also delayed in nearly all young females. The result of such delays is usually to reduce the numbers of young produced during that season; also, individuals of late litters may have too little time for maturation before the critical winter period and may suffer unusually high mortality. Breeding of muskrats *(Ondatra zibethica)* in Iowa continued through only part of the usual breeding season in a year of high populations (Errington, 1957), and wild rats *(Rattus norvegicus)* in the city of Baltimore had a markedly low pregnancy rate during population highs (Davis, 1951).

Survival rates of young also strongly affect population levels. Young are clearly the expendable part of the population and show the greatest fluctuations during population changes. A 74 per cent decline in pocket gopher density in western Colorado in 1958 was associated with an extraordi-

nary drop in survival of young (Hansen and Ward, 1966), and in southern Colorado Hansen (1962b) found that whereas high survival of young pocket gophers was characteristic of periods of high densities, low survival of young was associated with a declining population. Similarly, Krebs (1966) reported better survival for expanding than for declining populations of the California vole.

In mammals the contribution made to a population by reproduction clearly depends on a variety of factors, and is seldom constant within a species from year to year. As mentioned, variation in litter size, numbers of litters and length of breeding season, the age at which young animals breed, and survival of young are all important variables. In addition, the litter size (Table 15–11) and the percentage of females that become pregnant changes with age distribution of a species; the age composition of a population may therefore have a marked effect on the reproductive performance of that population.

Table 15–11. Litter Size of Consecutive Litters of the Montane Vole *(Microtus montanus)*

Litters	N	Mean Litter Size	Range
1	12	4.2	2–6
2	12	4.7	3–7
3	10	5.0	3–7
4	9	4.2	2–6
5	6	5.8	3–10
6	6	5.5	3–7
7	5	3.4	1–6

Note that young and old animals have relatively small litters. (After Negus and Pinter, 1965.)

Table 15–12. Life Table for Dall Sheep *(Ovis dalli)* in Mount McKinley National Park, Alaska

Age (years)	Age as Per Cent Deviation from Mean Length of Life	Number Dying in Age Interval per 1000 Born	Number Surviving at Beginning of Age Interval per 1000 Born	Mortality Rate per 1000 Alive at Beginning of Age Interval	Expectation of Life, or Mean Lifetime Remaining to Those Attaining Age Interval (years)
0–0.5	−100.0	54	1000	54.0	7.06
0.5–1	− 93.0	145	946	153.0	–
1–2	− 85.9	12	801	15.0	7.7
2–3	− 71.8	13	789	16.5	6.8
3–4	− 57.7	12	776	15.5	5.9
4–5	− 43.5	30	764	39.3	5.0
5–6	− 29.5	46	734	62.6	4.2
6–7	− 15.4	48	688	69.9	3.4
7–8	− 1.1	69	640	108.0	2.6
8–9	+ 13.0	132	571	231.0	1.9
9–10	+ 27.0	187	439	426.0	1.3
10–11	+ 41.0	156	252	619.0	0.9
11–12	+ 55.0	90	96	937.0	0.6
12–13	+ 69.0	3	6	500.0	1.2
13–14	+ 84.0	3	3	1000.0	0.7

(After Deevey, 1947.)

PREDATION

Rates and age distributions of mortality are important population characteristics. Mortality varies with age, and has been carefully studied in some mammals. Specific mortality is the number of individuals of a population that have died by the end of a given time span. Specific mortality at given ages can be expressed in a life table. The life table in Table 15–12 is based on data assembled by Murie (1944) during his classic study of Alaskan wolves. Ages of some small mammals are difficult to establish, and a life table for females may be based on numbers of litters that females have had (Table 15–13).

Of the mortality factors to which mammals are susceptible, predation looms high in importance. There has been much heated debate on the ability of predators to control or influence densities of prey species or to influence population cycles, and the final word has not been heard. The degrees of impact that predators have on prey have been summed up by Pearson (1971:41): "The effectiveness of predation varies from the relatively ineffective predation of rats on man, in which rats are

Table 15–13. Life Table for Female African Root Rats *(Tachyoryctes splendens)*

Litter/Age Groups	Actual Number of Moles	Survivorship (lx)	Death Rate (dx)	Mortality per 1000 Living (qx)
0	412	1000	750	760
1	99	240	19	79
2	91	221	29	131
3	79	192	136	708
4	23	56	41	732
5	6	15	8	553
6	3	7	5	714
7	1	2	2	1000

lx = the number surviving at the beginning of each age interval out of 1000 born.
dx = the number dying in each age interval out of 1000 born.
qx = the mortality rate per thousand alive at the beginning of each age interval. The number of mole rats in litter/age group 0 is an estimate.
(From Jarvis, 1973.)

able occasionally to kill infants or incapac-
itated adults, through the mink-muskrat
system described by Errington (1967), in
which mink take a significant proportion
of homeless and stressed muskrats, to the
almost total effectiveness of carnivore predation on *Microtus* until almost the last
one has been killed." A predator-prey relationship must obviously have some stability; as indicated by Lack (1966:301), "Only
those predatory species which have not
exterminated their prey survive today,
hence we observe in nature only those systems which have proved sufficiently stable
to persist, and many others were presumably terminated in the past by extinction."
Although the effectiveness of even a single
species of predator seems to vary according to the specific situation considered,
some general predator-prey relationships
that apply to mammals as well as to other
animals can be recognized.

The observed responses of a predator to
changes in the density of a prey species indicate that predation is density influenced.
The numbers of a preferred prey taken by a
carnivore increase as the density of the
prey increases (Fig. 15–41), because the
greater the number of prey animals per
unit area, the greater the opportunity for
predators to encounter and capture them.
This is a functional response on the part of
the predator (Holling, 1959, 1961). Errington (1937) noted such a response in the
predators of muskrats and suggested that

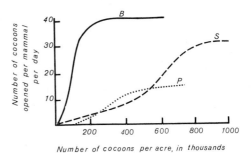

Figure 15–41. Functional responses on the part of
three predators, the short-tailed shrew, *Blarina brevicauda* (B), the masked shrew, *Sorex cinereus* (S), and
the deer mouse, *Peromyscus maniculatus* (P), to density of prey. The prey are sawfly larvae, which are
removed from their cocoons by the predators. (After
Holling, 1959.)

the intensity of predation is a function of
population levels of the prey. There may
also be a numerical response, involving an
increase in the predator density with a rise
in the prey population. The numerical
response may be the result of immigration
of predators, as in the case of the striking
responses to lemming abundance on the
part of the pomarine jaeger (Table 15–14),
or may be due to increased breeding success, as in the case of the masked shrew
(Fig. 15–42).

The relative populations of predators
and their prey have often been regarded as
being in dynamic equilibrium. The degree
to which a balance between prey populations and predator populations is reached,
and the extent to which the relationship is

Table 15–14. Densities of Breeding Pomarine Jaegers *(Stercorarius pomarinus)*
and Nesting Success Near Point Barrow, Alaska

Year	Spring Lemmus Density (no./acre)	No. of Pairs of Jaegers	Census Area (square miles)	Density (pairs/square mile)	Maximum Density (pairs/square mile)	Breeding Success (per cent of eggs)
1952	15–20	34	9	3.8	5–6	30–35
1953	70–80	128	7	18.3	25–26	20–25
1954	<1	0	—	—	—	—
1955	1–5	2	15±	0.13	—	0
1956	40–50	114	6	19.0	22–23	4
1957	<1	0	—	—	—	—
1958	<1	0	—	—	—	—
1959	1–5	3	15±	0.20	—	0
1960	70–80	118	5.75	20.5	25	55

Note the correlation between high densities of lemmings *(Lemmus trimucronatus)* and high populations of
nesting jaegers. (After Maher, 1970.)

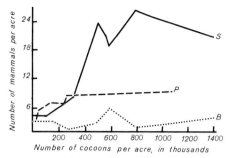

Figure 15–42. Numerical responses on the part of the short-tailed shrew (B), the masked shrew (S) and the deer mouse (P) to high densities of sawfly larvae. (After Holling, 1959.)

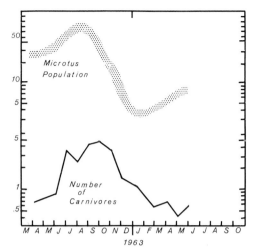

Figure 15–43. Densities of California voles (*Microtus californicus*) compared to densities of predators (feral cats, raccoons, foxes, and skunks) during a population cycle of voles. Multiply vertical scale by 100 to obtain numbers of voles present. (After Pearson, 1971.)

dynamic, varies widely in situations involving mammals. It depends in part on the ratio of predator density to prey density, on the relative sizes of the predator and the prey and the ease with which prey can be captured, and on the degree to which the prey populations are cyclic. Studies by Mech (1966) on Isle Royale in Lake Superior during the period from 1959 to 1961 indicated that the ratio of moose to wolves was roughly 30 to 1; 20 wolves were supported by approximately 600 moose, the wolves' primary food. The weight differential between an adult moose and an adult wolf was roughly 14 to 1 (approximately 450 kg to 33 kg), and the wolves had difficulty killing moose. A strongly contrasting situation was studied in California by Pearson (1971), who found that the ratio of numbers of predators varied from 72 to 1 in 1962, during a period of low vole populations, to 5410 to 1 during a peak in vole numbers (Fig. 15–43). In this case the prey was a cyclic vole (*M. californicus*) with an adult weight of roughly 45 gm. The predators — feral cats, raccoons, gray foxes and skunks — averaged perhaps 2.25 kg in weight, yielding a rough estimate of prey to predator weight of .02 to 1. The predators could catch voles easily.

These examples are based on two very different patterns of predator-prey interaction. The wolf-moose interaction resulted in relatively stable predator and prey densities, whereas the situation involving the California vole was one of great instability.

Because of the difficulty with which wolves bring down moose, the pressure they exert on the moose population is highly selective in that primarily young or old animals are taken; adult moose in the prime of their reproductive life are not killed (Mech, 1966). The predators of the vole, on the other hand, show a high preference for this prey and find it easy to catch; their kill is far more nearly random and they are able to kill almost every vole during times of vole scarcity (Pearson, 1966). As suggested by the above examples, predator-prey interactions are complex; so many variables are involved that few generalizations relating such interactions can be universally applied.

Some predators tend primarily to remove vulnerable individuals from the prey population. These are individuals that because of inexperience, old age, injury, or sickness are readily captured; or they are animals that are forced by intraspecific competition for space into marginal habitats in which their vulnerability to predators is increased. Work on wolves by Murie (1944), Crisler (1956), and Mech (1966) has shown that vulnerable individuals of the prey species, in these cases Dall sheep, caribou, and moose, were taken more frequently than were healthy, mature individuals. Careful field studies by Hornocker (1970a, 1970b) on the mountain lion in Idaho showed that of 53 lion-killed elk and 46 lion-killed deer, 75 per cent of the elk were young (less than 1.5 years) or old (more than 8.5 years), and 62 per cent of the deer were young or old. These percentages of young and old animals were considerably greater than would be expected if the lions had killed randomly. On Isle Royale, 18 of 51 wolf-killed moose were calves; most of the remainder of the animals killed were from 8 to 15 years old, and 39 per cent of these old animals showed evidence of debilitating conditions (Mech, 1966). The wolves had killed no adults from 1 to 6 years old. Errington (1943, 1946, 1963) found that in Iowa as muskrat populations rose above a "threshold of security" a number of animals were forced into marginal habitats by intraspecific competition for space. This "vulnerable surplus" was preyed upon heavily by mink and red foxes, which made a marked functional response to this available food source. This pattern of predation has been called compensatory predation.

But can predators control densities of mammals, or are they simply killing individuals that would quickly be removed from the population by other means? The answer clearly depends on the specific predator-prey interaction considered. Murie (1944:230) concluded that, in Alaska, wolf predation on Dall sheep lambs was the most important factor limiting numbers of sheep, and on Isle Royale the wolves seemingly kept moose densities below the level at which food supply would be the limiting factor (Mech, 1966:167). In Idaho, however, predation by mountain lions had little impact on the populations of deer and elk; here the densities of deer and elk were controlled by the winter food supply (Hornocker, 1970b).

Heavy predation on populations of small mammals has been shown to affect population levels. Data reflecting the severity of predation on lemmings demonstrate how important this source of mortality may be locally. On the coastal plain near Point Barrow, Alaska, during times of high lemming densities, the combined impact of several predators deals a staggering blow to lemming populations and is an important factor causing the population crash (Pitelka et al., 1955; Pitelka, 1957a). The combined kill of lemmings by the major predators amounted to at least 49 per acre during a cycle of abundance (Table 15–15). Studies in California by Pearson (1963, 1964 and 1966) on *Microtus* have demon-

Table 15–15. Impact by Predators on High Population of Lemmings *(Lemmus trimucronatus)* near Point Barrow, Alaska

Predator	Age Class	Density (ind./ square mile)	Daily Food Consumption (g/ind.)	Season's Lemming Consumption			
					(per acre)		
				(per ind.)	25 May to 15 July	16 July to 31 Aug.	Total
Pomarine jaeger	Adult	38	250	338	10	21	31
	Young	38	200	167	–	–	
Snowy owl	Adult	2	250	350	1.3	1.6	3
	Young	7	150	160	–	–	
Least Weasel		64	50	100	5	5	10
Glaucous gull		20	250	125	0.7	–	1
Waste					4	–	4
Totals					21	28	49

Data for the least weasel from Thompson, 1955; for the snowy owl from Watson, 1958. Table after Maher, 1970.

strated that predators preying upon these cyclic rodents are unable to control a rising prey population, but that "carnivore predation during a crash and especially during the early stages of the subsequent population low determines to a large extent the amplitude and timing of the microtine cycle of abundance" (Pearson, 1971:41).

Stability in a predator-prey system seems to be due generally to environmental heterogeneity (Murdoch and Oaten, 1975:117–120); the predator and prey occupy separate niches and are therefore partially separated. Refuges allow a segment of the prey population to escape predation. Carefully controlled laboratory experiments using a protozoan predator and a bacterium as prey suggest that, even with extremely limited environmental heterogeneity, the predator is unable to exploit the prey to the extent of causing the extirpation of either species (Van den Ende, 1973). The bacteria in the liquid culture could be captured by the protozoan, but, because the bacteria adhering to the walls of the culture vessel could not be dislodged by the ciliary action of the protozoan, the two species persisted and their populations approached stability.

DISEASE

Parasitism and disease are known to be a significant cause of mortality (Elton, 1942) among mammals, and may occasionally cause dramatic population crashes, as in the case of a die-off of prairie dogs (Cynomys gunnisoni) in Colorado (Lechleitner et al., 1962) caused by bubonic plague. Talbot and Talbot (1963) estimated that 47 per cent of the total mortality suffered by wildebeest was caused by diseases, of which the disease rinderpest seemed most important. The blood parasite Babesia is a source of mortality among African lions. Disease in relation to population regulation, however, has been a difficult factor to assess (Chitty, 1954). Disease as the single cause of death may be relatively unimportant, but it may be important in contributing to the vulnerability of an animal to predation or to stressful environmental conditions.

Parasitism has been regarded periodically as an important cause of mortality, but careful observation indicates that otherwise healthy animals can often tolerate a moderately heavy parasite load. Heavy parasitism has been found to accompany a general decline in health in rodents and rabbits during or following times of high density (Batzli and Pitelka, 1971; Erickson, 1944). Work in Colorado by Woodard et al. (1974) indicated that high late-summer mortality among young bighorn sheep (Ovis canadensis) was caused by heavy lungworm infestations and by bronchopneumonia.

WEATHER

Certain unusual weather-caused conditions, such as flooding, are known to result in significant mortality among mammals. A series of beaver colonies were decimated during a flood in Colorado (Rutherford, 1953), populations of small mammals in Oklahoma were reduced by stream valley flooding (Blair, 1939), and a 70 per cent decrease in the combined population of golden mice (Ochrotomys nuttalli) and cotton mice (Peromyscus gossypinus) was caused by a flood of three weeks' duration in eastern Texas (McCarley, 1959). Small mammals may face considerable cold stress and mortality during autumns when extreme cold descends and an insulating snow cover is late in developing; and rapid snowmelt with resultant flooding is an annual "catastrophe" that faces small mammals in areas with heavy snowfall.

Weather may have strong but indirect effects on mortality. In 1971 in Norway, for example, an estimated 41 per cent of the calf crop of red deer (Cervus elaphus, called elk in the United States) in the Aure area died, apparently as a result of poor nutrition of the pregnant cows because of unusually heavy late winter and spring precipitation (Wegge, 1975).

POPULATION CYCLES

Mammalian population cycles are among the most impressive biological phenomena. Striking changes in density occur primarily in temperate, subarctic, and arctic areas, but are not known to occur in tropical or subtropical regions. This difference is probably related to differences in species diversity between these areas. High latitude areas are characterized by biotic assemblages and food webs that are simple compared to those of tropical areas. The typical boreal community has a limited biota and supports few species of vertebrates, but some species may, at least periodically, be remarkably abundant. The simplicity of the northern community is seemingly partly responsible for its instability, for, where so few kinds of organisms exist, any marked fluctuation in the density of one species seems to disrupt the entire community. In tropical habitats, by contrast, there is an enormous diversity of plants and animals that supports many species of vertebrates. However, few of the many species of vertebrates have high population densities. In the complex tropical community the heterogeneity of the environment, the diversity of carnivores, the intricate patterns of niche displacement and potential competition, and the relatively small percentage of the energy resources available to any one species provide a buffer system against population outbreaks by any species. A complex food web, involving many pathways of energy flow, also provides a cushion against drastic population declines.

Even in northern Alaska, population cycles are more pronounced in coastal areas, where only two microtines occur, than in the foothills, where there are five species. Pitelka (1957a:85) states, "Similarities in their feeding and sheltering activities strongly suggest that where more than one species is important, competition may act to depress their respective populations and hence to depress the likelihood of strong fluctuations."

CHARACTERISTICS OF POPULATION CYCLES

In areas where well-marked cycles of microtine rodents occur, population peaks may occur at three- to four-year intervals (Fig. 15–44). On the coastal slope of northern Alaska, oscillations of populations of lemmings *(Lemmus trimucronatus)*, the

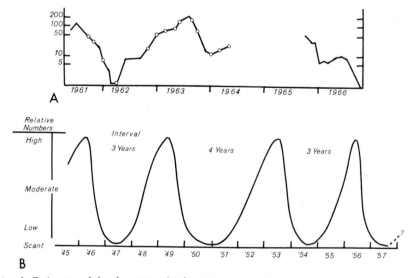

Figure 15–44. A, Estimates of the densities of voles *(Microtus californicus)* in an area near Berkeley, California, during a six-year period. (After Pearson, 1971.) B, Generalized curves (the amplitudes of successive cycles are actually not the same) showing fluctuations in brown lemming *(Lemmus trimucronatus)* populations near Barrow, Alaska. (After Pitelka, 1957b.)

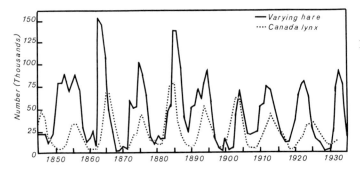

Figure 15–45. Cycles of population density of the varying hare (*Lepus americanus*) and the Canadian lynx (*Lynx canadensis*). The figure is based on the numbers of pelts purchased by the Hudson Bay Company. (After Mac Lulich, 1937.)

chief herbivore, are characterized by: a precipitous drop in density in the late summer and winter following a population peak; a period of one or two years of extremely low populations and localized distribution; an upsurge in the population in the winters following the low population, with peak numbers occurring early in the third or fourth summer (Pitelka, 1957b:85, 86). The short-term lemming cycles also occur in temperate and boreal parts of the Old World (Elton, 1942; Siivonen, 1954). Longer-term cycles, from eight to twelve years, are known in populations of the varying hare and the Canada lynx (Fig. 15–45). Mammalian population cycles occur in some temperate areas, where perhaps the most striking cycles involve microtines, which occasionally reach amazingly high densities (Table 15–16).

Microtine cycles are of special interest, both because they are extremely dramatic and because they have been carefully studied. Population cycles in lemmings and voles can be divided into four phases: (1) the increase phase, (2) the peak phase, (3) the decline phase, and (4) the phase of low densities. Each of these phases is to some extent distinct. Krebs and Myers (1974) present an excellent treatment of population cycles in small mammals. I have used their work repeatedly in preparing the following material on microtine population cycles.

The increase phase is a time when densities rise markedly from one spring to the next. This increase may continue over several years and may be interrupted annually by short-term population declines (Hamilton, 1937); or, more typically, it may occur within one year, with extremely sharp increases over a period of three or four months (Fig. 15–46). Microtines are flexible with regard to the duration of breeding, and during the increase phase breeding begins early in the spring and often con-

Table 15–16. Population Densities of Several Species of Microtines

Density (per acre)	Species	Region	Reference
1–20	*M. pennsylvanicus*	N. Minn.	Beer et al., 1954
6–67	*M. pennsylvanicus*	New York	Townsend, 1935
3000	*M. montanus*	N.W. U.S.A.	Spencer, 1958a
200–4000	*M. montanus*	Oregon	Spencer, 1958b
25–81	*M. californicus*	N. Cal.	Greenwald, 1957
425	*M. californicus*	N. Cal.	Lidicker & Anderson, 1962
25–145	*M. ochrogaster*	Kansas	Martin, 1956
250–300	*M. agrestis*	England	Chitty & Chitty, 1962
1900	*M. arvalis*	France	Spitz, 1963
1004	*M. guentheri*	Israel	Bodenheimer, 1949
2400	*M.* sp.	U.S.S.R.	Hamilton, 1937
50–100	*L. trimucronatus*	Alaska	Rausch, 1950
200–300	*L. lemmus*	Sweden	Curry-Lindahl, 1962

(After Aumann, 1965.)

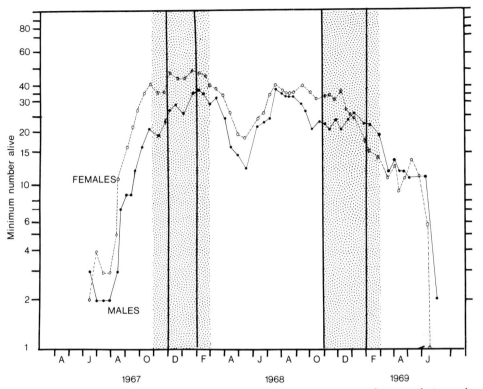

Figure 15–46. Changes in the densities of *Microtus pennsylvanicus* during a population cycle in southern Indiana. The shaded period is the winter. (After Gaines and Krebs, 1971.)

tinues into the winter. At this phase animals quickly reach sexual maturity. Some Norwegian lemmings *(lemmus lemmus)* during an increase phase were found by Koshkina and Kholansky (1962) to be pregnant at 20 days of age. Koshkina (1965) found that the rate of sexual maturation was affected by population density, with early maturation being typical of an increasing population. Survival is generally high during the increase phase, but there is much dispersal.

The peak phase is a time of relatively little change in density. The population increase ceases, and the population may remain fairly stable for at least a year, or may abruptly swing into a decline. During the peak phase the breeding season in summer is typically brief and no winter breeding occurs. Animals attain sexual maturity late, and young born at peak times may not mature sufficiently to breed during their first summer. Growth rates are relatively high at this time, and animals

are generally 20 to 50 per cent larger at the peak phase than at other times (Krebs and Myers, 1974). Mortality rates are relatively low during this phase.

The decline phase varies widely, from precipitous drops in density (population "crashes") to uneven declines lasting a year or more. As during the peak phase, the decline phase is typified by brief summer breeding and no winter breeding, and animals reach sexual maturity late. Little dispersal occurs during the decline, but mortality rates are high within the relatively sedentary population (Fig. 15–47).

The phase of low densities may last for one to three years, and annual shifts in abundance may occur (Krebs, 1966). At this time the breeding season is short, animals seem not to reach sexual maturity early, and mortality is high.

Mention should be made of the famous mass wanderings of Norwegian lemmings that accompany population highs and population declines. During these wanderings

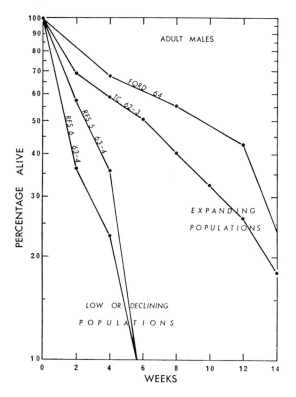

Figure 15-47. Survivorship curves for two expanding and two low or declining populations of *Microtus californicus* from time of first capture in live traps. Data for adult males only. (After Krebs, 1966.)

lemmings appear far from their preferred habitats. One such movement occurred in Norway in the summer and fall of 1963 and is well described by Clough (1965). The movement began in mid-July, when as many as 40 lemmings per hour passed an observation point; all animals were heading downhill. Lemmings also began to appear on the streets of a nearby town and in pine forests 12 miles down the valley. Clough was struck by the complete intolerance of one lemming for another. Each animal moved alone. Every time lemmings met one another, some aggressive action or strong avoidance behavior occurred, and large, mature females were dominant in all conflicts involving them. The wandering lemmings, however, were chiefly middle-sized or small, and often sexually immature individuals. By October lemmings were living in virtually all habitats, including those that were highly unsuitable, and the population was on the decline. The following March, Clough did not find a single lemming, nor were any found during his intensive work in May and June in areas where lemmings had been abundant

the previous summer. The winter's die-off had seemingly been complete.

FACTORS CONTROLLING POPULATION CYCLES IN SMALL MAMMALS

Although population cycles in mammals have been known for many years, and the patterns of fluctuation have been carefully described for some species, considerable uncertainty and controversy remains as to what factors control cycles. At present there is no definitive answer to this question, but a number of careful studies have suggested which hypotheses can be discarded and which merit further consideration. At least the search is narrowing.

Some factors seem not to be responsible for population cycles. In populations of lemmings and voles, litter size, rates of pregnancy, and sex ratios are rather constant and seldom vary during the phases of a cycle (Krebs and Myers, 1974). Lidicker (1973), however, found striking seasonal differences in litter size and sex ratios in *M. californicus*. Weather also seems not to be responsible, although Fuller (1967,

1969) and others have considered weather to have important effects on microtine populations. Such reports as that of Elliot (1969), which describes the association of a decline in a population of *Clethrionomys gapperi* in Alberta, Canada, with unusually severe weather, have been cited as evidence of the importance of weather. But not all microtines respond similarly to a given weather pattern, and no clear evidence indicates that weather has a major influence on microtine cycles.

Food has been considered by some as a dominant factor influencing microtine populations. Lack (1954b) regarded overexploitation of food as a major factor triggering changes in microtine densities, and Pitelka (1958) suggested that reduced availability of food could cause poor reproductive success and population declines. Because microtines are known to prefer certain foods and generally do not feed randomly on the most abundant plants, food might be expected to be limiting. But a number of studies have shown that voles and lemmings seldom consume more than 5 per cent of the energy available from the vegetation. Schultz (1965, 1969) showed that neither the quantity nor quality of tundra vegetation limited microtine populations, and he observed greater vegetative productivity in places with microtines than in experimental plots without them. Pieper (1964) regarded the nutrient-cycling activities of microtines to be important in influencing their densities. He hypothesized that the nutrients in the large amounts of urine and feces deposited beneath the snow by high numbers of lemmings become available rapidly to plants during snowmelt in the spring. The plants respond with high nutrient levels, which, at least temporarily, support high densities of lemmings. Although microtines unquestionably influence nutrient levels in plants, the nutrient levels in the plants appear not to influence microtine densities.

Studies in separate areas on different species of *Microtus* by Hoffmann (1958) and Krebs and DeLong (1965) involved fertilization of the vegetation and supplementary feeding of the rodents. Neither study demonstrated that the amount of food or its quality markedly affected the population densities. Furthermore, Batzli and Pitelka (1970, 1971) found that although in one area the favorite food plant of *M. californicus* was ten times more abundant than in another area, the two populations of voles underwent similar declines at the same time. Judging from available evidence, then, neither food availability nor nutrient levels strongly influence microtine densities.

As discussed in the section on predation, population cycles of some microtines are thought to be influenced by predation. The population crash that terminates the brown lemming cycle in northern Alaska is influenced by the concentrated efforts of a number of predators (Pitelka, 1957a:88). Pearson (1966, 1971) has presented data suggesting that the amplitude and timing of the microtine cycle are determined by intense predation during and immediately after the population crash. Lidicker (1973) thought that the "quasi" two-year cycles of *Microtus californicus* on an island occurred because no mammalian predator was present to depress low populations and to retard their recovery. The impact of even a single avian predator (the pomarine jaeger, *Stercorarius pomarinus*) on high lemming populations in summer has been shown to be heavy (Maher, 1970), and intense winter predation by a weasel (*Mustela nivalis*) may be responsible for nearly wiping out lemmings and delaying the recovery of the population until the weasels themselves decline (Maher, 1967; Thompson, 1955:173). Fitzgerald (1972) estimated that, during one winter in a study area in the Sierra Nevadas, weasels killed 40 per cent of the declining population of *Microtus montanus*.

Although predators clearly affect populations, they seem not to be the primary cause of population declines, nor is even intense predation able to stop population increases. Krebs et al. (1969) removed one third of the adults every two weeks from a fenced population of two species of *Microtus* and found that the population continued to increase. Even the stabilization of a population may require severe cropping: a population of M. californicus in an outdoor pen was not stabilized until over half the animals were removed each month (Houlihan, 1963). The effects of predators in modifying the timing and amplitude of cycles may well be important, but predation seems not to be the primary factor controlling cycles.

Stress is known to have pronounced effects on the mental and physical health of man (witness the high incidence of mental illness and ulcers among people living for long periods under stressful conditions), and is now known to have pronounced physiological effects on some other mammals. Since such pioneer works as those of Selye and Collip (1936, 1955), Green and his co-workers (1938, 1939), and Christian (1950) on the physiological responses of mammals to stress and to high populations, much attention has been focused on this subject and on the relationships between endocrine reactions to stress and population cycles in mammals.

The "stress syndrome" theory relies basically on the considerable body of evidence amassed in recent years indicating that, as populations rise and interactions between individuals of populations become progressively more frequent, adaptive physiological responses tend to reduce the population levels. A consistent pattern seems to be followed. As populations grow, social pressures, such as intraspecific competition and social strife, mount. These increasing pressures provide progressively stronger stimuli to the central nervous system (hypothalamus), which (through neuroendocrine mechanisms) affect the function of the anterior pituitary. The anterior pituitary responds by reducing the output of both growth hormones and gonadotropins (hormones that stimulate the adrenal cortex). The alteration of the hormonal balance causes certain behavioral changes, such as unusually high activity and heightened intolerance between individuals.

Of greater importance is suppression of normal reproductive function in the adult, decreased intrauterine and postnatal sur-

vival of young, unusual metabolic patterns, and lowered resistance to disease. As a result, mortality increases and reproduction may virtually cease. Increased adrenocortical activity, as indicated by increased adrenal weight, was found to be associated with high populations of Norway rats (Christian and Davis, 1956; Christian, 1959a), voles (Christian and Davis, 1966; Christian, 1959b; Louch, 1958), short-tailed shrews (Christian, 1954), muskrats (Beer and Meyer, 1951), and sika deer (Christian et al., 1960).

Some reliable data on wild mammals, however, do not support the stress syndrome theory. Adrenal glands of Canadian lemmings were found not to enlarge in times of high density (Krebs, 1963), and Clough (1965) observed no difference in adrenal weights or resistance to stress in voles from high populations compared to those from low populations. Much controversy still surrounds the question of the importance of the stress syndrome in governing population cycles of mammals.

Changing behaviors during population cycles are well documented, and may strongly influence population densities. Paired experimental encounters between voles on neutral ground showed that the most aggressive animals were from peak populations (Krebs, 1970). Such populations live under conditions of small home ranges and large individuals; these conditions were found by Turner (1971) to be related to aggressiveness in male voles. Conley (1971) studied the behavior of Microtus longicaudus in New Mexico and found that both male and female voles of a peak population were more aggressive than were animals from a declining population. The tendency of voles to disperse, on the other hand, is reduced at peak densities, but is strong during the increase phase (Myers and Krebs, 1971). Losses from a population are primarily due to emigration during the increase phase, but in the peak and decline phases seem to result in mortality of sedentary animals. The evidence leaves no doubt about the reality of behavioral changes during population cycles—but what causes these changes?

Well before the studies cited in the above paragraph appeared in print, Chitty (1958, 1960) hypothesized that in microtines selection of individuals for genetically determined behavioral features changed in association with changes in density; thus, microtine behavior shifted during a cycle and influenced the cycle itself. The complexity of such a system, however, made Chitty's ideas unattractive to some, and the rather rapid phenotypic and genotypic changes suggested by the hypothesis were questioned.

But critically important support was forthcoming. Krebs (1964b) found differences in skull size in relation to body size to be associated with different densities in some lemmings, and Chitty and Chitty (1962) recorded animals of both high and low growth potentials in a declining population of voles. Shifts in gene frequencies during changes in population densities were first demonstrated by Semeonoff and Robertson (1968) for declining populations of *Microtus agrestis* in Scotland, and results of careful experimental work on *Microtus pennsylvanicus* and *Microtus ochrogaster* by Myers and Krebs (1971) showed genetic differences between dispersing voles and resident animals. During the increase phase females that had just reached sexual maturity—individuals with the highest reproduction potential—were dispersing. Electrophoretic analyses of polymorphic plasma proteins (shown by Gaines and Krebs, 1971, to be under the genetic control of a single locus with multiple alleles) demonstrated a genetic component to the dispersal: both male and female dispersers differed from residents not only behaviorally but genetically.

Hilborn (1975) has further shown that during the increase phase of the populations of four species of *Microtus*, families tend to disperse as units; significant differences in the probability of dispersal occurred between families. Lidicker (1973), who carried out careful long-term studies on island populations of *M. californicus*, regarded the greater opportunities for emigration in mainland situations as a factor

important in extending the mainland vole cycles to three years. Chitty's hypothesis of genetic-behavioral control of microtine cycles has thus gained respectability through the support of an impressive body of evidence.

How, then, do biologists view the matter of the regulation of population cycles in small mammals? There is no universal agreement about the mechanisms of regulation, but the weight of considerable evidence is behind the hypothesis that in microtines, at least, some self-regulatory mechanism has evolved. Krebs et al. (1973) have presented a modification of Chitty's hypothesis that takes into account recently accumulated evidence (Fig. 15–48). They stress the importance of dispersal in affecting genetic change. During the increase phase of the cycle animals intolerant of crowding disperse and establish refuge populations. The rising population comes to consist more and more of density-tolerant and aggressive individuals,

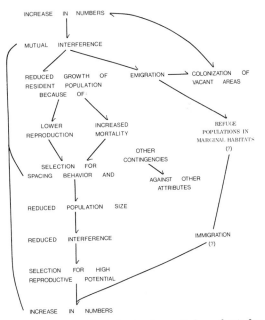

Figure 15–48. The behavioral-genetic hypothesis for the control of microtine cycles as presented by Krebs and Myers (1974). These authors explain that "Dispersal is viewed as being a more important aspect than originally proposed by Chitty. Central to the hypothesis is selection acting through behavioral interactions and changing the genetic composition of the population with fluctuating densities."

and these form the peak population. Extreme aggressiveness results in both reduced reproduction and poor survival, and the population declines. When the population reaches low densities, the density-intolerant and "docile" voles from refuge populations can immigrate, reproductive rates increase, and densities can increase.

The different behavioral patterns of the voles—density intolerance and the tendency to disperse *versus* aggressiveness and the tendency to be sedentary—are thought to be genetically controlled. Dispersal, therefore, generates a selective force that changes the genetic composition and behavior of the expanding population.

ZOOGEOGRAPHY

16

One of the most familiar kinds of biological information concerns zoogeography. Children learn that lions and zebras live in Africa and not in North America, and that kangaroos are typical only of Australia. This same type of knowledge of the presence or absence of various kinds of animals in different parts of the world is the substance of zoogeography, the study of animal distribution.

Considerations of zoogeography include two major approaches. The first is descriptive and static, and seeks to delineate the distributions of living species. Such information can be gained by field work and careful observation; it can be dealt with directly by using presently available evidence. The second approach is ecological or historical, and attempts to explain the observed distributions. Such inquiry often involves syntheses based on diverse lines of evidence. The ecologist, for example, may try to explain past or present distributions of animals on the basis of their environmental requirements. But scientists studying what Udvardy (1969:6, 7) calls "dynamic zoogeography" ask the most demanding question: How, when, and from where did animals reach the areas they now occupy? Virtually every fauna consists of animals that reached the area at different times, from different regions, and by different means. Our knowledge of the complex history of a fauna depends basically on the completeness of the worldwide fossil record and on our understanding of the geological history of the major land masses. Regrettably, however, our knowledge in these areas is incomplete, and even some of the major questions can only be answered tentatively.

Mammals occupy all continents, from far beyond the arctic circle in the north to the southernmost parts of the continents and large islands in the south. (Antarctica has no land mammals.) In the New World, the northernmost lands, the northern coasts of Greenland and of Ellesmere Island, are inhabited by the arctic hare (*Lepus arcticus*), collared lemming (*Dicrostonyx groenlandicus*), wolf (*Canis lupus*), arctic fox (*Alopex lagopus*), polar bear (*Ursus maritimus*), short-tailed weasel (*Mustela erminea*), caribou (*Rangifer tarandus*), and musk ox (*Ovibos moschatus*). A similar group of mammals, but lacking the musk ox, lives on the north coast of the Taymyr Peninsula (Soviet Union), the northernmost coast of Asia (Berg, 1950:19). The southernmost part of Africa has a rich mammalian fauna. Tasmania, the southernmost part of the Australian region, supports two monotremes, many marsupials, several native rodents, and several bats. On Tierra del Fuego, at the southern tip of South America, occur a bat, several rodents, a fox, otters, and a llama. The chiropteran family Vespertilionidae occurs almost everywhere there is land except in arctic areas, and the families Leporidae, Cricetidae, Sciuridae, Canidae, Mustelidae, and Felidae are native to all continents except Australia. All oceans, and seas connected to the oceans, are inhabited by cetaceans; odontocetes (toothed whales and porpoises) also live in some large rivers and lakes.

DISPERSAL AND FAUNAL INTERCHANGE

ANIMAL DISPERSAL. Dispersal occurs when an individual or a population moves

329

from its place of origin to a new area. The ability to disperse is as basic as the ability to reproduce, and is as necessary to the survival of a species. A spacing of members of a population so that each individual can satisfy its environmental needs is critical to all organisms. Territoriality is one familiar means by which this spacing is insured, and young of territorial species usually establish home ranges largely separate from those of other individuals, including their parents. The pressures exerted by reproduction and the necessity for the spacing of individuals create a tendency of populations to occupy ever-increasing areas, to colonize unoccupied localities, and to repopulate areas where they were previously extirpated. The more widespread a species, the less likely it is to be forced into extinction by local mortality, and as a result natural selection has usually favored those species that have broad distributions. A high adaptive premium is placed on dispersal ability. Udvardy (1969:12) has stated that "without evolved means of dispersal most animal populations would have succumbed, over a period of time, to the vicissitudes of the environment."

The ability of a population to expand into new areas depends on its innate dispersal ability (which is greater, for example, in fliers than in burrowers), on the breadth of environmental conditions that it can tolerate, and on the presence of barriers. Barriers may be ecological, with environmental conditions under which a species cannot survive, or more simply physical, such as bodies of water, precipitous cliffs or mountains, or rough lava formations. If enough information were available, much of the story of zoogeography could be told by considering the patterns of dispersal of animals as modified by the locations, effectiveness, and longevity of barriers.

MIGRATION AND FAUNAL INTERCHANGE. Certain regions have apparently been major centers of origin of mammalian groups. Many families first appear in the fossil record in the Eurasian area, and North America seems also to

have been the place of origin for many groups. The present mammalian faunas of regions such as Africa and South America were partly derived from migrations of mammals from northern continents. Despite uncertainty as to the place of origin of many mammalian groups (where a group first appears in the fossil record is generally taken as its place of origin), movements of mammals from place to place are in some cases well documented by the fossil record.

Simpson (1940) recognized several avenues of faunal interchange. The *corridor* is a pathway that offers relatively little resistance to mammalian migration and along which considerable faunal interchange would be expected to occur. Such a continuous corridor now exists across Eurasia; interchange of animals between Europe and Asia is highly probable and has apparently occurred frequently. A *filter route* has the effect of allowing passage of certain animals, but stopping others. Selective filtering has occurred at times along the land bridge that has periodically connected Siberia and Alaska. When this bridge was present in the Pleistocene, as an example, conditions were such that only animals adapted to cold climates could migrate between these two continents; mammals intolerant of cold conditions were denied use of this route. Mountain ranges, deserts, or tropical areas may also form filter barriers. The third and most restrictive route is the *sweepstakes route*. This is a pathway that will probably not be crossed by large numbers of any given type of animal, but is a route that an occasional individual may follow. Such a pathway is that between New Guinea and Australia or between Africa and Madagascar. Dispersal via a sweepstakes route must occur by swimming or flying, or by such uncertain means as rafting between one continent and another or between islands ("island hopping") on floating vegetation or debris. The probability that an animal will follow a sweepstakes route is extremely low if the route is long, as, for example, from North America to Hawaii, but is increased if an animal is small and can cling to floating material, is aquatic, or can fly. (The only land mammals that reached Hawaii without the help of man were bats). Despite the unlikelihood of a mammal's dispersal via a sweepstakes route, such dispersal has occurred, and has been the means by which some mammals originally reached Australia and South America.

MAMMALS OF THE ZOOGEOGRAPHIC REGIONS

The zoogeographical realms shown in Figure 16–1, which are the basis for the organization of the following section, were proposed by Wallace (1876) and have been widely used in discussions of zoogeography.

PALEARCTIC REGION. This region includes much of the northern part of the Old World, and is the largest of the zoogeographic regions. Included within this vast area are Europe, North Africa, Asia (except the Indian subcontinent and Southeast Asia), and the Near East (Fig. 16–1). The climate is largely temperate, but contrasting conditions exist, from the intense heat of North Africa to the arctic cold of northern Siberia. Broad areas of coniferous forests, comparable in many ways

to those of northern North America, are typical of much of the northern Palearctic Region, and deserts are widespread in the south. The Palearctic is separated from the Ethiopian Region by deserts, from the Oriental Region by the Himalayas, and from the Nearctic by the Bering Strait.

The Palearctic mammalian fauna is fairly rich, including some 40 families. Roughly 78 per cent of the Palearctic families also occur in the Ethiopian Region, and 70 per cent reach the Oriental Region (Table 16–1). The Palearctic shares 50 per cent of its mammalian families with the Nearctic. Many genera, and a few species, within the families Soricidae, Vespertilionidae, Cricetidae, Canidae, Ursidae, Mustelidae, Felidae, and Cervidae occur in both regions. Only two small families are restricted to the Palearctic (Spalacidae and Seleviniidae).

NEARCTIC REGION. This area includes nearly all of the New World north of the tropical sections of Mexico (Fig. 16–1). Habitats ranging from semitropical thorn forest to arctic tundra occur within the area. The mammalian fauna is correspondingly diverse, and includes some families that are mostly tropical in distribution (for example, Emballonuridae, Desmodontidae, and Tayassuidae) together

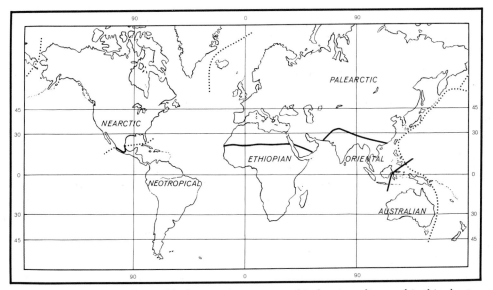

Figure 16–1. A map of the world, showing the zoogeographical regions discussed in this chapter.

Table 16–1. Approximate Percentages of the Families Occurring in Each Faunal Region Shared by Each Other Faunal Region

	Pal.	Near.	Neo.	Eth.	Or.	Aust.	Percentage of Endemic Families
Palearctic		50	38	78	70	40	5
Nearctic	75		78	38	49	31	3
Neotropical	33	52		25	30	20	43
Ethiopian	61	25	22		61	33	27
Oriental	70	38	35	78		48	12
Australian	52	32	28	55	60		39

with some primarily boreal families (Zapodidae, Castoridae, and Ursidae). Only one Nearctic family (Aplodontidae) is endemic. (An animal is endemic to an area if it lives nowhere else.) The mammalian fauna of the Nearctic resembles most closely that of the Neotropical.

NEOTROPICAL REGION. This region features great climatic and biotic diversity and includes all of the New World from tropical Mexico south. Much of the area is tropical or subtropical, and broad areas are covered with spectacular evergreen rain forest. Tropical savanna and grasslands occupy parts of the southern half of South America, and there are deserts in the south and along the west coast. The higher parts of the Andes support montane forests and alpine tundra. The South American part of the Neotropics was isolated from the rest of the world through most of the Cenozoic, but the Isthmus of Panama has provided a connection between South America and North America since the late Pliocene.

This region is second only to the Ethiopian Region in diversity of mammals. The Neotropical supports 46 families of mammals, and has the largest number of endemic families, 20. Especially characteristic of the Neotropical are marsupials, bats (including three endemic families), primates (two endemic families), edentates (two endemic families), and histricomorph rodents (eleven endemic or nearly endemic families). Two species of the genus *Lama* live in South America and are the only New World representatives of the family Camelidae. (Wild Old World camelids occur only in the Gobi Desert of Mongolia.) Tapirs are restricted to the Neotropical and Oriental Regions. The Neotropical mammalian fauna most strongly resembles that of the Nearctic, but it also shares over 33 per cent of its families with the Palearctic (Table 16–1).

ETHIOPIAN REGION. This region takes in Madagascar and Africa north to the Atlas Mountains, the Sahara Desert, and southern Arabia. The area is joined to the Palearctic by a land bridge (now broken by the Suez Canal) in northeastern Egypt. Deserts, tropical savannas, tropical forests, montane forests, and alpine tundra are all represented, and the most extensive tropical savannas in the world occur in Africa.

Next to the Neotropics, the Ethiopian region has the greatest number of endemic families of mammals. The impressive array of ungulates that inhabits the savannas of Africa is unmatched elsewhere, and Africa is the last stronghold of the families Equidae, Rhinocerotidae, Elephantidae, and Hippopotamidae. Although the only endemic artiodactylan family is the Giraffidae, nearly all of the African genera of antelope (Bovidae) are endemic. The primitive lemuroid primates of Madagascar and the diverse group of cercopithecid primates of Africa are especially typical of the region, and two of the four genera of great apes live in Africa. Apart from South America, Africa is the only area with a fairly diverse histricomorph rodent fauna. Viverrid carnivores reach their greatest diversity in the Ethiopian Region, where about 23 of the 25 genera are endemic. Over 60 per cent of the Ethiopian families of mammals also occur in the Oriental Region (Table 16–1).

ORIENTAL REGION. Included in this region are India, Indochina, southern China, Malaya, the Philippine Islands, and the islands of Indonesia east to a line (imaginary and controversial) between Borneo and Celebes and between Java and Lombok (Fig. 16–1). The area is dominated by tropical climates and once, before extensive clearing of lands by man, supported almost continuous tropical forests. Deserts occur in the Pakistan area. The Oriental Region is partly isolated from the Palearctic by deserts in the west and by the Himalaya Mountains to the north and northeast.

The mammalian fauna of the Oriental Region resembles most strongly that of the Ethiopian area, with which it shares 78 per cent of its families of mammals (Table 16–1). Many (70 per cent) of the Oriental families of mammals also occur in the Palearctic Region. The most distinctive elements of the Oriental mammalian fauna are all of tropical affinities. Four families of primates occur in this region. Four families of mammals—Tupaiidae (tree shrews), Cynocephalidae (flying lemurs), Tarsiidae (tarsiers), and Platacanthomyidae (spiny dormice)—are endemic, and each occupies forested tropical areas. Some 15 per cent of the Oriental families occur elsewhere only in the Ethiopian Region (Lorisidae, Pongidae, Manidae, Elephantidae, Rhinocerotidae, and Tragulidae). The presence in both areas of rhinos and elephants, great apes and lorises, and a diversity of viverrid carnivores makes the mammalian faunas of the Oriental and the Ethiopian regions seem much alike, but between these areas there are some striking faunal differences. Lacking in the Oriental Region are lemuroid primates, the distinctive African histricomorph rodents, and the diverse assemblage of antelope so typical of African savannas.

AUSTRALIAN REGION. This region includes Australia, Tasmania, New Guinea, Celebes, and many of the small islands of Indonesia (New Zealand and the Pacific area are not included). Within the area are islands of varying sizes and degrees of isolation. The island continent of Australia comes closer to New Guinea than to any other large island, but these land masses are separated by the Torres Strait, some 100 miles wide. The northern part of the area, including New Guinea and parts of the east coast of Australia, are covered with tropical forest, but much of Australia is tropical savanna or desert. Some of the most arid deserts in the world occur in the interior of Australia.

The Australian region is famous for its unusual mammalian fauna, and to the popular mind Australia itself is an area supporting marsupials almost exclusively. Actually, even Australia has nearly as many placental families (nine) as marsupial families (ten). Roughly 32 per cent of the families of the Australian Region are marsupials, and about 42 per cent (the monotremes and marsupials) are endemic. Some 61 per cent of Australia's families of mammals also occur in the Oriental Region.

OCEANIC REGION. The oceans of the world compose the Oceanic Region. Within this region live the whales and porpoises, most of the seals, the sea lions and walruses, and the inhabitants of isolated oceanic islands (usually bats and introduced murid rodents, and other mammals associated with man). The large islands are included within the regions with which their faunas have the most in common. Greenland and Iceland, for instance, are included in the Nearctic and Palearctic Regions, respectively.

HISTORICAL ZOOGEOGRAPHY OF MAMMALS

MAMMALIAN EVOLUTION AND FAUNAL SUCCESSION. Since the beginning of their spectacular adaptive radiation in the late Cretaceous and Paleocene, mammals have followed an evolutionary pattern typical of most plant and animal groups. In general, mammals have become progressively more diverse and better able to completely and efficiently exploit the niches available to animals with the basic mammalian structural plan. Along with this trend toward full occupation of the environment has gone a tendency toward increasing specialization. Whereas many Paleocene mammals were generalists and could probably rather inefficiently utilize a wide variety of foods, relatively few living mammals have this mode of life. A modern herbivore, for example, typically eats only a certain type of plant material, but its teeth and digestive system are adapted to efficient utilization of the food. Behavioral adaptations favoring selective foraging have also developed. The "average" modern mammal, compared to its Paleocene or Eocene counterpart, is able to find its food with less expenditure of energy, and to derive more energy from the breakdown of the food.

As mentioned in Chapter 15, biotic communities evolve just as do the interacting organisms that compose them. Whenever two or more organisms have attempted to play the same role in a community, to occupy the same ecological niche, the unstable situation that developed resulted in faunal change. One organism became master of the niche, and the others either moved away, evolved the ability to occupy a different niche, or became extinct; natural selection favored those structural or behavioral modifications that allowed an animal to be the stronger competitor, to most efficiently occupy its niche. Thus, a succession of mammalian faunas have occupied the earth, each better able to efficiently exploit the environmental possibilities of the times than were former faunas (Table 16–2). A number of factors, such as weather patterns and vegetation,

have importantly "guided" the evolution of mammalian faunas. In addition, a factor of major importance has been the timing of migrations.

FAUNAL STRATIFICATION AND FAUNAL ORIGINS. All large land masses support stratified faunas: not all of the animals have occupied the areas for the same length of time, nor have they all come from the same place. As an example, the mammalian fauna of Africa consists of animals representing families or genera that evolved elsewhere and dispersed to Africa, together with representatives of groups that apparently evolved on the African continent and have occupied the area throughout much of the Cenozoic. Horses had their origin in North America in the Paleocene (Jepsen and Woodburne, 1969) and probably did not reach Africa until the Pleistocene. Today horses (zebras) still form a conspicuous part of the African scene, but are completely absent from the New World. Old World monkeys (Cercopithecidae), on the other hand, probably originated in North Africa, and have seemingly occupied Africa continuously since the Oligocene. The Old World monkeys clearly represent part of an early African faunal stratum, whereas horses are part of a late one.

CONTINENTAL DRIFT, MAMMALIAN EVOLUTION, AND DISPERSAL

CONTINENTAL DRIFT. Teaching in geology and paleontology in North America has been dominated for many years by the view that the positions of the

Table 16–2. Ecological Replacement of Older Genera by Younger Genera as Shown by Pleistocene Fossil Record

Older Genera	Younger Genera
Nannippus (3-toed horse) — replaced by	*Equus* (modern horse)
Stegomastodon (mastodon) — replaced by	*Mammuthus*, mammoth (elephant)
Capromeryx (pronghorn) — replaced by	*Antilocapra* (pronghorn)
Hypolagus — replaced by	{ *Lepus* (hare)
	{ *Sylvilagus* (rabbit)
Pliophenacomys — replaced by	*Microtus* (meadow vole)
Arctodus — replaced by	*Ursus* (brown and grizzly bears)

(After Hibbard et al., 1965.)

continents and the intervening oceans are fixed, that they have remained immutable back through the vast sweep of geological time. Because these tenets were accepted by most North American paleontologists, they were forced to rely on often tenuous intercontinental land bridges or sweepstakes dispersal to account for intercontinental movements of terrestrial animals. But within the last decade our geological, paleontological, and biogeographical perspective has been drastically transformed by convincing evidence in favor of the theory of continental drift. As put by Gould (1973), acceptance of the theory of continental drift "has created the greatest revolution in man's thinking about his earth since the Copernican Revolution."

Since 1756, when the German minister Theodor Lilienthal noted that "facing coasts of many countries, though separated by the sea, have a congruent shape . . ." (Calder, 1972:42), the possibility that continents have not been immovable has repeatedly been considered. The proposition that continents have drifted over the earth was formulated in detail by the German Alfred Wegener in 1912, but our modern views of continental drift are based more on the careful work of the African geologist A. L. DuToit (1957). These men and others observed that the shapes of the eastern coastlines of the New World could be fit in jigsaw puzzle fashion against the western coastlines of the Old World, and took this evidence to indicate that the continents of the two hemispheres had originally formed one great land mass that had split apart. Recent computer analysis has shown the fit between the continental coastlines to be amazingly precise, and virtually identical rocks and fossils on presently widely separated continents have provided further evidence that these continents have not always been apart.

The discovery by the American paleontologist Edwin H. Colbert of the ancient reptile *Lystrosaurus* in Antarctica, on a "summer" day (in December) in 1969, put the capstone on the pyramid of evidence supporting continental drift. This Triassic, nonaquatic reptile had previously been found on other southern continents; its distribution could only be explained by assuming that the continents had once been connected. Wegener's wild-eyed theory was vindicated—the weight of the evidence became too much for even the staunchest "antidrifters" to bear. (References on continental drift appear in the bibliography under Colbert, 1973; Dietz and Holden, 1970; and Menard, 1969.)

One present view of the history of the major land masses, based on an acceptance of continental drift, is that some 200 million years ago there was a single great land mass, Pangaea (Wegener's term). (An earlier cycle of continental drift, during the Paleozoic, is less well understood than the Mesozoic-Cenozoic cycle under consideration.) This supercontinent was divided by a series of rifts that by the end of the Triassic (some 180 million years ago) had split Pangaea into a northern land mass, Laurasia, and a southern series of land masses, collectively called Gondwana (Fig. 16–2). By the end of the Cretaceous (roughly 65 million years ago). South America had moved westward, well away from Africa, which was separated from Laurasia by a narrow sea (Fig. 16–3). The east coast of North America and the west coast of Europe were presumably still in contact, but further drifting of the continents throughout the Cenozoic led to the existing arrangements of these land masses.

The force behind the slow, inexorable movement of continents is the spreading of the sea floor. The lithosphere of the earth—its crust, roughly 70 km thick—is divided into a series of huge plates (Fig. 16–4). These largely rigid plates are constantly but slowly changed by the addition at one border of upwelling molten rock and the destruction at the other border where the plates plunge down into the earth and again become part of the deep, molten core of the earth (Fig. 16–5). This vast system of crustal movement is called plate tectonics, and is a major focal point of modern geological research.

Continents float passively on the lithosphere and are carried along with the movements of the plates. Geological and paleon-

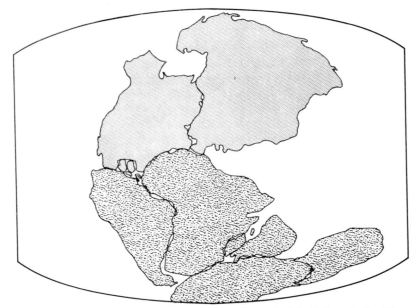

Figure 16-2. A hypothetical reconstruction of Pangaea near the end of the Paleozoic Era. The northern land masses (stippled) represent Laurasia, and the southern lands, Gondwanaland. (From Colbert, 1973.)

tological evidence not only documents the divisions of the land masses and their movements away from each other, but also suggests that collisions have occurred between continents carried on different, adjacent plates. Thus India was carried northward on the Indian Plate and collided in the Cenozoic with the Eurasian Plate. The concept of continental drift and plate tectonics has given great unity to the geologists' views of the creation of major land forms: the major mountain chains of the world are formed in part by deformation of the earth's crust at the leading edges of moving plates; earthquake zones are concentrated along these lines of tension; and extensive ocean trenches occur where the basaltic crust and the lithosphere of a plate plunge deep into the earth (Fig. 16-5).

The revolution of geological thought is obvious; but to biologists also the shift from the static-continent to the dynamic-plate concept is of tremendous importance: just as continents have separated, collided, or drifted progressively further apart, so terrestrial biotas have been isolated or brought together, entire patterns of distribution of marine biotas have been profoundly altered, and global ecological

diversity has shifted. To the evolutionary biologist, the movements of the earth's crust provide "the stage for all biological activity." In the publication from which this quotation was taken, McKenna (1972) reviewed relationships between organic evolution and plate tectonics. Several of his points form an appropriate background for discussions later in this chapter.

1. The effects of rising oceanic ridges. At times of rapid rise of an oceanic ridge, when there is maximum sea floor spreading, more water is displaced than during "quieter" times. This results in a global rise in sea level and simultaneous flooding of low-lying continental areas throughout the world. The flooded regions of shallow water greatly increase the space available to marine organisms, but reduce land area and may isolate sections of a continent and alter global climatic patterns. The evolutionary consequences of these drastic environmental changes seem to be heightened rates of extinction and speciation. If the sea floor spreading is episodic, these pulses may well be associated with intervals of rapid organic evolution. The extensive inland seas of the Cretaceous that divided eastern from western United

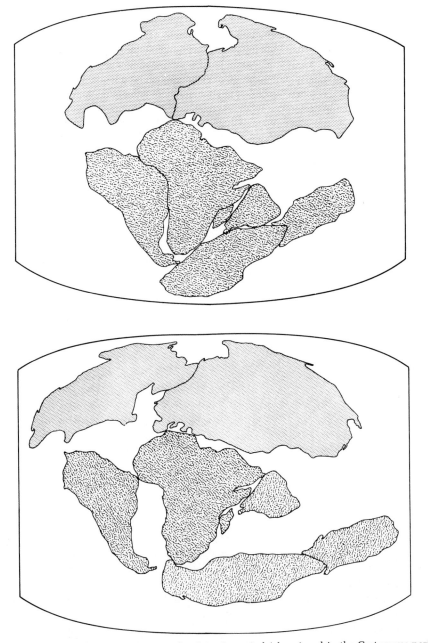

Figure 16-3.   Pangaea as it may have been in the Triassic period (above) and in the Cretaceous period (below). (From Colbert, 1973.)

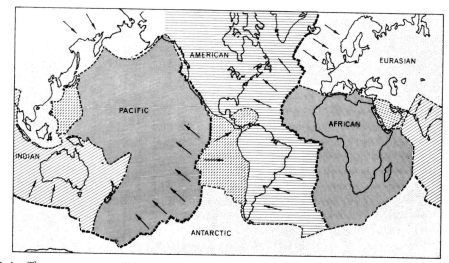

Figure 16-4. The present extent of the major tectonic plates and the directions of their movement. Movements of plates are influenced by upwelling of molten rock from deep within the earth along rift lines between two tectonic plates (see Fig. 16-5). (From Colbert, 1973.)

States and eastern from western Eurasia might well have been the result of an episode of maximum sea floor spreading, and these divisions may well have influenced the early evolution of eutherian and marsupial mammals (see page 35).

2. The effect of the separation of continents. A division of a land mass is accompanied by a great increase in the extent of shoreline; new continental shelves and slopes develop, and new circulations of

marine waters follow the newly developed waterway. As the land masses move apart into new climatic zones, their biotas are faced with new selective pressures. An increase in the total diversity of species is the result (Valentine and Moores, 1970). A case in point is Australia: the evolution of this island continent's biota has been strongly influenced by the northward movement of Australia away from Antarctica, from which it separated in the early

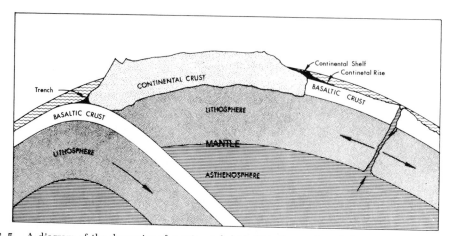

Figure 16-5. A diagram of the dynamics of continental drift. This figure, from Colbert (1973), is explained by him as follows: "Rifting between two tectonic plates (American and African) results in the welling up of molten rock from the depths, to form the midoceanic ridge (right). The westward drift of the American plate, carrying the continental block, caused a collision with the Pacific plate, this latter being forced down to be destroyed in the mantle. The zone of collisions is marked by the pushing up of the Andes and the formation of a deep trench, with resultant deep earthquakes."

Tertiary. As it moved northward toward Asia, Australia entered tropical and subtropical climatic zones favorable to a biotic diversity that could never have evolved if Australia had remained part of the vast Antarctic continent.

3. The effect of collisions of land masses. A long-isolated land mass tends to evolve a unique biota, and if this land approaches and finally contacts another mass the mixing of the biotas proceeds through various stages: during the approach fortuitous exchange of organisms occurs by rafting across the narrowing intervening waterway; when the lands actually touch, an avenue of dispersal that filters out organisms not adapted to shoreline conditions develops; finally, contact on a broad front provides a corridor for biotic interchange. Associated with these events is a destruction of the local marine environments and climatic alteration. Mountains formed in response to the collision of the land masses may form a partial barrier to faunal and floral mixing. But the final outcome is competition between ecological counterparts, widespread ecological replacement, the eventual evolution of a single biota, and a reduction in the diversity of world species.

4. The effects of "Noah's Arks." Because large and small land masses are carried slowly by underlying plates, biotas can be transported from one land mass to another over long distances during long periods of time. A one-way corridor is developed when the inhabitants of the ark land in a new port: although the inhabitants of the ark can thus gain access to a new land where they may or may not survive, the inhabitants of the new land can occupy the ark but have no means of transportation back to the ark's port of departure. The welcoming committee for the ark may be a heterogeneous one: the inhabitants of the ark may be met upon landing not only by the native inhabitants but by descendants of adventuresome ark inhabitants that made an earlier landing via floating vegetation. In addition, the biota of the ark would be occupied by portside inhabitants considerably before the ark docked. In the case of Australia, for example, rodents invaded this ark long before it closely approached their Asian place of origin, and the competition the rodents have offered Australian marsupials has markedly reduced the evolutionary options open to this group.

It is obvious, then, that when one considers the biogeography of individual species or of entire biotas, one must take into account plate tectonics and continental drift. The beauty of these concepts is that they often provide explanations for biogeographic patterns that have long appeared inexplicable.

CONTINENTAL DRIFT AND MAMMALIAN DIVERSITY. As pointed out by Kurten (1969), the fact that mammals evolved during a span of the earth's history when continents were moving apart may be a key to mammalian diversity. The Cenozoic radiation of mammals has occupied a shorter span of time than that available to reptiles for their radiation during the age of reptiles (65 compared to 200 million years), but mammals have diversified to a much greater extent, as reflected by a greater number of orders (about 30 orders of mammals to 20 orders of reptiles). Kurten believes that the greater diversity of mammals is a result of continental drift. Mammals evolved on several land masses under conditions of isolation or semi-isolation, whereas reptiles developed before the continents had moved far apart and therefore developed under conditions allowing freer faunal interchange between evolving stocks. A striking feature of mammalian evolution has been duplication of functional, and to some extent, structural types in separate groups. Examples of such convergent evolution are abundant: the members of several orders specialize in eating ants and termites (Fig. 16-6); the orders Marsupialia, Rodentia, Lagomorpha, Artiodactyla, and Perissodactyla all contain herbivorous, cursorial mammals that pursue basically similar modes of life; and small, terrestrial, insect-eating mammals have developed within at least four orders (Insectivora, Marsupialia, Edentata, and Rodentia). The greatest du-

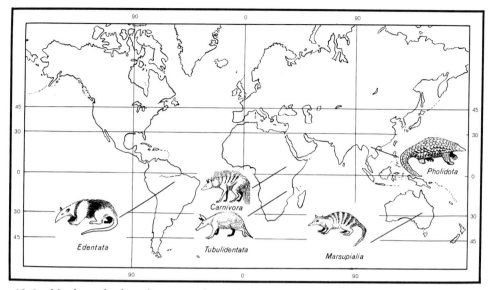

Figure 16–6. Members of at least five mammalian orders that occur in southern continents are adapted to eating ants and termites. Edentata—all members of the family Myrmecophagidae; Carnivora—the hyaenid *Proteles*; Tubulidentata—the aardvark *Orycteropus*; Marsupialia—the numbat *Myrmecobius*; Pholidota—the pangolin *Manis*.

plication has occurred in southern land masses, which have been longer and more completely isolated than have the Nearctic and Palearctic areas. Mammalian diversity, then, may be as much a result of the progressive Mesozoic and Cenozoic separation of the continents as of the structural and functional adaptability of the mammals themselves.

NEARCTIC AND PALEARCTIC MAMMALIAN FAUNAS. The high percentage (roughly 65 per cent) of Nearctic mammalian families that also occur in the Palearctic has long suggested Cenozoic avenues of dispersal between these regions. Although presently separated by a seaway across the now submerged Bering-Chukchi platform, Alaska and northeastern Asia are parts of the same tectonic plate. During much of the Cenozoic, these land masses were connected by a dry land bridge across what is now the Bering Strait. (This bridge has been termed Beringia.) Although this bridge has allowed repeated faunal interchange between Asia and the Nearctic, it was not the key to the faunal similarities between Europe and the Nearctic in the early Tertiary.

From the Cretaceous until roughly the end of the Eocene, a marine seaway, the Turgai Strait, stretched from the Mediterranean to the Arctic Ocean, thus separating Europe from Asia and interposing a barrier between the terrestrial faunas of these areas. Recent geological evidence based on studies of plate tectonics points toward persistence of connections between northern Europe and North America until the mid-Eocene, some 50 million years ago. These connections allowed dispersal between Europe and North America and account for early Tertiary faunal similarities between the areas (McKenna, 1975).

Since mid-Eocene, because of the drifting apart of Europe and North America, there has not been a continuous and direct North Atlantic land bridge. From the time of the severing of this dispersal route until the Turgai Strait was drained in the Oligocene, Europe was isolated not only from North America but from Asia. Whereas the early Tertiary was clearly a time when North America and Europe had broad faunal affinities, after the Eocene the only route of Palearctic-Nearctic biotic exchange was Beringia, and since this time

North America has had its closest affinities with Asia.

The degree of faunal interchange between the Nearctic and Palearctic fluctuated during the Cenozoic, with peaks in the early Eocene, early Oligocene, late Miocene, and Pleistocene (Fig. 16–7). No faunal mingling seems to have occurred in the Recent, by which time the North Atlantic dispersal route was long since inundated and interchange across Beringia was barred by the Bering Strait. Because Beringia is at a high latitude, even before the climatic cooling of the Pleistocene this dispersal route probably functioned as a filter barrier. The rodent families Heteromyidae and Geomyidae evolved in mid-Tertiary in response to the arid and semiarid conditions that developed in the southern part of the United States; these families and those with southern distributions, such as the Phyllostomatidae and Procyonidae, were apparently unable to reach or cross the Bering land bridge.

The derivation of the Nearctic mammalian fauna is complex. There are a number of families now established in North America, and widespread in Eurasia, that had their origins in Europe or Asia and reached North America via Beringia or, in the early Tertiary, across the North Atlantic route. The Talpidae and Soricidae, for example, first originated in Europe in the Eocene and appear in the North American fossil record in the Oligocene. Their route to North America must have been across the North Atlantic bridge, for these families do not appear in Asia until the Recent (Talpidae) or the Pliocene (Soricidae). The Felidae first appears in the Paleocene of Asia, and occurs in both Europe and America in the Eocene. In this case North America must have served as an immigration route to Europe, for at the time of appearance of felids in Asia the Turgai Strait separated this land mass from Europe. The Zapodidae appeared in the Oligocene in Europe, but did not reach North America until the Miocene; the Cervidae arose in Asia in the Oligocene and arrived in North America in the Miocene. These families clearly used the Beringian dispersal route. North America was also the source of some families that spread throughout the Palearctic: the Vespertilionidae and the Equidae appeared first in North America and emigrated to the Palearctic, the Vespertilionidae via the North Atlantic route and the Equidae probably by way of both this and the Beringian route.

The Nearctic mammalian fauna has been augmented to a very limited extent by species of Neotropical origin. Except for some bats, members of only two families that originated in South America were able to become established in the Nearctic. A porcupine (Erethizontidae) has become widely established in the Nearctic, and

Figure 16–7. The intensity of interchange of land mammals between Eurasia and North America in most of the Cenozoic, as indicated by the numbers of closely related mammals on the two continents. (Based upon Figure 28 from *The Geography of Evolution* by George Gaylord Simpson. © 1965 by the author. By permission of the publisher, Chilton Book Company, Philadelphia.)

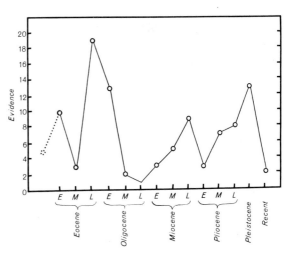

during the Pleistocene large relatives of the armadillo—ground sloths and glyptodonts—were far more widespread in North America than the armadillo is now.

THE ETHIOPIAN MAMMALIAN FAUNA. As previously mentioned in the discussion of the Palearctic Region, Ethiopian-Palearctic faunal interchange took place periodically in the Tertiary, but one-way transfers also seem to have occurred. The vertebrate fossil record indicates that, in the Eocene and again in the Oligocene, Africa received mammalian immigrants from Eurasia; however, there was no reciprocal movement of African mammals back to Eurasia. Such one-way dispersal was perhaps associated with the separation from Europe of land masses less than continent-sized (microplates) and their movement to Africa, thus "inoculating" Africa with northern mammals (Van Couvering and Van Couvering, 1975). Such Tertiary microplate movement in the western Mediterranean area has been described by a number of geologists. These microplates may well have functioned as classic "Noah's Arks."

The present African mammalian fauna has clearly had a complex derivation. Tertiary immigrants from the north have contributed importantly, and adaptive radiation of some of these groups within Africa gave rise to a number of endemic orders and families. In the late Tertiary, extensive faunal interchange with Eurasia reduced the distinctiveness of the African mammalian fauna. In the Pleistocene, however, the narrowing of the dispersal route between Africa and Eurasia and the development of extreme desert conditions throughout much of North Africa formed a strong filter barrier; the Ethiopian mammalian fauna again tended toward endemism. Africa, together with some parts of Asia, today serves as a refuge for a diverse Pleistocene world fauna that has disappeared elsewhere.

A high percentage of Palearctic families (78 per cent) also occurs in the Ethiopian Region, a surprising situation in view of the present narrow and inhospitable single link between Africa and Eurasia at the Isthmus of Suez. (This link is now broken by the Suez Canal.) At various times during the Cenozoic, however, broad dispersal routes existed between North Africa and Eurasia. In the Paleocene, Africa was linked to Asia via Arabia and to Europe via the area that is now Turkey and Bulgaria; at least temporary links were seemingly present in the Oligocene, Miocene, and Pliocene. Pleistocene and recent connections have been much restricted and have been via the Isthmus of Suez, the breadth of which changed with fluctuations in sea level. Seemingly, then, there have been periodic opportunities since the Cretaceous for faunal exchange between Eurasia and Africa when equable climates did not exert the strong filtering effect typical of the inhospitable Recent route through Egypt and the Sahara Desert.

NEOTROPICAL MAMMALS. The origins of the Neotropical mammalian fauna have long held the interest of scientists. The classic works of Simpson (1950, 1965, 1969) laid the groundwork for an understanding of Neotropical historical biogeography, and recent advances in our knowledge of plate tectonics and the dispersion of continents have refined previously held views. Controversy remains, particularly about the importance of the Isthmus of Panama in controlling mammalian dispersal between North and South America (see, for example, Hershkovitz, 1966, 1969, and Patterson and Pascual, 1968, 1972). Central to a discussion of the derivation of the Neotropical mammalian fauna is the history of the "Isthmian Link." Many paleontologists and biogeographers accept the view that South America was an island continent through all but the end of the Tertiary. The following points made by Savage (1974) summarize the probable history of the Isthmus of Panama: (1) A major seaway, some 400 to 100 km wide, separated North from South America from the Cretaceous through part of the Miocene; (2) The seaway was closed progressively from north to south during the Miocene; (3) Crustal uplift by early Pliocene formed a continuous land corridor between North and South America.

The present mammalian fauna of the Neotropics was seemingly derived primarily from four sources (Savage, 1974). The oldest, termed the *South American* (the terminology is that of Savage), includes marsupials, xenarthral edentates, condylarths, protoungulates and bats. Members of these groups reached South America by the early Tertiary; they radiated there, and in some cases their descendants became extinct without dispersing elsewhere. Because there is no compelling evidence in favor of a land connection between North and South America in the Cretaceous or early Tertiary, the means by which this early contingent arrived is in doubt; fortuitous overwater transport, perhaps by rafting, must have occurred. A second source (the *Young Southern*) is composed of primates, caviomorph rodents, manatees, bats and cricetid mice. These arrived in South America at various times in the Tertiary, seemingly by sweepstakes dispersal across the inter-American seaway. A third group (the *North Tropical* unit), including mammals that evolved in mid-Tertiary in middle America (the area that is now Mexico and the northern part of Central America), moved into South America when the

Isthmus of Panama allowed dry land dispersal in early Pliocene. The fourth unit (the *North American*), representing the most recent invaders of South America, is composed of a variety of North American mammals that entered via the Isthmus since early Pliocene. A diagrammatic illustration of the effectiveness of Middle America as a filter barrier is shown in Figure 16–8.

The Tertiary marsupial radiation in South America has been discussed previously (see page 50), but the extremely impressive ungulate radiation also deserves mention. The condylarth stock that reached South America very early in the Tertiary radiated rapidly, and by the end of the Paleocene a diverse series of evolutionary lines were established. In isolation from North American ungulate stocks, the South American ungulates went their unique evolutionary ways. Although they clearly filled many of the same niches occupied by other lineages of ungulates in other parts of the world, many of the South American ungulates were anomalous-looking beasts unlike any ungulates elsewhere. Some had unhandsome arched profiles with the snouts lengthened into

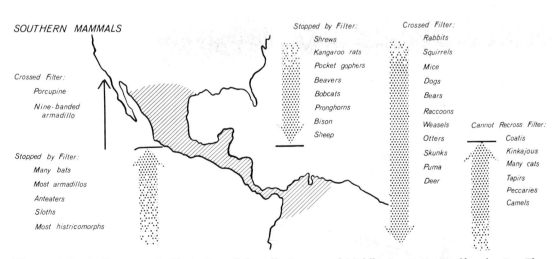

Figure 16–8. A diagrammatic illustration of the effectiveness of Middle America as a filter barrier. The crosshatched area is the filter zone, and the success or failure of animals in crossing the barrier is shown. (Based upon Figure 45 from *The Geography of Evolution* by George Gaylord Simpson. © 1965 by the author. By permission of the publisher, Chilton Book Company, Philadelphia.)

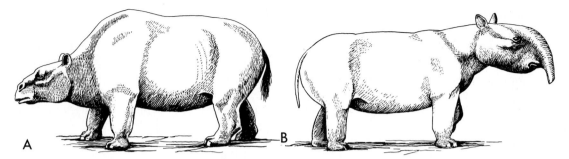

Figure 16–9. Restorations of two South American notoungulates. A, *Toxodon* (Toxodontidae), a Pleistocene form; B, *Pyrotherium* (Pyrotheriidae), an Oligocene form. (After Patterson and Pascual, 1972.)

proboscises, low-slung bodies, and short limbs (Fig. 16–9), and one can imagine that they moved ponderously.

These South American ungulates spanned a considerable size range. There were rat-sized little ones and giants approaching the size of an elephant. Especially successful was the order Notoungulata. This order included various herbivorous species, one of the largest of which was toxodon, a stubby-legged, rhinoceros-like beast some nine feet in length (Fig. 16–9A). Another important order, Litopterna, included a number of cursorial ungulates of smaller size. One advanced Miocene genus (*Thoatherium*) had one-toed feet that not only were much more specialized than those of the contemporary North American horses but were even

more specialized than the feet of present-day horses (Figs. 16–10, 16–11).

These distinctive South American ungulates reached their peak of diversity and numbers in the Oligocene and Miocene, but they declined in the Pliocene and by the end of the Pleistocene only fossils remained. Patterson and Pascual (1972) stressed that the decline of the South American ungulates was not due to the invasion of South America by Nearctic ungulates and carnivores. Major faunal shifts occurred before the emergence of the Isthmus of Panama. By this time over half of the mammals occupying the adaptive zone of the large herbivores were not ungulates: of the ten families of large Pliocene herbivores, four were edentates and two were giant caviomorph rodents.

Figure 16–10. A restoration of *Thoatherium* (Prototheriidae), a highly cursorial South American ungulate from the Miocene. (After Patterson and Pascual, 1972.)

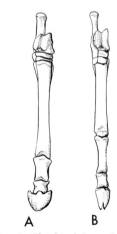

Figure 16–11. A, The hind foot of a modern horse (*Equus*) and B, that of a Miocene South American litoptern (*Thoatherium*). Note that the vestiges of digits two and four are more strongly reduced in the litoptern than in the horse.

The faunal interchange between North and South America across the emergent Isthmus in the Pliocene has long impressed scientists. In 1893, in his *Handbuch der Paleontologie*, von Zettel wrote that ". . . there was thus accomplished, toward the end of the Pliocene, one of the most remarkable migrations of faunas that geology has been able to record." At the time of the linking of North and South America by the emergence of the Isthmus, the mammalian faunas of these areas differed sharply. Taxonomic diversity, however, was similar. North America supported 8 orders and 26 families, whereas South America had 7 orders and 26 families. But the two faunas had different degrees of success in dispersing from their homelands. Thirteen northern families moved far into South America, beyond latitude 25°S, while only seven southern families moved northward beyond 25°N. The relatively small area of tropics north of the Isthmus and the semiarid and arid lands to the north of it limited the extent of the northward dispersal of some southern groups. The periodic glacial advances in the Pleistocene and the accompanying spread southward of boreal conditions may also have limited the access to North America by South American mammals. On the other hand, Nearctic mammals adapted to warm or tropical environments could spread widely into South America. Further, the more limited glaciation in South America and the dispersal route provided by the temperate flanks of the extensive north-south-trending Andean chain probably favored the southward movement of northern species (Patterson and Pascual, 1972).

When northern mammals entered South America in the Pliocene, there was apparently some duplication of functional roles between them and the native mammals. The intensity of the competition cannot be clearly read in the fossil record, but the abrupt disappearance of the saber-toothed marsupial carnivores (Thylacosmilidae) with the appearance in the fossil record of saber-toothed felids suggests competitive replacement in at least this case. But not all southerners suffered at the hands of the northerners. Some South American groups not only withstood the onslaught from the north but were able to move counter to its flow. Notably successful were the edentates, which migrated northward and occupied much of the Nearctic in the Pleistocene. Some of the large ground sloths outlasted the large Pleistocene ungulates and subungulates and did not become extinct in North America until some 11,000 years ago. The camels and tapirs are also of special interest. These animals have persisted in the Neotropical region, but have become extinct in the Nearctic, whence they originally came.

It is clear, then, that the South American mammalian fauna has a complex derivation. The unusually large number of endemic species is a reflection of the degree and duration of separation of South America from other continents through most of the Cenozoic.

HISTORY OF THE AUSTRALIAN MAMMALS. The native mammalian fauna of the Australian Region, and especially of Australia itself (Table 16–4), is famous for its uniqueness. Many of the functional roles played elsewhere by carnivores or ungulates are pursued by marsupials that are restricted either to Australia itself or to the Australian zoogeographical region.

For many years, the origin of the Australian marsupial fauna has been debated, but recent studies of continental drift and of the position of Australia in late Mesozoic and early Tertiary strongly suggest a southern route. Marsupials originated in the New World, and by early Tertiary were present in both North and South America. Australia was situated far south, near Antarctica, until as late as the Cretaceous or perhaps the Eocene. The discovery of Triassic reptiles on the Antarctic mainland demonstrates that this continent supported vertebrate life during this part of the Mesozoic, and indicates that Antarctica could have been the stepping stone used by marsupials to enter Australia from South America in the late Cretaceous or early Tertiary. This theory is supported by many biologists, and will gain virtually universal

Table 16–3. Earliest Fossil Records in North America of Eurasian Immigrants That Crossed the Bering Land Bridge

Genera	Earliest Records
Ursus, bear	middle Pleistocene (late Blancan)
Mammuthus, mammoth elephant	middle Pleistocene (late Kansan)
Bison, bison	middle Pleistocene (Illinoisan)
Smilodon, saber-toothed cat	middle Pleistocene (Irvingtonian)
Gulo, wolverine	middle Pleistocene (Irvingtonian)
Cervlaces, extinct moose	upper Pleistocene (Rancholabrean)
Rangifer, caribou	upper Pleistocene (Wisconsin)
Oreamnos, mountain goat	upper Pleistocene (Wisconsin)
Ovibos, musk ox	upper Pleistocene (Wisconsin)
Ovis, sheep	upper Pleistocene (Wisconsin)
Alces, moose	upper Pleistocene (Wisconsin)
Bos, yak	upper Pleistocene (Wisconsin)
Saiga, Asiatic antelope	upper Pleistocene (Wisconsin)
Bootherium, extinct bovid	upper Pleistocene (Wisconsin)
Symbos, woodland musk ox	upper Pleistocene (Wisconsin)

Note that all of the mammals are boreal types.
(After Hibbard et al., 1965.)

acceptance if future paleontological work in Antarctica yields marsupial fossils.

Although marsupials and eutherians both reached South America by early Tertiary, only marsupials entered Australia at this time. Why did eutherians fail to accompany the marsupials? Perhaps eutherians did not yet occur in the specific parts of South America or in the habitats from which marsupials dispersed to Australia, or perhaps marsupials had superior ability to cross water barriers. Possibly marsupials were more numerous than eutherians in the parts of South America closest to Australia, and by chance the mammalian stock that reached Australia was marsupial. In any case, Australia drifted northward, away from Antarctica, in early Tertiary. Australia thus provided a refuge where for a considerable time marsupials underwent an adaptive radiation free from competition with terrestrial eutherians, the first of which probably reached Australia in the Pliocene.

Monotremes, the other "original" group of Australian mammals, perhaps evolved on the island from advanced therapsid reptilian ancestry.

Nine of the twenty-one families of Australian mammals are placentals (Table 16–4). These groups arrived in Australia at various times, and all can easily be assigned to families that occur in other areas. Some have undergone little change since their arrival in Australia.

Bats are a group that have remained nearly unchanged. Apparently, bats entered Australia at various times in the Ter-

Table 16–4. Native Recent Mammals of Australia

	Families		Genera	
	NUMBER	% ENDEMIC	NUMBER	% ENDEMIC
Monotremata	2	100	2	100
Marsupialia	10	100	68	100
Chiroptera	7	0	21	29
Rodentia	1	0	13	77
Carnivora	1	0	1	0
Totals	21	57	105	82

tiary, and of the 21 Australian genera of bats, only 2 are restricted to Australia. Interchanges of bats between New Guinea and Australia, which were perhaps frequent because of bats' ability to fly, have kept the Australian bats from differentiating markedly from those of the Oriental Region.

Rodents are abundant and diverse (about 13 genera) in Australia, and some species have undergone great specialization, but all of these rodents are clearly assignable to the widespread family Muridae. Simpson (1961) has divided the Australian murids into four groups, according to their history on this island continent. (1) The "Rattus group" consists in part of species commensal with man that were introduced by European man, but also includes species endemic to Australia that perhaps developed from pre-Pleistocene immigrants. (2) The "old Papuan" group contains genera that evolved in Australia from murid ancestors that arrived probably no later than the Pliocene. (3) The "old Australians" are a fairly diverse group of, in some cases, highly specialized rodents (remarkable adaptations to dry climates by two members of this group are discussed on p. 447). Their ancestors were probably the first rodents to reach Australia, probably in the Pliocene. (4) The "hydromyine" group (of the murid subfamily Hydromyinae) consists of two semiaquatic genera. One apparently evolved in Australia from ancestors that came from New Guinea, and the other came recently from New Guinea.

The family Canidae is represented in Australia by the feral dingo (*Canis dingo*), which was probably brought to Australia by aborigines.

Perhaps the most remarkable feature of the Australian mammalian fauna is the presence of a marsupial assemblage that is fairly balanced, in the sense that it fills most of the terrestrial niches. Kangaroos and wallabies take the place of ungulates, dasyurids take the place of shrews and to some extent of rodents, phalangerids take the place of squirrels, and so on (see Chapter 5). Placental mammals that

reached Australia without man's help have in large part either filled adaptive zones that marsupials could not occupy, as in the case of the bats, or have occupied niches that they could perhaps fill more effectively than could marsupials, as in the case of the murid rodents. Such recently man-introduced placentals as dingos and, more recently, rabbits and foxes have had the unfortunate effect of displacing native marsupials or preying heavily on them.

THE UNUSUAL MAMMALIAN FAUNA OF MADAGASCAR. Islands long isolated from continents frequently have unusual mammalian faunas. These faunas may be dominated by a group equally important nowhere else, as in the case of the marsupials of Australia, or they may be extremely poor in mammals, as in the case of New Zealand where the only native mammals are bats. Madagascar is an interesting example of an area supporting a mammalian fauna with little ordinal diversity, many endemics, and a seemingly incomplete exploitation of habitats.

Madagascar is a large island some 995 miles in length, with a maximum width of 350 miles. It lies 260 miles east of the east coast of Africa. Madagascar separated from Africa by the start of the Cenozoic, and has been isolated from other land masses since that time. The island has supported six orders of mammals in Recent times (Table 16–5). Most of the mammals are endemic, and the most highly diversified groups, the lemuroid primates and the tenrecid insectivores, probably reached Madagascar from Africa by island hopping and rafting early in the Tertiary. Many of the Malagasy mammals occupy niches filled elsewhere by mammals of different taxa (Table 16–6). This phenomenon is called complementarity by Darlington (1957:23). The only artiodactyl that was present before man's arrival was the now extinct hippopotamus (*Hippopotamus lemelii*), and today the introduced river hog (*Potomochoerus*) is the only wild artiodactyl. The ungulate niche has largely gone unfilled, although a group of large Pleistocene lemurs, now extinct, may have been terrestrial herbivores.

THE ISLAND SYNDROME. Mammals

Table 16-5. Comparison of the Mammalian Fauna of the Panama Canal Zone and Madagascar

	Panama Canal Zone	Madagascar
Marsupialia	6	0
Insectivora	0	10
Chiroptera	40	12
Primates	5	10
Edentata	7	0
Lagomorpha	1	0
Rodentia	19	8
Carnivora	11	6
Perissodactyla	1	0
Artiodactyla*	3	1

*The artiodactyl from Madagascar, a hippopotamus, is now extinct. (After Eisenberg J. F., and E. Gould, 1970. *The Tenrecs: A Study in Mammalian Behavior and Evolution*, Smithsonian Institution Press.)

isolated on islands typically face different selective pressures than do members of parent mainland stocks. On islands, competition is usually reduced or may be absent, predators are often absent or few kinds are present, and the flora may be depauperate. In some cases, as on some small desert islands off the east coast of Baja California, one or two species of rodents are the only mammalian inhabitants, and just a fraction of the number of species of plants that occur on the mainland is present. Through time island mammals tend to diverge from parent mainland stocks, but the pattern of divergence is not consistent for all species. Island mammals typically differ in size from mainland relatives, and Foster (1964) pointed out that whereas some mammals become larger on islands, some become smaller. Island rodents and marsupials are generally larger; insectivores, lagomorphs, carnivores, and artiodactyls, however, are usually smaller. Examples are numerous. The *Peromyscus* inhabiting islands off the coast of British Columbia are unusually large, but the caribou that inhabited one of these islands (but is now extinct) was a dwarfed form. The gray fox (*Urocyon littoralis*) that lives on the Channel Islands off the coast of California is substantially smaller than the mainland *Urocyon cinereoargenteus*, and

the elephant that lived on these islands in the Pleistocene was a dwarf. However, evolutionary patterns on islands are not completely consistent. Counter to the usual trend, not all island insectivores are small. Unusually large insectivores (solenodonts; see page 84) evolved on some of the islands of the West Indies, and the largest insectivore of all time (*Deinogalerix koenigswaldi*) lived on the Mediterranean island of Gargano. This insectivore was larger than a fox, and probably fed on rodents (Freudenthal, 1972).

The remarkable dwarfed Pleistocene mammals of the Mediterranean islands have been discussed by Sondaar (1977). In the Pleistocene, elephants (*Elephas*) and deer (*Cervus*) lived on many of these islands, and some supported hippos (*Hippopotamus*). These mammals must have reached the islands by sweepstakes routes, for the extent to which they diverged morphologically from the mainland stocks and the fact that generally not all types occurred on an island suggest that access to the islands was across water. All of the three types listed above are known to be strong swimmers, and a single pregnant female could have founded a population on an island. Of special interest are the similar evolutionary trends exhibited by large mammals on a number of islands between which passage of terrestrial mammals would have been impossible. Elephants on the islands became strongly dwarfed compared to the parent mainland stock of *Elephas namadicus*. *Elephas falconeri* of Sicily, an example of extreme dwarfism, was roughly a meter high, some one quarter the size of its mainland progenitor and, compared to mainland elephants, *E. falconeri* had short distal segments of the limbs, cheek teeth with fewer enamel ridges, and the skull was relatively much lower (Fig. 16-12), with a reduction of the elaborate system of air sinuses. Short leggedness was especially pronounced in the island deer, but the pig-sized island hippos also became short-legged.

These patterns of parallel evolution probably resulted from similar selective

Table 16–6. Some Major Feeding Niches and the Mammals that Occupy them in Panama and Madagascar

	Anteaters		Primary Insectivore and Secondary Frugivore		Carnivore	
	ARBOREAL	TERRESTRIAL	ARBOREAL	TERRESTRIAL	ARBOREAL	TERRESTRIAL
PANAMA (major genera only)	Edentata: Myrmecophagidae *Cyclopes* *Tamandua*	Edentata: Myrmecophagidae *Myrmecophaga*	Marsupialia: Didelphidae *Marmosa* Primates: Callithricidae *Saguinus* Cebidae *Aotes*	Edentata: Dasypodidae *Cabassous* *Dasypus*	Carnivora: Mustelidae *Eira* Felidae *Felis*	Carnivora: Mustelidae *Mustela* *Galictus* Canidae *Urocyon*
MADAGASCAR			Insectivora: Tenrecidae *Echinops* Primates: Lemuridae *Microcebus* *Cheirogaleus* Daubentoniidae *Daubentonia*	Insectivora: Tenrecidae *Centetes* *Hemicentetes* *Oryzorictes* *Geogale*	Carnivora: Viverridae *Cryptoprocta* *Galidia*	Carnivora: Viverridae *Cryptoprocta* *Galidia* *Fossa* *Viverricula*

(After Eisenberg, J. F., and E. Gould, 1970. *The Tenrecs: A Study in Mammalian Behavior and Evolution*, Smithsonian Institution Press.)

Figure 16–12. Differences between the skull morphology of the widespread mainland species *Elephas namadicus* (A) and the insular species *E. falconeri* (B) from Crete. Approximate actual skull lengths are 768 mm for *E. namadicius* and 288 mm for *E. falconeri*. (After Maglio, 1973.)

pressures on the many isolated islands. No large predators were on the islands. Large size is an extremely effective adaptation to avoid predation, and without large predators great size was no longer of advantage. An unreliable food supply for herbivores may have favored smaller size and, in the absence of predators, overpopulation might have triggered periodic heavy mortality. Beds of deer bones found on the island of Crete are interpreted by some paleontologists as evidence of mass mortality, and abnormalities of the bones suggest starvation as the cause of death. In the dwarf elephant *E. falconeri*, the reduc-

tion of the crest of the skull and the reduced number of enamel ridges on the molars were related to the general dwarfing (Maglio, 1973:94). The marked shortening of the limbs of the deer is thought by Sondaar (1977) to have been due to two factors: the absence of predators and the consequent lack of need for speed, and the need for sturdy and well-braced limbs with which to negotiate mountainous terrain.

Just as mammals on islands are divergent structurally, some have changed behaviorally. Mammals that live on islands and have no mammalian predators often show little fear of man. As an example, Blake (1887:49) found gray foxes on Santa Cruz Island to be remarkably tame. They commonly approached to within a meter of a man, and regularly visited Blake's camp for scraps of food. These foxes even approached sleeping persons and pulled at their blankets.

CLIMATE AND MAMMALIAN DISTRIBUTION. Climate had and still has a pronounced effect on animal distribution. The present patterns of distribution of many mammals can be explained in terms of climatic changes in the Pleistocene and Recent.

The Pleistocene was a time of pronounced climatic shifts, when periods of lowered temperatures alternated with periods of relative warmth. Temperatures during the cool periods may have been 4° to 8°C below present temperatures, and temperatures in the intervening warm periods were probably higher than those now. Geological and paleontological evidence both indicate four major episodes of cool climates separated by three warm intervals. Accompanying the periods of cooling, which were apparently worldwide, were a number of spectacular environmental changes. Precipitation increased everywhere, and with increased snowfall continental glaciers developed and pushed southward. At one time in the Pleistocene over 25 per cent of the land surface was covered with glaciers: Eurasia had 3,200,000 square miles of ice; the Nearctic ice sheet covered 4,500,000 square miles, and during its greatest push southward reached what is now Kansas. Glaciers on Mount Kenya in Africa extended about 1700 m below the present vestigial snow fields (at some 4500 m), and New Guinea and Madagascar had montane glaciers. The distributions of floras were changed. Boreal vegetational zones were pushed downward on mountains, and coniferous forests spread southward over areas that previously supported less boreal floras. Concurrently, tropical floras receded toward the equator, and deserts became far more restricted than they are today.

The Pleistocene began 2 to 3 million years ago (Evernden et al., 1964) and ended about 10,000 years ago with the extinction in the Nearctic and Palearctic of such common Pleistocene mammals as elephants, camels, woodland musk ox, ground sloths, horses, and the giant beaver. Following the retreat of the last continental glacier at the close of the Pleistocene was a warm, moist climatic phase (perhaps 8000 to 6000 years before the present), followed by a warm, dry phase (about 6000 to 4000 years before the present), when temperatures were higher than those today and precipitation was lower.

Considerable fossil evidence documents the southward extensions of the ranges of boreal mammals during glacial advances. Remains of the musk ox (*Ovibos*), arctic shrew (*Sorex arcticus*), and collared lemming (*Dicrostonyx*) have been found well south of their present northern ranges (Fig. 16–13). Two voles that today have separate ranges that extend far north occurred together in the Pleistocene in the area that is now Virginia and Pennsylvania (Hibbard et al., 1965). Abundant evidence verifies the occurrence of northern assemblages of mammals during the Pleistocene as far south in the United States as Kansas and Oklahoma. There were reciprocal northward movements of subtropical or desert mammals during interglacial times, as indicated by the fossil occurrence of such animals as the hog-nosed skunk (*Conepatus*) and jaguar (*Panthera onca*) far north of their present ranges. A fossil record of the jaguar, for example, is from northern Nebraska, some 1300 km north of the animal's present range.

One of the most common and obvious patterns of mammalian distribution – the occurrence of isolated or semi-isolated populations of northern mammals on mountain ranges at fairly low latitudes – is the result of Pleistocene southward migrations of boreal faunas. During glacial advances, assemblages of boreal mammals were widespread in lowlands well south of their present ranges. Concurrent with the movements of these mammals northward during the retreat of cool climates were

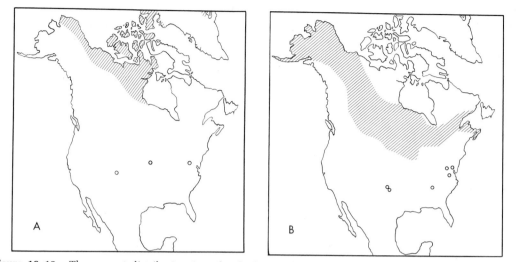

Figure 16–13. The present distribution (crosshatched) and the Pleistocene records (circles) of the musk ox *Ovibos* (A), and the arctic shrew *Sorex arcticus* (B). (After Hibbard et al., 1965.)

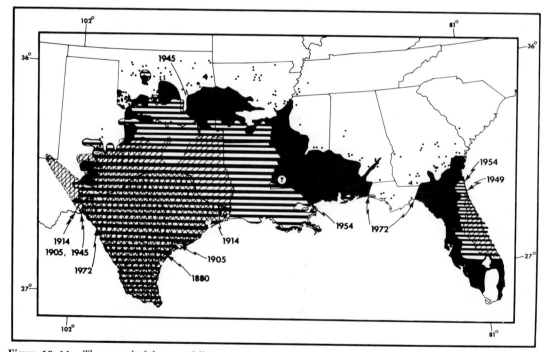

Figure 16–14. The spread of the armadillo in the United States in about the last 100 years. The dates and different patterns of hatching roughly indicate the spread at various times. Scattered single records during a 1972 survey are indicated by dots. (From Humphrey, 1974.)

movements of boreal mammals into montane areas. Here, because of the effect of elevation on climate, cool refuges were available. Many of these montane populations have persisted in "boreal islands" far south of the northern stronghold of their closest relatives, and the zonation of mammalian distributions on some mountain ranges in the southwestern United States has perhaps resulted from Pleistocene faunal movements (Findley, 1969). The White Mountains of east central Arizona constitute a "boreal island," where several species having northern affinities are isolated from other populations of their species. Populations of least chipmunks (*Eutamias minimus*) and redbacked mice (*Clethrionomys gapperi*) are isolated in this way, and the water shrews (*Sorex palustris*) in the White Mountains are separated by some 300 km of inhospitable (for this shrew) habitat from the nearest other water shrew populations, which occur in northern New Mexico.

Some relict populations, left behind after northward and eastward retreats of boreal conditions, also occur in some moist lowland localities. One such isolated population of the southern bog lemming (*Synaptomys cooperi*) occurs in a restricted marshy area in western Kansas (Hibbard and Rinker, 1942); another population, that occupies an area of moist habitat only 100 m wide and about 1.5 km long, is at a fish hatchery in extreme southwestern Nebraska (Jones, 1964:220,221).

Man has obviously reduced the ranges of many mammals, but some species are extending their ranges northward today, perhaps in response to the present warm climatic cycle. The armadillo has extended its range from southern Texas into much of central and southern United States in recent years (Fig. 16–14). The cotton rat (*Sigmodon hispidus*) and opossum (*Didelphis virginiana*) are also moving northward (Hall, 1958). Many ranchers have told me that the raccoon (*Procyon lotor*) has extended its range within the last 50 years from the river valleys in the plains of eastern Colorado deep into the foothills of the Rocky Mountains, where it is common today.

17   BEHAVIOR

The behavior of any animal is of great interest because it is, so to speak, "the proof of the pudding." In the case of the pronghorn, for example, great running speed became part of a unified functional system only because of a complex of behaviors that evolved in association with this ability. The formation of herds and systems of social behavior, the preference for open situations, the "flashing" of the white rump patch as a danger signal to other antelope, and the remarkable ability to detect enemies at a distance all allow the pronghorn to utilize its great speed effectively to escape from predators. How an animal uses its morphological and physiological equipment is of vital adaptive importance, and forms the substance of behavior.

The behavior of mammals is of particular interest because of its flexibility and variability. Compared to other vertebrates, mammals learn extremely rapidly and can modify behavior on the basis of past experience. This ability, superimposed on a rich array of innate (instinctive and unlearned) responses or behaviors, makes for complex behavioral patterns that often differ widely between species. Remarkably well-developed sense organs, coupled with a brain capable of rapid evaluation of complex sensory information, have enlarged the perceptual sphere of mammals and have facilitated the evolution of communication and rich social behavior.

THE ETHOLOGICAL APPROACH

This chapter will deal largely with ethology, the study of behavior in relation to structure and mode of life, or, as put by Tinbergen (1963), "the biological study of behavior." One might suppose that behavior could be more readily observed and analyzed than could other aspects of biology, and that detailed behavioral information on many species would have been assembled relatively early. But this is not the case; indeed, little is known of the behavior of many animals that are well known morphologically. A number of recent field studies under natural conditions (such as those of Geist, 1971; Kruuk, 1972; McCullough, 1969; and King, 1955) have provided a foundation of information on the behavior of some mammals, but a substantial frame of knowledge is yet to be built on this foundation. The stage of synthesis and the formulation of comprehensive theories have already begun, however, as in the fine theoretical treatment of social evolution in ungulates by Geist (1974), and by Jarman (1974).

Although mammals are remarkable in their ability to learn and to profit from experience, built-in patterns of behavior form an important part of the behavioral repertoire of mammals. Such innate behavior is not individually variable within a species, but is unlearned and is a part of an individual's heritage that is shared with other members of his species. These built-in behaviors are best regarded as simple sequences of movement elicited by specific stimuli. The term *Erbkoordination*, coined by Konrad Lorenz (1950), seems especially appropriate, and refers to a simple but specific hereditary movement or pattern of coordination. Such behavior in canids or hyaenids seems to be the lowering of the head, rump, and tail—assuming the "submissive posture"—in response to the menacing jaws, high head, cocked ears, and high tail of a dominant individual. In the parlance of some ethologists, the visual impression presented by the dominant animal is the innate releasing mechanism (IRM) that triggers the innate submissive-posture response in the other individual. Such innate movements are inherited just as are structural features, and are favored by selection because of their adaptive importance.

Clearly, mammals are not completely unique behaviorally: they are set apart from other vertebrates by their superior ability to learn, remember, and innovate, but they resemble other vertebrates in their wide use of innate behaviors.

NONSOCIAL BEHAVIOR

A number of behaviors relate basically not to social interactions but to such vital activities as feeding and seeking or preparing shelter. Feeding behavior is highly variable and in some species is a nonsocial activity, and in some species the preparation of shelter involves remarkably complex nonsocial behavior.

FEEDING BEHAVIOR. Herbivores utilize food that is unable to escape, and therefore are spared some of the problems that face carnivores. But a herbivore is generally equipped to utilize efficiently only specific types of vegetation, and must often face seasonal shortages of food and cope with plant material that is difficult to digest. Specialized feeding behaviors, together with specializations of the dentition and digestive system, tend to maximize the return of energy from food relative to expenditures of energy in securing it.

Some of the most specialized foraging behaviors occur among rodents. Pocket gophers (Geomyidae), mole rats (Bathyergidae), and other fossorial rodents dig complex burrow systems, and part of their diet consists of underground parts of plants they encounter. Naked mole rats (*Heterocephalus glaber*) of East Africa

were found by Jarvis and Sale (1971) to return repeatedly to feed on large tubers exposed in the burrows; some of the tubers that had been heavily fed upon remained alive and were sprouting. Pocket gophers frequently feed on aboveground vegetation, which is gathered during brief forays from an open entrance to the usually closed burrow. The animal scurries a few cm from the burrow to a plant, clips it hurriedly, and retreats quickly into the burrow tail first, as if yanked back forcibly by a string tied to the tail. Seldom does the animal venture more than a few cm from the security of its burrow.

The closely related kangaroo rats and pocket mice (Heteromyidae), by contrast, travel far from their burrows and gather seeds from the soil by using the long claws of the small forefeet. These rodents collect seeds in the cheek pouches rapidly and rather indiscriminately, and dash back to the burrow with them. Typically, many loads are taken to the burrow in an evening. Later, in the safety of the burrow, the animals selectively eat the seeds that yield the most energy (Reichman, 1975). This style of foraging enables the animals to gather food rapidly, but to minimize aboveground exposure to predators. This delayed-eating behavior departs from the pattern of many rodents, which consume food as soon as it is found. Some kangaroo rats can apparently discriminate between dry seeds and those that are moist and would spoil in the burrow. Moist seeds are placed in shallow pits near the burrow, covered with soil, and allowed to dry before being transferred to underground caches (Shaw, 1934). In some desert areas the Merriam's kangaroo rat (*Dipodomys merriami*) seems to use the shallow pits dug when the animals prospect for seeds as "seed traps." In the summer in southern Arizona, where afternoon winds drift seeds into depressions in the soil, these kangaroo rats regularly visit the shallow pits and remove blown-in seeds. The persistent checking of these seed traps probably increases the foraging efficiency of the kangaroo rats in seasons when seeds are not being produced.

Hoarding behavior is highly developed in a number of mammals. Some rodents, some shrews, some carnivores, and the pika (a lagomorph) store food. Hoarding behaviors seemingly evolved under selective pressures imposed by seasonal food shortages or by seasonal difficulties in foraging. An example of the latter situation was reported by Kenagy (1973) for the little pocket mouse (Perognathus longimembris), a tiny heteromyid that cannot forage during the cold part of the year because of the extremely high energetic cost of maintaining homeothermy. This rodent stays in its burrow continuously for up to five months during the winter, alternating between feeding on seeds it has stored and periods of dormancy. Many sciurid, cricetid, and heteromyid rodents store food in underground or protected caches. "Scatter hoarding" is practiced by some squirrels, which bury food at scattered points within the home range.

The caching behavior of the North American red squirrels (Tamiasciurus hudsonicus) is particularly notable. These squirrels depend for food on seeds of fir and pine. Cones are cached in holes in large middens formed by the litter of cone fragments that accumulates beneath a squirrel's favorite feeding sites. The middens are frequently 20 to 30 feet in diameter and contain from 2 to 10 bushels of cones, and are in shady situations where the moisture retained within the midden aids in the preservation of the green cones (Finley, 1969). Small numbers of cones are commonly cached in logs or pools of water. The cones are harvested in late summer and fall, and are cut on occasion at the rate of 29 per minute (Shaw, 1936:348). The squirrels are such effective harvesters that one pine in northern California lost 93 per cent of its 926 cones to them (Schubert, 1953). Seeds from the cached cones are eaten during winter and, when snow is deep, access burrows through the snow into the midden are maintained.

The dextrous forefeet of rodents are often used in food handling. The bipedal squirrel-like feeding posture frees the hands, which grip and manipulate the food while it is gnawed by the incisors. This behavior is seemingly innate, for hand-reared rodents were found to handle food in an essentially adult manner when they were still too poorly coordinated to maintain their balance (Ewer, 1968:32). The small forefeet of kangaroo rats and pocket mice skillfully sift seeds from the soil and put them in the cheek pouches. Even some rodents with forelimbs highly specialized for digging retain considerable manual dexterity. The plains pocket gopher (Geomys bursarius), which has powerfully built forelimbs, with large forefeet equipped with long claws, holds food in typical rodent fashion when it is eating. This fossorial rodent has several specialized feeding behaviors that serve to avoid the ingestion of soil and that demand considerable dexterity. Dirt or water is carefully stripped from leaves by the claws, and unusually wet or sandy food is often held in both forepaws and rapidly shaken (Vaughan, 1966b). Probably the importance of the forelimbs of rodents for food handling has limited the extent to which the limbs have become specialized for digging or locomotion.

The efficient capture of food by predaceous mammals is only possible because of complex behavioral modifications. The capture of flying insects in darkness by bats involves incredibly specialized behaviors (Chapter 21). To all predators, however, the pursuit, capture, and killing of live prey, which in many cases have their own finely tuned behaviors adjusted to defense or to escaping predators, presents a considerable problem.

Except for the use of cooperative action, the canid style of predation is one of the least specialized. Most canids are quite cursorial, and are able to capture prey by virtue of speed or by group effort. A pack of wolves will often bring down prey much larger than a single animal could handle. Wolves develop by experience a remarkable ability to recognize vulnerable prey (Crisler, 1958:106; Murie, 1944:109, 166–174). Mech (1966:121, 124) observed that moose that showed signs of weakness by a lack of defense or by being easily overtaken were attacked, whereas wolves repeatedly abandoned the pursuit of a moose that put up an aggressive defense. Wolves occasionally chase prey for several miles, seemingly testing the animal's stamina. Mech (1966:126–138) found that wolves first slashed at the rump of a moose, perhaps because this is the least dangerous site to attack, and when injuries to this area had partially immobilized the moose the wolves tore at the throat and shoulders and often grasped the nose. Similar styles of killing by wolves are used on deer (Stenlund, 1955:31), on caribou (Banfield, 1954:47), and on elk (Cowan, 1947:159). Hyenas often bring down prey by attacking the hind limbs. Canids often kill small prey, such as rabbits, by grabbing the animal across its back and shaking it violently, and this method is used widely by other carnivores.

The cats employ more specialized hunting and killing techniques. Cats are not long distance runners, but usually depend on short rushes directed against surprised prey. The sudden rushes of lions seldom cover more than 50 to 100 yards, and leopards and smaller felids frequently only make several bounds to reach their prey. The cheetah, an exceptional felid, may chase an antelope several hundred yards at speeds up to 112 km per hr (70 miles per hour!).

In order to use the typical feline hunting technique effectively, a cat must get close to its prey. The stalking of prey by felids involves a series of beautifully coordinated behaviors described in detail by Leyhausen (1956). When prey is sighted, the cat crouches low to the ground, and approaches using the "slink-run" and taking advantage of every object offering concealment. As the distance to the prey is reduced the cat moves more slowly and cautiously. At the last available cover the cat stops and "ambushes." The body is held low and just before the attack the heels are raised from the ground and the weight is shifted forward, just as a sprinter readies himself in the last tense instant before the sound of the gun. The brief rush to the prey ends in a spring; the forefeet clutch the animal but the hind feet remain planted and steady the cat for the possible struggle. Some of the components of this total behavior pattern may be observed in kittens as they stalk an insect or a scrap of paper. The cat makes the kill not by belaboring the prey as do many canids, but either by a powerful bite at the base of the skull or the neck, which crushes the back of the skull or some of the cervical vertebrae and the spinal cord, or, in the case of large prey, by strangulation. The shortening of the felid jaws is a specialization that contributes to the power of the bite. The tiger may kill prey as large as buffalo by gripping the throat and waiting for strangulation (Schaller, 1967). The cheetah also uses this style of killing.

Most cats are solitary, and cooperative effort in killing prey is rare. An exception is the African lion, the only truly social felid, which often hunts in groups in which there is some cooperation between individuals, with adult females doing most of the killing. The lion typically deals with large prey, often as heavy or heavier than the predator itself. Schaller (1972), who observed many kills, found that when attacking prey the size of zebra or wildebeest the lion attempts to bring the prey to the ground by clutching the rump, hind legs, or shoulders with the forepaws and throwing the prey off balance. When the prey falls, the lion grabs with its jaws for the neck or nose and maintains a grip until the prey is suffocated. On occasion a lion puts its mouth over the muzzle of the prey, surely a specific and effective means of strangulation. Schaller points out that by centering the bite on the neck or nose the lion immobilizes the horns, remains clear of thrashing hooves, and can easily keep the victim on the ground. This specialized killing behavior enables a single animal to kill large prey, and, of great importance, reduces the risk of serious injury from the powerful prey.

A wide variety of both placental and marsupial carnivores has been observed to use the neck bite to kill prey. This killing behavior was studied in the house cat by

Leyhausen (1956), who presented the predators with normal and headless rats, and with rats with the head fastened to the tail end. The cats aimed their bite at any constriction in the body; with normal prey this obviously results in the neck bite.

Unusual behavior patterns enable some carnivores to break the exoskeletons or shells of invertebrates and eggs. Some mongooses use the forefeet to throw objects against hard surfaces (Ewer, 1968: 48; Dücker, 1957). Clomerid millipedes have an unusually hard exoskeletal armor, roll into a sphere when disturbed, and are thus invulnerable to many predators. The banded mongoose (Mungos mungo), however, smashes a millipede by using both front feet to throw it between the hind legs aginst a rock (Eisner, 1967). The spotted skunk (Spilogale putorius) uses a technique for breaking eggs that involves kicking the egg with a hind leg and sending the egg against an object such as a rock (Van Gelder, 1953). The sea otter (Enhydra lutris) smashes the sturdy shells of mollusks by using a tool: the otter floats on its back with a flat stone on its chest and pounds the mollusk against the stone (Fisher, 1939). An individual was observed to pound mussels (Mytilus) on a stone a total of 2237 times during a feeding period lasting 86 minutes (Hall and Schaller, 1964:290). These otters are clearly selective in their choice of stones, and may use the same one repeatedly.

SHELTER-BUILDING BEHAVIOR. Many mammals have evolved elaborate shelter-building behaviors that aid them in maintaining homeostasis. The nests, burrows, or houses of mammals provide insulation that augments the animal's own pelage and reduces the rate of thermal conductance from the animal to the external environment or vice versa (Fig. 15–6; p. 273). Insulation from extremes of both heat and cold is important in reducing the energy expended in thermoregulation. The woodrat (Neotoma) collects a variety of materials with which it builds houses or improves the shelter provided by rock crevices or vegetation (Fig. 17–1). At the center of the woodrat house, or in a bur-

row beneath it, is a carefully constructed nest (Fig. 17–1). The nest may be globular or cup-shaped, and is formed of grasses and plant fibers. Many terrestrial rodents construct similar nests beneath logs or rocks or in burrows. Arboreal rodents frequently build nests in the branches of trees or within hollows or holes in trees. Some nest-building behaviors are perhaps common to many rodents, but the choice of nesting site seems to be species specific. For example, red tree mice (Phenacomys longicaudus) of the humid coast belt of Oregon and California build their nests only in Douglas firs (Pseudotsuga menziesii), the needles of which provide their primary food (Benson and Borell, 1931).

Fossorial rodents follow rather complex patterns of movement when digging. Probably many of the specific components of the total digging sequence are innate behaviors. Pocket gophers (Geomyidae) use the forefeet to loosen the soil by powerful downward sweeps, and the hindlimbs kick the accumulated soil backward from beneath the animal. Pocket gophers periodically eject soil from a burrow entrance by pushing it with the chin and forelimbs. Careful studies of the pocket gopher by Kennerly (1971) have shown that the long and complex series of behavior patterns associated with mound building are basically innate, but may be modified by learning. "Autoformulated releasers" probably play important roles in guiding mound building in rodents (and probably many other behavioral sequences in other mammals). Such a releaser is any alteration by an animal of its perceptual environment that acts to release the animal's subsequent behavior. Thus, the mound of earth itself, and changes in the mound due to the pocket gopher's activity, release subsequent behaviors associated with mound building. The animal characteristically alternates direction in pushing soil from the burrow; it pushes a series of 5 to 20 loads to the right, a similar series to the left, and so on. The frequency distribution of directions of pushing soil from the mouth of the burrow to the rim of the mound indicates that efforts are mainly in three directions:

Figure 17-1. Above, the house of a Stephen's woodrat (*Neotoma stephensi*) at the base of a juniper tree *(Juniperus sp.)*. Note the pile of sticks to the right of the tree and in the crotch between two trunks. Below, the nest of a Stephen's woodrat. This nest, about 180 mm in diameter and composed of grass and shredded juniper bark, was exposed when a woodrat house was dismantled. (Photographs by David M. Kuch.)

the pocket gopher usually pushes soil either directly in front of the mouth of the burrow or at an angle of 90° to either side. This results in the fan-shaped mound so typical of pocket gophers. That learning plays a part in burrowing and mound-building activity is suggested by the fact that young animals are less successful in plugging the openings of burrows than are older animals.

The burrowing and mound-building behaviors of some African mole rats (Bathyergidae) differ markedly from those of pocket gophers. *Heliophobius* for example, uses its incisors to excavate soil, and pushes the dislodged soil in back of its body with its feet. The animal transports a load of soil to the surface by backing up against it with the rump and large hind feet. The forefeet push the animal backwards, and the head and upper incisors are braced against the roof of the burrow to gain purchase. Unlike pocket gophers, which appear briefly at the surface each time they push a load of soil onto the mound, *Heliophobius* pushes a core of soil onto the mound without appearing on the surface.

SOCIAL BURROWING BEHAVIOR

Especially remarkable is the cooperative burrowing behavior of the naked mole rat. Jarvis and Sale (1971) described and photographed this "digging chain" (Fig. 17–2). The lead member of the chain chisels soil away with its protruding incisors and kicks soil back to the second member, which takes the load and backs toward the mouth of the burrow. When it reaches the mouth of the burrow it kicks the soil out of the burrow with a powerful sweep of the hind feet. The soil spews out as it would from a miniature volcano, and the mound takes the form of a small volcanic crater (Fig. 17–3). After kicking the soil out, this member of the chain straddles other members and returns to a position in back of the digger. The digger is replaced by another mole rat periodically. The new digger often uses its incisors to pull the animal it is relieving back from its position, frequently to the accompaniment of loud squeaking by the latter. On occasion the labor is divided differently between members of the chain, and each member in back of the digger passes soil back to the next. The well-coordinated, social digging behavior of the naked mole rats enables the animals to extend their burrow systems

Figure 17–3. The mound formed when naked mole rats kick soil from the mouth of their burrow. (From Jarvis and Sale, 1971.)

rapidly during the wet seasons when the soil is friable, and the spewing of the soil from the burrow is seemingly the only way the animals can bring the fine, loose soil to the surface. As might be expected, burrow systems of colonies of naked mole rats are extensive and extremely complex.

COMMUNICATION

Communication has often been broadly defined to include all interactions between animals that serve to transmit information between them; but, if all types of stimulus-reception sequences are regarded as communication, then essentially all behavior of one animal that can be perceived by another must be regarded as communication. For the purposes of discus-

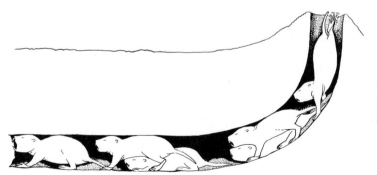

Figure 17–2. The digging chain of the naked mole rat (*Heterocephalus glaber*) of East Africa. See text for explanation. (After Jarvis and Sale, 1971.)

sion here, the definition of communication proposed by Otte (1974) will be used. A small segment of the multitude of stimuli received when an animal "views" its environment with its receptors is produced by other organisms and, through natural selection, has become modified to convey information to certain other animals. Only these stimulus-reception sequences involve communication.

Communication signals, then, are "behavioral, physiological, or morphological characteristics fashioned or maintained by natural selection because they convey information to other organisms" (Otte, 1974:385). "By far the greatest part of the whole system of communication seems to be devoted to the organization of social behavior. . . ." (Marler, 1965:584). This comment was applied to primates, but may be equally valid for all mammals. Each type of communication—visual, olfactory, auditory, and tactile—will be considered separately, but it should be stressed that usually a complex of several kinds of communication signals passes between animals.

VISUAL COMMUNICATION. Visual signals involving displays were perhaps derived in vertebrates from movements showing intention (to flee, for example), from displacement activities (seemingly inappropriate actions that typically occur when two opposing "desires," such as to escape or to attack, are in conflict), and from such autonomic responses as the erection of hair (Hinde, 1970:668). The evolution of displays is in the direction of reduced ambiguity. They have tended to become simplified, exaggerated, and stereotyped (repeated without variation). Highly developed facial musculature, the ability of the body and ears to assume a variety of postures, the control many mammals have over the local erection of hair, and large and conspicuous secondary sexual characteristics, such as horns, allow mammals a remarkable breadth of visual communication. Some visual signals are familiar: the dog wags its tail as a sign of friendship, and the cat arches its back, erects its fur, and raises its tail in a defen-

sive threat. Such displays can be observed readily and their functions or messages can often be recognized, but only a start toward an understanding of the broad area of visual communication in mammals has been made.

Facial expressions are of great importance in communication, and natural selection has often favored the development of distinctive facial markings that focus attention on the head. As described by Lorenz (1963), the facial expressions and postures of the ears of dogs signal degrees of aggressiveness or submissiveness. The posture of the head and the facial expression of many ungulates provide visual signals to other members of the herd or to territorial or sexual rivals (Fig. 17-4). An elk (Cervus elaphus) ready to run from danger elevates its nose and opens its mouth (McCullough, 1969); the Grant's gazelle (Gazella granti) holds its head high, elevates its nose, and pricks its ears forward when challenging another male (Estes, 1967). The heads of both of these animals are conspicuous: the elk's because the dark brown head and neck contrast strongly with the pale body, and the gazelle's because of bold black patterns. The intricate facial expressions of primates are frequently made more obvious by distinctive and species-specific patterns of pelage coloration and by brightly colored skin.

The body is used for signaling in many species. This type of signaling is particularly well developed in ungulates that inhabit open areas and that gain an advantage from coordinated herd action. The Grant's gazelle and the Thompson's gazelle (Gazella thompsonii) of Africa, which have two warning displays (Estes, 1967), twitch the flank skin (conspicuously marked in G. thompsonii) just as they begin to run from a predator that has entered the minimum flight distance (the minimum distance at which an approaching enemy causes the animals to run). The most effective display is a stiff-legged bounding gait, called "stotting," used as the gazelles begin to run. The conspicuousness of this display is enhanced by the erection and flaring of the hair of the white

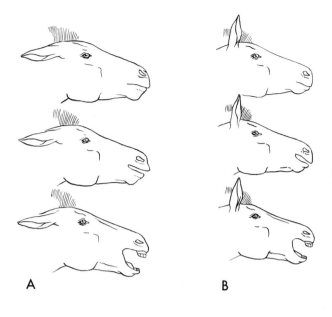

Figure 17–4. Facial expressions of horses: series A, three intensities of threat expressions; series B, three successive stages of greeting expressions. (After Trumler, 1959.)

A   B

rump patch. In some monkeys and apes the presentation of the hind quarters as if inviting copulation is a social gesture symbolic of friendship, and is accepted by a brief "token" mounting (Heinroth-Berger, 1959). Kangaroos threaten one another by standing bipedally at their maximum height. An understanding of this display helps to explain the extremely aggressive attitude of adult male kangaroos toward people. Because of his bipedal, erect stance, man is constantly assuming a posture interpreted by the kangaroo as a hostile threat.

The effectiveness of visual signaling is heightened in many species of mammals by weapon automimicry in the form of striking (and to our eyes handsome) markings (Guthrie and Petocz, 1970). The ears are commonly used signaling devices in mammals and, seemingly because of the proximity of the ears to the horns of artiodactyls which are universally used in signaling, the ears in many species are marked or adorned with hair in such a way as to mimic the horns. This probably strengthens the visual signal given by the horns, as well as making the posture of the ears extremely obvious. Several examples include the prong of the pronghorn horn (Fig. 14–3), which gives a hooked effect that is mimicked by the black and hooked ear tip; the horns of the roan antelope (Fig. 17–5), which are mimicked by the tufted ears; and the marks inside the ears of the klipspringer (Fig. 14–35A), which give the

Figure 17–5. The drooping ear tips of the roan antelope *(Hippotragus equinus)* are an example of automimicry. (Photograph taken in Kenya, East Africa, by Richard G. Bowker.)

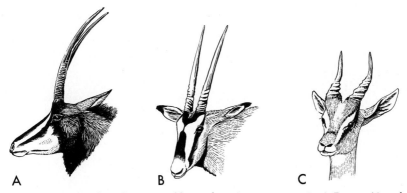

Figure 17–6.   Facial markings of antelope. A, Sable antelope (*Hippotragus niger*); B, oryx (*Oryx beisa*); C, young Grant's gazelle (*Gazella granti*).

impression of a halo of short horns. Facial markings may also play a role in automimicry by accentuating the horns. In the sable antelope, oryx, and Thompson's gazelle, black markings create a design that extends the contours of the horns (Fig. 17–6).

OLFACTORY COMMUNICATION. Eisenberg and Kleiman (1972:1) define olfactory communication as "the process whereby a chemical signal is generated by a presumptive sender and transmitted (generally through the air) to a presumptive receiver who by means of adequate receptors can identify, integrate, and respond (either behaviorally or physiologically) to the signal." The chemical signal serving to elicit a response in a conspecific receiver is known as a *pheromone*, whereas an *allomone* conveys a message to a receiver of a different species. Olfactory communication is effective because specific chemicals can convey very specific messages, and a scent mark on an object will persist and yield a message long after it is deposited. Because scent is released into the air and disperses rapidly, however, a receiver must have a sense of smell acute enough to find the source by detecting concentration gradients. Also, olfactory signals "broadcast" in the air are as available to a predator as to a conspecific. Pheromones of mammals have a variety of sources.

Urine and feces contain metabolic wastes that serve as chemical signals. Many kinds of mammals are highly spe-

cific in their choice of sites for urination and defecation, and in some species a stereotyped routine is associated with urination and defecation. The dik-dik (*Madoqua kirki*), a small, brush-dwelling, African bovid, deposits its feces in conspicuous piles at the borders of its territory, urinates on the piles, and makes scratch marks around the pile with its hoofs (Fig. 17–7). These obvious piles provide both olfactory and visual signals announcing territorial boundaries (Hendrichs and Hendrichs, 1971). A beautiful example of scent marking of territories by wolf packs was presented by Peters and Mech (1975), who found that scent marking with urine was concentrated along the borders of territories.

Urine and feces seemingly convey a considerable amount of information. Males of most species of mammals can recognize by the smell of her urine when a female is in estrus, and in many species copulation will not be attempted until this time. Male urine is also used in signaling. Mature bull elk during the rut (the breeding season) do not urinate normally; instead, the penis is extended and the animal squirts urine on the belly and the thick hair on the chest. The smell of the urine may have an important communication function (McCullough, 1969:82, 99, 110). By "self marking" with his metabolic wastes the bull is probably advertising his general physical condition. While he is in excellent condition his urine advertises this information

Figure 17-7. A male dik-dik *(Madoqua kirki)* carefully smelling its dung pile (left) and then marking the pile by scratching the soil around it. (Photograph taken in Kenya, East Africa, by Margaret H. Bowker.)

to rival bulls; when his condition declines, this is also conveyed to other bulls via the smell of the urine. This type of communication may avoid disruption of breeding activity and of the harem by delaying attempts at deposing the harem master until his exhaustion allows a fresh bull to replace him. This system assures that the breeding is carried out by a series of fresh, dominant bulls that are ready to service cows when they are in estrus.

Scent urination in goats *(Capra hircus)* was studied by Coblentz (1976), who suggested that in addition to advertising age dominance and physical condition to other males, the scent urination or self-marking behavior may have evolved because of its importance in male-female interactions. The behavior perhaps serves to hasten estrus and to synchronize it with the period of peak physical condition in the male. Thus scent urination would increase the fitness of the dominant male by enabling him to make the most, reproductively, of the brief time of his peak condition.

The signals provided by urine and feces may also be valuable to such refuging species as bats. To some bats, the urine and feces deposited in retreats may pro-

vide valuable olfactory signals that aid the animals in finding optimal roosts.

Also important as sources of pheromones are a variety of glands. Glands associated with the mouth, eyes, sex organs, anus, and skin are known to produce chemicals used in olfactory communication. Secretions from five locations on the body of the Australian honey glider *(Petaurus breviceps)* serve functions ranging from attracting newborn young, in the case of the pouch gland of the female, to contributing to a community odor within the social group, in the cases of the frontal and sternal glands (Schultze-Westrum, 1965). Müller-Schwarze (1971) described a number of pathways of social odors in mule deer *(Odocoileus hemionus)*; these are shown in Fig. 17-8.

Scent marking is used in a variety of behavioral situations, but appears frequently to be associated with an expression of dominance, as is illustrated by the behavior of a molossid bat observed by Langworthy and Horst (1971). Dominant males of a captive colony of *Molossus ater* often mark subordinate males with secretions from well-developed throat glands.

Reproductive behavior in some, and perhaps most, terrestrial mammals is strongly

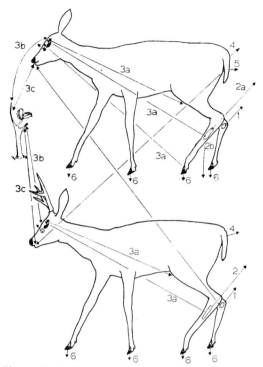

Figure 17–8. Sources of scents used in intraspecific communication and pathways of social odors in the mule deer *(Odocoileus hemionus).* The scents of the following are transmitted through the air: tarsal organ (1), metatarsal gland (2a), tail (4), and urine (5). When the animal lies down the metatarsal gland marks the ground (2b). The hind leg is rubbed against the forehead (3a), and the forehead is rubbed against twigs (3b). Marked objects are sniffed and licked (3c). The interdigital glands (6) deposit scent on the ground. (From Müller-Schwarze, 1971.)

influenced by the sense of smell. Experimental removal of the olfactory bulbs of laboratory rats impairs male reproductive behavior and, in male laboratory mice and hamsters *(Cricetus)* so treated, copulatory behavior was abolished. The vomeronasal organ has been regarded by many as playing an important sexual role in the male. After smelling the genital area and urine of a female, some mammals, notably perissodactyls, artiodactyls, and some carnivores, make a characteristic facial expression involving the upward curling of the upper lip and often the lifting of the head. This distinctive behavior, called *Flehmen* by Schneider (1930), has been thought by some to be important in activating the vomeronasal organ and in perceiving sex-

ual pheromones. This hypothesis has gained support from laboratory studies on the hamster by Powers and Winans (1975), who found that destruction of the afferent nerves of the vomeronasal organ produced a disruption of copulatory behavior in one third of the altered animals. This procedure, coupled with destruction of afferent nerves of the olfactory bulbs, completely eliminated copulatory behavior in all experimental animals. Powers and Winans suggest that inputs from both the vomeronasal organ and the olfactory bulbs are necessary for the arousal of sexual activity. Olfactory communication may also play an important role in the reproductive cycles of some primates. The smell of vaginal secretions of rhesus monkeys *(Macaca rhesus)* in estrus is sexually stimulating to males and promotes copulation (Michael et al., 1971).

Self anointing, an unusual signaling behavior shared by a number of species of hedgehogs (Erinaceidae), involves the smearing of saliva over the quills and results in a pungent smell easily detectable by man. This behavior has been studied in the European hedgehog *(Erinaceus europaeus)* by Brockie (1976). When removed from their nests, nestlings self anoint; seemingly this aids the mother in recovering them. Self anointing occurs in adults only during the breeding season, and is assumed to serve as a sexual signal.

Some kinds of mammals are known to be able to discriminate between individuals of their species entirely by scent, and this ability may be very widespread. Even in the primitive primate *Lemur fulvus*, this ability is well developed (Harrington, 1976). The mechanism for this recognition in the small Indian mongoose *Herpestes auropunctatus* has been studied by Gorman et al. (1974) and Gorman (1976). The anal pockets of this mongoose produce acids used in scent marking objects within the animals' home ranges. The glands contain a series of six volatile carboxylic acids derived from bacterial decomposition of sebaceous and apocrine secretions within the pocket. The six acids occur in different relative amounts in different individuals, and animals recognize each other by scent on the basis of the unique carboxylic acid profiles. Bacterial production of these acids has been demonstrated in a number of mammals. Sebum and, to a limited extent, apocrine secretions are substances deposited on the pelage of mammals. These substances seemingly serve as waterproofing agents, and when metabolized by skin bacteria produce odor-yielding carboxylic acids. Gorman (1976) suggests that selection has favored the concentration of sebaceous and apocrine glands into discrete organs, where bacteria could produce in relatively large amounts the carboxylic scents used in olfactory signaling.

Although the specific message conveyed from one animal to another by scent marking is difficult to establish, careful studies, using not only observational but also neurological and biochemical techniques, have yielded an approach to an understanding of mammalian scent-marking behavior. In her careful consideration of scent marking, Ralls (1971:449) indicated that it is used by mammals "in any situation where they are both intolerant of and dominant to other members of their species." Ewer (1973:243–252), in her discussions of scent marking by carnivores, makes a similar point, and the work on European rabbits (Oryctolagus cuniculus) by Mykytowycz (1968) and on Mongolian gerbils (Meriones unguiculatus) by Thiessen et al. (1971) raises several major points: (1) the maturation and use of scent glands is controlled by gonadal hormones produced at sexual maturity; (2) most scent marking is done by dominant males; (3) scent marking is associated with the possession of a territory. But scent is used under widely differing behavioral contexts, not only in association with territorial defense and certainly by both sexes.

We clearly lack many of the final answers, and more definitive statements must await further research. A reasonable summary statement, in my judgment, is that of Eisenberg and Kleiman (1972:24), who regard scent "as a means of exchanging information, orienting the movements of individuals, and integrating social and reproductive behavior."

AUDITORY COMMUNICATION. The sense of hearing in mammals is acute, and auditory communication is of great importance. Indeed, the sounds of some mammals that are rarely seen are commonly heard. Impressive choruses of howls of the coyote may be heard nightly in some parts of the western United States where the animals themselves are only occasionally seen. The importance of nearly constant auditory communication to a herd animal is difficult for one to imagine. Virtually continuous noises made by the members of a herd serve to integrate the group by keeping individuals apprised of the locations of one another. When he was very close to a herd of tule elk, McCullough (1969:71) observed that "there is a continuous array of sounds—foot bone creaking, stomach rumbling, teeth grinding, and others." The crunching of vegetation by McCullough was instantly distinguished by the elk as distinct from similar sounds made by their own feeding activities. In caribou (Rangifer tarandus), also, the creaking and snapping of foot bones can be heard for considerable distances and enable scattered members of a herd to remain in auditory contact (Kelsall, 1970).

Vocal communication is widely used by mammals. In man, of course, this type of communication reaches its most complicated development, but even in other primates some type of "language" can be recognized. The Japanese macaque (Macaca fuscata) has a repertoire of some 25 sound signals (Mizuhara, 1957). The more basic sounds used by the rhesus monkey may be linked by a series of intermediate sounds, and one basic sound may grade independently into other calls (Rowell, 1962). This yields a remarkably complex and rich vocal repertoire. The functional importance of some sounds can be recognized. The quiet "grunt," for example, is used by many primates to maintain contact with each other (Marler, 1965:568), and vocal sounds are used by many mammals to maintain or re-establish contact with one another. This seems to be the function of howling choruses in canids and the calls of young in a variety of species. The function of the complex vocal noises of cetaceans is as yet poorly understood, but is being intensively studied (see, for example, Dreher, 1966, and Reysenbach de Haan, 1966).

The vocal repertoire of some rodents consists basically of ultrasonic signals. Laboratory rats (*Rattus norvegicus*) are known to produce several such vocalizations: a 50-kHz call is associated with aggression, and a call of the same frequency is emitted during mounting; a 22-kHz signal is typical of a defeated and subordinate male. Given the appropriate behavioral setting, some calls apparently convey a rather specific message. A 22-kHz call is given after ejaculation by a male rat during its refractory period (this is the period during which the animal cannot initiate copulation spontaneously). The male's female partner ceases "courtship" behavior during this vocalization and usually stays away from the male. Similar calls are given by females resisting the mounting behavior of males. Barfield and Geyer (1972) suggest that this is a desist-contact signal given by an animal seeking social withdrawal. Parent-young communication is also based on ultrasonics. Parents respond to the ultrasonic distress vocalizations of helpless young by bringing them back to the nest, and it has been found that a decrease in the acoustical energy of the calls as the young grow older is associated with the development of homeothermy (and an attending decrease in the vulnerability to cold) by the young (Noirot, 1969). During mating, ultrasonic pulses made by rodents may function to reduce aggressive behavior, and some ultrasonic vocalizations may even serve as territorial announcements (Sewell, 1968).

Vocalizations may facilitate individual recognition in many kinds of mammals. Brown (1976) found that female pallid bats (*Antrozous pallidus*) and their young recognize each other on the basis of distinctive vocal signatures. The "who-oop" call of the spotted hyena (*Crocuta crocuta*) differs between individuals to the attentive human ear, and may well provide for individual recognition among the hyenas themselves.

Compared to what remains to be learned, we know little about vocal communication in mammals under natural conditions. Here, again, is a promising area for research.

TACTILE COMMUNICATION. Little is known about tactile communication in mammals, but its use is probably widespread. The sexual behavior of many mammals includes precopulatory activities by the male such as laying the chin on the rump, nuzzling the genital area, or touching various parts of the female's body. These behaviors are presumably sexually stimulating to the female or at least cause her to accept mounting by the male. Perhaps tactile stimuli are of greatest importance in connection with sexual activities in most mammals, but in primates they have assumed other roles.

Mutual grooming by primates, which clearly provides tactile stimuli, serves to cleanse a partner's fur in places inaccessible to self grooming. But mutual grooming may also have an important social function in promoting social contact and allowing familiarity between individuals. In anthropoid primates, embracing and touching of hands or of the genital areas are types of "friendly behavior," and formalized neck biting is a gesture of domination in some species.

Mutual grooming occurs also in collared peccaries (*Dicotyles tajacu*) and is described by Sowls (1974). Animals rub their heads against each other's flanks and the rump, which bears a much-used scent gland, in a ritual that seems to serve as a greeting ceremony, but may also have a scent-marking component. African dik-diks use the nose to touch various parts of another's body. Often the female rejoining her young or a male rejoining a female will perform this behavior. Seemingly this is primarily a tactile reassurance and a reassertion of familiarity, perhaps akin to mutual grooming (Bowker, 1977).

SOLITARY MAMMALS

Many mammals, including some marsupials, rodents, and insectivores and many carnivores, are essentially solitary and territorial. The home range of a pocket gopher (Geomyidae), for example, contains its burrow system, which, except in the breeding season, is occupied by a single animal and is apparently defended against interlopers. Although the burrow system may be extended and some sections may be plugged, most individuals probably occupy the same area throughout most of their lives. These animals may be colonial, but are never social. Concentrations of burrow systems often occur locally, but each burrow system is exclusively "owned." Similarly, in the case of two species of ground squirrel (Spermophilus mexicanus and S. armatus), the colony consists of a number of solitary individuals living in close proximity (Edwards, 1946; Balph and Stokes, 1963).

Leyhausen (1964) regards some carnivores, such as various kinds of felids, as territorial but not truly solitary. Individuals occupying neighboring territories probably meet on occasion and gain some familiarity with one another, and establish a social order featuring mutual respect for one another's territorial rights. Bailey (1974) found that in Idaho female bobcats occupied virtually exclusive, but relatively small ranges, whereas those of males were larger and overlapped ranges of males and females. Few interactions between adults were observed, and signaling by scent marking was the primary means by which an animal advertised its presence. For many solitary species, little is known of the brief periods of social life, but an awareness of the identities and positions of their nearest conspecific neighbors is probably a critical part of their perceptual world.

SOCIAL BEHAVIOR

THE EVOLUTION OF SOCIAL PATTERNS. For many years, much of the literature on social behavior reflected a belief that there is some pervasive benefit derived from sociality, some automatic increase in survival and fitness resulting from a social life. In his thoughtful consideration of evolution of social behavior, Alexander (1974) stressed that, to the contrary, social living has important disadvantages. Competition for food, mates, and space is heightened, and the conspicuousness gained is disadvantageous to prey and predator alike. An additional liability for social animals that assemble in very large groups is the possibility that disease or parasites will be spread widely and rapidly in the population. One must, then, look for advantages attending group living that more than compensate for the disadvantages.

Alexander discusses three broad advantages. (1) An individual's vulnerability to predation may be reduced by effective group defense or herd behavior. Defense of the group by dominant males is an important antipredator strategy of baboons, and a cohesive, running herd of ungulates presents a problem for predators. "The safety of the herd consists of the cohesive mass of animals running in an organized manner. The animals exposed are only those on the outside, and even these are protected by the number of flying hoofs and the ebbs and surges within the group. The vast array of movement has a disorienting effect on the observer's vision" (McCullough, 1969:72). In this latter situation, selection against the straggler or the individual who breaks from the herd is intense; the conspicuous misanthrope is the animal most easily singled out and killed by predators. (2) The cooperative effort of a predatory group (such as hyenas or wolves) may be effective in bringing down large prey that could not be killed by a single predator. With baboons, scattered but rich sources of food can be found more often by many searchers than by a single animal. (3) A paucity of safe nocturnal or diurnal retreats may have forced a partly social life on such animals as baboons and some bats.

After animal groups form, refinements in

social behavior evolve. Alexander views these refinements as serving several functions. They may increase the advantages gained by group living. Such a behavior as the formation of the defensive ring of animals by musk oxen (*Ovibos moschatus*) tends to increase the invulnerability of the herd to predation by wolves. Further advantages may also be gained by groups of predators: the precise positions and spacing maintained by individuals of some foraging groups of cetaceans may increase the ability of the group to perceive and capture prey. Selection in some primates has seemingly favored individuals that show great fidelity to a restricted social group; this behavior would limit intergroup transmission of disease and would favor the development of immunity to diseases common to many members of the group, thus reducing mortality. The stress to which a baboon is subjected when it tries to enter a new troop would be difficult for a diseased or heavily parasitized animal to survive, and may thus protect the troop from the introduction of "new" pathogens. These behaviors and many others may have evolved because they allowed primates to avoid diseases to some extent.

Most importantly, in Alexander's view, the evolution of social behavior affects reproductive competition between group members and reproductive performance of the population at large. The social system of the hamadryas baboon (*Papio hamadryas*) described by Kummer (1968), for example, is based on the one-male unit. An adult male maintains a group of from one to several females, which are threatened or punished when they stray. This is a stable unit, and the male copulates with his females only. But by keeping his females with him constantly and by being aware as they come into estrus, he ensures his fitness. Young males are often unit followers, have opportunities to copulate with estrus females on occasion, and become familiar with the social behaviors that may later be used in gaining and maintaining their own units. During the evolution of this system, the fitness of the socially in-

tegrated individuals was presumably greater than that of the individual who was solitary or did not learn the behavioral tactics associated with the social life.

Many ethologists agree that in animals the evolution of social systems is associated with increased fitness of individuals. But a social system in any group of animals is tested against the constraints imposed by a specific environment: perhaps as strongly as any factor, the abundance and distribution of food limits the evolutionary options. Indeed, the evolution of some social systems may have been influenced primarily by selective pressures imposed by the distribution of food in time and space. Using baboons as examples again, the one-male social unit of the hamadryas baboon seems well adapted to dry areas where productivity of the habitat is low and food has a patchy distribution but is nowhere abundant. The savanna baboons, however, occupy a more productive area where food is scattered but a patch may provide abundant food; these animals forage in large social groups.

Wilson (1975:456) regards milk as the key to sociality in mammals. Mothers must invest considerably more energy in the care of postnatal young than in the growth and development of intrauterine young. Young are associated with their mothers during much of their early lives. This mother-young group is the basic social unit in mammals, and even in solitary species the bond between the young and the mother is close. Because the care and nourishment of the young demand from the females a tremendous amount of time and energy, the females "are the limiting resource in sexual selection" (Wilson, 1975:426). Males, on the other hand, typically invest little time and energy in the young; the males are thus free to make behavioral adjustments, such as polygynous breeding or the holding of harems, that increase their fitness.

Just as the distribution of food influences the evolution of social behavior, so it affects the evolution of mating systems. Females tend to compete for food and, in the case of bats, for shelter, whereas males compete for females (Bradbury, 1977). By excluding other conspecific male bats from an area with resources sought after by females, a male can gain exclusive access to females. If it is not feasible to protect an area and its resources, a male can either attach itself to a group of females and drive away other males (as in the harem system of some ungulates) or, if females do not assemble in communal groups, males and females can meet and copulate at a mating area advertised by the male. These strategies will be considered in the next section.

Viewing the entire class Mammalia, highly evolved social behavior is the exception rather than the rule. Monotremes are solitary, as are most marsupials, but some members of the marsupial family Phalangeridae and some kangaroos and wallabies (Macropodidae) are highly social. Among eutherian mammals a wide variation in "sociobiology" occurs, with highly evolved social behavior occurring in a number of orders.

SOCIAL BEHAVIOR AND MATING SYSTEMS. The separation of social behavior and mating systems in discussions of mammalian behavior is difficult and artificial, and for this reason they will be discussed together in the following section. Patterns of social behavior in mammals are so diverse that broad summary statements are hard to frame. An additional problem is the incompleteness of our knowledge of the social behavior of many mammals. As might be expected, the large, more spectacular mammals and those hunted for sport have been most thoroughly studied. But some of the most important groups of mammals remain poorly understood; although rodents and bats together comprise over half of the known species of mammals, we have only a rudimentary understanding of social behavior in these groups. This section gives a necessarily cursory overview of the sociobiology of mammals. Excellent treatments of this subject are by Wilson (1975), Eisenberg (1966), and Ewer (1968, 1973).

Social behavior has evolved in two marsupial families. In the Petauridae several species are social to some extent, but so-

ciality is perhaps best developed in the honey glider (Petaurus breviceps), in which cohesive family units are dominated by males (see page 67). In the family Macropodidae, sociality is developed to varying degrees, and probably the most highly evolved marsupial social behavior occurs in the whiptail wallaby (Macropus parryi). The population studied by Kaufmann (1974b) consisted of subunits called mobs. The members of a mob occupy a home range to the near exclusion of members of other mobs, but the area is not defended. The social organization of a mob is loose, but some structure is provided by a rather flexible dominance hierarchy between the males that is maintained by nonviolent ritualized fighting. (A dominance hierarchy is a fairly permanent social system based on dominance. Each individual recognizes its "position"; that is to say, it recognizes the animals that it can dominate and those that dominate it.) An estrous female is typically accompanied for from one to three days by her dominant-male consort, who has exclusive mating rights.

The facial markings of the whiptail wallaby are of interest. These markings are exceptionally striking, resembling those of some social antelope, and may play an important role in signaling and orienting attention in whiptail wallabies during the complex ritualized courtship, and aggressive behaviors.

Among insectivores, sociality is rare. Eisenberg and Gould (1970) found that some species of the family Tenrecidae exhibit limited social behavior. In the most social tenrecid, Hemicentetes semispinosus (Fig. 6–12), several females and their young and a single male occupy the same den. Among members of the Soricidae, the largest family of insectivores, the greatest display of sociality occurs when a number of shrews occupy the same nest, presumably in the interest of lowering the energy cost of thermoregulation. Solitary lives, and the use of olfactory signals to bring the sexes together in the breeding season, are characteristic of most insectivores.

Although our knowledge of the sociobiology of bats is extremely fragmentary, there is no doubt that within this group a wide array of social systems occur, and some species are known to have complex social behavior. A few species are completely solitary except during copulation and when the mother-young bond is maintained briefly. A more common pattern, typical of vespertilionid bats and some species in other families, involves the separation of the sexes at the time of parturition. Females form nursery colonies exclusive of males, with the sexes associating again when the young can fly and forage. Monogamous family groups are formed by a few bats in the families Emballonuridae, Nycteridae, Rhinolophidae, and Vespertilionidae. Of great interest as examples of complicated social behavior are the several species of harem-forming bats that have been studied.

Thanks to the careful work of Bradbury and Emmons (1974), the elaborate social organization of some Neotropical bats is well known. The sociobiology of *Saccopteryx bilineata* can serve as a basis for comparing social behavior in several closely related bats of the family Emballonuridae. This small insectivorous bat (Fig. 17–9) occurs widely in the Neotropics, where it lives by day in colonies in the buttress cavities at the bases of large tropi-

Figure 17–9. The hovering display of the neotropical emballonurid bat *Saccopteryx bilineata*. The complex social behavior of this bat is discussed in the text. Photograph by Bruce Dale © National Geographic Society.

cal trees. Each colony is organized into a number of harems. Each harem and its territory is maintained as a discrete social unit by a single male, and contains from one to eight adult females. The harem territories are from about 0.10 to 0.36 m^2, and females are regularly spaced from 5–8 cm apart. This individual distance seems inviolate, except by young, which up to an age of some two months can approach their mothers. Territorial defense by males involves a remarkably intricate series of displays. Frequently, males face each other across a common territorial boundary and give brief, high frequency "barks." A barking session generally ends when each male withdraws from the boundary. A mutual approach by two males often leads to the males moving along near the boundary opposite each other, with periodic violations of the minimum distance, leading to a wavy pattern of movement. During the least frequent but most intense type of border dispute, one male charges across the boundary and strikes at the other with closed wings. The other male makes a brief flight and lands forcefully near the first, causing it to retreat in turn. Often four or five attack-supplant sequences occur before each male retreats from the boundary. A quite different display, called "salting," seemingly includes both visual and olfactory signals. The male approaches another bat with forearm extended but with the digits flexed and the chiropatagium folded, opens the orifice of the wing gland, and shakes the wing as a salt cellar would be shaken. Males salt females in their own territories, or another male's females across a territorial boundary. A male will also salt another male across a boundary during a territorial dispute, and alternate reciprocal saltings by a pair of males also occur. Never did Bradbury and Emmons observe the antagonists sustain injury during agonistic displays.

In a different behavioral context, during interactions with females, territorial males use a diverse repertoire of displays. Territorial males are first to return to the colony sites at dawn, and when females and their young begin to return, males begin singing. Each song is a long series of chirps, and some components of the song are audible to man. But as the females and young begin alighting in numbers in their respective territories, the males stop singing and turn to other displays. Considerable salting of females in bordering territories occurs at this time, and each female that lands in a harem is greeted by a hover display by the harem master. This display consists of a brief flight by the male, during which he hovers just above the female. During the hovering both animals usually vocalize, and the male may open the orifices of the wing glands. The hovering display seems to be a greeting, perhaps a means of retaining females, whereas the long song may serve to attract females. The period of displaying and sorting of animals into their territories generally occupies several hours, after which the relative positions of the bats in the colony remain fairly constant. A summary of behavior during a 60-minute period is shown in Figure 17–10. Females are highly aggressive toward one

another and rigidly maintain their individual distances; this tendency is probably a major factor limiting harem size. Females vocalize frequently and spontaneously, and the vocalizations are highly variable.

There are surprisingly sharp contrasts between the social behavior of *Saccopteryx bilineata* and that of the closely related *Saccopteryx leptura* and *Rhynchonycteris naso*. All of these emballonurid bats are sympatric in many areas. Whereas *S. bilineata* forms large colonies and males perform a complex series of conspicuous displays to attract and maintain females in harems, *S. leptura* occurs in small groups of twos and threes, males seem tolerant of one another and, except for some vocalizing, there is none of the wide array of social behaviors typical of *S. bilineata*. Perhaps *S. leptura* is monogamous and forms enduring pair bonds. The use of relatively unprotected roosting sites by *S. leptura* may preclude the occurrence of large colonies and conspicuous behaviors, whereas the protected colony sites of *S. bilineata* will allow these patterns of behavior. The third species, *R. naso*, roughly resembles *S. leptura* in social structure.

Detailed observations of the behavior of a laboratory colony of *Carollia perspicillata* have been made by Porter (1975). This frugivorous phyllostomatid bat is common and widespread in the Neotropics. The laboratory colony contained six males and four females, and was maintained in a large flight cage. Early in the study the animals established a stable pattern of roosting; the four females roosted in one corner of the cage and were guarded by one male; the other five males roosted singly or in pairs elsewhere. None of these five males interacted with females, nor did they fly within 1 m of the females' corner. Two months after the observations began, two females bore young. Within six weeks a male that had previously roosted with other males some 5 m from the harem began roosting alone within 3 m of the females. Soon the interloper began roosting about 1 m from the females and became aggressive toward the harem male. On the first day of confrontation the males period-

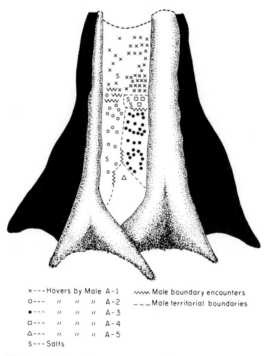

x --- Hovers by Male A-1   ⋀⋀⋀ Male boundary encounters
o ---   "   "   "   A-2   _ _ _ Male territorial boundaries
● ---   "   "   "   A-3
□ ---   "   "   "   A-4
△ ---   "   "   "   A-5
s --- Salts

Figure 17–10. The occurrence of various behaviors (hovers, saltings, and boundary displays) by the adult males of a colony of *Saccopteryx bilineata* during a 60-minute observation period. (From Bradbury and Emmons, 1974.)

ically approached each other, the harem male closest to the harem, and hung close to each other with necks extended until they were nearly nose to nose. Their mouths were partly open and their wings were partly spread, but the bats maintained this "frozen posture" except for the rapid flicking of their tongues. The bats then rocked back and forth from foot to foot and occasionally struck at each other with their wings. After about a week of repeated confrontations, the nose-to-nose posture was used progressively less and wing shaking and rocking became more intense and came to resemble a boxing match (Fig. 17–11). The boxing activity was only briefly interrupted during the night by feeding or grooming.

Figure 17–11. Two male *Carollia perspicillata* performing the "boxing" ceremony. See text for discussion. (Photograph by Fran Lang Porter.)

Thirteen days after the initial confrontation, the interloper began roosting nearer the females while the harem male began leaving the harem more frequently than usual. On the seventeenth day the new male displaced the original harem male and assumed control of the harem; he began roosting with the females and defended their immediate area. At this time, the former harem male began roosting some 1.5 m from the harem. Neither competing male was injured, but both seemed exhausted. During the week following the displacement the boxing displays became progressively less frequent and were finally abandoned. The stereotyped nature of the displays and the rather orderly progression of events suggest the importance of these behaviors. Although in the laboratory more time was perhaps spent than could have been spared under natural conditions for the behaviors associated with the displacement, the same behavioral progression probably occurs under both situations.

The amazingly specialized hammerheaded bat has been described previously and its use of leks has been mentioned (see page 106). The raucous vocalizations of the male are used not only for attracting females but for establishing dominance ranking and for maintaining a small territory within the lek. The males that are able to occupy the areas in the lek most favored by females probably do most of the breeding. Because the males may take several years to establish themselves in these favored areas, the genetic constitution of the population is influenced strongly by the hardiest and most aggressive males.

Primates are primarily social mammals, and some of the most complex mammalian social systems occur in them. A major trend in primates is toward the abandonment of the sense of smell and the refinement of visual depth perception. The eyes have broadly overlapping visual fields, and stereoscopic vision allows the precise discrimination of distance necessary for rapid arboreal locomotion and for dextrous manipulation of objects with the hands. As might be expected, visual and auditory signals are of prime importance, manual dexterity in the form of mutual grooming is widely used in social situations, and olfaction is of relatively little importance, especially in anthropoid primates. An impressive body of literature is available on primate social behavior, and only brief comments on this subject are appropriate here. (An excellent overview of the sociobiology of nonhuman primates is given by Wilson, 1975).

Even in the most primitive primates, the Lemuridae, sociality is well developed; but compared to those of many other primates the lemurid social systems are rather simple. The mouse lemur (*Microcebus murinus*), studied by Petter (1962) and Martin (1973), occurs on Madagascar in dispersed "population nuclei." The proportion of females to males in these nuclei is four to one, and as many as fifteen females may occupy the same nest. Surplus males not accompanying groups of females often occupy nests on the periphery of the area. Although mouse lemurs occupy nests together, perhaps because nest sites are at a premium, there is no organized social life, and animals forage alone. This primate, therefore, must be regarded as a basically solitary creature that is flexible enough to be able to occupy nests communally.

The ring-tailed lemur (*Lemur catta*), studied by Jolly (1966, 1972), is far more advanced socially, being perhaps the most socially progressive of all prosimian primates. This lemur lives in troops that range in size from some 10 to 20 or more animals. Adult males and females are equally represented in the troop, and their total numbers are usually equalled by the numbers of young. Troops occupy exclusive areas and there is little intertroop contact. Social organization within a troop is based on dominance patterns. Females are dominant over males, a reversal of the usual primate system. A male-dominance hierarchy is established, and dominant males seem to be able to remain for long periods with a troop, but (surprisingly) do not have first access to estrous females.

As befits their rather lowly position in the primate taxonomic scheme, lemurs make wide use of olfactory signals. Both sexes mark branches with secretions from the genitals and, using the palms of the hands, males mark branches with other secretions. Scent glands occur on the chests and forearms of males. During aggressive confrontations the tail is pulled between the forearms and annointed with scent and is then lifted high and waved to disperse the scent. Males indulge in "stink fights," which involve palmar marking, tail marking and tail waving, and often displacement of one animal by the other.

The animals face each other when performing the scent marking, and the visual displays by the two animals, each using the conspicuously banded tail, seem to be mirror images of each other. The dominant animal moves forward while the other retreats. Vocalizations are also important during social interactions and a variety of vocal signals is used.

Although a variety of types of social organization occur among anthropoid primates, most higher primates must learn to be responsive to a complex *social field*. An individual remains aware simultaneously of the attitudes and displays of a number of members of the social group and of the social ranks of these animals. Manipulation of the social field becomes an important aspect of the behavior of many primates, and even the ranking of an individual may depend in part upon the animal's effectiveness as a manipulator. As a result of enduring social bonds between female baboons, two "friends" can put up an intimidating united front when one is threatened by a third individual. In some primate social systems, the rank of a female depends partly on her close association with a dominant male and her ability to depend upon his help or protection during aggressive confrontations. Her status may abruptly decline if the male is deposed from his dominant position. In some baboon troops and in other primate societies in which structure and cohesiveness of the group are maintained by strong dominance patterns, the dominant male is the focal point of attention. An individual's behavior and the behavior of the entire social group are geared to the responses of this leader.

Savanna baboons (*Papio* spp.) that pursue an almost entirely terrestrial life are often vulnerable to attack by a predator in places where immediate escape to trees is impossible. Food is frequently scattered, and a troop must forage over wide areas, thus increasing the chance of encounters with predators. A large and tightly organized social group has evolved in these baboons, perhaps largely in response to this pressure. These groups include from roughly a dozen to over 150 individuals. Each group occupies a largely exclusive home range; although home range boundaries are usually respected, they are seemingly not defended. When a group is moving, the less dominant males move to the side or ahead, where they protect the group from a surprise attack by giving the alarm when a predator is sighted. The positions of the members of the group are uniform, with mothers carrying young near the center together with the dominant males (Fig. 17–12; Hall and DeVore, 1965:70). These males quickly respond to threats from any quarter and their united action provides the primary defense of the troop.

Sexual dimorphism is pronounced in baboons, and enhances the male's intimidating appearance as well as his fighting ability. The male of the anubis baboon (*Papio anubis*), for example, is roughly twice as large as the female, is more powerfully built, and has comparatively huge canines. The long fur over the crown, neck, and shoulders accentuates the impression of size and ferocity (Fig. 17–13).

In baboons in both East and South Africa, the mating pattern seems to have a consistent relationship to dominance rank-

ing. Whereas subadult, juvenile, and less dominant males copulate with females in the early states of estrus, dominant males have exclusive rights to females during the period of maximal sexual swelling (the time when ovulation occurs). In some groups only the highest ranking male copulated with females during the height of the swelling (Fig. 17–14), and in a group observed by DeVore in Kenya not one dominant male attempted copulation until the swelling was at its peak. This system seems to ensure that most young are sired by the most powerful and aggressive males.

Most anthropoid primates retain a largely arboreal style of life, and among these species there is considerable diversity in the social organization. The simple social system of the dusky titi monkey (*Callicebus moloch*; Cebidae) of South America occurs in several other cebids and in marmosets (Callithricidae). The social unit of the titi is the closely knit family group, which includes a mated pair and one or two young. The pair is probably mated for life. Members of a family unit stay close together when foraging, and frequent confrontations between two neighboring units seem to confirm territo-

Figure 17–12. A diagrammatic representation of the positions of the members of a moving troop of baboons (*Papio anubis*). (After Hall and DeVore, 1965.)

Figure 17–13. The threat "yawn" of a male baboon *(Papio anubis)* displays the large canines. The eyes are closed during the threat, and the whitish lids are conspicuous. (Photograph by Irven DeVore.)

rial boundaries. Despite their simple social organization, these animals use a broad array of signals and their vocal repertoire is one of the richest known. Moynihan (1966) believes that in the acoustically confusing forest community they occupy, where an extraordinary diversity of birds are calling and other cebids are vocalizing,

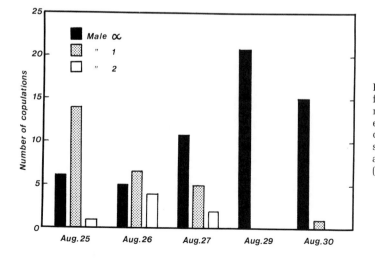

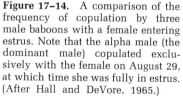

Figure 17–14. A comparison of the frequency of copulation by three male baboons with a female entering estrus. Note that the alpha male (the dominant male) copulated exclusively with the female on August 29, at which time she was fully in estrus. (After Hall and DeVore, 1965.)

the complex vocabulary of the titi facilitates precise, unambiguous, and "private" communication.

Studies of the African red colobus monkey *(Colobus badius tephrosceles)* by Struhsaker (1976) provide evidence that a social system involving large multimale social groups has evolved within rain forest as well as savanna or desert environments. This colobus monkey is primarily a leaf eater and lives in tropical forests. Its social units are large multimale groups that average some 45 animals. These groups are not territorial and home ranges overlap broadly, but spacing calls tend to keep neighboring groups apart.

The great apes (Pongidae) have social systems that display no radical departures from basic primate patterns, but there are some unique features. The mountain gorilla *(Gorilla gorilla beringei)* groups include from 2 to 30 animals. Social interplay between individuals is surprisingly amiable (Schaller, 1963, 1965a, 1965b; Fossey, 1972). Particularly notable is the age-graded male troop, with the nucleus of the group consisting of the dominant silver-backed male (ten years of age or older) and adult females and their young. Additional males, including less dominant silver backs and black-backed males, attach themselves to the periphery of the group. Of further interest is the low-keyed aggressive behavior associated with assertions of dominance. The great size and formidable presence of gorillas belie their peaceful, nonviolent natures.

The chimpanzee *(Pan troglodytes)* has in recent years been the subject of considerable observation in the field by a number of workers, including Van Lawick-Goodall (1968) and a group of Japanese scientists (Izawa and Itani, 1966; Izawa, 1970; Nishida and Kawanaka, 1972; Sugiyama, 1973). The basic social unit of chimpanzees is an often dispersed group of from 30 to 80 animals that show considerable fidelity to a large home range. Particularly unusual is the looseness of the organization of the social group, with intricate patterns of establishment and dissolution of small parties. Highly evolved visual, tactile, and vocal communications are used by chimpanzees. When a party of chimpanzees finds trees bearing fruit, their almost manic vocalizations and actions attract other parties to the bonanza.

Primates are clearly set apart from other mammals by the complexity of their social systems, the closeness and permanence of their social bonds, and by the high level of development of vocal and visual communication. Much remains to be learned of primate sociobiology, and of special importance is the behavior of forest-dwelling species.

Social behavior is not widespread in the Lagomorpha, Pikas *(Ochotona* spp.) live in colonies, but individuals are solitary, and most rabbits are also solitary except during the breeding season. The European rabbit *(Oryctolagus cuniculus)* is an exception. Certain males have a number of females in their territories, and the females establish an order of dominance. Some social groups are dominant to others and maintain larger territories.

Rodents are predominantly solitary and many are strongly territorial, but within the order diverse social patterns occur, ranging through simple family groups to large social aggregations forming "towns," in the case of the prairie dog. Social evolution in rodents has been influenced by many factors, important among which are patterns of diel and yearly activity, distribution and abundance of food, distribution of favorable habitat, and length of the growing season.

The beaver *(Castor canadensis)* illustrates a simple level of social development. A beaver colony is a family group in which mutual tolerance is the rule, and apparently no social integration exists (Tevis, 1950). In wild colonies of Norway rats *(Rattus norvegicus)* a similar colonial system prevails, but a dominance hierarchy gives order to the system (Steiniger, 1950).

The genus *Marmota* provides a striking and remarkably interesting example of different social systems evolving under differing environmental pressures. Barash (1974) discussed the contrasting social behavior of the woodchuck (*Marmota monax*), the yellow-bellied marmot (*Marmota flaviventris*), and the Olympic marmot (*Marmota olympus*). The woodchuck is common in fields and edge situations in parts of eastern United States. Here growing seasons are reasonably long, some 150 days in southern Pennsylvania. Except at the time of copulation and during the period before weaning when mothers and young live together, woodchucks are solitary and are aggressive toward one another. The Olympic marmot lives in the state of Washington in alpine meadows near timberline in the Olympic Mountains. Winters are long here, and the growing seasons are short, from about 40 to 70 days. Olympic marmots are highly social; colonies are tightly organized social units including, usually, a male, two females, young of the year, and young of the two previous summers. Social contact is frequent and amiable, and during the early morning individuals enter occupied burrows and exchange ritualized greetings with burrow residents; there are no individual territories, with all areas occupied by a colony being equally accessible to all colony members. No clear nor strongly enforced dominance hierarchy exists. The yellow-bellied marmot, which usually lives in montane areas with growing seasons intermediate between the two extremes described above, has, neatly enough, an intermediate social system. Colonies contain a number of adults, but individuals maintain discrete home ranges, and greetings are infrequent and have aggressive overtones. Barash summarizes the degrees of sociality by indicating that an individual Olympic marmot averages one greeting per hour, the average greeting rate for the yellow-bellied marmot is roughly one tenth this rate, whereas greeting behavior is yet to be described for the woodchuck.

Studies on other species of marmots have provided further evidence that the shorter the growing season, the higher the degree of sociality (Barash, 1976).

Among rodents, the most complex social system known is that of a prairie dog (*Cynomys ludovicianus*), which has been carefully studied by King (1955, 1959). Prairie dogs formerly occupied many parts of the western United States, where they occurred in large "towns," often including over 1000 animals and covering many acres. (Now, lamentably, prairie dog towns are rare in many parts of the west, owing partly to intensive, government-sponsored poisoning campaigns.) The functional social units are *coteries*, which generally consist of an adult male, several adult females, and a group of young. No dominance hierarchy exists within the coterie. The paths, burrows, and food within the area held by a coterie are shared by its members, but hostility between coteries is the universal pattern.

Members of the coterie become familiar with each other in part by grooming, playing, and "kissing" behaviors. During kissing the mouths are open and the incisors are bared. This behavior is seemingly a ritualized method of distinguishing between friend and foe. Faced with the threatening expression of open mouth and bared teeth, a trespasser retreats, while a fellow coterie member meets its "friend" and kisses. A two-syllable territorial call is used to proclaim ownership of territory. A repetitive, high-pitched yelp is a warning of danger. During the spring period, when females are pregnant or are lactating, the coterie system partially dissolves, and some yearlings and adults establish themselves beyond the territorial limits of their coteries. The personnel of coteries thus may change, but the territory itself is stable.

An individual gains several advantages from this social system. Many eyes are watchful for danger, and many voices are ready to sound a warning. The effect of the foraging of the animals is to keep vegetation low over a wide area and to provide terrestrial carnivores with little concealment. Perhaps equally effective in providing for long-term occupancy of an area, the animals are kept spaced so that overuse of food plants is generally avoided.

The correlation between severity of climate and social behavior is associated also with distinctive patterns of dispersal of young. Returning to the genus *Marmota*, because of the long growing season available to the woodchuck, young animals attain considerable growth during their first summer and can disperse in the year when they were born. Much less growth occurs the first summer in the yellow-bellied and Olympic marmots (Fig. 17–15), and their survival would be jeopardized if they were to disperse during their first season. These two species have responded to the more severe climates and slow growth of young by delaying the dispersal of young. Yellow-bellied marmots disperse their second summer (as yearlings), whereas Olympic marmots disperse their third summer (as two-year-olds). These delays, which are highly adaptive because they allow greater time for growth and favor survival of dispersing animals, can only occur when the social organization fosters extended tolerance of young on the part of adults.

A rather inhospitable environment has exerted selective pressures that in the case of the naked mole rat have influenced the animals' social organization. Cooperative digging behavior in these animals (see page 360) has enabled them to burrow rapidly in soil in which digging is extremely difficult except during the brief rainy seasons.

Our knowledge of the behavior of cetaceans is as yet extremely incomplete, but current evidence indicates that most species are social. Not only do some cetaceans travel and forage in social groups in which some consistent spatial organization is evident, but cooperative behavior is known. Several dolphins (*Delphinus delphis*) were observed lifting an injured individual to the surface where it could breathe, and females without young will help a mother tend its young. Cetaceans are known to produce a wide range of vocalizations. This ability, and the large brain typical of cetaceans, have inspired some writers to make extravagant claims of high intelligence for these animals (see, for example, Lilly, 1961, 1967). Careful considerations of vocalizations of several cetaceans by Dreher and Evans (1964) and by Caldwell and Caldwell (1972) indicate a repertoire of from 16 to 19 vocal signals, plus a number of signals produced by snapping the jaws shut or hitting the water with the flukes. But, until free-ranging cetaceans are carefully studied, uncertainty will remain about the diversity and complexity of their communication systems. The reasonable suggestion has been made that the intricate and oddly melodious songs of the humpback whale (*Megaptera*) are used for long-range communication in the open sea.

Although among carnivores sociality is probably the exception rather than the rule, highly organized social systems have evolved in some species. The lion "pride," the spotted hyena "clan," and packs of wolves (*Canis lupus*) or African hunting dogs (*Lycaon pictus*) are examples. Some diurnal viverrids are also social, with the

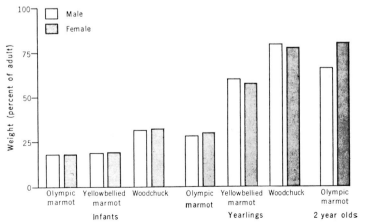

Figure 17–15. Growth rates of three species of marmots expressed as percentages of adult weights that young marmots attained at different ages. The Olympic marmot (*Marmota olympus*) lives at high elevations where growing seasons are short (40 to 70 days); the yellow-bellied marmot (*M. flaviventris*) lives at intermediate elevations where growing seasons are from 70 to 100 days; and the woodchuck (*M. monax*) lives at low elevations where growing seasons are some 150 days. (After Barash, 1973.)

packs of the banded mongoose (*Mungos mungo*), which may consist of some 40 or more animals, representing the extreme.

Kruuk (1972) studied spotted hyenas from mid-1964 to early 1968 in the Ngorongoro Crater and the Serengeti Plain of Tanzania, East Africa. His excellent work revealed that, contrary to popular opinion, hyenas kill much of their own food. Of particular interest, he observed a remarkable social system. The basic social unit of the spotted hyena is the clan, which may contain as many as 80 animals. The clan system is periodically disrupted in the Serengeti, where seasonal migrations of wildebeest and zebras result in drastic shifts in food supply. Each clan defends a territory, the boundaries of which are maintained in part by systematic scent marking. Territorial disputes are often violent, and individuals are occasionally killed during border warfare. Females are larger than males and are dominant to them. A rather complex dominance hierarchy exists within a clan, and strong bonds of "friendship" develop between females. At a kill, dominance is often not asserted, and competition is based largely on the ability to eat extremely rapidly rather than on fighting prowess or size. Females tend their cubs at a central denning area, and here young receive early training (or practice) in the social rituals of the species.

Spotted hyenas offer an interesting example of intraspecific mimicry. The external genitalia of the female hyena mimic the male penis and scrotum: the female's clitoris is very large, resembles a penis, and is erectile, and two sacs filled with fibrous tissue form a false scrotum. So closely alike are the external genitalia of the two sexes that, beginning in Aristotle's time (384–322 B.C.), hyenas have been considered by many to be hermaphroditic.

When two hyenas meet, they typically go through a meeting ritual, part of which involves the external genitalia. Kruuk describes one such greeting: "From several hundred meters distance, a male comes walking up, without hesitating, straight toward the female. He shows signs of ner-

vousness, tail flicking, teeth slightly bared, and penis erected. The female gazes straight ahead of her, seemingly ignoring the male. When the male reaches the female, he quickly touches the side of her head with his muzzle, then lifts his leg, whereupon the female does the same, both now showing full erection. They spend about 8 seconds sniffing each other's genitals, then separate, each walking away in opposite directions." Kruuk concludes that the intraspecific mimicry evolved because of its importance during meeting ceremonies. This ritual probably enables individuals to be close together briefly and to "identify" each other while attention is attracted to the genitals and to a course of action alternative to that of fighting. This cooling-off period perhaps allows aggressive tendencies to subside. The leg-lifting action seems to be an appeasement gesture, hence its initiation typically by a subordinate animal. The adaptive value of such behavior to a spotted hyena is perhaps heightened by the flexible social life of these animals. Although often social, hyenas may be solitary for varying lengths of time, and peaceful meeting behavior and recognition of individuals is of frequent importance.

The social system of the spotted hyena seems highly adaptive when considered in relation to the animal's environment. This predator is well adapted both for scavenging and for killing its own prey. Scavenging is largely a solitary enterprise, whereas success in killing large animals such as wildebeest is greatly improved by group action. Hyenas, therefore, must have a social organization that favors social hunting and peaceful social feeding, while also having behaviors enabling individuals to safely meet one another after being separated. In predators with such powerful offensive weapons as the teeth of hyenas, control of aggression is of critical importance, and the primary function of the great array of scents, displays, and vocal signals is probably to restrict the use of these weapons (Figs. 17–16, 17–17). Regarding the clan system, by defending a territory a clan not only gains the advan-

Figure 17-16. Two spotted hyenas doing the "parallel walk," a threat behavior directed toward a third hyena. (From Kruuk, 1972.)

tage of exclusive use of a familiar area, but the limited numbers of animals feeding on a kill ensure opportunities for feeding for the lowest-ranking members, the juveniles.

Careful, long-term field studies on the lion (*Panthera leo*) in Serengeti National Park of Tanzania by Schaller (1972) and Bertram (1973, 1975) have provided a fascinating picture of the social life of this animal. The social unit of the lion is the pride. This permanent and fairly stable social unit has a nucleus of from about 3 to 12 adult females. There is virtually no recruitment of outside females: all of the females are born and grow up within the pride. A pride probably lasts for many decades, and its female members are all closely related. When a young female reaches some three years of age she is either accepted as a member of the pride or is driven from it. These rejected females, and males that are not attached to prides, become nomads, tend to follow migratory movements of their prey, and constitute about 15 per cent of the total lion population in the Serengeti. A pride is usually controlled by several adult males, which defend the pride against outside males. Each pride occupies a territory that is largely exclusive of those of other prides. Although the boundaries of the territory shift to some extent, essentially the same area is held by a pride year after year. All members of a pride are not together at all times. Individuals may hunt alone or part of a pride may separate from the rest; but members of a pride are familiar with each other and social contacts are peaceful or seemingly affectionate (Fig. 17-18). A member of the central sisterhood of the pride leads a stable, if at times violent, life, and her reproductive life is some 13 years.

A male, on the other hand, does not associate consistently with a single pride throughout life, and his reproductive life may be only two or three years. The life cycle of the male can be divided into several periods. Young males stay with their pride until some three years of age, when

Figure 17-17. Two hyenas attacking a third. Note the tail-high, ears-forward, aggressive posture of the attackers and the "fleeing posture" of the animal being attacked. (Photograph copyright Hans Kruuk.)

Figure 17-18. Two female members of a pride nuzzle one another after having killed and fed on a zebra. (Photograph taken in Amboseli National Park, Kenya, East Africa.)

they are either forced from the pride or leave it voluntarily. Often several males leave the pride together; these males may be brothers or, because the females of a pride are all grandmothers, mothers, sisters or daughters, the young males are at least closely related. The young male outcasts become members of the nomadic population, and for the first time in their lives are unable to depend on the hunting prowess of the experienced females of their pride for providing food. A good deal of scavenging is typically done by these nomads. After roughly two years of nomadic life the members of the group of males are approaching the prime of life, and are sexually mature, and are sufficiently formidable, in terms of strength and aggressiveness, to take over a pride. They virtually always take over a different pride than the one into which they were born. A pride lacking males may be taken over peacefully, or several males past the prime of life may be displaced and their pride taken over. On occasion violent fighting accompanies the displacement, and because a group of males can successfully challenge males holding a pride whereas one or two males cannot, selection favors tight social ties between males. The new owners disrupt the life of the pride: females may abort fetuses they carry, the cycle of females coming into estrus is interrupted, and the newcomer males may even kill cubs. But after a few months females again begin coming into estrus, they are bred by the new males, and males often enter into the care of the cubs.

After only two or three years, however, when the males are aging and perhaps some have died or have been injured, they are driven out by a new group of prime males. The displaced males, their reproductive life terminated, are again part of the nomadic segment of the population. Because they have become accustomed to being provided for by the females, and because of their declining physical condition, their life expectancy is not great.

Reproductively, lions seem to be highly inefficient. That is to say, an extremely low percentage of copulations results in offspring. In addition, the mortality rate of cubs is some 80 per cent, and the killing of cubs by lions is an important cause of death. Bertram (1975) details the situation as follows: "Assuming that lions mate every fifteen minutes for three days, that only one in five three-day mating periods results in cubs, that the mean size of litters is two and a half cubs and that the mortality among cubs is 80 per cent, then a male must mate on the average some 3000 times for each of his offspring reared to the next generation." Because each copulation is so relatively unimportant, and because the males of a pride are typically closely related and are thus genetically similar, there seems to be little pressure on the males to fight for the chance to copulate with an estrous female. This inefficiency and lack of competition between males may be closely related to the rather unique mode of life of a top predator. The lion has no important predator except man, and the size of the pride is perhaps controlled largely by periodic food shortages. The life span of lions is fairly long; reproductive inefficiency, therefore, does not prejudice the survival of the pride. Under these conditions there is seemingly no strong selective pressure favoring highly efficient reproduction, but the critical factor may be the reduction of aggression within a pride to a level permitting the survival of at least some young. Bertram suggests that reproductive inefficiency and reduction of competition between males results in increased stability of the pride, with less frequent changes of the male guard, and hence greater chances for the survival of cubs.

In the pinnipeds that are polygynous—the otariids, some phocids, and the walrus—the males are extremely vocal, are much larger than the females, and maintain breeding territories (Fig. 17–19). In the California sea lion (*Zalophus californianus californianus*) studied by Peterson and Bartholomew (1967), large bulls establish territories at sites adjacent to the water that are favored as hauling out places by females, which arrive at the rookery a few days before they give birth.

Nonterritorial bulls usually form aggregations apart from the breeding rookery. Because the same females do not continuously occupy a male's territory, and males make no effective effort to herd females into territories, the term "harem" cannot be applied to the females within a territory. Females enter estrus and copulation occurs roughly two weeks after parturition. Fighting between males occurs during the establishment of territories, and males on established territories signal their possession by incessant barking. Little actual fighting occurs after territories are established, but a "boundary ceremony" between males on adjoining territories periodically reaffirms territorial boundaries. These ceremonies involve an initial charge toward one another, followed by open-mouthed head shaking as the animals confront each other at close quarters (Fig. 17–20), and a final standoff in which the bulls stare obliquely at each other. The ceremony is so carefully ritualized that should animals get uncomfortably close to one another they adroitly avoid actual contact. Females are aggressive toward one another through much of the breeding sea-

Figure 17–19. Harems of northern fur seals (*Callorhinus ursinus*) on St. Paul Island in the Bering Sea. (From Orr, 1976.)

Figure 17–20. A boundary ceremony between two male California sea lions; the animals are confronting each other and "head shaking." (From Peterson and Bartholomew, 1967.)

son; again, however, injury is avoided by ritualized aggressive threats. Although males may be on territories in a rookery from May through August, each male maintains a territory for only a week or two; territories are thus occupied by a succession of males.

The northern fur seal (*Callorhinus ursinus*) has a breeding season that extends from June to December, and large numbers of animals assemble on the Pribilof Islands of the North Pacific. Nonbreeding males form bachelor "cohorts" around the edges of the breeding grounds. Territorial males herd females into their territories and maintain fairly stable harems (Peterson, 1965).

In contrast to the pinnipeds mentioned above, male elephant seals (*Mirounga angustirostris*) establish a social hierarchy on the breeding ground but are not territorial (Le Boeuf and Peterson, 1969). The highest ranking males stay close to breeding females, and breeding success is closely correlated with social rank. In one study area on Año Nuevo Island off the coast of California, 4 of the highest ranking males, which comprised but 6 per cent of the 71 bulls in the area, copulated with 88 per cent of the 120 females. At another study area on the same island, the alpha bull (the bull at the top of the hierarchy) maintained its rank throughout the breeding season and was involved in 73 per cent of the observed copulations.

The social life of elephants is known through the work of Laws and Parker (1968), Hendrichs and Hendrichs (1971), and Douglas-Hamilton (1972, 1973). The elephant social system is structured at several levels. The first level is that of the family group, including an old matriarch and some 10 to 20 females and their offspring. Because of the long life span of elephants, the family unit generally includes grandmothers, mothers, sons, daughters, grandsons, and granddaughters. Tight social bonds between females may last 50 years or more. The second level is the kinship group, consisting of several family groups that remain in the same vicinity and on occasion mingle peaceably. Under some conditions, as during migration, many kinship groups may band together to form "clans," containing on occasion 100 or more animals. The clan probably has no social cohesion at any level above the kinship group.

Males leave the family units when they become sexually mature and assemble in all-male groups in which males establish dominance by ritualized fighting and sparring. Dominant males are temporarily attached to family units with females in estrus. Especially remarkable is the importance of cooperative and altruistic behavior within the family unit. A nursing elephant is allowed to suckle from any lactating female, young females approaching sexual maturity are solicitous of the well-being of small calves, and the safety of a calf seems to be the concern of the entire family unit. When threatened, the adult members of the family form a united defensive phalanx.

Ungulate social behavior is of particular interest for several reasons. Many species are large, occupy open situations and can be observed easily, and have therefore been well studied. Because these open-country dwellers are probably the most highly social of all ungulates, we have a reasonably good understanding of ungulate social behavior. Further, a growing knowledge of the environments occupied by a variety of ungulates has provided a basis for a theoretical approach to the relationships between ecology and the evolution of social behavior, morphological features, and color patterns. Thoughtful work by Jarman (1974), Estes (1974), and Geist (1974), among others, has provided a conceptual base from which the broad and fascinating diversity of ungulate behavior can be viewed.

In some ungulates a small group of one or more families may form the social unit. The warthog (Phacochoerus africanus) group does not defend its home range, but male hippopotami (Hippopotamus amphibius) that live in "schools" have a hierarchy that determines which animals are to occupy favored sites near the females. Males defend their paths to foraging areas.

Many ungulates have breeding cycles that feature harems of females, each maintained by a dominant male. The area occupied by the females is the strongly defended territory or, in some cases, the male maintains a "breeding territory" that is defended even when females are not present. The essence of this territorial behavior is that it persists only through the breeding season and apparently promotes breeding largely by mature, vigorous, and aggressive males. McCullough (1969:99) reported that in a herd of elk he studied, only 12 per cent of the bulls—the largest individuals—did 84 per cent of the copulation he observed.

The rutting behavior of the elk is especially well known owing to the studies of Darling (1937), Graf (1955), Struhsaker (1967), McCullough (1969), and others.

McCullough recognized four main categories of bulls during the breeding season. Primary bulls are powerful, mature individuals that shed the velvet from their antlers early, and are the first to establish harems. Secondary bulls are large individuals that take over the harems by defeating the primary bulls as the latter become exhausted. Tertiary bulls assume control of the harems after the secondary bulls decline. Opportunist bulls are those whose only contact with cows is by chance. When a bull becomes exhausted through constantly keeping cows herded together, driving rival bulls away, and copulating, all while unable to obtain adequate food and rest (Figs. 17–21, 17–22), it is beaten in a fight with a fresh bull, who takes over the harem from the deposed "master."

Of particular interest in the present discussion are ritualized social behaviors evident during the rut. Bulls advertise their location, vigor, and sexual readiness by several acts. The most obvious and characteristic of these is "bugling," a stirring, high-pitched call with considerable carrying quality. In addition, a bull frequently thrashes low vegetation and rakes the ground with his antlers while spurting urine onto his venter. Bulls also wallow in boggy places and gouge the soil with the antlers, and frequently rub the antlers and scrape the incisors against trunks of trees. The master bull drives competitors away from the harem by a ritualized charge: with the head and neck extended, the bull jogs stiff-legged at the intruder while grinding the teeth and lifting the upper lip to display the upper canines. If the interloper does not retreat, a series of preliminary behaviors lead to the bulls facing each other and smashing their antlers together. Such an encounter typically occurs after the physical condition of the primary bull is on the wane, and is a serious test of strength, with each contestant attempting to lunge forward and catch its adversary off balance. Such encounters may end indecisively, but if the master bull is clearly defeated his reign as harem master is ended. In the case of one bull observed by McCullough (1969:89), the ani-

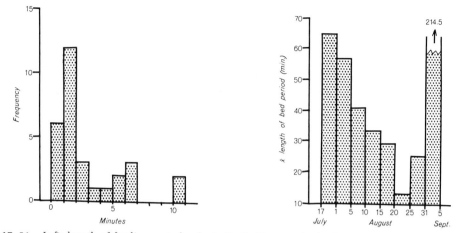

Figure 17–21. Left: length of feeding periods of a bull elk *(Cervus elaphus)* at an intense rutting stage. Right: changes in bedding (resting) periods of a bull elk as the rut progressed; this bull was defeated and gave up his harem on 31 August. (After McCullough, D. R.: *The Tule Elk; Behavior and Ecology,* 1969. Originally published by the University of California Press; redrawn by permission of The Regents of the University of California.)

mal was deposed from the position of harem master one month after shedding the velvet, and exhibited no further rutting behavior; he even ignored cows in estrus.

Jarman (1974) has given order to our view of the behavior of bovid ungulates by relating the sociobiology of these animals to their ecology. Using carefully assembled data on vegetation and on ungulate densities and distributions in Africa, Jarman has shown that open woodland, savannas, and grasslands support the highest diversity and biomass of bovids. In such areas grass forms the bulk of the vegetation available to ungulates. A high percentage of each grass plant is usable by ungulates, whereas much less of a tree or a shrub is edible. Grasslands, therefore, have high productivity of plant material available to ungulates, but in these habitats growing seasons are short. The pattern is one of high productivity of nutritious vegetation during seasonal rains and long intervening dry spells when vegetation has low nutritional content. By contrast, moisture and plant productivity are more uniform throughout the year in forested areas.

How do bovids respond to these different patterns of plant composition and seasonal productivity? Grass provides great quantities of forage, but its energy content is lower than that of many small

forbs. Because of the relatively high metabolic rates of small antelope, these animals have higher energy needs than do the large species and tend to be selective feeders, utilizing forbs or only the most nutritious parts of grasses. The small antelope, therefore, because they are depending on a food base that is far less abundant and more dispersed than grasses, do not attain the densities of the larger grass-eating bovids.

Associated with the differences in density between large and small bovids are sharp differences in social organization. The five categories of Jarman (1974), as outlined by Wilson (1975), are shown in Table 17–1, and indicate the relationships between habitat, feeding style, and social organization in African bovids.

Estes (1974) has stressed a major structural and behavioral dichotomy within the family Bovidae. The ancestral bovids were probably small forest dwellers, perhaps resembling today's forest-dwelling duikers (Fig. 17–23). The expansion of grasslands in the Miocene and Pliocene of Eurasia and Africa set the stage for the movement of bovids into open grassland or savanna habitats. This major evolutionary step brought some bovids under the influence of new suites of selective forces, and led to their structural and behavioral divergence

Figure 17–22. A bull elk during the rut. Top, the bull bugling; the coat of this animal is caked with mud after wallowing. Middle, the bull on the right (the same animal whose feeding and bedding periods are shown in Figure 17–21) is in poor condition, has broken antlers, and is close to the end of his reign as harem master. Bottom, the same bull on 12 September, nearly two weeks after his defeat. Note the scars and broken antlers. (Photographs from McCullough, D. R.: *The Tule Elk: Behavior and Ecology*, 1969. Originally published by the University of California Press; reprinted by permission of The Regents of the University of California.)

from the persistently forest-dwelling species. Viewed today, the dichotomy is between, on the one hand, the forest-dwelling browsers that are generally small, cryptically marked, and simple-horned, and escape from predators by hiding and, on the other hand, the open country grazers that are of medium or large size, are conspicuously marked, possess large and often complex horns, and use their speed in the open to escape predators. The behavioral dichotomy is also clearly delineated: the forest dwellers are either solitary or live in small family groups, and scent marking is the primary means of communication; the open country grazers

Table 17-1. Behavioral/Ecological Classification of Some African Bovids

Social Organization and Feeding Style	Size (in kg)	Antipredator Behavior	Examples
Class A Solitary or in pairs or family groups; small, permanent home range; highly diversified diet, but selective feeders.	1–20	Freeze, dash to cover and freeze, or lie down. Do not outrun or counterattack predators.	Dik-dik (*Madoqua*), duiker (*Cephalophus*)
Class B Several female offspring units associate; group size 1 to 12; permanent home range; males solitary; diversified diet.	15–100	Similar to class A, but with some outrunning of predators for short distances.	Reedbucks (*Redunca*), vaal rhebuck (*Pelea*), oribi (*Ourebia*), lesser kudu (*Tragelaphus imberbis*)
Class C Larger herds, of six to hundreds of members; males have breeding territories; selective browsers and grazers.	20–200	Diverse. Hiding used in heavy cover, running used in open situations; communication of alarm behavior important.	Kob, waterbuck, lechwe (*Kobus*), gazelles (*Gazella*), impala (*Aepyceros*), greater kudu (*Tragelaphus strepsiceros*)
Class D During sedentary times societies as in Class C species; gigantic herds develop during migration; feeders on diversity of grass; selective as to plant parts eaten.	100–250	Run from large predators or may mount unified counterattack on smaller predators.	Wildebeest (*Connochaetes*), hartebeest (*Alcelaphus*), topi (*Damaliscus*)
Class E Large, stable herds of females and young with males organized into dominance hierarchies; herd size up to one or two thousand; no coalescing of herds during migration; unselective grazers or browsers.	200–700	Run from predators or mount unified counterattacks even on large predators. Groups respond to distress calls of young.	Buffalo (*Syncerus caffer*), beisa oryx (*Oryx beisa*), gemsbok (*Oryx gazella*), probably eland (*Taurotragus*)

(Based on Jarman, 1974, and Wilson, 1975.)

Figure 17–23. A red duiker (*Cephalophus natalensis*) in dense bush in southern Kenya, East Africa.

are typically highly gregarious and use primarily visual signals.

As noted and described in summary fashion in Table 17–1, the social organization of the classes of bovids recognized by Jarman and Wilson forms a progression from the small social units of the selective feeders of the bush and forest to the very large herds of the unselective feeders of the grassland. To flesh out the skeleton of

this outline, comments on the social behavior of several species follow.

Although both the impala and the Uganda kob (*Adenota kob*) have somewhat similar social organizations (Class C in Table 17–1), their breeding behavior differs. The impala has been studied by Jarman (1970), and by Jarman and Jarman (1974). Dominant male impala, comprising roughly one third of the population of adult males, maintain territories in the breeding season. These territories form a mosaic of adjoining areas, and each dominant male defends his area against other males of comparable social status. Females and bachelor males occupy home ranges that typically include a number of territories. A territorial male attempts to round up females that enter his area and keep them within it. In bachelor herds, the hierarchy is based partly on age distinctions, with older, larger-horned animals dominating younger ones. Adult males of different ages are not set apart by obvious clues such as body size and horn size, and among these animals a social hierarchy is established by frequent aggressive encounters featuring displays and fighting (Fig. 17–24). Males at or near the top of the bachelor hierarchy challenge territorial males, and repeated encounters between a territorial male and his challenger might span several weeks. A male holding a prime territory much frequented by impala is kept busy herding females, checking for females in estrus, and keeping bachelor

Figure 17–24. Sparring between two pairs of male impala (*Aepyceros melampus*) that are members of a bachelor herd. See text for explanation. (Photograph taken in Tarangire National Park, Tanzania, East Africa.)

and competitive adult males at a distance. These males become exhausted and lose their territories more quickly than do males holding areas less preferred by impala. In areas with seasonal precipitation, the territorial system is abandoned during the dry seasons.

Both male and female Uganda kob return to traditional and permanent territorial breeding grounds called leks to breed, and breeding occurs throughout the year (Buechner, 1961, 1963, 1974; Buechner and Roth, 1974). Each lek is some 200 m in diameter, and is defended by a single male through displays and ritualized fighting. The turnover of males on territories is high. Males on territories with the most traffic of females and of invading males are generally displaced within ten days. As they enter estrus, the females move into the leks and pass freely through the territories of the males, who defer to one another as the females cross territorial boundaries. Both sexes show great fidelity for their "home" leks, and males are especially faithful to the vicinity of the lek. Over half of the mature males tagged by Buechner (1974) remained within 50 m of the home lek from 90 to 100 per cent of the time.

The advanced social organization of the African buffalo (Synceros caffer) has been studied by Sinclair (1970, 1974) in Tanzania. This nonterritorial animal forms herds of from 50 to 2000 animals. The size of a given herd is rather constant, with a mean herd size of about 350 animals. For the first three years of its life a young buffalo tends to remain near its mother, and bonds between mothers and daughters seem closer than those between mothers and sons. In the third year of life, males begin to leave their mothers, and when they are four and five years old they form subgroups within the herd. Older, adult males that remain with the herd establish a linear dominance hierarchy. The repeated sparring typical of immature males may result in the formation of this hierarchy. The less dominant males are driven from the herd and form small bachelor groups that remain separate from the mixed herds. Old males, over about ten years of age, leave the herd and become extremely sedentary; they are either solitary or form small social units. The breeding is done largely by the dominant males of the herd, with the highest ranking males having the greatest access to estrous females.

Especially remarkable is the way in which the herd functions as a tightly knit unit. An entire herd will rally to the defense of an animal in distress, and their formidable united front will discourage even the largest predators. A herd also moves and feeds as a closely massed unit. There is little attempt to maintain individual distance, and the bodies of herd members may on occasion be touching. The newborn young of some kinds of antelope, such as wildebeest, are very quickly able to run fast enough to keep up with the running herd, but Sinclair noted that for several weeks after birth young buffalo are unable to stay with a running herd. Perhaps the coordinated defense behavior of the herd is especially important in protecting the young, slow-moving calves.

Geist (1974) points out that the widespread substitution by bovids of ritualized combat and aggressive displays for damaging physical contact has probably evolved under selection exerted by high densities and high diversities of predators. Bovids that attack and injure others invite damaging counterattack and are likely to be wounded, whereas those that use nondamaging, ritualized fighting are less likely to sustain injury. Because predators often concentrate on conspicuously wounded animals, selection by predators strongly favors the adherence by bovids to ritualized intraspecific contests. This great development of ritualized combat in African bovids, which must face many diverse predator populations, contrasts with the more damaging encounters between members of northern species, such as bighorn sheep (Ovis canadensis), that are under far less pressure from predators.

The matter of fighting among horned or antlered artiodactyls merits further comment. An interesting correlation seems to exist between the degree of ritualization of fighting and the structure of the head weaponry. In elk the pattern of branching of the antlers is such that antlers of combatants interlock during a fight; there is little chance that tines will inflict injury. These animals can afford the luxury of forceful contact during tests of strength; injuries seem to occur largely when the antlers are disengaged and one animal is attempting to escape. In the genus *Oryx* the horns are long, unbranched, and pointed, and are extremely dangerous weapons. Fighting is highly stylized, with the pushing occurring with the foreheads together but with little contact between the horns (Walther, 1958). The test of strength occurs within the limits of a pattern of behaviors that avoids injury from the dangerous horns. Encounters between male Grant's gazelles, which have quite dangerous horns, usually go no further than intimidation by a neck display (Fig. 17–25) whereas, in the Thompson's gazelle, Estes (1967:189) found that "natural selection has operated on horn configuration and fighting style to produce a relatively safe type of parry-thrust combat, thus obviating the need for a display substitute." In most cases mutual adaptations between head weapons and fighting style have insured that opponents come together and measure each other's strength without the use of the full offensive or defensive potential of the weaponry. The pattern of rings around the horns of many African antelope may possibly serve as "nonslip" devices that reduce the danger of injury to fighting males.

THREAT AND APPEASEMENT. Threat behaviors are among the most familiar activities of mammals. A dog lifts its upper lip to expose the length of its upper canines, a cat opens its mouth and hisses, and some rodents grind their teeth. These actions all signal a readiness to fight or to attack if the antagonist does not retreat or take other appropriate action. A threat is typical of a situation in which conflicting tendencies preclude either an immediate attack or a hasty retreat.

Threats can be simple, as in animals that merely open the mouth widely to display the teeth (Fig. 17–26), or complex, as in some horned artiodactyls in which both distinctive postures and movements are involved, but usually seem to advertise the most important offensive weapons (Fig. 17–13). Visual threats may be made more impressive or startling by audible threats such as explosive hisses or growls. Strictly defensive threats are often used by animals that are under pressure from an aggressor but that would gladly escape. The cat's threat with the back arched and the side of the body confronting an aggressor is such a behavior, and the "oblique stare" used by bull sea lions on adjacent territories is seemingly a ritualized defensive threat. In an extreme defensive posture, many carnivores such as cats and weasels lie on their backs with the teeth and claws

Figure 17–25. The neck display of the Grant's gazelle (*Gazella granti*). (From Estes, 1967.)

Figure 17-26. Open-mouthed threat by a pouched "mouse" (*Dasyuroides byrnei*, Marsupialia). (Photograph by Jeffrey Hudson.)

ready for action. The animal has retained its intention to defend itself while abandoning any inclination to attack.

Some mammals have carried this type of defensive behavior a step further by discarding the pretense of defense. Such an appeasement posture or behavior is a complete surrender, and contains no elements that are likely to trigger an opponent's aggression. Complete vulnerability is emphasized, and the response on the part of the dominant animal is to cease its hostile activity. Wolves and many other mammals appease their dominant opponents by lying on the back with the vulnerable throat and underside unprotected. In several artiodactyls, lying down serves as appeasement (Walther, 1966; Burckhardt, 1958), and a subordinate black wildebeest (*Connochaetes gnou*) may roll on its side with its belly towards its superior and the side of its head on the ground (Ewer, 1968:177). In the Grant's gazelle, a lowering of the head, the reverse of the high-headed threat posture, is adopted by a submissive animal (Walther, 1965). In some primates the presenting of the rump as the female does prior to copulation is an appeasement gesture. The brightly colored skin on the rump of some Old World monkeys may serve in part to make the rump conspicuous and thus to make "presenting" appeasement gestures more effec-

tive. A "grin" serves as an appeasement in some anthropoid primates—perhaps this is akin to responding to the cowboy's demand to "smile when you say that, podner."

Appeasement behavior clearly serves several purposes. It allows an animal being defeated in a fight to avoid further injury, and in many cases allows a subordinate animal to avoid a contest. In highly social species, threat and appeasement behaviors foster the peaceful perpetuation of a dominance hierarchy, and allow animals to be close to one another with a minimum of energy wasted on aggressive interactions. Appeasement behavior may even be important in permitting a subordinate animal to seek social contact without risking attack (Schenkel, 1967). The evolution of appeasement behavior has followed the same course toward ritualization as have other behavioral patterns. Ritualization serves to make signals as simple, obvious, and as unambiguous as possible.

FRIENDLY BEHAVIOR. In many social species patterns of friendly behavior, and often close bonds between individuals, help maintain social structure. Grooming is the most common type of friendly behavior. This may involve the grooming of infants by the mother, mutual grooming of the fur by adults as in primates, or grooming as part of courtship behavior. Other friendly or recognition behaviors include smelling of the mouth and the anal and genital region in dogs, the mutual embrace of chimpanzees *(Pan troglodytes)* described by Goodall (1965:471, 472), and the mutual pressing together by duiker antelope *(Cephalophus maxwelli)* of the maxillary scent gland (Ralls, 1971:446). The choral howling so characteristic of such social canids as wolves may keep members of a group apprised of the locations of other members and may strengthen familiarity and bonds between individuals of the group. The nuzzling and tail wagging of wolves preparing for a hunt may serve to create a common mood in preparation for a cooperative effort (Lorenz, 1963).

Accompanying social behavior and strong friendly ties between individuals is often extreme pugnacity towards "outsiders." An individual recognized as not belonging to the social group may be attacked and driven away. Smell is an important clue used in such recognition (Ewer, 1968; Crowcroft, 1966, Ralls, 1971), but to some species visual signals are also important.

ACTIVITY RHYTHMS

A striking aspect of the behavior of animals is the rhythmic or cyclic pattern of activity. Some species are active at night *(nocturnal)* and some during the day *(diurnal)*; some are active primarily at dawn and dusk *(crepuscular)*. The activity periods tend to be at regular intervals—the time of emergence of a particular species of bat may differ by no more than two or three minutes night after night. Animals also exhibit other kinds of cyclic behavior. The timing of reproduction is cyclic and, in some mammals such as some rodents or bats, daily or seasonal shifts occur between highly active and torpid states. Migratory movements are also cyclic. Daily activity rhythms, those based on the 24-hour cycle, are termed *circadian rhythms*, and are better understood than are other types of rhythms.

Circadian rhythms differ markedly between species. Most mammals are nocturnal, but even between two nocturnal species there are contrasts between the patterns of activity (Fig. 17–27). In general, small mammals that are especially vulnerable to predation, such as rodents, tend to be nocturnal (chipmunks and ground squirrels are exceptions), whereas less vulnerable species, such as many ungulates, may be more or less active during the day. The activity cycles of carnivores seem to be geared to the circadian cycles of their prey or to the period when hunting is most rewarding. Martens *(Martes americana)* frequently forage by day, when red squirrels *(Tasmasciurus hudsonicus)* are active, whereas coyotes *(Canis latrans)* hunt at dusk and at night, when rabbits and rodents are feeding.

Circadian rhythms are also influenced by interactions between species with similar environmental needs; in some cases competition between species is reduced or eliminated because their activity cycles are

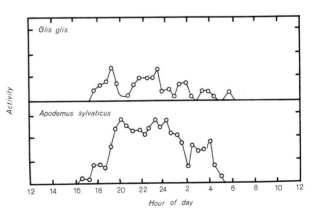

Figure 17–27. The autumn activity cycles of two nocturnal rodents. (After Eibl-Eibesfeldt, 1958.)

out of phase. Two species of fishing bats (*Noctilio*), both of which feed over water, avoid interference partly by foraging at different times of the night (Hooper and Brown, 1968). Clearly, the circadian rhythm of an animal is part of its total adaptation to its particular mode of life and environment, and has evolved just as have morphological characters.

The question of whether circadian cycles are endogenous (internally controlled) or exogenous (ultimately regulated by external stimuli) has occupied the attention of many biologists. Clearly, some strong endogenous control is present in many species. As an example, careful work on the flying squirrel (*Glaucomys volans*) by DeCoursey (1961) showed that even under constant environmental conditions, including continuous darkness, flying squirrels maintained regular activity periods that only deviated ±2 minutes from the mean value for activity periods under natural conditions. When a laboratory animal whose circadian cycle is not in phase with the natural 24-hour light-dark cycle is again exposed to normal day and

night conditions, however, its cycle rapidly shifts and becomes "synchronized." The cycle becomes adjusted and locked — en-*trained* — to the 24-hour cycle (Bruce, 1960). Circadian cycles, and other animal behaviors, are seemingly regulated by intricate and as yet poorly understood interactions between endogenous and exogenous factors.

As might be expected if circadian cycles are adaptive, they shift seasonally in some species (Fig. 17–28). Attending the seasonal changes in environmental temperatures are changing metabolic demands put on small mammals, and some shifts in circadian rhythms may allow the animals to avoid activity during times of most intense temperature stress. Studies by Wirtz (1971) have shown that in the deserts of California the antelope ground squirrel (*Ammospermophilus leucurus*) is most active at midday in the winter, whereas in the summer the greatest activity is in the morning and late afternoon; the midday heat is then avoided. The shift from nocturnal activity in the summer to diurnal activity in the winter by a bank vole (Fig. 17–28) proba-

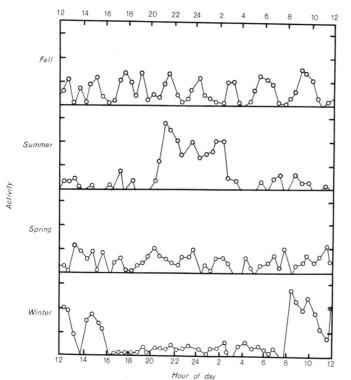

Figure 17–28. Seasonal changes in the daily activity cycles of a vole (*Clethrionomys glareolus*). (After Eibl-Eibesfeldt, 1958.)

bly results in a considerable saving of energy.

Activity cycles are also clearly geared to the basic metabolic demands of animals. In shrews, activity periods are distributed more or less evenly through the day, and the smaller shrews have shorter and more frequent bursts of activity (Crowcroft, 1953). The high metabolic rates of the smaller shrews require frequent periods of feeding and short intervals between feedings. Larger mammals such as rabbits and rats, which have much lower metabolic rates than those of small shrews, can meet their energy needs by feeding at dusk and at night, and some rabbits probably have only two feeding periods per 24-hour cycle.

Although less thoroughly studied than circadian rhythms, *circannian rhythms* play an equally prominent role in the lives of some mammals. Among vertebrates, such vital activities as breeding, migration, and hibernation are phased on an annual cycle or a circannian rhythm. Pengelley (1967) defined a circannian rhythm as an endogenous cycle that has a length of approximately one year. Among mammals, such rhythms have been documented by Pengelley and Fisher (1963) for the golden-mantled ground squirrel (*Spermophilus lateralis*), by Davis (1967) for the woodchuck (*Marmota monax*), and by Heller and Poulson (1970) for two species of ground squirrels (*Spermophilus*) and four species of chipmunks (*Eutamias*).

Circannian rhythms are a major key to the survival of some temperate zone and arctic mammals. As said by Heller and Poulson (1970), these rhythms allow an organism "to anticipate, and thus prepare for, a future, annually occurring environmental condition such as cold weather, drought, food scarcity or optimal breeding time." The rhythm also ensures some flexibility of response to cyclic environments that may differ markedly from year to year. In addition, circannian rhythms enable "the organism to integrate a large number of environmental cues and through phasing the rhythm respond most favorably to

conserve energy and to ensure reproductive success."

In hibernators in temperate regions, the rhythms make the animals sensitive in the autumn to falling temperatures and declining body weight so that the onset of hibernation may be hastened by unfavorable conditions or delayed by favorable temperatures and food supplies. In arctic areas, the extremely harsh environment and the sudden onset of winter, coupled with the brief time available for breeding and for putting on fat in preparation for hibernation, make flexibility nonadaptive. Here the adaptive premium shifts to a precise, inflexible, optimal schedule. Breeding at the optimal time each year, regardless of climatic conditions, probably ensures the greatest reproductive success, and precision in the onset of hibernation ensures maximum over-winter survival. But even in tropical areas, circannian rhythms may be highly adaptive. In Kenya, East Africa, for example, the African false vampire bat (*Cardioderma cor*) and the giant leaf-nosed bat (*Hipposideros commersoni*) become pregnant well before the onset of the late March–April rainy season, seemingly in anticipation of the burst of insect abundance that accompanies the rains. This pattern of breeding is probably controlled by a circannian rhythm.

The factors controlling these remarkable rhythms are as yet poorly understood. For arctic hibernators, Heller and Poulson (1970) suggest that photoperiod is the environmental factor of primary importance in phasing the underlying circannian rhythm. In temperate zone hibernators, in contrast, although the timing of breeding seems inflexible, the animals are responsive to environmental conditions in the fall and may delay or hasten their entrance into hibernation. The situation is not entirely simple, however, for even temperate zone hibernators that occupy the same area do not follow the same circannian rhythms. The activity of the golden-mantled ground squirrel, an inhabitant of mountains in the western United States, is controlled largely by an endogenous rhythm. This animal feeds into the fall and stores seeds, a relatively nonperishable food, in its burrow, but its entrance into hibernation is relatively tightly scheduled. The Belding ground squirrel (*Spermophilus beldingi*), that often lives almost side-by-side with the golden-mantled ground squirrel, feeds on green material that decomposes quickly if stored underground; this squirrel feeds as long as possible in the fall, putting on more and more fat, but stores no food. Whereas the golden-mantled ground squirrel can perhaps afford greater rigidity in the timing of its hibernation because of the cushion of stored food in the hibernaculum, the Belding ground squirrel must depend entirely on food stored in the form of body fat, and thus feeds as long as it is energetically feasible.

REPRODUCTION

18

Compared to the types of reproduction in all other vertebrate classes, the reproductive pattern typical of mammals departs furthest from that of primitive vertebrates. Primitive, ancestral vertebrates were presumably egg layers, and this style of reproduction, or fairly modest variations on this theme, is typical of all classes of vertebrates but the Mammalia. In all mammals but prototherians, young remain during their embryonic and fetal life within the uterus, and here embryonic differentiation of tissues and organs and growth of the fetus occur. Nourishment and protection for the intrauterine young are provided by the mother, and under most conditions survival rates of the fetuses are high. After birth all young mammals are nourished by milk from the mother, and parental care, or in most cases maternal care, lasts until young are reasonably capable of caring for themselves. The young of some mammals stay with their parents through an additional period of learning that increases the chances for survival when they become independent. In sharp contrast, in most nonmammalian vertebrates (the birds are an exception) the young have little or no parental care after hatching, or in the case of ovoviviparous animals, after birth. In mammals, the combined effect of the high survival rate of fetuses and extended postpartum care is an increase in the efficiency of reproduction in terms of expenditure of energy per young that reaches reproductive maturity. Most lower vertebrates lay great numbers of eggs at tremendous metabolic expense, and the success of the species depends on the survival of an extremely small percentage of the young; considering any given young of a lower vertebrate, survival is un-

likely. In mammals, on the contrary, relatively few young are produced, but the likelihood for survival of any given young is fairly high.

The following sections of this chapter consider unique features and typical major patterns of mammalian reproduction. No attempt has been made to catalogue exhaustively the reproductive cycles of all families of mammals, but tables are included that review features of the reproductive cycles of selected mammals representing most orders. Descriptions of reproduction in many mammalian species are given by Asdell (1964), and the tables are partly based on his work.

THE MAMMALIAN PLACENTA

One of the most distinctive and important structures associated with the reproduction of therian mammals is the placenta. Differences among the major placental types have been used in distinguishing some of the higher taxonomic categories of mammals (subclasses and infraclasses); and some primary contrasts between reproductive patterns in mammals depend partly on placental differences. Further, the relative competitive abilities of marsupials and eutherians may be strongly influenced by basic differences among the structures and functions of the placentas of these groups.

A functional connection between the embryo and the uterus is necessary in animals in which development of the fetus occurs within the uterus, and in which nutrients for the fetus come directly from the uterus rather than from yolk stored in the ovum. This connecting structure, the placenta, allows for nutritional, respiratory, and excretory interchange of material by diffusion between the embryonic and the maternal circulatory systems, and consists of both embryonic and uterine tissues. The placenta also functions as a barrier that excludes from the embryonic circulation bacteria and many large molecules. In addition, in eutherians the placenta produces certain food materials and synthesizes hormones important for the maintenance of pregnancy. Mammals are not unique in having a placenta, for certain fishes and reptiles establish placenta-like connections allowing diffusion of materials between the vascularized oviduct and the embryo. Among mammals, the major types of placentae differ sharply in structure and in the efficiency with which they facilitate the nourishment of the embryo.

CHORIOVITELLINE PLACENTA. This, the most primitive type of mammalian placenta, occurs in all marsupials except those of the family Peramelidae, the bandicoots. In marsupials with a choriovitelline placenta, the yolk sac is greatly enlarged to form a placenta. The blastocyst does not actually implant itself deep in the uterine mucosa, as is the case in eutherians, but merely sinks into a shallow

depression made by erosion of the mucosa. The contact is strengthened by the wrinkling of the wall of the blastocyst that lies against the uterus, and this wrinkling serves to increase the absorptive surface of the blastocyst. The embryo is nourished largely by "uterine milk," a nutritive substance secreted by the uterine mucosa and absorbed by the blastocyst. The embryo also derives nourishment from limited diffusion of substances between the maternal blood in the eroded depression in the mucosa and the blood vessels within the large yolk sac of the blastocyst (Fig. 18–1).

CHORIOALLANTOIC PLACENTA. This type of placenta occurs in the bandicoots and in all eutherian mammals. Although similar to the eutherian chorioallantoic placenta in basic structure, the peramelid placenta achieves less effective transfer of substances between the fetal and maternal circulations. In peramelids the allantois is fairly large and becomes highly vascularized; the blastocyst rests against the endometrium on the side where the allantois contacts the chorion. At the point of contact with the blastocyst, the uterus is highly vascularized, and the part of the chorion against the vascularized endometrium is more or less lost. At this point of approximation of the maternal blood

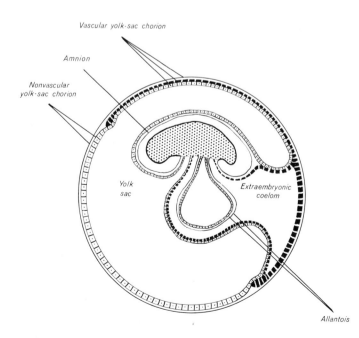

Figure 18–1. A diagram of the embryo and extraembryonic membranes of a marsupial (*Didelphis*). (After Torrey, T. W.: *Morphogenesis of the Vertebrates*, 3rd ed., John Wiley & Sons, Inc., 1971. By permission of John Wiley & Sons, Inc.)

stream and the allantois, exchange of materials occurs across the allantoic membranes. Because the peramelid allantois lacks villi and only its corrugations serve to increase its absorptive surface, a limited surface area is available for exchanges of material between the maternal and fetal blood streams; supplementary nutrition is supplied by uterine milk. Probably due partly to the lack of villi and the resulting lack of absorptive efficiency of the allantois, the gestation period of peramelids is fairly short, and the suckling period is long (Table 18–1).

In eutherian mammals the chorioallantoic placenta reaches its most advanced condition with regard to facilitating rapid diffusion of materials between the fetal and uterine blood streams. In eutherians the blastocyst first adheres to the uterus and then sinks into the endometrium. The mechanisms by which implantation occurs are not fully understood. Proteolytic enzymes secreted by the chorion have been thought to erode the endometrium and allow the blastocyst to sink into the cavity thus formed, but little definite evidence supports this view. Studies by Boving (1959) have shown that in the domestic rabbit dissociation of uterine epithelial cells overlying blood vessels facilitates the implantation of the blastocyst. This dissociation is initiated by a local rise in pH caused by a bicarbonate compound produced by the blastocyst, and by a reciprocal reaction on the part of the endometrium to remove the bicarbonate. How widespread this reaction is among mammals, and therefore how important it is as an implantation-furthering device is not known.

As implantation proceeds, chorionic villi grow rapidly and push farther into the endometrium as local breakdown of uterine tissue continues. The resulting tissue "debris" is often called embryotroph; this nutritive substance is absorbed by the blastocyst and nourishes the embryo until the villi are fully developed and the embryonic vascular system becomes functional. In response to the presence of the blastocyst, the uterus becomes highly vascularized at the site of implantation. When the eutherian placenta is fully formed, the complex and highly vascularized villi provide a remarkably large surface area through which rapid interchange of materials between the maternal and fetal circulations can occur (Fig. 18–2). The extent to which the villi increase the surface area available for diffusion is difficult to imagine; the extent of this increase is suggested by the fact that the total length of the villi in the human placenta is roughly 30 miles (Bodemer, 1968).

Among eutherians the degree to which the maternal and fetal blood streams are separated in the placenta varies widely. Lemurs, some ungulates (suids and equids), and cetaceans have an *epitheliochorial placenta*, in which the epithelium of the chorion is in contact with the uterine epithelium and the villi rest in pockets in the endometrium. Under these structural conditions, oxygen and nutrients must pass through the walls of the uterine blood vessels and through layers of connective tissue and epithelium before entering the fetal blood stream. In ruminant artiodactyls, the uterine epithelium is eroded locally, and contact between the chorionic ectoderm and the vas-

Table 18–1. Breeding Cycles of Several Families of Marsupials

Family	Breeding Season; Polyestrous or Monestrous	Litter Size	Gestation Period in Days	Suckling Period in Days
Didelphidae	March–Oct.; monestrous or polyestrous	2–25	13	70–80
Dasyuridae	April–Dec.; monestrous	3–12	8–34	49–150
Peramelidae	March–June; polyestrous	1–7	15	59
Phalangeridae	all year; 1 litter/2 yrs. or polyestrous	1–6	17–35	42–165
Macropodidae	all year; monestrous or polyestrous	1–2	24–43	64–270

(Data on suckling period from Sharman, 1970.)

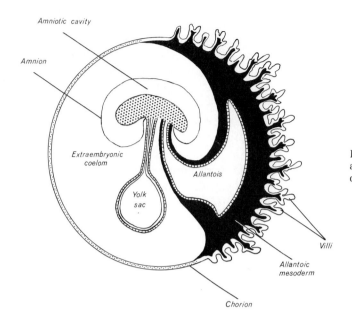

Amniotic cavity

Amnion

Extraembryonic coelom

Allantois

Yolk sac

Villi

Allantoic mesoderm

Chorion

Figure 18-2. A diagram of the embryo and extraembryonic membranes of a placental mammal. (After Balinsky, 1975.)

cular uterine connective tissue occurs. This is a *syndesmochorial placenta*. In carnivores, erosion of the endometrium is carried further and the epithelium of the chorion is in contact with the endothelial lining of the uterine capillaries. This is called an *endotheliochorial placenta*. Destruction of the endometrium in some mammals may involve even the endothelium of the uterine blood vessels, allowing blood sinuses to develop in the endometrium; the chorionic villi may then be in direct contact with maternal blood. This *hemochorial placenta* occurs in some insectivores, bats, anthropoid primates, and some rodents. In rabbits and some rodents, the destruction of placental tissue is so extreme that only the endothelial lining of the blood vessels in the villi separates the fetal blood from the surrounding maternal blood sinuses (Arey, 1974:147). In this case a *hemoendothelial placenta* results.

The shape of the placenta is governed by the distribution of villi over the chorion. Several different distributions of villi occur in mammals. The lemurs, some artiodactyls, and perissodactyls have a *diffuse placenta*; this type of placenta has a large surface area because the villi occur over the entire chorion. Ruminant artiodactyls have *cotyledonary placentae*, consisting of more or less evenly spaced

groups of villi scattered over the most avillous chorion. Carnivores have a *zonary placenta*; a continuous band of villi encircles the equator of the chorion. In the *discoidal placenta*, villi occupy one or two disc-shaped areas on the chorion; this type occurs in insectivores, bats, some primates, rabbits, and rodents.

At birth the fetal contribution to the placenta is always expelled as part of the "afterbirth," but the maternal part of the placenta may or may not be lost at this time. In mammals with the *epitheliochorial* type of placenta, the villi pull out of the uterine pits in which they fit, none of the endometrium is pulled away, and no bleeding occurs at birth. This placenta is termed *nondeciduous*. In mammals with placentae allowing more intimate approximation of uterine and fetal blood streams, because of the extensive erosion of the uterine tissue and the extensive intermingling of uterine and chorionic tissue, at birth the uterine part of the placenta is torn away, resulting in some bleeding. This type of placenta is *deciduous*. The hemorrhaging after birth is soon stopped by the collapse of the uterus and by contractions of the myometrium, which tend to constrict the blood vessels, and by clotting of blood.

The rate of movement of substances

from the maternal to the fetal blood stream in the placenta is of course increased by the reduction of the number of interposed membrane barriers. Because of the difference between the number of such barriers in man and the pig, for example, sodium is transferred 250 times more efficiently by the human placenta than by that of the pig (Flexner et al., 1948). The remarkable absorptive ability of the allantoic placentae of such mammals as insectivores, bats, primates, rabbits, and rodents is due largely to the great surface area afforded by the complex system of villi, to the extensive erosion of uterine mucosa and the resulting development of blood sinuses into which the villi extend, and to the loss of nearly all of the membranes separating uterine from fetal blood.

THE ESTROUS CYCLE, PREGNANCY, AND PARTURITION

In mammals reproduction is characterized by a series of cyclic events that are under nervous and hormonal control. As with many complex functions of the vertebrate body, the regulation of the reproductive cycle is maintained through reciprocal controls between endocrine organs and their secretions. The events characterizing the stages in the reproductive cycle in mammals are well known, but details of the hormonal regulation of these events are not completely understood. The ovarian cycle results in the development of ova, their release from the ovary, and their passage into the uterus; the uterine cycle involves a series of cyclic changes in the uterus.

The ovarian cycle includes two major phases: (1) the growth of the follicle and the release of the ovum; and (2) the development of the corpus luteum from the ruptured follicle. This cycle seems to be controlled largely by pituitary and ovarian hormones. The pituitary produces FSH (follicle-stimulating hormone) and LH (luteinizing hormone), which act together to stimulate growth of the follicle and to initiate the secretion of estrogen by the ovary.

Estrogen acts on the pituitary to stimulate increased production of LH, to initiate production of LTH (luteotropic hormone), and to reduce the secretion of FSH; under the influence of interactions between these hormones ovulation occurs (ova are released from follicles) and the corpus luteum forms from the ruptured follicle. Maintained by LTH from the pituitary, the corpus luteum produces progesterone, which sensitizes the uterus for implantation. If fertilization of ova does not occur, the corpora lutea regress and estrogen and progesterone production is reduced. The pituitary responds by resuming production of FSH, and another ovarian cycle is initiated.

This pattern, involving spontaneous ovulation, is the typical situation in mammals, but some deviations are known. In the cat and some rodents, the follicles develop, but ovulation does not occur until after copulation. In rabbits the follicles do not develop fully before copulation and a long estrus may occur; ripening of the follicles and ovulation are initiated by copulation.

As the ovarian cycle proceeds, a series of cyclic changes occur in the uterus. The uterus consists of an outer peritoneal layer; an intermediate layer of smooth muscle, the *myometrium;* and an inner layer of uterine mucosa, the *endometrium.* Just before ovulation, the endometrium becomes thicker; this uterine stage is called the *proliferation phase.* In many mammals a period of "heat," during which the female is receptive to the male, occurs at the end of the proliferation phase. This time of receptivity immediately prior to ovulation is termed *estrus.* After ovulation the endometrium develops further and becomes highly vascularized; this is the *progestational phase* of the uterine cycle.

In most mammals, if the ova are not fertilized the endometrium shrinks and the vascularization is reduced. In nonprimate mammals, no extensive bleeding occurs during the regression of the endometrium and the period of receptivity is short; in these animals the uterine cycle is referred to as the *estrous cycle.* Some mammals have repeated estrous cycles during a year and are said to be *polyestrous;* others have a single estrous cycle per year and are *monestrous.*

In some species sexual cycles are seemingly influenced by day length. The eyes receive light over a progressively longer period of time each day during the spring and initiate a nervous reflex transmitted by the optic nerves to the brain, where certain centers stimulate the release of gonadotropic hormones by the pituitary. The ovarian cycle may be triggered in some rodents by vitamins or nutritional factors present in the green vegetation that appears in the spring (Negus and Pinter, 1966). Although considerable information relating to mechanisms controlling reproductive cycles is scattered through the literature, the importance of these mechanisms remains obscure. It seems doubtful that in most mammals any single factor or simple combination of factors is responsible for controlling the estrous cycle.

The ovarian and uterine cycles in primates are different, to some extent, from those in other mammals. This primate cycle is called the *menstrual cycle.* In man and most other primates considerable bleeding is typical of the time of endometrial breakdown, ovulation occurs at regular intervals throughout the year, and females may be receptive over an extended period. Even in such advanced primates as the gorilla (*Gorilla gorilla*), however, copulation is cyclic and occurs rarely except when the vulva of the female is swollen in association with ovulation (Nadler, 1975).

When copulation occurs the sperm reach the oviducts within a matter of minutes in some species, and fertilization of the ova usually occurs within 24 hours of ovulation. The zygotes move down the oviducts, aided by contractions of the muscles of the oviducts, and usually reach the uterus and implant within a few days.

The delicate hormonal control of pregnancy is exerted in eutherians by interactions between hormones produced by the pituitary, the ovary, and the uterus. In the early part of pregnancy, chorionic gonadotropin is of critical importance in preserving the corpora lutea and in preventing regression of the endometrium. This hormone is produced first by the trophoblast during its implantation in the endometrium, and then by the chorion, which develops from the trophoblast. During early pregnancy the corpus luteum, because of its production of progesterone, is important in maintaining pregnancy by keeping the endometrium in a thickened and highly vascularized condition and by altering the ability of the myometrium to perform coordinated contractions that might expel the embryo. In some species progesterone sensitizes the endometrium and increases the efficiency of implantation of the blastocyst.

The maintenance of pregnancy in many mammals is not entirely under the control of the corpus luteum; instead, as pregnancy continues, hormones are produced progressively more by the placenta and less by the ovary. In man the placenta is thought to produce chorionic gonadotropin and estrogens, and is probably the most important source of progesterone. During the latter stages of pregnancy the placenta seems to be a nearly independent endocrine gland, which in man, at least, takes over the functions of the pituitary gland and the corpus luteum.

An important hormone of pregnancy, but one whose function is important mainly at parturition, is *relaxin.* This hormone is known to occur in a variety of mammals, and may be universal among mammals. The concentration of relaxin in the blood stream increases toward the end of pregnancy; this hormone causes relaxation of the pelvic ligaments and the pubic symphysis in preparation for parturition. In some mammals, such as pocket gophers (Geomyidae), the connective tissue joining the pubic bones is resorbed at puberty and a gap between the pubic bones remains during the rest of the life of the animal (Hisaw, 1924). Relaxin may be produced by the uterus or by the placenta, and in man is known to be produced by the ovaries during pregnancy (Guyton, 1971).

Birth is accomplished by rhythmic and powerful contractions of the uterine myometrium, aided by the abdominal muscles.

Continued contractions force the placenta from the uterus and the vagina. The uterine contractions are seemingly under the control of interacting hormones. *Oxytocin*, produced by the hypothalamus and stored in the posterior lobe of the pituitary, occurs in increasingly higher concentrations in the maternal blood stream toward the end of pregnancy; oxytocin can initiate contractions of the uterus. Apparently the reduced concentrations of progesterone late in pregnancy are insufficient to block the effects of the increasing levels of oxytocin, with resulting contractions of the myometrium and parturition. The amount of estrogen also increases late in pregnancy, and the sensitizing of the uterus by estrogen shortly before parturition may allow oxytocin to initiate uterine contractions.

The newborn mammal is nourished by milk produced by the female's mammary glands. Under the influence of estrogen and progesterone from the placenta, these glands undergo considerable growth during pregnancy. Milk secretion is stimulated and regulated by *prolactin*, produced by the anterior lobe of the pituitary. Prolactin is secreted in progressively larger amounts during the latter part of pregnancy and after parturition, and when its inhibition by placental hormones is removed after birth, milk secretion begins. Milk production is partly under neural control and continues only as long as the suckling stimulus persists.

THE MARSUPIAL-PLACENTAL DICHOTOMY

The recognition of the marsupial-placental dichotomy is based on a number of biological differences; primary among these are the contrasting reproductive patterns. Whereas marsupials bear virtually "embryonic" young after a brief gestation period, many placentals bear highly precocious young and all placentals bear young that are anatomically "complete," after a relatively long gestation period. Why this difference? In one of his excellent papers on the marsupial-placental dichotomy, Lillegraven (1976) considered evidence bearing on this question. The following remarks are based largely on his discussions.

Vertebrates have an extremely effective immune-response system serving to destroy invading foreign antigenic materials. During prolonged gestation in vertebrates lacking substantial stored energy reserves (such as yolk), an intimate contact between mother and fetus must insure efficient physiological exchange, especially throughout the period of organogenesis and rapid growth. But in this critical period, because its paternal antigens may be recognized as foreign by the maternal system, the fetus risks destruction during its extended "parasitism" of the mother. The mechanisms tending to avoid this immune response are not understood for most vertebrates, but for mammals some information is available (see Anderson, 1972; Hughes, 1974).

In those marsupials that have been studied, the eggshell membranes are retained through the early two thirds of the gestation period. Although these membranes permit the passage of nutrients and probably of enzymes and antibodies, they interpose a barrier between the antigen-bearing parts of the embryo and the lymphocytes in the uterine fluid that initiate the immunological reaction. Immune-response rejection of the early fetus is thus avoided. Late in the gestation period, however, after the shedding of the shell membranes occurs and the fetal and maternal tissues are in close contact, the stage is set for an immunological attack. The apparent precipitous birth of the rudimentary young after the short gestation period is seemingly an adaptation enabling marsupials to avoid this attack.

Although eutherian mammals lack the shell membranes, the early stages of the zygote are separated from maternal tissues by the zona pellucida (a noncellular layer surrounding the zygote) and do not elicit an immune response. Later embryological stages are protected by the trophoblast, the epithelial covering of the blastocyst that develops prior to implantation, and by the placental trophoblast and its noncellular external secretion (Wynn, 1971). Adcock et al. (1973) found that in man chorionic gonadotropin blocks the action of maternal lymphocytes, protects the surface of the trophoblast, and allows the fetus to be accepted. Throughout the gestation period, despite the close apposition of the uterine tissues and the fetus, at least one layer or trophoblast constantly provides a barrier between fetal and maternal tissues. Because of its remarkable capacity for isolating the fetus from maternal tissues that would initiate an immunological response, while at the same time not hindering efficient transport of essential materials, the trophoblast has allowed eutherians the luxury of a long gestation period during which embryogenesis and rapid growth occur and an anatomically "complete" young is formed. Lillegraven (1976) states: "The 'invention' of the trophoblastic tissues by primaeval eutherians was probably the single most important event in the history of the subclass."

This event may well have been important in allowing a structural diversity among eutherians (from forms with grasping hands to those with hoofs or wings to marine forms with flippers) that marsupials could never match. Because all newborn marsupials must have forelimbs at birth that enable them to make their way to the marsupium, diversity of forelimb structure is impossible. Limitations imposed by the necessarily brief gestation period and long extrauterine development while attached to a nipple in the pouch have thus greatly restricted the adaptive options open to marsupials.

MAJOR REPRODUCTIVE PATTERNS

Although several deviations from the usual scheme occur, most mammals follow a similar reproductive pattern with regard to development of the embryo. After ovulation the ovum passes down the oviduct, where it is fertilized. Early cell cleavages occur during the several days occupied by passage of the zygote down the oviduct,

and by approximately the time the zygote enters the uterus it has become a hollow ball of cells enclosing a fluid-filled cavity. This stage is called the *blastocyst*. After further enlargement, the blastocyst implants in the endometrium. Implantation occurs between the fifth and fourteenth day after copulation, and the timing of implantation varies little within a species. After implantation, the embryo develops a system of membranes and blood vessels in the placenta that allows diffusion of nutrients and waste materials between the uterine and embryonic blood vessels.

In eutherian mammals, the length of time from fertilization to implantation is considerably shorter than the period between implantation and birth. Typically, fertilization occurs shortly after ovulation, and the development of the embryo from fertilization to birth is an uninterrupted process. Perhaps in response to specialized activity cycles, some mammals have abandoned this usual pattern of continuous development. One departure involves a delay of ovulation and fertilization until long after copulation (delayed fertilization); another is typified by normal fertilization and early cell cleavages, but also by an arresting of embryonic development at the blastocyst stage (delayed implantation).

DELAYED FERTILIZATION. This unusual pattern of development occurs in a number of bats inhabiting north temperate regions. As early as 1879 Fries recognized that males of some species of the families Rhinolophidae and Vespertilionidae had the unusual ability to store viable sperm through the winter, long after spermatogenesis had ceased; later studies detailed the reproductive cycles of the females of these species (see Guthrie, 1933; Hartman, 1933; Wimsatt, 1944, 1945). These remarkable reproductive cycles are seemingly adaptations to continuous or periodic winter dormancy, and occur in a number of New World and Old World species included in the following genera: *Rhinolophus, Myotis, Pipistrellus, Eptesicus, Nycticeius, Lasiurus, Plecotus, Miniopterus,* and *Antrozous.* Dealyed fertilization may be the typical pattern in all but the tropical members of the family Vespertilionidae. Excellent papers by Wimsatt (1944, 1945) and by Pearson and his colleagues (1952) describe delayed fertilization as it occurs in vespertilionids, and the following remarks will be based largely on those studies.

The reproductive cycle of the big-eared bat (*Plecotus townsendii*) in California follows a timetable similar to that of many vespertilionids of temperate zones. The testes descend into the scrotum in the spring. This migration is caused mostly by increased production of testosterone, which is cyclical in bats. The males become reproductively active in August. The testes begin to enlarge in the spring and are largest in September; spermatogenesis occurs mostly in late August and September. The testes regress and spermatogenesis ends before winter, but the caudal epididymides retain motile sperm through February, and the accessory reproductive organs remain enlarged throughout the winter. Young males are not fertile in their first autumn. In the females a single graafian follicle enlarges in the autumn, but remains in the ovary throughout the winter. A female may be inseminated repeatedly in the fall and winter, and males frequently copulate with hibernating females, although usually all females are inseminated by the end of November. The most typical vespertilionid pattern is for most copulation to occur before hibernation. The sperm are stored in the uterus, where they remain motile for at least 76 days. In *P. townsendii*, ovulation usually occurs in late February or March, either while the females are still at the winter roost or shortly after they leave. In many species inhabiting cold regions, ovulation occurs shortly after the females emerge from the hibernacula. Implantation is nearly always in the right horn of the uterus in *P. townsendii*, but ovulation occurs with equal frequency in each ovary. The gestation period in this species is highly variable, ranging from 56 to 100 days. This variation is probably due to regional differences in ambient temperatures and hence to the different body-temperature routines that

occur in bats of widely separated colonies. Periodic torpor or low body temperatures after the beginning of gestation slow the development of the embryo.

Several features of the unique cycle are especially noteworthy: (1) The development of the male reproductive organs is out of phase; that is, the testes have regressed when the caudal epididymides and accessory organs are most enlarged and when breeding activity is at its peak. (2) Males retain viable sperm in the caudal epididymides long after spermatogenesis has ceased. (3) Females do not ovulate until long after they have been inseminated, but are able to store viable sperm for several months. (4) Because of differing metabolic routines in different individuals, the rate of development of the embryo is highly variable.

Delayed fertilization is seemingly a highly advantageous adaptation in mammals with long periods of dormancy. Spermatogenesis, enlargement of reproductive organs, and copulation require considerable energy. In species that practice delayed fertilization, these activities occur in the late summer and autumn, when males are in excellent condition and have abundant food, rather than in the spring, when the animals are in their poorest condition and when food (insects) may not yet be abundant. Ovulation and zygote formation occur almost immediately upon emergence from dormancy, rather than being delayed until after males attain breeding condition and copulation occurs. The female can therefore channel more energy into nourishment of the embryo than would be available if copulation were occurring immediately after hibernation. Perhaps the major advantage is that of hastening the time of parturition and allowing the longest possible time for development of young before the winter period of dormancy.

DELAYED DEVELOPMENT. Fleming (1971) showed that in the Jamaican fruit bat (*Artibeus jamaicensis*) development of the blastocyst is delayed. Blastocysts conceived after the birth of young in July or August soon implant in the uterus as in most mammals; the blastocyst then becomes dormant, however, and further development is delayed until mid-November. This delayed development allows the young resulting from late summer matings to be born in early spring, when fruit is abundant.

DELAYED IMPLANTATION. This deviation from the "normal" reproductive pattern occurs in a variety of mammals, representing the orders Chiroptera, Edentata, Carnivora, and Artiodactyla (Tables 18–2 and 18–3). These mammals obviously do not share a common heritage; in addition, they occupy a wide variety of habitats, and pursue differing modes of life. Delayed implantation in each group, therefore, has probably evolved separately and in response to different selective pressures. Delayed implantation is either *obligate*, and constitutes a consistent part of the reproductive cycle, or is *facultative*, and provides for a delay of implantation on occasions when an animal is nursing a large litter. A good discussion of delayed implantation is given by Daniel (1970).

In mammals with obligate delayed implantation, ovulation, fertilization and

Table 18–2. Periods During Which Blastocysts Remain Dormant in Some Mammals with Obligate Delayed Implantation or Delayed Development

Species	Dormancy of Blastocyst (in months)
Order Chiroptera	
Equatorial fruit bat (*Eidolon helvum*)	3+
Jamaican fruit bat (*Artibeus jamaicensis*)	2½
Order Edentata	
Nine-banded armadillo (*Dasypus novemcinctus*)	3½–4½
Order Carnivora	
Grizzly bear (*U. arctos*)	6+
Polar bear (*U. maritimus*)	8
River otter (*Lontra canadensis*)	9–11
Harbor seal (*Phoca vitulina*)	2–3
Gray seal (*Halichoerus grypus*)	5–6
Walrus (*Odobenus rosmarus*)	3–4
Order Artiodactyla	
Roe deer (*Capreolus capreolus*)	4–5

(Data mostly from Daniel, 1970; data on *Artibeus jamaicensis* from Fleming, 1971.)

early cleavages up to the blastocyst stage occur normally, but further development of the blastocyst is arrested and it does not implant in the uterine endometrium. The blastocyst remains dormant in the uterus for periods of from 12 days to 11 months. Little growth of the blastocyst occurs during its dormancy, which begins generally when the embryo consists of from roughly 100 to 400 cells. The western spotted skunk (*Spilogale gracilis*) studied by Mead (1968) follows a reproductive pattern fairly typical of mammals with delayed implantation. Males become fertile in the summer, and copulation and fertilization of the ova occur in September. The zygote undergoes normal cleavage, but stops at the blastocyst stage; the blastocysts float freely in the uterus for 180 to 200 days. After implantation, the gestation period is about 30 days, and the young are usually born in May. During the period of dormancy, each blastocyst is covered by a thick and durable *zona pellucida*, a noncellular protective layer. This general pattern of delayed implantation occurs in a number of carnivores, but the timing of the cycle varies between species (Table 18–3).

The adaptive advantage of delayed implantation is not understood for all species. In *Macrotus waterhoussi* (Bradshaw, 1962; Wimsatt, 1969b) and in *Miniopterus schreibersii*, in which delayed implantation is known, this specialization may confer the same advantages as those resulting from delayed fertilization in vespertilionids of temperate zones.

Facultative delayed implantation occurs in some species in which the female is inseminated soon after the birth of a litter. This type of delay is known in some marsupials, some insectivores, and some rodents. In certain rodents that have postpartum estrus, implantation of blastocysts is delayed when the female is suckling a large litter.

Our understanding of the factors controlling normal blastocyst development or dormancy of the blastocyst in eutherian mammals is incomplete. Present evidence suggests that estrogen causes the uterine endometrium to form proteins essential for

Table 18–3. Reproductive Cycles of Some North American Mammals with Delayed Implantation

	Breeding Season	Time of Implantation	Length of Delay Period	Time of Parturition	Litter Size	Gestation Period
Long-tailed weasel (*Mustela frenata*)	July	March	8 mo.	April–May	6–10	9 mo.
Ermine (*Mustela erminea*)	June–July	March	8½–9 mo.	April–May	6–10	9½–10 mo.
Mink (*Mustela vison*)	Feb.–March	March	0–1 mo.	April–May	3–8	40–70 days
Marten (*Martes americana*)	July–August	March–April	8 mo.	May	2–3	9 mo.
Fisher (*Martes pennanti*)	March–April	Feb.–March	11 mo.	March–April	2–4	11½–12 mo.
Wolverine (*Gulo gulo*)	Spring (?)–Summer	Jan.–Feb.	5 mo. +	March–April	2–4	8–9 mo. +
Badger (*Taxidea taxus*)	July–Aug.	Feb.	6 mo.	March–April	2–3	8 mo.
Western spotted skunk (*Spilogale gracilis*)	Sept.	April	6–7 mo.	May–June	4–7	8 mo.
Black bear (*Ursus americanus*)	June	Nov.	6 mo.	Jan.–Feb.	1–4	7 mo.
Northern fur seal (*Callorhinus ursinus*)	Late July	Nov.–Dec.	3½–4½ mo.	Late July	1	12 mo.

(Courtesy of Philip L. Wright.)

rapid growth of the blastocyst, and that a deficiency of these proteins results in dormancy of the blastocyst (Daniel, 1970). One protein seemingly responsible for regulation of the differentiation and growth of the blastocyst was named "blastokinin" (Krishnan and Daniel, 1967). Experimentally administered doses of estrogen or progesterone or both have been used in an attempt to initiate growth of the dormant blastocyst in mammals with the obligate type of delayed implantation. These procedures have not been successful in renewing growth of the blastocyst, indicating that in such animals some blocking of the action of estrogen in the endometrium must occur (Daniel, 1970). McLaren (1970) has proposed that during lactation in mice (Mus) implantation is delayed by an initial inability of the blastocyst to "hatch" from the zona pellucida, which must be shed before implantation can occur.

Delayed implantation is an important part of the reproductive cycles of many marsupials. In most marsupials, the suckling of the young in the pouch inhibits estrus and ovulation; but in some kangaroos and wallabies (Macropodidae) a type of delay occurs that is termed embryonic diapause by Sharman (1970). In most macropodids for which embryonic diapause is known, the mother undergoes postpartum estrus; copulation occurs and the ovum is fertilized early in the life of the young she carries in the pouch. The suckling of the pouch young initiates neural and hormonal responses that arrest the activity of the corpus luteum and induce dormancy of the blastocyst; cell division in the blastocyst ceases and it does not implant. In contrast to the dormant eutherian blastocyst, which consists of an inner cell mass that gives rise to the embryo and a hollow sphere of cells that gives rise to extraembryonic membranes, the marsupial blastocyst consists of only 70 to 100 cells that form a single spherical layer of cells of one type (protoderm). The marsupial blastocyst is surrounded by protective coverings consisting of an albumin layer and a shell membrane. When the young leaves the pouch, development of

the corpus luteum and growth of the blastocyst resume, the blastocyst implants, and rapid growth of the embryo resumes.

In marsupials, the young suckle after they leave the pouch for a period roughly comparable to the suckling period in eutherian mammals of similar size, but the intrauterine period for the marsupial fetus is often short (Table 18–1). As a result, a newborn young may be attached to a nipple and suckling while a much older young is returning periodically to suckle from a separate nipple. In both marsupials and eutherians the composition of the milk changes during pregnancy. In marsupials, the milk secreted early in the suckling period contains little or no fat, whereas milk secreted later in the period may contain as much as 20 per cent fat. During "double suckling" in kangaroos a remarkable thing occurs: separate mammary glands concurrently produce vastly different milks, although both glands are seemingly under the same hormonal influences. Compared to the low-fat–content milk produced by the gland supporting the pouch young, the milk produced by the gland supporting the advanced young has three times as much fat. The physiological basis for this remarkable arrangement is not known.

The available evidence suggests that in marsupials no extraovarian hormones are secreted during pregnancy. The placenta, important as an endocrine organ in eutherians, does not serve this function in marsupials. The reproductive physiology of marsupials is reviewed by Sharman (1970), who believes that the differences between marsupial and eutherian reproduction point to a separate evolution of viviparity in these two groups after they diverged from a common oviparous ancestral stock.

This view is taken to task effectively by Lillegraven (1976), who presents strong evidence in favor of the opposing view. Lillegraven suggests that the common ancestor of marsupials and placentals was viviparous, and bore virtually "embryonic" young; he hypothesizes that viviparity did not evolve independently in marsupials and placentals. These groups of mammals seemingly diverged in Early Cretaceous, and compelling paleontological and anatomical evidence indicates an origin from a common ancestral stock. Temporary closures of the eyes and ears in newborn marsupials guard against desiccation, and partial closure of the mouth ensures secure attachment to the nipple and immovable jaws in which the dentary-squamosal jaw joint (absent at birth) can develop. In placental mammals that bear precocious young, these closures have no function, yet they still occur briefly during intrauterine development, and perhaps represent a developmental stage inherited from a common ancestor with marsupials. This ancestor may have borne rudimentary young after a brief gestation period, just as do marsupials today.

REPRODUCTIVE CYCLES OF MAMMALS

Tremendous variation occurs among the reproductive cycles of different species of mammals, and this variation is reflected by differences in the lengths of the gestation periods in different mammals. The duration of gestation depends in part on the size of the animal, on whether or not delayed fertilization or delayed development occurs, and on the rate of intrauterine development of young. The longest gestation periods are those of elephants (up to 22 months). But gestation periods cannot be predicted on the basis of size of the mammal alone. The blue whale (*Balaenoptera musculus*), for example, is the largest living animal, and probably the largest animal that has ever lived, but its gestation period is only 11 months. For some unknown reason, growth of the whale fetus is amazingly fast. As might be expected, the gestation periods in species with delays in fertilization or development are characteristically long. The gestation period of the fisher (*Martes pennanti*), a small carnivore with delayed implantation, is roughly the same length as that of the blue whale. At the other extreme, many

marsupials have remarkably short gestation periods (Table 18–1). This is a result of the unique marsupial reproductive pattern typified by brief intrauterine development of young, birth while the young are still poorly developed, and a long period of suckling while the young are in the pouch. In the common opossum (*Didelphis marsupialis*), the gestation period is but 12 or 13 days, considerably less than that of some tiny shrews.

The frequency of breeding and the size of the litters in any mammalian species are adaptive features that have doubtless evolved in response to a great number of factors. Among these are longevity of individuals, duration of suckling period of young, time during which the young are dependent on the parents, death rate of young, annual activity cycles of adults, and such environmental factors as availability of food supply and severity of seasonal changes in temperature and precipitation. Under the influences of these and other factors, a reproductive pattern has evolved in each species that is geared to the greatest possible success in rearing the young. According to Lack (1948, 1954a), litter size is a result of natural selection that tends to favor the most consistently successful litter size in terms of survival of young. Williams (1967) offered a refinement of Lack's principle that takes into account the total reproductive performance of adult animals. Because the metabolic cost of raising large and well-nourished litters is paid by a lowering of future reproduction, in any population litter size will represent the best reproductive investment for the environmental situation under which the population is operating.

This "best investment," as represented by litter size, differs from area to area, even within a species (Spencer and Steinhoff, 1968). Within a species, litter size generally becomes larger at northern latitudes and at higher elevations because the severe winters and brief growing seasons in these areas limit the number of litters an animal can have during its lifetime (Table 18–4). The most adaptive pattern in

Table 18–4. Geographic Variation in Deer Mouse Litter Size (*Peromyscus maniculatus*) as Indicated by Embryo Counts

Area	Elevation in Feet	Number of Females	Mean Litter Size
Plains:			
Larimer Co., Colorado	5100–5300	56	4.0
Mountains:			
Plumas Co., California	3500–5000	96	4.6
Larimer Co., Colorado	5500–6500	37	4.4
Larimer Co., Colorado	8000–11,000	47	5.4
Routt Co., Colorado	10,500	111	5.6

(Data for California from Jameson, 1953:48; for Larimer Co., Colorado, from Spencer and Steinhoff, 1968:283; for Routt Co., Colorado, from Vaughan, 1969:60.)

boreal areas, then, is one involving a few large litters; in less severe climates, more but smaller litters are produced. In some tropical areas, breeding continues through much of the year, but the pronounced wet and dry seasons in many tropical regions partially restrict breeding to periods of high productivity of food.

Pat statements regarding timing of breeding can seldom be made, however, for considerable interspecific variation in reproductive patterns can occur in mammals of a single locality. In the Panama Canal Zone, most rodents breed throughout the year, but in some species breeding is restricted to certain seasons (Fleming, 1970); and in a subalpine locality in Colorado some rodents are polyestrous and breed over a several month period, whereas others are strictly monestrous (Vaughan, 1969).

Especially critical is the seasonal timing of the gestation period, the suckling period of the young, and the period during which young are becoming independent of the parents. During the intrauterine development of the young and the period of lactation, unusually heavy metabolic demands are put on the female; also, the survival of the young depends on their ability to obtain sufficient food when they are becoming independent. The reproductive patterns of most mammals are timed so that these critical parts of the cycle occur during times of high productivity of food. Most mammals of temperate or

boreal areas bear young in the spring or summer, when food is abundant and optimal weather conditions for survival of young occur. Because the summer snow-free period is short in some northern or mountain regions, the young of many small mammals are frequently born soon after snowmelt, and the periods of lactation, weaning, and early independence of young occur during the brief period of rapid growth and flowering of the plants and the time of maximum activity of insects.

For some hibernating species, the summer period in boreal areas seems barely long enough for growth sufficient to prepare young for dormancy. Some populations of least chipmunks (*Eutamias minimus*) are probably under such environmental pressure. Spring populations of this species in the high mountains of northern Colorado often contain some nonreproductive individuals. These animals are considerably lighter in weight than the average members of the population and are probably individuals born the previous summer that were unable to gain sufficient growth and develop sufficient fat to provide energy for both hibernation and rapid gonadal development prior to emergence from hibernation (Vaughan, 1969:60–62).

Within the order Chiroptera a wide variety of reproductive patterns is known. The absence or scarcity of insects during the winter in temperate regions has forced the bats of these areas to become inactive in the winter and to bear their young in spring or summer, when food is abundant. All of these bats are monestrous, and synchronous breeding is the rule.

Among tropical bats, however, there is no such ubiquitous pattern. In both the Neotropics and the Paleotropics, some species of bats do not face seasonal food shortages that limit reproduction. In the neotropical Desmodontidae, and in many members of the Emballonuridae, Phyllostomatidae, and Molossidae, breeding is asynchronous but continues through the year. *Myotis nigricans*, studied in Panama by Wilson and Findley (1970), breed continually for much of the year, with a cessation of reproduction for several months (October through December) toward the end of the rainy season. In some Paleotropical regions the situation is especially complex.

Seemingly of major importance have been the effects of the wet and dry seasons on patterns of insect abundance and flowering and fruiting of plants. As shown in Table 18–5, within a family, or even within a species, reproductive cycles show great geographic variation. The Egyptian fruit bat (*Rousettus aegyptiacus*), which occurs in Egypt and virtually throughout Africa south of the Sahara Desert, in East Africa gives birth semiannually, in March and September, at the start and toward the end of the long rainy period; in Egypt, however, reproduction continues throughout the year. Some kinds of bats retain the same reproductive pattern over broad areas, but the timing of the cycle in different areas is out of phase. The giant leafnosed bat (*Hipposideros commersoni*), an insectivorous species, bears young in Zambia in October, at the start of the long rainy season. By contrast, in southern Kenya, where two brief rainy seasons occur, one in March and April and one in November and December, copulation probably occurs in November and births are in April. The

period of births in each area thus coincides with the rains and with great insect abundance, but, according to the calendar, the cycles are out of phase.

A remarkable situation in *Hipposideros caffer* has been reported by Brosset (1968), who found that in Gabon birth occurs in some colonies in March and in others in October. The broad pattern in this species is for births in October in southern latitudes and in March north of the equator. Leaving details and individual species aside, at least four major reproductive patterns have evolved among African bats (Table 18–5).

In deserts, similarly, the reproductive cycles of small mammals are timed to take advantage of periods of plant growth. In the deserts of Arizona, for example, a burst of growth of small forbs occurs in the spring, after the winter precipitation, and another period of plant growth occurs in late summer, in response to the summer thunderstorms. Reichman and Van De Graaff (1975) found that during these periods of plant growth both sexes of the kangaroo rat *Dipodomys merriami* ate more green plant material than during the rest of the year, and that immediately following these periods there were surges in reproduction. A similar relationship was found for female Arizona pocket mice

Table 18–5. Reproductive Patterns in Selected Paleotropical (African) Bats

Polyestrous: Yearlong Asynchronous Breeding	Polyestrous: Asynchronous Breeding for Part of Year	Diestrous: Two Synchronous Breeding Periods	Monestrous: One Synchronous Breeding Period
Rousettus aegyptiacus (P)	*Epomophorus wahlbergi* (P)	*Rousettus aegyptiacus* (P)	*Eidolon helvum** (P)
Rousettus lanosus (P)	*Lavia frons* (MG)	*Myonycteris torquata* (P)	*Myonycteris torquata* (P)
Epomophorus labiatus (P)		*Taphozous mauritianus* (E)	*Rhinopoma hardwickei* (RM)
Epomops franqueti (P)		*Nycteris hispida* (N)	*Coelura afra* (E)
Tadarida pumila (M)		*N. thebaica* (N)	*Taphozous nudiventris* (E)
		Cardioderma cor (MG)	*Hipposideros commersoni* (R)
		Rhinolophus landeri (R)	*H. cylops* (R)
		Tadarida condylura (M)	*H. caffer* (R)
			Pipistrellus nanus (V)
			Eptesicus somalicus (V)
			Tadarida pumila (M)

*This species has delayed fertilization; copulation and parturition seem not to be synchronous within a population.

E, Emballonuridae; M, Molossidae; MG, Megadermatidae; N, Nycteridae; P, Pteropodidae; R, Rhinolophidae; RM, Rhinopomatidae; V, Vespertilionidae.

(Data mostly from Kingdon, 1974; some data taken by T. J. O'Shea and T. A. Vaughan in Kenya.)

(*Perognathus amplus*) at times of the year when these rodents were active. These authors suggested that the availability of green vegetation may serve as a physiological signal for reproduction at times when sufficient moisture is available to females for lactation and a new seed crop insures increasing food for an expanding population.

Corroborative evidence was presented by Van De Graaf and Balda (1973), who found striking differences between the reproductive activity of two populations of *D. merriami* in Arizona at sites only 92 miles apart; one site had received three times more rain in the autumn than had

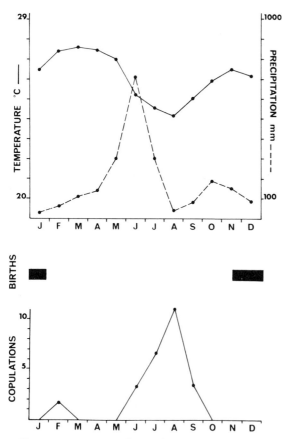

Figure 18–3. Seasonal reproductive behavior in an Old World monkey (Lowe's guenon, *Cercopithecus campbelli lowei*) inhabiting a tropical forest in Ivory Coast, West Africa. Mean monthly temperature is shown in the upper graph by the solid line and mean monthly precipitation by the dotted line. The lower graph shows the number of copulations per 100 hours of observation each month. (From Bourliere, 1973.)

the other site, and had vastly more green vegetation. At the "wet" site 90 per cent of the adult kangaroo rats were reproductively active, whereas but 14 per cent of the males and none of the females sampled at the "dry" site were in reproductive condition. In addition, the kangaroo rats from the "wet" site were significantly heavier (an average of some 7 g) than those from the other site. These studies offer no proof that reproduction is triggered by green vegetation, but it seems obvious that this influences general body condition and reproduction.

In Panama the two periods of natality of the Jamaican fruit bat (*Artibeus jamaicensis*) are at times when fruit is available (Fleming, 1971). In western Mexico four insectivorous bats of the family Mormoopidae have their young within a brief period in the early part of July at the time when the rains begin and insects suddenly become abundant (Bateman and Vaughan, 1972). Many, perhaps the majority, of tropical rain forest mammals breed seasonally (Bourliere, 1973). This is the typical pattern for Old World monkeys of the genus *Cercopithecus* (Fig. 18–3). For rodents, seasonal breeding at times of peak rainfall has been recorded in Zaire by Rahm (1970), in Uganda by Delany (1971), and in Gabon by Dubost (1968).

Reproductive rates vary markedly between species and are dependent on many factors such as litter size, duration of the suckling period, and the time required for young to reach sexual maturity. As might be expected, the reproductive rate is low in large animals that have long gestation periods and a single young, that suckle their young for long periods, and that have young that take many years to reach sexual maturity. Because of the lengths of the periods of gestation and suckling in the elephant, for example, a female may bear a young once every four years or even less often, and an average female may have only four young in her lifetime. By comparison, some small rodents have amazingly high reproductive potentials because of polyestrous reproductive cycles and rapid growth and maturation of the young.

The montaine vole (*Microtus montanus*) of the western United States is a good example of a small rodent with a high reproductive rate. This vole is polyestrous and often breeds throughout the late spring, summer, and early autumn. The gestation period is only 21 days, and a postpartum estrus occurs. A female may have three or four litters in the breeding season, the litter size averaging four to six. Young females can breed at 21 days of age, and males are fertile at roughly twice this age. In a summer period, therefore, a female can bear 20 young, and, as a conservative estimate, these young might produce an additional 20 young before winter. Thus, a roughly 20-fold increase from the original pair of mice could occur. In the European field vole (*M. agrestis*) breeding may continue through all but the midwinter months (Baker and Ranson, 1933), and the California vole (*M. californicus*) may breed nearly all year (Greenwald, 1956). It is not surprising that occasionally, under conditions that favor high survival of young, tremendously high population densities of microtine rodents occur.

The early age at which some small mammals can reproduce is especially noteworthy and clearly has an effect on reproductive rates and population increases. Kalela (1961) reported conception in a lemming (*Lemmus lemmus*) at an age of only 15 days, and birth of the litter at 38 days. The pigmy mouse (*Baiomys taylori*) is reported by Hudson (1974) to breed as early as 28 days after birth. The age at puberty in six species of *Peromyscus* ranges from 36 to 51 days (Asdell, 1964).

GROWTH OF YOUNG

Growth and development of young are remarkably rapid in some mammals. The young least shrew (*Cryptotis parva*) roughly doubles its birth weight at the end of four days (Conaway, 1958). Young evening bats (*Nycticeius humeralis*), which weigh 2 g at birth, roughly double their weight by 18 days of age, and by this time the wings more than double their length (Jones, 1967). These bats can fly when 18 days old, but adult weight is not attained until roughly 60 days. Similar growth rates occur in the cave myotis (*Myotis velifer*; Kunz, 1973); such rates are probably typical of many small bats. Growth rates are also quite rapid in rodents, but rates may differ sharply between closely related animals. One half of adult body weight was attained by several species of *Peromyscus* (white-footed mice) at from 23 to 48 days after birth (Table 18–6). Characteristically, the early growth of small mammals is rapid, but the rate declines shortly after weaning. As an example, in three species of kangaroo rats that are weaned between 15 and 25 days of age, young attained roughly half the adult weight within 30 days, but full adult weight is not reached in two of these species until from 150 to 180 days after birth (Fig. 18–4).

Some of the most astounding growth rates known in mammals have been recorded for the southern elephant seal (*Mirounga leonina*). This is a huge animal; the males reach weights of over 3000 kg (more than three tons). An average female

Table 18–6. Rate of Postnatal Growth of Deer Mice (*Peromyscus*) as Indicated by Percentages of Mature Weight and Age at One-Half Growth

| Species | Age in Weeks | | | | | | One-Half Growth in Days |
	1	2	4	6	8	10	
P. maniculatus	24	38	60	81	92	99	23
p. truei	19	30	51	72	85	89	27
P. megalops	9	14	27	41	51	75	48
P. floridanus	19	27	51	68	74	78	28
P. californicus	23	34	55	73	85	90	25

(After Layne, 1968.)

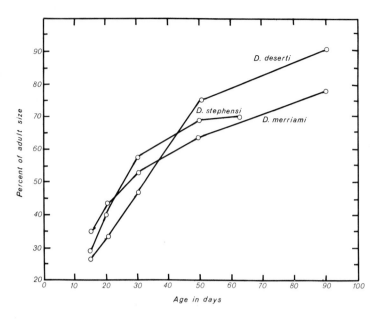

Figure 18–4. Growth rates of three species of kangaroo rats (*Dipodomys*). (After Lackey, 1967.)

weighs 46 kg at birth. The weight doubles by 11 days of age and quadruples by 21 days (Laws, 1953:33). The young Weddell seal (*Leptonychotes weddelli*) doubles its weight within two weeks after birth (Bertram, 1940:32). The rapid growth of pinnipeds is facilitated by high-energy milk that is up to 53 per cent fat (Amoroso and Matthews, 1951). The period of suckling is usually long in pinnipeds, lasting for a year and a half in the walrus (*Odobenus*; Chapskiy, 1936:120) and for a year in some California sea lions (*Zalophus californianus*; Peterson and Bartholomew, 1967:44). According to Scheffer (1958:25), the long suckling period and rich milk are adaptations allowing considerable growth and fat deposition before the young animals must face the stresses associated with their first winter at sea.

Some newborn mammals are helpless and poorly developed (altricial), whereas others are well developed and capable of taking care of themselves to some extent (precocial). Most mammals are of the former type, and many are born hairless, with the eyes closed and the external ear opening sealed. Many small rodents are not covered with fur until at least one week of age, and the eyes open between the 10th and 20th day. Newborn cricetid rodents are typically uncoordinated and lack locomotor ability, and such ability is not developed to any extent until the eyes open. Newborn carnivores, although usually fully furred, are helpless, and the eyes and ears are not open. At the other extreme, some mammals are remarkably well developed and alert at birth or soon thereafter. Perhaps the prime examples of such young are those of perissodactyls and artiodactyls. Newborn of these animals are fully furred, their eyes are open, and they are soon able to run. Within five minutes of its birth a calf wildebeest (*Connochaetes taurinus*) rises to its feet and begins to follow its mother. The extremely long limbs of most young antelope enable them within a few days after birth to keep pace with adults.

METABOLISM AND TEMPERATURE REGULATION IN MAMMALS

19

Major barriers to mammalian distribution are easily recognized. Bodies of water, arid lands, or mountains may be absolute barriers to dispersal, depending on the environmental tolerances of the specific mammals. Equally limiting, however, are environmental temperatures; the distributions of some mammals—Neotropical sloths, for example—might be described most precisely by reference to the extreme temperatures and the seasonal patterns of temperature change that can be tolerated. Air temperatures from −50° to 50° C may be encountered at various times and places on the earth, but at best mammals can only survive body temperatures of approximately 45° to 0° C, and can be normally active only within the range of body temperatures between roughly 45° and perhaps 30° C. In mammals, interspecific differences in the ability to withstand temperature extremes occur even between closely related species (Fig. 19–1), and it is not surprising that no one species is adapted to facing the full range of temperature extremes known for mammals as a group. Just as some mammals are adapted to a few food sources or to a restricted type of habitat, some can live only within a narrow range of temperatures. Knowledge of metabolism and temperature regulation in mammals is essential to an understanding of their ability to adapt to the great variety of ecological settings they occupy.

Most animals are *ectothermic*. In these animals body temperature is regulated by heat gained from the environment rather than by heat produced by the animals'

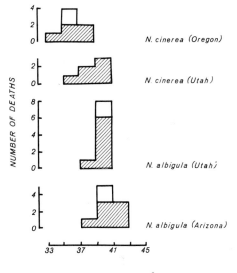

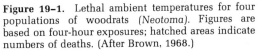

Figure 19–1. Lethal ambient temperatures for four populations of woodrats *(Neotoma)*. Figures are based on four-hour exposures; hatched areas indicate numbers of deaths. (After Brown, 1968.)

own metabolic processes. Mammals and birds are unusual in being *endothermic*; their body temperatures are controlled largely by metabolic activity and by modifications that carefully regulate the rate of heat exchange with the environment. The most obvious advantage of endothermy is the freedom it allows from dependence on environmental temperatures. Mammals and birds, whether terrestrial or aquatic, remain active and typically retain a remarkably constant body temperature under an imposing variety of temperature extremes, ranging from intense desert heat to extreme arctic cold. This type of thermal regulation allows these endotherms to have broad ecological and geographical distributions. Some mammals and birds maintain constant body temperatures year round, even in areas with drastic seasonal changes in temperature. Of advantage to some mammals and birds, however, is the ability to maintain a constant body temperature at some times but to allow the body temperature to fluctuate at other times. Such animals are said to be *heterothermic*. This pattern of regulation is of critical adaptive importance to some animals by facilitating metabolic economy, particularly during energetically stressful periods.

Endotherms have gained an advantage over ectotherms by developing partial independence of environmental temperatures, but this advantage is maintained at considerable metabolic expense. Some 80 to 90 per cent of the oxidative energy produced by endotherms is used for maintaining thermal homeostasis. To maintain a constant body temperature when the ambient temperature is below the body temperature, an animal must keep a balance between heat lost to the environment and heat produced by metabolism. When the environmental temperature exceeds the body temperature, heat gained from the environment must be dissipated by some cooling device. Both heat production and heat dissipation demand some outlay of energy and, in some extreme environments, the energy demands of endothermy are extremely high.

A major adaptive trend in some mammals is toward the reduction of the metabolic cost of endothermy. This is often done by decreasing *thermal conductance*. This is expressed as the metabolic cost (in cubic centimeters of oxygen per gram of body weight) for a given time interval per °C difference between the body temperature and the environmental temperature. (The units of this quantity are thus cm^3 $O_2/g/hr/°C$.) In some mammals in extremely cold environments, the difference between body temperature and environmental temperature may be 70° C or more; because the rate of loss or gain of

heat in a body is proportional to the difference between the body's temperature and that of the environment, reduction of loss of heat by the lowering of thermal conductance is essential in these mammals. Insulation, in the form of fur or blubber (or feathers in birds), is the primary means of reducing thermal conductance. Control of the blood supply to peripheral parts of the body is also important.

Each endotherm has a *thermal neutral zone* within which little or no energy is expended on temperature regulation. The thermal neutral zone is "the range of temperatures over which a homoiotherm can vary its thermal conductance in an energetically inexpensive manner and on a short time scale" and keep a constant body temperature (Bartholomew, 1968). Within this zone the fluffing or compressing of the fur, local vascular changes, or shifts in posture suffice to maintain thermal homeostasis.

At the lower limit of the thermal neutral zone is the *lower critical temperature*, the point below which the balance between metabolic heat production and heat loss to the environment cannot be maintained by metabolically inexpensive variations in thermal conductance. Below the lower critical temperature, at which the rate of heat flow is minimal, oxidative metabolism must be increased to keep the body temperature constant. Obviously, if a constant body temperature is to be maintained over a wide variety of ambient temperatures, adjustments of both thermal conductance (through changes in insulation) and heat production (through metabolic changes) are necessary.

The *upper critical temperature* is the point above which a constant body temperature can only be maintained by an increase in metabolic work above the resting level to dissipate heat. This temperature is far less variable than is the lower critical temperature, but is of great importance to desert mammals. These mammals usually do not have access to water and must strictly minimize water loss. Animals faced with temperatures above the upper critical temperature dissipate heat by evaporative cooling, which involves considerable water loss. Because such loss in desert species is extremely disadvantageous, these animals generally avoid temperatures above the upper critical temperature. In some mammals the upper critical temperature is unusually high or may even be difficult to detect, as in the anomalous case of the African rock hyrax (Heterohyrax brucei), studied by Bartholomew and Rainy (1971). The body temperatures of experimental hyraxes rose as the environmental temperatures rose, and at an environmental temperature of 42.5° C, and body temperatures in the vicinity of 41° C (105° F), the rate of oxygen consumption was actually lower than at temperatures nearer the mean body temperature of animals not under heat stress (36.4° C). Figure 19-2 summarizes the usual relationship between oxygen consumption and environmental temperatures in a mammal.

The following considerations of the reactions of different mammals to conditions of thermal stress provide a basis for a general understanding of metabolism and temperature regulation in mammals.

HOMOIOTHERMY AND COLD STRESS

HOMOIOTHERMY IN TERRESTRIAL MAMMALS. A primary adaptation of mammals inhabiting cold terrestrial environments is the development of highly effective insulation. This insulation may be so remarkably effective that the thermal neutral zone of an animal may extend downward to −40° C, as in the case of the arctic fox (Alopex lagopus). The length of the woolly underfur and the longer guard hairs varies seasonally; the summer pelage, which is acquired in the spring, has reduced insulating qualities, but the longer winter coat, which replaces the summer pelage in the fall, has great insulating ability and allows the fox to keep a constant body temperature in extreme cold with relatively little metabolic drain. This reduction in energy loss may be of critical importance to arctic foxes, for they are at the top of the food chain; they are thus in an energetically precarious position because they depend on such prey species as lemmings, which fluctuate widely in density and in availability.

Additional metabolic savings may be gained by reductions in peripheral circulation. During cold stress, extremities such as legs and ears, which dissipate heat rapidly because of their relatively poor insulation, receive a reduced blood supply and are allowed to become cool, thereby reducing the temperature differential between these parts and the environment. Such regional heterothermy is common among mammals of cold areas, and differences between temperatures of different parts of the body are surprisingly high. In an Eskimo dog with a deep body temperature of 38° C, for example, the temperature of the foot

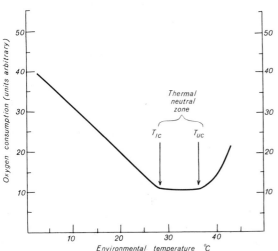

Figure 19-2. Oxygen consumption in relation to environmental temperatures in a hypothetical mammal. Abbreviations. T_{lc}, lower critical temperature; T_{uc}, upper critical temperature. (After Bartholomew, 1968.)

pads may be 0° C, that of the dorsum of the foot, 8° C, and that of the carpal area of the forelimb, 14° C (Irving, 1966).

The hollow hair of some large ungulates is remarkable insulation, and allows the animals to modify their activities relatively little in the winter. Pronghorns (Antilocapra americana) commonly remain in open and often windswept situations in temperatures far below 0° C. The metabolic saving in the winter resulting from decreased thermal conductance is of great importance in some species that must endure not only extreme cold but also reduced availability of food.

As might be expected, behavior plays a part in reducing cold stress. Many small mammals curl up in a ball or hunch the body so that the overall shape is nearly spherical. By minimizing the ratio of the outer surface to the volume, the most advantageous shape in terms of retention of heat is obtained, and lightly insulated surfaces (feet, face, and parts of the venter) are well protected. Even the tree-roosting red bat (Lasiurus borealis), which seemingly goes into short-term hibernation during winter cold spells in the central United States, assumes a posture with the furred tail membrane covering the venter and the head tucked downward; the total form approaches the shape of a sphere (Davis, 1970). The response of seeking shelter may also be of great importance. An animal burrowed deeply into the snow faces an ambient temperature of roughly 0° C, whereas the ambient temperature above the snow might be many degrees below zero. Wintering herds of deer and elk often frequent ridge tops or south-facing slopes, where the cold of the night is quickly moderated by the first rays of the sun, and, during the Alaskan winter, moose abandon many basins into which cold air drains from the surrounding mountains.

As mentioned above, when cold stress forces an animal to conserve heat, a spherical shape is most advantageous. But some small mammals that must be active under conditions of extreme temperatures have shapes that allow rapid heat transfer. Some ground squirrels and weasels have long, slim bodies, and thus have greater surface area relative to body mass than do mammals with "normal" shapes. For desert ground squirrels this form is perhaps of advantage in allowing for rapid dissipation of heat; however, inasmuch as weasels inhabit chiefly areas that are cold in winter, their shape would seem to be maladaptive with respect to thermoregulation. And so it has been found to be by Brown and Lasiewski (1972). Body surface relative to body weight is some 15 per cent greater in the long-tailed weasel (Mustela frenata) than in normally shaped woodrats (Neotoma). Under cold stress, woodrats curl up and assume a spherical shape, whereas weasels assume more nearly the shape of a flattened disc (Fig. 19–3). As a result, cold-stressed weasels have metabolic rates 50–100 per cent greater than those of normally shaped mammals of the same weight. Weasels are thus paying an extremely high energetic cost for being long and thin. Brown and Lasiewski suggest that the long, thin form has so increased the predatory ability of weasels by allowing them to follow small prey into their retreats that the increase in food intake more than repays the metabolic cost incurred by the elongate form.

Of further interest, this body shape may have forced weasels to develop sexual dimorphism (such dimorphism in weasels is discussed on page 287). Weasels are typically active periodically both day and night, and their intense foraging activity would be expected to lead to strong intraspecific competition. By adopting sexual dimorphism, and by each pair of weasels maintaining an exclusive foraging area, this competition may be partially alleviated. It may be significant that small mustelids that are not unusually long bodied are not sexually dimorphic.

SIZE IN RELATION TO HOMOIOTHERMY. Small size is disadvantageous in terms of heat conservation, but favors heat dissipation. In general, the smaller the animal, the greater the surface area relative to volume; the surface area is proportional to the square of the body length and the volume is proportional to

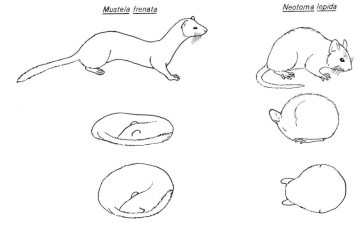

Mustela frenata   *Neotoma lepida*

Figure 19–3. Postures of weasels and woodrats under normal resting conditions (top) and under cold stress (middle, lateral view; bottom, vertical view). (After Brown and Lasiewski, 1972.)

the cube of the length. The surface-to-volume ratio, then, varies as the two-thirds power of the weight. The empirical relationship between body temperature (T_B), lower critical temperature (T_{LC}), and body weight (W) in mammals is represented by the expression $T_B - T_{LC} = 4W^{0.25}$. "Because T_B is essentially independent of body weight (W) in mammals, as weight decreases T_{LC} approaches T_B" (Bartholomew, 1968:322).

The calculated lower critical temperature for a mouse weighing 20 g is 29° C, a temperature considerably higher than that usually encountered by nocturnal mammals. Basal metabolic rate (the metabolic rate necessary for simply the maintenance

of life in a resting organism), lower critical temperature, and thermal conductance all vary inversely with body size, and all are intimately related. The metabolic rate, as measured in oxygen consumption per gram of body weight per hour, climbs so precipitously with decreasing body weight that a mammal weighing less than about 3.5 g would be unable to eat sufficient food to sustain activity. This is shown in Figure 19–4, which also illustrates the fact that rates of oxygen consumption differ strikingly even between small mammals: the tiny masked shrew consumes oxygen at a rate over four times that of the larger deer mouse. Carrying the comparisons further, the metabolic rate of a horse is only

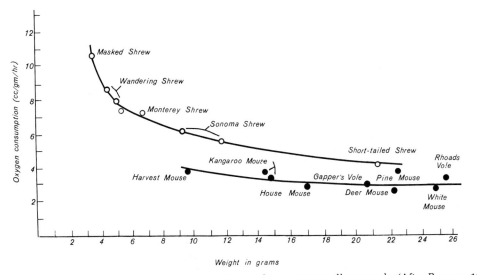

Figure 19–4. Oxygen consumption in relation to body weight in some small mammals. (After Pearson, 1948.)

approximately one tenth that of a mouse (Krebs, 1950). Small mammals must face an additional acute problem related to insulation, for they are limited as to the length that hairs of the pelage can attain, and the pelage is correspondingly limited in insulative effectiveness.

Most small mammals (such as mice and shrews) that inhabit cold areas and are active throughout the winter remain beneath the snow pack, thereby avoiding the intense cold. These animals are often active at the soil-snow interface, in the zone of "depth hoar." This is a stratum of loose snow that develops beneath a fairly deep snow pack and through which small animals can readily travel. These mammals forage in winter in a subnivean environment (an environment beneath the snow) with a constant temperature near 0° C, and when resting presumably seek refuge in nests, which provide insulation that augments the animal's pelage. Small mammals that are intermittently active above the snow in the winter typically maintain thermal homeostasis by increasing heat production via increases in oxidative metabolism. The metabolic cost to small mammals of activity in the open in winter is extremely high. Field workers soon learn that these animals, when captured in live traps in cold weather, will die in a very short time if not supplied with ample food and nesting material.

At the other end of the size scale, it might be expected that exceptionally large mammals would commonly be faced with the problem of heat dissipation because of a more limited surface area relative to volume. This seems to be the case. In whales, for example, the flippers and flukes and at least parts of the skin are important as heat-radiating surfaces, even when these animals are immersed in cold water.

HOMOIOTHERMY IN AQUATIC AND SEMIAQUATIC MAMMALS. Temperature regulation is a demanding problem to mammals that inhabit cold water. The rate of heat loss by an endotherm in water is some 10 to 100 times as great as the rate of loss in air of the same temperature (Kanwisher and Sundnes, 1966:398). Arctic and antarctic waters are near 0° C year round, and northern lakes and rivers approach this temperature in the winter. Consequently, a temperature differential of some 35° C between deep body temperature and ambient temperature is usual in mammals swimming in these waters. Despite the thermal inhospitality of this environment, cold waters are temporarily or permanently inhabited by a variety of mammals, some of which are highly specialized for coping with cold. The muskrat (Ondatra), the beaver (Castor), some shrews (Sorex), the river otter (Lontra), and the mink (Mustela) spend considerable time in cold water.

In these animals heat is lost to the water most rapidly by the foot pads, the nose, and other bare surfaces, for most of the body is insulated by a layer of air that is entrapped by the fur. Although air entrapped in the fur retards heat dissipation, heat is lost by an animal far more rapidly during immersion in water than when the animal is in air. Calder (1969) found that in two species of shrews (Sorex palustris and S. cinereus) and two species of mice (Zapus princeps and Peromyscus maniculatus) thermal conductance in the water when the fur had entrapped air was four and a half times that in air; when the fur was wet to the skin the conductance rose to nine times that in air. Calder also measured heat loss in the water shrew (S. palustris), the smallest homoiothermic diver. The body temperature of water shrews with air entrapped in the fur dropped an average of 1.4° C in 30 seconds during dives beneath the surface of the water, whereas a shrew with fur wet to the skin had a temperature drop of 4.5° C in the same time. The meticulous grooming and drying of the fur performed by a water shrew after a dive is clearly highly adaptive. Some nearly permanent inhabitants of marine waters, such as otariid seals, similarly use entrapped air as insulation. The cetaceans, phocids (earless seals), and walruses (Odobenus), however, lack appreciable fur, and the surfaces of their bodies are in contact with water that may in extreme cases be 40° C below the deep body temperature. How they maintain a constant body temperature under such demanding conditions is of considerable interest.

These cold-water, hairless marine mammals have thick layers of subcutaneous blubber that form an insulating envelope around the deep, vital parts of the body. A substantial amount of the weight of a marine mammal may be contributed by the blubber. For example, blubber constitutes 25 per cent of the weight of the Weddell seal (*Leptonychotes weddelli*; Bruce, 1915). In the small (75 kg) harbor porpoise (*Phocoena*), 40 to 45 per cent of the weight is blubber, and only 20 to 25 per cent is muscle (Kanwisher and Sundnes, 1966:405). Studies of seals by Irving and Hart (1957) have shown that the skin temperature varies directly with the water temperature down to 0° C. The cooled surface of the body and the thick blubber are seemingly an effective insulation, as indicated by the fact that the lower critical temperature of some seals is 0° C. The skin of seals has a well-developed vascular supply, and the temperature gradient between the deep parts of the body and the skin is controlled largely by changes in the blood supply to the skin.

Some of the most extreme thermal demands faced by endotherms are those met by cetaceans. Whales and porpoises live their entire lives in the water, and some species continuously occupy water at or near the freezing point. All cetaceans have insulating layers of blubber, but an extreme situation is faced by a small porpoise that must maintain a deep body temperature some 40° C higher than that of the sea, from which it is insulated by only 2 cm of blubber. No inflexible pattern of thermoregulation is adequate even in inhabitants of the sea, which offers a relatively constant thermal environment. Some cetaceans migrate seasonally from cold waters to warm tropical seas. Because of the high thermal conductivity of water, skin temperatures generally equal water temperatures, and variations in water and skin temperatures of roughly 20 to 30° C may occur seasonally. The temperature of

the body core, however, remains constant, and insulation requirements, therefore, may vary fivefold. Heat production by mammals varies tremendously as a result of changes in metabolic level. Metabolic activity and heat production increase roughly ten times in an animal going from a resting state to one of maximum exertion. It has been estimated that a cetacean at rest in cold water needs roughly 25 to 50 times the insulation it needs when swimming at high speed in tropical waters (Kanwisher and Sundnes, 1966:399).

Gigantic differences in the ability of cetaceans to keep warm result from differences in size and in thickness of blubber. The biggest whale is some 10,000 times as heavy as the smallest porpoise, has roughly a 10 times more advantageous mass-to-surface ratio with regard to heat retention, and has a much thicker shell of blubber. Because of these differences, the whale has approximately a 100-fold advantage over the small porpoise in its ability to keep warm. The very factors working in favor of heat retention in the large cetaceans, however, are obviously disadvantageous under conditions of great activity or warm waters. Because of the vast bulk of large cetaceans, dissipation of heat is an acute problem. As an example of the slowness of diffusion of heat in large cetaceans, the deep muscle temperature of a dead and eviscerated fin whale (*Balaenoptera*) dropped only 1° C in twenty-eight hours (Kanwisher and Leivestad, 1957). Clearly, cetaceans must have considerable "thermal versatility." How is this versatility achieved?

As is usual in considerations of biological problems, no single answer is appropriate, nor has sufficient research been done on the problem to suggest even most of the probable answers. Although much remains to be learned, several points seem well established. First, metabolic rates of cetaceans differ markedly in different species. The small porpoises have much greater basal metabolic rates than do large whales, as could be expected because of the inverse relationship between basal metabolic rate and size, and the former thus have a much more rapid rate of heat production even when resting. Metabolic rate seems to be in part an adaptive feature, for certain animals that have difficulty keeping warm, such as small porpoises, have even higher metabolic rates than would be expected on the basis of size alone.

Second, blood flow through the well-developed vascular system in the flippers, dorsal fin, and flukes of cetaceans allows these structures to function effectively as heat dissipators under conditions of heat stress. The flow can apparently be shut down during cold stress, allowing for a minimum of heat loss from these surfaces.

Third, a remarkable series of vascular specializations allow for great variations in the thermal resistance offered by the blubber. A system of countercurrent heat exchange in the vascular network supplying the blubber minimizes heat loss to the blubber and skin, and hence to the environment. This system, utilized by many mammals, involves arterioles and venules that lie against one another, often in a complex network. Heat diffuses from the arterioles to the venules and serves to heat the venous blood before it enters the body core; much of the heat of the arterial blood is thus returned to the body core before it is lost to the environment from such poorly insulated surfaces as bare skin or appendages. In cetaceans, a second venous system in the blubber bypasses the countercurrent system during heat stress and allows considerable heat loss to the environment when heat dissipation is of prime importance. Similar countercurrent and bypass systems occur in the flippers and fins. The extremities and much of the surface of the body can thus serve to dissipate heat, or can be maintained under an altered vascular supply that provides for maximal heat retention.

The longitudinal folds of blubber on the throat of the rorqual (Balaenoptera) probably function partly as a cooling device (Gilmore, 1961) by providing increased surface area for heat dissipation. The highly vascularized skin at the bottom of these grooves can be exposed to the water. Morrison (1962) found that these grooved

anterior parts of the humpback whale (Megaptera) were slightly cooler than were other parts of the body, suggesting their importance in heat dissipation.

The great quantities of blubber on large whales (up to 20 cm in thickness) are seemingly not primarily useful as insulation; these animals could probably maintain a constant deep body temperature, without increased heat production, with much less insulation. These fat deposits may be useful primarily as food stores that can support an animal during periods of migration and fasting. The consumption of only half of a whale's blubber could support the animal at a basal metabolic rate for from four to six months (Parry, 1949).

REACTIONS OF MAMMALS TO HEAT STRESS

Some of the most severe problems in thermoregulation are those faced by mammals living in hot regions. In many low-latitude deserts, daytime surface and air temperatures in the summer rise well above the body temperatures of most mammals. Under such conditions heat from the environment is absorbed, while at the same time the animals themselves are producing considerable metabolic heat. In order to maintain thermal homeostasis these animals must avoid as much as possible the absorption of heat from the environment, dissipate such heat as is absorbed, and lose endogenous heat. Unless the body temperature is elevated, these heat transfers must occur against a thermal gradient, from the relatively cool animal to the relatively hot environment. Such heat transfers invariably involve evaporative cooling, a luxury that most desert organisms cannot afford since they live in a region where water is in critically short supply. Nonetheless, even extremely hot and arid deserts are occupied by mammals, and some kinds, notably rodents, are quite common in such areas. A variety of physiological, anatomical, and behavioral adaptations have allowed mammals to inhabit these seemingly inhospitable regions.

AVOIDANCE OF HIGH TEMPERATURES.
Most desert animals are never subjected to
the extremely high diurnal temperatures,
nor are they able to survive them; their
success is based on the ability to avoid extremely high temperatures rather than to
cope with them. Perhaps the saving grace
of the desert is the typically great daily
and seasonal fluctuation in temperature.
Temperatures frequently drop markedly at
night, and winters are usually cool or cold.
As a result, soil temperatures well below
the surface are never high (Fig. 19–5), even
in the summer, and to this refuge of coolness and relatively high humidity nearly
all desert rodents retreat during the day.
All but a very few desert rodents are strictly
nocturnal, and all are more or less fossorial;
these animals are active above ground in
the part of the diel cycle when temperatures are lowest and humidities are highest. The studies of McNab (1966) and Mac-
Millen and Lee (1970) suggest that most
fossorial and nocturnal desert rodents have
metabolic rates below those predictable on
the basis of body size. The low rates are
associated with lowered metabolic heat
production while the animals are in the
humid burrows during the daytime summer heat. This lowered heat production
probably precludes the use of wasteful (in
terms of water loss) evaporative cooling in
order to dissipate heat while the rodents
are below ground.

Most desert carnivores are also largely
nocturnal, some are fossorial, and all seek
shelter during the hottest part of the day.
Several other means of avoiding the extremes of daytime heat are used by some
mammals of the deserts of southwestern
United States. The white-throated woodrat
(Neotoma albigula) remains during the
day in burrows insulated by piles of sticks
and other debris. Frequently these houses
are built beneath the shade of low-growing
vegetation or under mesquite. In the burrow beneath a woodrat house near Yuma,
Arizona, the temperature fluctuated only
2.5°C throughout a 24-hour cycle and
never exceeded 34.4°C; temperatures outside the house varied 13.7°C and reached
41.5°C (Brown, 1968:24). Jackrabbits
(Lepus californicus and L. alleni) stay in
"forms" in the shade of bushes in the day,
and are reluctant to run far in the open in
midday. Steady or pulsating vasodilation
of vessels in the huge ears of L. californicus allows for rapid dissipation of heat
at ambient temperatures below body temperature (Hill and Veghte, 1976). Bighorn
sheep (Ovis canadensis) and javelinas
(Dicotyles tajacu) often take shelter in rock
grottos or in the shade of steep rock outcrops, where for much of the day their
body temperatures are above air temperature and they can dissipate heat.
TERRESTRIAL THERMOREGULATION
IN PINNIPEDS. As might be expected,

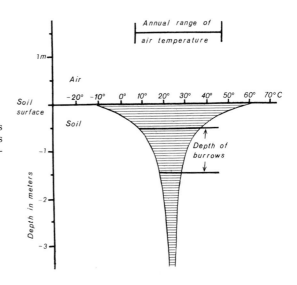

Figure 19–5. Annual range of subsoil temperatures
in a desert in Arizona. Note that most rodent burrows
are at depths at which heat or cold stress is never encountered. (After Misonne, 1959.)

the very adaptations that enable a seal or sea lion to reduce heat dissipation in cold water make the animals unable to stand high temperatures. Whittow (1974) and his coworkers at the University of Hawaii studied the terrestrial thermal budget of California sea lions (Zalophus californianus), which inhabit some coasts of Mexico and the Galapagos Islands where high temperatures occur regularly. At air temperatures of about 30°C (86°F), after the sea water had evaporated from their body surfaces sea lions were unable to maintain a constant body temperature, and with continued exposure under experimental conditions their temperatures rose to slightly over 40°C (about 104°F). When the sea lions slept their heat production dropped some 24 per cent, an obvious advantage for energy conservation and temperature regulation, but they were unable to dissipate sufficient heat in direct sunlight to avoid heat stress. Under stress the animals fanned their flippers, which are known to sweat, thus increasing evaporative cooling. They also urinated and wet the undersides of their bodies, thus further increasing evaporation. But under experimental conditions, these behaviors were inadequate and the animals were increasingly hyperthermic. Only when they were able to wet their bodies in the sea did body temperatures drop and stabilize.

Terrestrial heat production by a sea lion is dissipated approximately as follows: 2 per cent is lost by respiratory evaporative cooling, 12 per cent by evaporation from the skin, 52 per cent by nonevaporative heat loss (conduction and convection) from the skin, and 15 per cent by conduction from the parts of the body against the sand. Nineteen per cent of the metabolic heat is stored, leading eventually to an elevation of body temperature. It is obvious why sea lions and other pinnipeds have difficulty staying out of the water for long periods on a warm day, and increased activity on land at night is understandable. Although their physiological makeup limits the time they can spend hauled out on a warm or hot day, by choosing windy sides of islands and by basking at sites

where spray from breaking waves repeatedly wets their bodies and increases evaporative cooling, sea lions can considerably extend their resting times on land.

HEAT STRESS IN MAMMALS UNABLE TO AVOID HIGH TEMPERATURES. A few mammals are able to tolerate exposure to extreme desert heat. Most ungulates are obviously unable to use the types of shelter available to small mammals, and often occupy open areas where there is little shelter. These mammals must be able to withstand hours of exposure to environmental temperatures in excess of their body temperatures.

Large size itself is advantageous to mammals that must tolerate high temperatures. Because of the weight-surface ratio discussed earlier, the larger the animal, the greater will be its ability to withstand exposure to high temperatures due to a relatively reduced surface area for heat gain. Stated differently, large animals have greater thermal inertia than do small ones. Of additional importance, just as insulation in the form of thick pelage slows the loss of body heat in low ambient temperatures, fur provides a partial heat barrier that slows the penetration of heat to the body surface when temperatures are high. Another advantage of fur under some circumstances is that it reduces water loss through the skin, an extremely critical advantage in arid environments. Although both large size and fairly thick fur are important aids to avoiding rapid heating in hot areas, under conditions of intense heat other factors usually help tip the delicate thermal balance away from lethally high body temperatures.

Studies of temperature regulation in the camel by Schmidt-Nielsen (1959) have revealed a carefully regulated and highly adaptive diel cycle of changes in body temperature. Camels in the Sahara desert in the winter, when cool temperatures (from roughly 0° to 20°C) prevailed, had fairly constant body temperatures that varied between 36° and 38°C. The fluctuations in body temperatures were not random, but followed the same pattern day after day, regardless of weather. In the summer, variations in body temperature were considerably greater; generally animals had temperatures in the morning between 34° and 35°C, and the body temperature reached a peak of approximately 40°C late in the day (Fig. 19–6). The camels were seemingly able to regulate their temperatures, but did so only above or below these extremes; when body temperatures reached 40.7°C, evaporative cooling in the form of sweating was used to dissipate heat, and body temperatures never exceeded 40.7°C. Thus, the camel accepted a heat load during the day that sharply elevated its temperature; but during the relative coolness of the desert night the heat stored during the day was passively dissipated, and the body temperature was allowed to drop to some extent (perhaps, according to Schmidt-Nielsen, to a critical lower limit below which metabolism is disrupted). Schmidt-Nielsen (1964:44) estimated that for a camel to dissipate by evaporative cooling the heat load accepted during a hot day would require the expenditure of some 5 liters of water. In an animal that does not have frequent access to water, such daily water loss would lead to fairly rapid dehydration. An additional advantage of high body temperature during the day results from the narrowing of the gap between environmental and body temperature; the smaller this temperature differential, the lower the rate of heat flow from the environment to the body.

As an adaptation to intense heat, similar patterns of temperature fluctuation occur in the oryx *(Oryx)*, the eland *(Taurotragus)*, and the gazelle *(Gazella)*, African antelope that occur in desert or savanna areas. The oryx frequently inhabits areas where no shade is available, and it does not seem to seek shelter in the day. The ability of this animal to withstand a diurnal heat load is exceptional. Under experimental conditions the oryx could withstand exposure to an ambient temperature of 45°C (113°F!) for 12 hours (Taylor, 1969a:91). During this period the body temperature rose above 45°C and was sustained at this level for up to eight hours with no injury to the animal. Hence, rather than gaining heat from the environment, the oryx was actually losing heat. Such high body temperatures would kill most mammals fairly quickly, but circulatory specializations apparently allow the oryx to survive such extreme "overheating."

The brain, seemingly the most heat-sensitive organ, is provided in this animal with a specialized countercurrent cooling system of its own in the sinus cavernosus beneath the brain (Taylor, 1969a:92). The external carotid artery, on its way to the brain, divides into many branches in this sinus, and these branches are in close proximity to veins returning from the nasal passages (Fig. 19–7). These veins carry relatively cool blood, because evaporative cooling of the nasal mucosa tends

Figure 19–6. Diagrammatic representation of daily patterns of temperature change in three mammals subjected to desert heat. Note that the antelope ground squirrel *(Ammospermophilus leucurus)* goes through a series of heating-cooling cycles during the day. (Mostly after Bartholomew, 1964.)

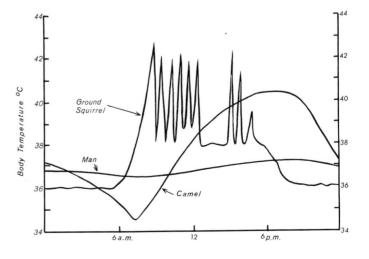

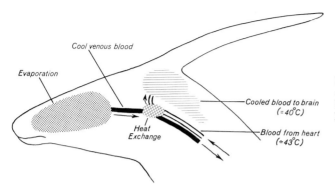

Figure 19–7. Diagrammatic representation of the countercurrent cooling system of the brain of a gazelle as proposed by Taylor and Lyman. (After Taylor and Lyman, 1972.)

to cool the blood supplying these surfaces. Heat exchange between this cool venous blood and the blood in the carotid artery network assures that the blood supply of the brain is cooler than that of most of the rest of the body (Fig. 19–8).

The reactions of ungulates to high temperatures may depend in part on their physiological conditions: dehydrated animals typically respond differently to heat stress than do animals with unlimited access to water. This is shown well by the tiny African dik-diks (*Madoqua kirki* and *M. guentheri*) studied by Maloiy (1973). At temperatures between 20 and 45°C patterns of change of body temperature, respiration rate, and cutaneous evaporation differed between dehydrated and hydrated animals, with differences between the responses of body temperature being most striking (Fig. 19–9). Until the ambient temperature became extremely high, about 45°C, the dehydrated dik-dik's body tem-

perature was above ambient temperature and the animal could lose some heat to its environment. At high temperatures, evaporative cooling, largely through panting, is the main route of heat loss. (Thermoregulation in relation to water economy in dik-diks is discussed on page 442.)

For an antelope running at high speed the uncoupling of brain and body temperatures is an important adaptation. During such exertion a Thompson's gazelle probably produces heat at some 40 times the resting rate. Taylor and Lyman (1972) found that during a gazelle's 7-minute run at 40 km per hour, the temperature of the brain remained below 40.5°C (Fig. 19–8) although the temperature of the blood in the carotid artery rose to nearly 44°C. Whereas the body stored great amounts of heat during running, the countercurrent system in the sinus cavernosus assured that blood supplying the brain was cooled and did not reach lethal levels.

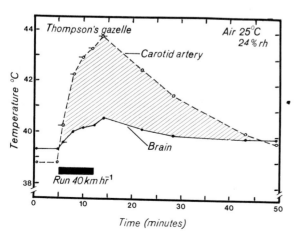

Figure 19–8. Temperatures of the brain and carotid artery of a Thompson's gazelle (*Gazella thompsonii*) before, during, and after a run of 7 minutes at 40 km per hour. The temperature of the carotid artery rose precipitously after the gazelle began to run and exceeded the brain temperature until some 40 minutes after the run. (After Taylor and Lyman, 1972.)

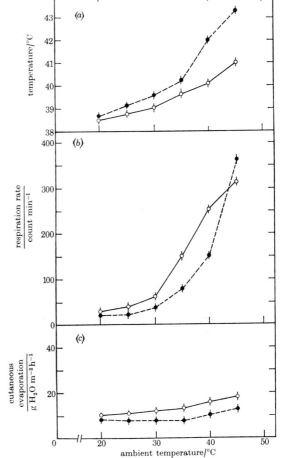

Figure 19–9. Physiological responses of hydrated and dehydrated dik-diks *(Madoqua kirki)* to different temperatures. Rectal temperature is shown in (a), respiratory rate in (b), and cutaneous evaporation in (c). (From Maloiy, 1973.)

Tolerance to high body temperatures may require specializations (as yet, unknown) of various enzyme systems, such as those of the muscles. Many of the physiological reactions known to occur in animals that are not subject to heat stress do not proceed properly at elevated temperatures; also, many enzymes are unstable or denatured at high temperatures.

Some reduction of heat load occurs in the eland and the oryx by metabolic adjustment that reduces internal heat production; this is most striking in the oryx. A dehydrated oryx reduces its metabolic rate at high temperatures sufficiently to reduce evaporative water loss to 17 per cent below that of an individual with free access to water. In the oryx, as in any species of mammal, temperature regulation is controlled by a variety of interrelated adaptations. The following features are known to contribute to heat tolerance in the oryx: diurnal storage of heat and reduced or negligible heat flow from the environment at high body temperatures; decreased metabolism at high temperatures; cooling of blood in arteries supplying the brain; pale-colored pelage that reflects considerable heat; and panting to produce evaporative cooling under extreme conditions.

Typical panting, involving rapid and shallow respiration, is used entirely for heat dissipation and is an effective aid to temperature regulation. Laboratory studies of dogs, for example, indicated a tolerance of an ambient temperature of 43°C for at least seven hours (Robinson and Lee, 1941). Panting utilizes evaporative cooling of the mouth, tongue, and, probably most important, the nasal mucosa (Schmidt-Nielsen et al., 1970). In the dog, and in many other mammals with long snouts and excellent olfactory ability, the turbinal bones of the nasal cavity are intricately rolled and provide a great surface area for olfactory epithelium in the nasal mucosa. This large, moist surface area is ideal for evaporative dissipation of heat. The tongue is probably also important as a site of heat loss during panting, for the blood flow to the tongue increases sharply at onset of panting and during heat stress increases six times over normal.

The resting respiratory rate of a dog is roughly 30 per minute, but this rate rises abruptly, with virtually no intermediate rate, to over 300 per minute during panting. As demonstrated by recent laboratory studies (Schmidt-Nielsen et al., 1970), air movement during panting in the dog is largely unidirectional: most of the air passes in the nose and out the mouth. This type of flow achieves the maximum effectiveness of evaporative cooling. The lateral nasal glands, which open some 2 cm inside the opening of each nostril, supply a major share of the water used in evaporative cooling during panting in dogs (Blatt et al., 1972). Under experimental conditions the rate of secretion of one of these glands in a dog rose from no secretion at 10°C to 9.6 g per hour at 50°C; between 20 and 40 per cent of the evaporative cooling during panting at high temperatures results from evaporation of the fluid from these glands. Because the glands are situated anterior to the turbinals, they tend to keep the nasal mucosa moist when air is drawn rapidly in through the nostrils during panting.

This system has some advantages over cooling effected by sweating. There is little loss of salt during panting, whereas salt loss during sweating (except probably in donkeys and camels) is always appreciable. In addition, adequate ventilation of evaporative surfaces always occurs during panting; in cool, still air, however, sweating is seemingly not equally efficient. One disadvantage of panting is that the increased activity increases metabolism, thereby contributing more heat to be dissipated. Studies of respiratory frequency

in dogs (Crawford, 1962) have indicated that these animals are panting at the resonant frequency of oscillation of the diaphragm (the natural frequency of vibration of this structure) and may therefore economize on energy output. Considering water loss relative to total body surface area of a mammal, the amounts of water loss in sweating and panting are probably similar. Both panting and sweating are obviously not effective means of cooling in high humidities.

Ground Squirrels. Hudson et al.(1972) compared thermoregulation in nine species of ground squirrels (*Spermophilus*) and found them to be physiologically remarkable in several ways. Ground squirrels have unusually low metabolic rates, down to 50 per cent below that expected for mammals of their size; and, as an odd accompaniment, their thermal conductance can be up to 50 per cent greater than expected. In other words, ground squirrels dissipate heat especially rapidly while producing heat at an unexpectedly low rate.

The maintenance of thermal stability by ground squirrels seems to depend largely on (1) their remarkable control over the rate of cutaneous heat loss, (2) their ability to raise sharply their heat production, and (3) their capacity to dissipate heat at high temperatures by evaporative cooling. Hudson and his coworkers found that in the laboratory thermal conductance typically changed markedly between different measurements on the same animal, and metabolic rate was also labile. Ground squirrels clearly have a high level of vasomotor control of the skin: they can increase the cutaneous blood flow when heat dissipation is necessary and decrease the flow when conservation of heat is of advantage. A delicate balance between metabolic rate and conductance thus maintains a fairly constant body temperature, while at very high temperatures panting occurs (see Table 19–1 for changes in breathing rates) and heat is dissipated rapidly via both the respiratory tract and the skin.

Hudson et al. (1972) present an interesting and reasonable hypothesis: The high

Table 19–1. Highest Tolerated Ambient Temperatures for Ground Squirrels and Some Responses

Species	N	Highest T_A (°C) Tolerated	30°C	Mean Breathing Rate at Highest T_A Tolerated	T_B (°C)	T_A at which 100% of Metabolic Heat is Evaporated (°C)
C. lateralis	4	<39	–	316	41.4	–
C. armatus	5	39	100	268	41.0	39.0
C. richardsoni	5	39	–	192	41.5	–
C. spilosoma	5	40	142	240	41.0	39.6
C. leucurus	7	41	–	–	41.3	41.0
C. townsendi	5	41	73	230	41.0	40.8
C. tereticaudus	7	46	–	–	41.2	–

(From Hudson et al., 1972.)

level of control of physical temperature regulation and the low basal metabolic rate of ground squirrels evolved as adaptations to hibernation, a strategy requiring rapid cooling and/or warming of the body. These same physiological abilities preadapted ground squirrels to life in seasonally hot areas by allowing the dissipation of heat when there is little disparity between body temperature and ambient temperature. Probably only minor quantitative changes in physiological performance fitted some species for the very high temperatures of deserts.

Especially remarkable in their ability to be active in desert heat are the desert ground squirrels. In the heat of the day, these are the only small mammals that are conspicuously active. In the case of the antelope ground squirrel (*Ammospermophilus leucurus*), hyperactive is a more appropriate descriptive term. This small (roughly 90 g) rodent appears extremely nervous, and dashes from enterprise to enterprise whether in the cool of winter or the searing heat of summer. It is often active when surface soil temperatures are 65°C or more. Studies of the antelope ground squirrel by Hudson (1962) have shown that, compared to most mammals, the thermoneutral zone of this squirrel is high, between roughly 33°C and 41°C. Throughout its thermoneutral zone, the body temperature is maintained at least slightly above the air temperature, allowing some dissipation of body heat to the environment even at these high tempera-

tures. This squirrel is able to store heat, and operates in a seemingly normal manner at body temperatures above 43.5° C. Because of the high temperatures at and slightly above ground level, the zone in which this squirrel is usually active, its body absorbs heat rapidly and the animal is forced periodically to unload heat; to do so it retreats periodically to the dense shade of a bush or rock, or into its burrow, presses its ventral surface against the ground, and loses heat to the relatively cool substrate (Fig. 19–6).

Hudson observed that under laboratory conditions an antelope ground squirrel could reduce its body temperature from approximately 41°C to 38°C within three minutes when transferred from high ambient temperatures to a chamber with an ambient temperature of 25°C. As a last resort when subjected to continued heat stress, ground squirrels salivate copiously and spread the saliva over the head, where evaporative cooling occurs. In desert ground squirrels, physiological adaptations alone were probably unable to ensure survival under stresses imposed by intense solar radiation and high air and soil temperatures. Occupancy of low deserts by some ground squirrels may well have been made possible by the evolution of a suite of behaviors that augmented the physiological abilities.

Marsupials. A number of marsupials, including some of the larger kangaroos (*Macropus*), the quokka (*Setonix brachyurus*), a rabbit-sized macropodid, and the Tasmanian devil (*Sarcophilus harrisii*), a raccoon-sized dasyurid, are excellent temperature regulators in the face of heat stress. Panting is the primary means of evaporative cooling in marsupials (Dawson, 1973b). Under conditions of extreme heat stress, however, both sweating and licking occur. Sweating is important mostly in the large kangaroos, in which it is used in addition to panting to dissipate heat during sustained exertion (Dawson, 1973b; Dawson et al., 1974). At high ambient temperatures many marsupials resort to salivating heavily and licking the appendages. This may be an extremely effective means of rapid heat loss, for vascular specializations of the forelimbs of the red kangaroo (*Megaleia rufa*), and very probably those of other large kangaroos, provide for increased blood flow to the forelimbs when heat stress initiates the licking behavior (Needham et al., 1974). The repertoire of thermoregulatory responses of some marsupials, the quokka for example, includes all of those just mentioned, but sweating is of major importance and licking is not used in some species.

An interesting example is the rat-kangaroo (*Potorous tridactylus*), studied by Hudson and Dawson (1975). Thermoregulation at high temperatures in this small and rather generalized macropodid involves a specialized system of evaporative cooling. This marsupial weighs about 1 kg. Its metabolic rate and body temperature (36°C) are low compared to those of eutherian mammals of similar size; thermal conductance from the well-furred body is low. (Studies by MacMillen and Nelson, 1969, and by Dawson and Hulbert, 1970, have shown that a number of marsupials have body temperatures equivalent to those of eutherians but have reduced metabolic rates that are about two thirds those of placentals of comparable size.) At ambient temperatures below body temperature, heat is dissipated by the rat-kangaroo primarily by panting, but at ambient temperatures approaching and exceeding body temperature the bare tail, which comprises 9.4 per cent of the total surface area, is a major route for heat loss.

Vasodilation in the skin of the tail allows for increased nonevaporative heat loss, and at temperatures near and above body temperature profuse sweating of the tail, but not of the body, produces rapid evaporative cooling. Constant side-to-side movement of the tail further facilitates evaporation. The maximum rate of sweating of the tail is extremely high, amounting to 620 to 650 $g/m^2/hour$, roughly double the highest measured rates in such eutherians as horses and cows.

Rat-kangaroos are nocturnal, prefer areas of dense ground cover, and thus probably avoid intense daytime heat, so sweating from the tail may be of importance primarily during exertion when the animals are hopping. Because of its low metabolic rate, good insulation of the body is seemingly essential for heat conservation under cool conditions, whereas the evolution of compensatory responses of the tail—delicate vasomotor control of the skin and profuse sweating—has provided the animal with a system for the rapid dissipation of heat at high temperatures or during exertion.

BEHAVIORAL THERMOREGULATION. Most discussions of temperature regulation in mammals center on physiological means of producing, conserving, or dissipating heat. Because physiological temperature regulation is extremely well developed in mammals, this is appropriate. But because this type of thermoregulation generally involves adjustments in metabolic rate, it is energetically costly. As a result, selection has favored the evolution of behavioral control of body temperature as a "cheap" means of thermoregulation. While it is more or less important to many mammals, in the lives of a few mammals behavioral temperature regulation plays a central role.

Behavioral thermoregulation is a conspicuous part of the daily routine of some African rock-dwelling hyraxes. Field studies by Sale (1970) have indicated the importance of behavior in adjusting to heat or coolness, and laboratory studies by Bartholomew and Rainy (1971) attest to the unusual system of regulation of body temperatures in the rock hyrax *Heterohyrax brucei.* The body temperatures of normally active individuals vary from 35° to 37° C, and are affected by ambient temperatures. The standard metabolic rate is some 20 per cent below that expected on the basis of body weight, and the mean minimum heart rate (118 per minute) is 52 per cent below the expected level.

Outside their nocturnal retreats, hyraxes adjust their postures and locations to exploit the environment to maintain appropriate body temperatures. I observed a colony of *H. brucei* on the Yatta Plateau, southern Kenya, in July, 1973. When hyraxes first emerge from their nocturnal retreats in deep rock crevices, they avoid extensive contact between their ventral surfaces and the cool rock, turn broadside to the first rays of the sun, and bask while maintaining a semispherical body form,

which presumably avoids excessive heat dissipation. As the air and the rock begin to warm, the hyraxes tend to sprawl on the rock, presenting a large surface area to the sun (Fig. 19–10A). Bartholomew and Rainy (1971) found that the lowest body temperatures of hyraxes were reached shortly before sunrise, despite their huddling together during the night. The basking utilizes solar radiation rather than metabolic heat production to raise body temperature. After basking, the hyraxes generally feed.

As the temperature rises abruptly during the morning, the hyraxes often move first to the dappled shade beneath the sparse foliage of trees or bushes (Fig. 19–10B); then later, when ambient temperatures reach 30°C or above, the animals move to deep shade. During these hot times the hyraxes often lie full length on the rock

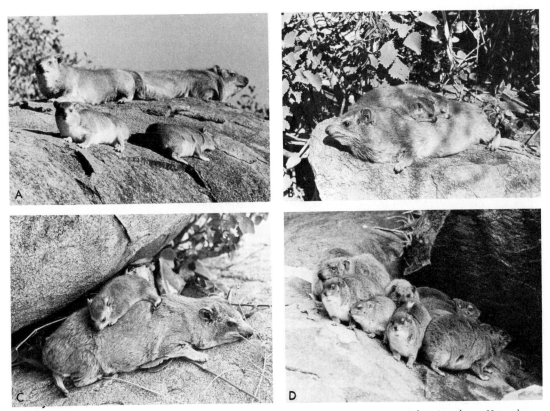

Figure 19–10 Hyraxes behaviorally thermoregulating. A, Two *Procavia johnstoni* (above) and two *Heterohyrax brucei* (below) basking with their bodies broadside to the early morning sun; B, a female *P. johnstoni* and her young basking in mottled light and shade; C, a female *P. johnstoni* and her young stretched out on deeply shaded rock during the heat of the day; D, both species huddled together on a cool day. (Photographs copyright H. N. Hoeck.)

(Fig. 19–10C). At peak ambient temperatures during the afternoon, they tend to remain on deeply shaded rocks. Before dark they may again be sprawled on rock, but this time on warm rock in the open. Solar radiation and shaded sites seemingly provide means of passively absorbing and dissipating heat, respectively.

Of particular importance, the rocks on which the hyraxes live provide an auxiliary means of adjusting body temperatures. The rock outcrops form massive heat sinks with vastly more thermal inertia than that of the air. During the day, shaded or partially shaded rock heats relatively slowly and heat transfer from the ventral surfaces of the hyraxes to the rock probably allows for cooling; fully shaded rock even during the hottest part of the day remains cooler than the air and may enable them to unload heat passively. On the other hand, on a breezy evening just before dark, air temperatures decline rapidly, but the surfaces of previously sunlit boulders are relatively warm, and reradiate heat absorbed during the mid-day. The hyraxes then use heat from the rock to compensate for body heat lost to the air by conduction and convection. On cool, cloudy days hyraxes huddle tightly together, thus reducing the surface area exposed to the air (Fig. 19–10D).

The regulation of body temperature at very low metabolic cost in hyraxes is reflected by a casual approach to feeding. On a day when I had 12 hyraxes under observation for most of the morning, 6 did not feed until at least three hours after they began basking. Behavioral thermoregulation seemingly took precedence over feeding.

HYPOTHERMIA AND METABOLIC ECONOMY

Hypothermia (the lowering of body temperature) is critically important in the lives of some small mammals. These animals are able to exploit efficiently such abundant food sources as tiny seeds and small insects, and can use a nearly limitless variety of shelters; but in terms of metabolism, small size is an extremely costly luxury. The weight-to-surface ratio of small mammals favors rapid dissipation of heat, and a small mammal must have a high metabolic rate sustained by frequent feeding to maintain thermal homeostasis. To a shrew that maintains a fairly constant body temperature, for example, a continuously available and rich food source is a necessity. Some areas are highly productive of adequate food for small mammals in some seasons, but are relatively unproductive at other times. Winters in the north and dry seasons in the deserts and in some tropical areas are typically times of potential food shortage for small mammals, and are also periods when conditions of temperature or moisture may limit their activity.

It is not surprising, therefore, that some small species have evolved means of surviving periods of food shortage and temperature stress and of taking advantage of times of moderate temperatures and high productivity of food for reproduction and storage of food or fat. Many small (up to woodchuck size) mammals periodically conserve energy by allowing the body temperature to drop to near that of the environment. This is not a primitive feature, a manifestation of some ancestral inability to sustain a steady temperature at all times, but is instead a highly adaptive ability. The body temperature in heterothermic mammals may vary widely, but within certain limits, depending on the species, is still under control. This adaptive hypothermia may well have been a factor important in furthering the success of the two largest mammalian orders, the Rodentia and the Chiroptera. Many small bats would be unable to forage only at night and fast throughout the day if they could not conserve energy in the day by hypothermia, and seasonally hostile areas would not be inhabited by some small rodents if these animals retained constant thermal homeostasis.

From evidence assembled in roughly the last 25 years, a rather complex picture of heterothermy in small mammals has emerged, but it is clear that the metabolic economy gained by hypothermia is of importance to many small mammals daily or seasonally. Periodic torpor occurs in a monotreme *(Tachyglossus aculeatus)*, in marsupials (Bartholomew and Hudson, 1962), in some members of the orders Insectivora, Chiroptera, Primates, and in many rodents.

A number of physiological changes, all favoring metabolic economy, occur during adaptive hypothermia or torpor. These include lowering of heart rate, progressive vasoconstriction, suppression of shivering, reduction in the rate of breathing, and lowered oxygen consumption, in some cases to levels 5 per cent below that of a resting, nonhibernating individual. These changes all occur during entry into torpor, and actually precede the sharp decline in body temperature. The stimuli for hibernation have been widely studied and, although many questions remain to be answered, it is clear that different mammals are responsive to different stimuli or respond differently to the same stimuli.

Some cricetid and heteromyid rodents become torpid rather quickly in response to low temperatures and lack of food, but some, the golden hamster *(Mesocricetus auratus)* for example, require two to three months of cold "preparation" before becoming torpid. Mammals with circannian rhythms, such as the arctic ground squirrel *(Spermophilus parryii)* enter torpor in response to diminishing photoperiods and falling temperatures in the autumn (see page 395). Laboratory studies by Dawe and Spurrier (1969), in which injection of blood of hibernating thirteen-lined ground squirrels *(Spermophilus tridecemlineatus)* into active animals was followed by hibernation by the latter, suggest that some "trigger substance" in the blood stream may initiate hibernation in squirrels.

Preparation for winter dormancy or hibernation often (but not always) involves great increases in body weight. This gain ranges in sciurids from 80 per cent of the fat-free weight in the golden-mantled ground squirrel *(Spermophilus lateralis)* to 30 per cent of this weight in the yellow-pine chipmunk *(Eutamias amoenus;* Jameson and Mead, 1964).

Levels of torpor differ between species and can be regarded as "shallow" or "deep." A number of small rodents, including the genera *Baiomys, Perognathus, Peromyscus,* and one species of *Reithrodontomys,* do not spontaneously become torpid but enter torpor when denied food. During the shallow torpor of these rodents, body temperatures stay above roughly 15°C, whereas during the deep torpor of some vespertilionid bats, body temperature drops to 1.4°C (Hock, 1951), and that of the birch mouse *(Sicista)* of Europe reaches 5°C (Johansen and Krog, 1959). Species differ in their "critical" body temperatures—the lowest body temperature they can withstand without arousing. This temperature is fairly high in some heteromyids that undergo shallow torpor (12°C in *Perognathus hispidus),* but, as would be expected, it is low in mammals using deep torpor. Critical body temperature is close to freezing in some bats of the genus *Myotis,* 2.8°C in the golden-mantled ground squirrel, and 4°C in the European hedgehog *(Erinaceus europaeus).*

Arousal from hibernation is energetically costly and is associated with metabolism of energy-rich "brown fat" (Chaffee and Roberts, 1971) in some species or with shivering in others. The high cost of arousal suggests that the most advantageous strategy for a hibernator would involve continuous deep torpor; but, surprisingly, periodic arousal during the hibernation period is the rule among those species studied. The broad pattern is one of progressively increasing periods of torpor through the early stages of hibernation and decreasing periods in the late stages. Among several rodents the maximum periods of torpor were from 12 to 33 days, and a period of 80 days was recorded for the little brown bat *(Myotis lucifugus).* The European hedgehog was found to be torpid 31 per cent of the time at 10°C and 80 per cent of the time at 4.5°C). There is considerable evidence that the length of the period of torpor is determined by the time it takes for the bladder to fill; in other words, it is determined by excretory requirements.

(For an excellent review of torpor in mammals, see Hudson, 1973.)

Periods of torpor are seemingly characteristic of the life cycles of some rodents inhabiting hot regions. In the cactus mouse (*Peromyscus eremicus*), an inhabitant of the deserts of the southwestern United States and northern Mexico, torpor may occur in both summer and winter (MacMillen, 1965). Cactus mice remain in their burrows for several weeks during the driest part of the summer, and laboratory animals entered torpor in the summer in response to a reduced food supply or, in some cases, to restricted access to water. In the winter, laboratory animals were torpid by day and active by night when their food was in short supply, and they were able to become torpid at ambient temperatures below 30°C (Fig. 19–11). The cactus mouse has a narrow thermoneutral zone (28° to 35°C) and a low basal metabolic rate for a mammal its size; these features are seemingly typical of small mammals able to enter torpor under moderate temperatures. According to MacMillen, the summer torpor in the cactus mouse is a device for reducing the use of food and water and for surviving periods of water and food shortage on the surface. The California pocket mouse (*Perognathus californicus*), an inhabitant of seasonally dry chaparral areas, undergoes a daily cycle of diurnal torpor in the laboratory when its food supply is reduced, and maintains a delicate balance between food supply and metabolic economy (Tucker, 1962, 1965). Its periods of torpor are adjusted so that the shorter the food supply, the longer the daily torpor becomes. Both the cactus mouse and the California pocket mouse are adapted to moderately high-temperature torpor; neither can arouse, and the animals will die, when the body temperature goes below about 15°C.

Torpor in response to heat, drought, food shortages, or combinations of these stresses, is called estivation. Among small mammals occupying hot areas, estivation is obviously an important and widespread device for avoiding hot-season stresses. In addition, periods of torpor in some species may strongly reduce competition for limited food between closely related and sympatric rodents during periods of food shortage.

The Mohave ground squirrel (*Spermophilus mohavensis*) remains in its burrow from August to March; this period spans both extremely hot and cold seasons. Laboratory studies indicate that during this

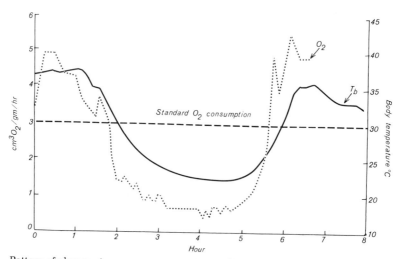

Figure 19–11. Pattern of changes in oxygen consumption and body temperature in the cactus mouse (*Peromyscus eremicus*) during entry and arousal from torpor at an ambient temperature of 19.5°C. The straight dashed line shows the standard rate of oxygen consumption in active mice at an ambient temperature of 20°C. The dotted line shows oxygen consumption and the solid line gives body temperature. The cycle of torpor was initiated by deprivation of food and water. (After MacMillen, 1965.)

period the animals are intermittently torpid for periods of from several hours to several days (Bartholomew and Hudson, 1960, 1961). This squirrel is able to elevate its temperature from 20° to 30°C in from 20 to 35 minutes, and even during its active period in spring and summer has an unusually variable body temperature (from 31° to 41.5°C).

TEMPERATURE REGULATION IN BATS

Recent studies of a variety of bats have clarified and, at the same time, complicated the picture of temperature regulation in these animals. Among different species, contrasting reactions to temperature changes occur, and within the Chiroptera most mammalian styles of temperature regulation are represented.

Seemingly the larger megachiropterans are homoiotherms. Those that have been studied are able to maintain body temperatures within fairly narrow limits (35° to 40°C) over a range of ambient temperatures from roughly 0° to 40°C. No diurnal torpor occurs in these bats, which are usually quite active during the day in their communal roosts. Many pteropodid bats react to cold stress by shivering and by enveloping their bodies with their wings (Fig. 19–12); the wings serve as blankets that provide considerable insulation for the body (Bartholomew et al., 1964). Many megachiropterans roost in trees where they are periodically exposed to direct sunlight.

Several devices for lowering body temperature were observed by Bartholomew and his coworkers in animals under heat stress. Vasodilation occurred in surfaces such as the scrotum, wing membranes, and ears; these naked surfaces are seemingly efficient heat dissipators. Other reactions to high temperatures were extension of the wings, fanning of the wings, and panting. Under intense heat stress, the animals salivated copiously and licked their bodies. Regular use of wing licking as an aid to evaporative cooling under natural conditions has been observed in southern Kenya, East Africa, by T. J. O'Shea. On a number of

Figure 19–12. An African epauletted bat (*Epomophorus wahlbergi*, Pteropodidae) with its wing membranes enshrouding its body. (Photograph by Richard G. Bowker.)

hot afternoons he observed roosting epauletted fruit bats (*Epomophorus wahlbergi*) partially spread their wings and lick the membranes, in which the engorged blood vessels could be seen clearly. These bats were hanging beneath the sparse canopy of an acacia tree. A different thermoregulatory pattern occurs in some megachiropterans; Bartholomew et al. (1970) have shown that some of the smaller species in New Guinea are capable of becoming torpid.

Compared to megachiropterans, microchiropteran bats are highly variable in their responses to temperature extremes. Among tropical or subtropical microchiropterans two extreme patterns of response have been found. The Australian species *Macroderma gigas* (Megadermatidae) is able to maintain a stable body temperature in the face of ambient temperatures as low as 0°C, and many of the reactions to temperature extremes in this bat are similar to those of megachiropterans (Leitner and Nelson, 1967). At the other extreme, the Neotropical species *Desmodus rotundus*, the vampire bat, seems unable to regulate its body temperature (Lyman and Wimsatt, 1966). The inability of this species to dissipate heat causes death at ambient temperatures as low as 33° C. Nor can this species withstand low temperatures: after an initial short-lived attempt at maintaining the body temperature by increasing metabolism, the body temperature drops and body temperatures between 17°C and 27°C are often lethal. Vampires have no ability to rewarm their bodies if the ambient temperature is not raised. The style of temperature regulation of most tropical bats probably lies somewhere between these extremes.

Many tropical microchiropteran bats from the Old World and from the Neotropics have similar diel activity cycles. These bats are active at night and inactive during the day, and this activity cycle is reflected by their temperature cycles. In the Neotropical phyllostomatid bats that have been studied, body temperatures are from 37° to 39°C during the night, and are two or three degrees lower during the day (Morrison and McNab, 1967). In general, these bats are able to maintain high body temperatures despite moderately low ambient temperatures; only one species cannot sustain "normal" body temperatures at ambient temperatures below 12°C. Some Old World members of the families Rhinopomatidae, Emballonuridae, Megadermatidae, and Rhinolophidae have responses to temperature extremes similar to those of phyllostomatids (Kulzer, 1965). But broadly speaking, cold stress can usually be tolerated by tropical bats for only fairly short periods, after which the body temperature falls uncontrollably. Body temperatures below 20°C are often fatal.

The Molossidae, a widespread but largely tropical and subtropical group, seem to have a pattern of temperature regulation intermediate between that of the tropical families and the pattern typical of microchiropterans inhabiting temperate areas. The strictly tropical molossids have some diel variation in body temperature, but cannot cope with low temperatures, and die when the body temperature goes below 20° C. Molossids that inhabit temperate areas, however, can tolerate body tempera-

tures as low as 10°C. In southern California in the winter, the western mastiff bat (Eumops perotis) becomes torpid during the day, when its body temperature drops to within 1° to 2°C of the ambient temperature at temperatures from 9° to 28°C. The metabolic rate is spontaneously elevated in the afternoon and evening, and the bats are active at night (Leitner, 1966).

Adaptive hypothermia, often involving (at different seasons) both daily torpor and hibernation, occurs in many vespertilionids of north temperate areas, and seems to be the key to the survival of some species in cool or cold regions. During the summer some microchiropterans of temperate zones undergo daily torpor ("Tagesschlaflethargie"; Eisentraut, 1934) at low ambient temperatures. Tremendous metabolic savings are realized by microchiropteran bats that become hypothermic at low temperatures. Under experimental conditions the average metabolic rate of six Myotis lucifugus kept at an ambient temperature of 35°C was some 33 times that of these bats when kept at 5°C (Henshaw, 1970:201, Fig. 18). In addition, strikingly abrupt rises in metabolic rate occur during flight. The metabolic rate of the phyllostomatid bat Phyllostomus hastatus was found by Thomas and Suthers (1972) to be about 30 times greater in a flying individual than in one resting at a body temperature of 36.5°C. It is obvious that by evolving foraging strategies that allow the briefest possible periods of flight and, in bats capable of adaptive hypothermia, by lengthening the daily periods of hypothermia, bats in cool areas can make significant energy savings.

Some microchiropterans, such as the pallid bat (Antrozous pallidus), are able to roost in summer in situations allowing the utmost energy savings. In many hot desert and semidesert areas, pallid bats roost in deep crevices in cliffs, and these cliffs function as tremendous heat sinks. In one area in Arizona where these bats were studied, temperatures deep within crevices remained close to 30° C throughout the diel cycles on summer days. The temperature gradient in several crevices late in a

July afternoon was from 31°C deep in the crevices to 46°C at their entrances. By adjusting their positions in the thermal gradient of a crevice the bats were able to dissipate heat early in the day, to maintain a low body temperature and a low metabolic rate until late afternoon, and to passively raise their body temperatures prior to emergence by moving toward the relatively hot mouth of the crevice (Vaughan and O'Shea; 1976). In early spring, when days and nights were cool, the pallid bats in the study area preferred to roost in shallow, south-facing, poorly insulated crevices, in which temperatures often rose from about 10°C in the early morning to some 32°C just before dark. Here at sundown the bats could passively attain body temperatures high enough to allow flight.

Some bats abandon homoiothermy well before winter. Fat deposition is known to occur in late summer or early fall in some species of vespertilionids that hibernate (Krzanowski, 1961; Baker et al., 1968; Weber and Findley, 1970; Ewing et al., 1970), and three species of *Myotis* studied by O'Farrell and Studier (1970) became nonhomoiothermic during this time. In *M. thysanodes* the metabolic rate for homoiothermic individuals at an ambient temperature of 20.5°C is 6.93 cm^3 O$_2$/g/hr; the rate drops to 0.59 cm^3 O$_2$/g/hr in nonhomoiothermic bats. This drop in metabolic rate results in the saving of 2.81 kcal/day as a bat becomes nonhomoiothermic. Fat is deposited in preparation for hibernation at the rate of 0.17 g/day in the period of maximum fat accumulation, which requires 1.60 kcal/day. This energy is available primarily because of the late summer–autumn shift to daily hypothermia (Krzanowski, 1961; Ewing et al., 1970). Some birds accumulate fat in preparation for migration by greatly increasing food intake, but this increased intake

does not occur during the period of fat deposition in *M. lucifugus* and *M. thysanodes*.

Winter hibernation in bats differs from short-term torpor largely in the length of dormancy and in the levels to which the metabolic rate and temperature drop. The duration of hibernation for bats differs widely among species and within a species, depending on the area. In the northeastern United States, *M. lucifugus* remains in hibernation for six or seven months, from September or October to April or May (Davis and Hitchcock, 1965). Periods of hibernation for bats in warmer areas are probably considerably shorter. Ewing et al. (1970) used amounts of fat accumulated by bats in the autumn as a basis for estimating durations of hibernation. The estimated lengths of hibernation for several New Mexican bats were as follows: *M. lucifugus*, 165 days; *M. yumanensis*, 192 days; *M. thysanodes*, 163 days. Because no allowance was made for metabolic drain occasioned by intermittent periods of activity, these estimates are probably too high. Hock (1951) estimated that the metabolic rate of hibernating bats at ambient temperatures not much above freezing was 0.1 cm^3 O$_2$/g/hr, which is 257 to 385 times lower than the rate for a flying bat estimated by Studier and Howell (1969). At ambient temperatures near 5°C, bats in deep hibernation maintain their body temperatures about 1°C above ambient temperature. These bats are responsive to certain stimuli, and will begin arousal when handled or when subjected to unusual air movement. As a defense against freezing to death, bats spontaneously raise their metabolic rates at dangerously low ambient temperatures (below roughly 5°C) and either arouse fully or regulate their temperatures and remain in hibernation.

20 WATER REGULATION IN MAMMALS

Roughly 35 per cent of the earth's land surface is desert, where water is the primary limiting factor for plant and animal life. These desert areas are characterized by intense daytime heat in the summer, intense solar radiation by day and maximal heat loss by night resulting in great daily changes in temperature (commonly up to 30° C), extremely low humidity through most of the year, and small amounts of precipitation, often at irregular intervals. To an animal abroad on the desert on a summer day, the searingly dry winds and the radiation and reflection of heat from the hot and pale-colored soil add to the harshness of the environment. Few equally hostile environments occur on earth, and to the casual observer the desert gives the impression of overwhelming sterility. This impression is deceptive, however, for in reality the desert supports a great variety of animal life.

The abundance of mammal life on the desert and the severity of this environment are well described by Hall (1946:1,2): In the morning "scores of burrow openings around sandy dunes attest the density of population of small mammals—a density equaled in few other habitats—and inspection discloses that in nearly every burrow, a short distance back from the entrance the occupant has snugly packed a plug of moist sand to shut him away from the dangers of day. Before a person's curiosity is half satisfied about the burrows and the dozens of stories told by the tracks, the sun is up—and with it the wind, the wind that obliterates every telltale mark and burrow opening, leaving only smooth sand in their places. Little by little the heat re-

turns." Each desert mammal has evolved means of coping with the extreme conditions of this environment. Of these conditions none is more acute than the lack of water.

Water is absolutely essential to life; to all mammals life depends on the maintenance of an internal *water balance* within fairly narrow limits. (Water balance occurs when intake, through drinking, eating, and the production of metabolic water equals the output through the skin, respiratory passages and surfaces, feces, and urine.) Most mammals are under intense discomfort when water loss reduces their body weight by as little as 10 or 15 per cent, and death occurs in many mammals when such loss reduces the body weight by 20 per cent. Loss of water occurs rapidly on the desert: water loss in man on a hot summer day in the southwestern deserts of the United States has been recorded as 1.41 per cent of body weight per hour; comparable figures for the donkey and dog were 1.24 and 2.62 respectively (Schmidt-Nielsen, 1964:27). Deprived of drinking water, a man or a dog can survive only a day or two of exposure to the desert in the summer. Nonetheless, some small desert rodents live for long periods on diets of dry seeds and no drinking water. Similarly, all of the mammals in some desert areas must maintain water balance with only occasional access to water. Although much remains to be learned about mammalian adaptations for water conservation in arid environments, excellent studies in this field have provided a solid base of knowledge.

A number of different solutions to the

438

problem of maintaining water balance without drinking water are used by desert mammals. These solutions depend on the size of the animal, the timing of activity cycles, the foods eaten, and a variety of behavioral, structural, and physiological features. It is highly unlikely that any two desert mammals have solved this problem in the same way. The following discussions will not cover the subject of water conservation in mammals exhaustively, but will consider the adaptations that permit several kinds of mammals to maintain water balance in dry environments. A good discussion of water regulation in desert mammals is presented by Schmidt-Nielsen (1964).

MAMMALS DEPENDENT ON "WET" FOOD

A number of mammals that occupy deserts or semiarid areas are no better adapted to surviving without considerable moisture in their diets than are mammals of fairly moist areas. Even in some areas with fairly high precipitation, small mammals do not have regular access to drinking water and, as in the case of some desert rodents, satisfy their water requirements by eating moist food.

Succulent desert plants provide water for some desert rodents such as the white-throated woodrat (Neotoma albigula), which occupies the hot deserts of southwestern United States and northern and central Mexico. Although this woodrat often inhabits extremely barren areas, its distribution is limited to that of the succulent plants from which it obtains water. Paradoxically, this desert rodent is dependent on large amounts of water for the maintenance of water balance; but it has an important adaptation that allows it to use cactus as a water source: it has the ability to cope with oxalic acid ($C_2H_2O_4$), which occurs in large amounts in prickly pear and cholla cactus (Opuntia). Oxalic acid is a highly toxic compound to most mammals. The white-throated woodrat, however, is able to metabolize oxalic acid, and eats quantities of cactus that contain sufficient oxalic acid to kill other mammals of equal size (Schmidt-Nielsen, 1964:146–149). MacMillen (1964a) found that the desert woodrat (N. lepida) and the cactus mouse (Peromyscus eremicus) also utilize large quantities of Opuntia as a source of both food and water; apparently these species have also evolved a means of metabolizing oxalic acid and excreting the by-products. In some arid habitats rodents feed so heavily on cactus as to destroy it locally (Fig. 20–1).

The ability to obtain water from cacti and to deal with oxalic acid metabolically is not limited to the rodents mentioned

Figure 20–1. A patch of cactus (Machaerocereus gummosus) in Baja California in the process of destruction by the desert woodrat (Neotoma lepida). For much of the year this plant is a prime source of water.

above, all of which belong to the family Cricetidae, but also occurs in the rodent family Geomyidae, the pocket gophers. The northern pocket gopher (Thomomys talpoides), inhabiting fairly dry shortgrass prairies of Colorado, obtains water by eating prickly pear cactus (Vaughan, 1967). During the dry midwinter period, this plant comprised 79 per cent of the diet of these gophers in one area. In arid parts of Arizona and Colorado, the valley pocket gopher (T. bottae) has also been observed to make heavy seasonal use of prickly pear cactus. Anyone who has had contact with prickly pear cacti must admit that the behavioral ability of small mammals to cope with the spiny armor of these plants is as impressive as the physiological ability to deal with oxalic acid.

Other plants provide water for rodents inhabiting arid regions. The juniper (Juniperus), which contains about 70 per cent water, provides water during all seasons for the Stephen's woodrat (Neotoma stephensi), an inhabitant of semiarid areas in Arizona and New Mexico (Hanson, 1971). The desert ground squirrels (Spermophilus and Ammospermophilus) also obtain moisture from their food, which is largely green vegetation and insects.

Some desert rodents obtain water from succulent plants that contain high salt concentrations; these animals have kidneys that are able to produce highly concentrated urine (urine that has little water relative to the contained solutes). The North African sand rat (Psammomys obesus, Cricetidae) is such an animal. This small rodent obtains water from the fleshy leaves of halophytic plants (plants that grow in salty soil), which grow along dry river beds in the desert (Schmidt-Nielsen, 1964:183). These leaves are 80 to 90 per cent water, but contain higher concentrations of salt than seawater does, and also have large amounts of oxalic acid. In order to utilize this water source, the sand rat must produce urine with extremely high concentrations of salt and must be able to metabolize large quantities of oxalic acid. An Australian hopping mouse, Notomys

cervinus (Muridae), has a remarkably well-developed ability to concentrate electrolytes in its urine, and probably also uses the succulent but highly saline leaves of halophytic plants as a water source (MacMillen and Lee, 1969). The tammar wallaby is known to be able to drink seawater (Kinnear et al., 1968). A similar ability occurs in the western harvest mouse (Reithrodontomys megalotis), a rodent that commonly inhabits semiarid or arid areas. This animal can produce highly concentrated urine, and is able to survive short periods of water deprivation. Some populations of harvest mice inhabit salt marshes, areas regarded as "physiological deserts" because of the physiological problems of utilizing the water from the highly saline sap of the plants or from seawater. MacMillen (1964b) thought that these populations of harvest mice might obtain water by drinking seawater or eating halophytes, but more recent work by Coulombe (1970) suggests that daily torpor and water from dew or fog precipitation may be of equal importance. The fact remains, however, that this rodent can drink seawater and maintain weight on this water source.

Most deserts support a number of carnivorous or insectivorous mammals in which moisture requirements are seemingly met by the water in their food. The grasshopper mouse (Onychomys), a small rodent widely distributed in the deserts and semiarid sections of western United States and Mexico, is almost exclusively insectivorous at some times of the year. This mouse has thrived in the laboratory on an entirely meat diet, with no drinking water (Schmidt-Nielsen, 1964:185). In the North American deserts, kit foxes (Vulpes macrotis), badgers (Taxidea taxus) and coyotes (Canis latrans) must also be able to derive sufficient water from their meat diets, for these animals often live in areas remote from drinking water. Schmidt-Nielsen (1964:126, 127) found that the desert hedgehog (Hemiechinus auritus, an insectivore) and the fennec (Fennecus zerda, a fox), both inhabitants of North African deserts, could get adequate water from a predominantly carnivorous diet, as could

Figure 20–2. Gemsbok *(Oryx gazella)* in the Kalahari Desert of Africa. These animals are able to go for long periods without drinking. (Photograph by Fritz C. Eloff.)

the mulgara *(Dasycercus cristicauda)*, an Australian dasyurid marsupial (Schmidt-Nielsen and Newsome, 1962).

Few large ungulates inhabit barren deserts where no drinking water or cover is available. One notable exception is the oryx or gemsbok *(Oryx;* Fig. 20–2), a large antelope that occurs in arid and semiarid sections of Africa, and even penetrates the borders of the Sahara Desert. The oryx has an amazing ability to withstand intense desert heat (as described in Chapter 19, p. 425); perhaps more remarkable is the animal's lack of dependence on drinking water. Careful studies by Taylor (1969a) have shown that the water needs of the oryx are probably satisfied by its food, which consists of leaves of grasses and shrubs that by day may contain as little as 1 per cent water. After nightfall, as temperatures drop and the humidity rises, these parched leaves absorb moisture from the air, and probably contain approximately 30 per cent water during much of the night (Fig. 20–3). By feeding at night, therefore, the oryx can manage a nightly intake of some 5 liters of water with its forage. This is at best a minimal amount of water for a roughly 200 kg mammal living in shelterless desert, and is sufficient for the oryx only because of a combination of mechanisms that favor water conservation.

Some of the major means for reducing water loss in a dehydrated oryx include:

(1) voluntary hypothermia and the sparing use of evaporative cooling except under extreme conditions, when panting but not sweating is used; (2) reduction of evaporative water loss by lowering of the metabolic rate at high ambient temperatures during the day; (3) reduced permeability of the skin, resulting in reduced water loss by diffusion; (4) reduced respiratory rates and greater extraction of oxygen from inspired air during the night, and hence reduced water loss *via* expired air; (5) lowering of metabolism and respiratory water loss by more than 30 per cent during the cool

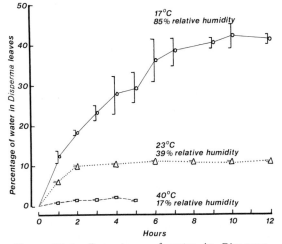

Figure 20–3. Percentages of water in *Disperma* leaves at various humidities and temperatures. The hygroscopic leaves may be an important water source for Grant's gazelles *(Gazella granti)* in some areas. (After Taylor, 1968a.)

night. Taylor also studied the eland (*Taurotragus*) and found that its diet of moist acacia leaves, together with physiological specializations to reduce water loss similar to those in the oryx, allowed this largest of African antelope to be independent of drinking water.

Even between quite similar ungulates, strategies for conserving water may differ, as indicated by studies of the Grant's gazelle (*Gazella granti*) and the Thompson's gazelle (*G. thompsoni*) by Taylor (1968b, 1972). These antelope occur together widely in East Africa, but the range of the Grant's gazelle extends into the harsh deserts of northern Kenya, while the Thompson's gazelle is restricted to less arid areas. Unexpectedly, however, under experimental conditions involving peak temperatures of about 40° C, the Grant's gazelle had a rate of pulmocutaneous water loss about one third higher per kg of body weight than that of the Thompson's gazelle. This apparent paradox was resolved when Taylor (1972) considered the performances of these antelope under extreme heat. At air temperatures over 42°C, the Thompson's gazelle used panting to increase evaporative cooling; at an air temperature of 45°C, it maintained a body temperature of 42.5°C. The Grant's gazelle, by contrast, did not resort to evaporative cooling, but allowed its body temperature to rise to 46°C, and was thus able to dissipate heat to the air. It is obvious that under desert conditions, with limited access to drinking water and intense heat at midday, the temperature regulation strategy of the Grant's gazelle would be highly adaptive. As in the case of the oryx, the major source of water for desert-dwelling Grant's gazelles may be leaves that absorb water at night.

The social behavior of a mammal may markedly influence its pattern of water regulation. Kirk's dik-dik (*Madoqua kirki*) is a sedentary, territorial antelope that lives in brushy, often hot and dry areas of East Africa. Because each pair or family group has a rigidly defined, exclusive area, extensive daily movements to and from sources of water would involve prohibi-

tively frequent aggressive interactions with other dik-diks. The social system of the dik-dik forces each animal to rely on water from plants within its own territory. This constraint, intensified in the dry seasons by high temperatures and reduced moisture content of the plants, provides strong selective pressures. These pressures have been effective: under stress from heat and restricted access to water, the dik-dik has the most highly concentrated urine of any ungulate studied (Table 20–1). Under these conditions the savings due to decreased urine volume and increased urine osmolality average about 20 to 40 g of water per day per animal (Maloiy, 1973).

Considerations of temperature regulation in a number of species of ungulates have led Taylor (1970) and Robertshaw and Taylor (1969) to conclude that there are two major systems of temperature regulation in ungulates. Small ungulates under heat stress allow their body temperatures to rise (to such levels as 46° C in extreme cases), avoid sweating, and often avoid panting. The strategy of the Grant's gazelle exemplifies this system. Large ungulates, on the other hand, have a relatively low surface-to-mass ratio and can afford to sweat. The correlation between body size and degree of use of cutaneous water loss proposed by Robertshaw and Taylor (1969) seems valid: under heat stress, the eland and the African buffalo (*Syncerus caffer*) dissipate heat by sweating, whereas the dik-dik and Thompson's gazelle dissipate heat from the respiratory tract by panting.

Table 20–1. Maximal Levels of Urine Concentration (Osmolar Concentration) in Dehydrated East African Mammals

Species	Urine Osmolality mOsm/kg H_2O
Dik-dik antelope	4300
One-humped camel	3100
Oryx	2900
Fat-tailed sheep	2900
African goat	2800
Impala antelope	2300
Donkey	1500
Zebu cattle	1400

(From Maloiy, 1973.)

Hypothermia during running may be an important adaptation in some mammals. Taylor et al. (1971) found that by comparison with that of the domestic dog, the body temperature of a running African hunting dog (*Lycaon pictus*) is high (41.2° versus 39.2° C) and the percentage of the total heat production during running lost by means of respiratory evaporation is low (25.1 versus 49.7 per cent). Taylor and his coworkers suggest that the greatly reduced pulmonary evaporation rate of the hunting dog may allow this animal to increase the distance it can chase prey. Lest the student assume that for the physiologist the course of research is all sweetness and light, I offer the following quotation from their paper: "The hunting dog is a quite intractable animal, and its odor alone makes it unsuitable for domestication. Nevertheless, at some inconvenience to our surroundings, a male hunting dog was hand-reared from birth and trained to run a treadmill while wearing a mask."

Our knowledge of the water needs of wild ungulates remains incomplete; adaptations that reduce water requirements must occur in all ungulates that occupy dry areas.

PERIODIC DRINKERS

In many arid or semiarid regions, scattered water holes or widely separated rivers offer water to mammals that can move long distances. The extensive grasslands of Africa form such an area, as did the North American Great Plains before the coming of white man. Most large mammals in such areas probably drink every day or two in hot weather, and seemingly are unable to survive for long periods without drinking. A few ungulates, however, occupy an intermediate position with regard to water needs. Although they can go for long periods without drinking, these mammals are not independent of drinking water, as are some desert rodents, and must drink water periodically. Such a mammal is the camel.

Our present knowledge of the water metabolism of the camel is largely a result of the work of Schmidt-Nielsen and his associates (Schmidt-Nielsen, Houpt, and Jarnum, 1956, 1957). Their work, done in the northwestern part of the Sahara desert on local domestic camels (*Camelus dromedarius*), substantiated the popular idea that camels can tolerate long periods without drinking water; but more important, Schmidt-Nielsen and his group explained the adaptations allowing this tolerance. The ability of their experimental animals to tolerate dehydration was remarkable. One camel went without water for 17 days in the winter on a diet of dry food; during this period it lost 16.2 per cent of its body weight. In some areas, camels that foraged on native vegetation in the winter were never watered. Two camels kept without water for seven days in the heat of the summer lost slightly over 25 per cent of their original body weights. All of these animals drank tremendous amounts of water after their periods of dehydration, and none showed ill effects.

The camel economizes on water in several ways. Because the body temperature of the camel drops sharply at night and, at the other extreme, the camel can tolerate considerable hyperthermia, the day is largely over before the body temperature rises to levels at which evaporative cooling, in the form of sweating, must occur. Thus, relative to man under similar conditions, for example, very little moisture is expended each day in cooling the camel (Fig. 20–4). As in the oryx, excess heat gained by day is lost passively at night. Further water economy results from the ability of the kidneys to concentrate urine, and from the absorption of water from fecal material. But despite these important water-saving devices, the camel loses water steadily through evaporation and in the urine and feces, and perhaps most remarkable is its ability to tolerate tremendous losses of weight (up to over 27 per cent of body weight) during intervals of dehydration.

Apparently the degrees of water loss from various parts of the body differ between man and the camel. When a man on

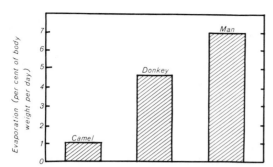

Figure 20–4. Amounts of water used for temperature regulation when the subjects were exposed to the sun in the Sahara Desert in June. The camel and the donkey were tested during periods when they were deprived of water. (After Schmidt-Nielsen et al., 1956.)

the desert has lost water equal to about 12 per cent of the body weight, the blood plasma has lost sufficient water to become viscous; as a result, the heart has difficulty moving the blood and the rate of the blood circulation to the skin and the rest of the body decreases. This leads to a marked reduction in the rate of dissipation of metabolic heat, to a sudden rise in body temperature, and to rapid death. Even in a camel that has been dehydrated to the extent of losing 20 per cent of its body weight, however, close to the normal plasma water content is retained, whereas large amounts of water are lost from interstitial fluids and from intracellular protoplasm. In a camel deprived of water for eight days, the loss of volume due to water loss was 38 per cent for interstitial fluids, and 24 per cent for intracellular water. However, only a 10 per cent reduction in water in the plasma occurred (Schmidt-

Nielsen, 1964:65). Although the camel becomes strikingly dehydrated and emaciated during periods without water, the blood apparently retains its fluidity and its ability to contribute to heat dissipation without straining the circulatory system. The donkey (*Equus asinus*) was also studied by Schmidt-Nielsen (1964:81–93) and proved to be as capable of tolerating weight loss due to dehydration as was the camel, but the donkey had a rate of water expenditure some 2½ times that of the camel and could therefore not be independent of water for intervals of more than a few days.

Both the camel and the donkey (Table 20–2) have an amazing ability to recoup their water losses rapidly. A camel that had lost 25 per cent of its body weight recovered its original weight within ten minutes by drinking, and a similarly dehydrated donkey was able to drink back its weight in just two minutes. The amount of water drunk in a brief interval is nearly unbelievable. Within a few minutes a dehydrated male camel studied by Schmidt-Nielsen drank 104 liters of water (weighing 228 pounds)!

NONDRINKERS DEPENDENT ON DRY FOOD

The most impressive feats of water conservation are those performed by granivorous desert rodents (seed-eating rodents). These small mammals are often common in barren sand dunes or rocky hills where drinking water is rarely avail-

Table 20–2. Water Consumption by a Donkey Deprived of Water in June in a Desert

Days Without Water	Initial Weight (kg)	Dehydr. Weight (kg)	Weight Loss (kg)	% of Body Weight	Water Consumed (liters)*	% of Dehydr. Body Weight
4	107.0	74.5	32.5	30.4	20.5 (23.5)	27.2 (31.5)
4	104.1	73.0	31.1	29.9	20.3 (26.9)	27.8 (36.8)

*Water taken within five minutes is shown without parentheses; water consumed within a two-hour period is shown in parentheses.
(Data from Schmidt-Nielsen, 1964.)

able, and their diets through much of the year include only dry seeds that are picked from the surface of the ground or are sifted from the soil. Some New World kangaroo rats and pocket mice (Heteromyidae), some jerboas (Dipodidae) and gerbils (Cricetidae, Gerbillinae) of the Old World, and some Australian mice (Muridae) are capable of living on dry diets. Work by Abbott (1971) has shown that the canyon mouse (Peromyscus crinitus), a cricetid rodent, can also live on a dry diet. As might be expected, the centers of distribution of these rodents are all desert regions. The following discussion is based largely on the work of Schmidt-Nielsen and his associates, whose publications dealing with water conservation in Merriam's kangaroo rat are cited in the bibliography.

Inasmuch as kangaroo rats do not drink water, water can only be obtained from food, either from free water in the food or from water formed by oxidation of food. Because the moisture content of the kangaroo rat's food is generally low, their chief source of water is probably "oxidation water." Channels of water loss from the body are evaporation and loss in urine and feces.

Water intake in the kangaroo rat can be accounted for fairly easily. The seeds eaten by this animal are high in carbohydrates, which yield a large amount of oxidation water. As an example, for every 100 g of dry pearled barley metabolized, a total of 53.7 g of oxidation water is produced (Schmidt-Nielsen, 1964:173). This may be augmented under natural conditions by free water in the food. Seeds scattered on or slightly beneath the surface of the ground absorb moisture at night as the temperature drops and the humidity rises, and seeds stored in humid burrows may contain as much as 30 per cent free water. If kangaroo rats show a preference for seeds that have been stored in burrows for a time, rather than for those freshly collected, the amount of moisture taken in with the food could be appreciable. This preference has not yet been demonstrated, but might be expected.

Water loss in kangaroo rats is minimized by several devices. Evaporative water loss in kangaroo rats is unusually low, due in part to a nearly total absence of evaporation from the skin. But equally important is the reduction of respiratory water loss. This economy results from the nasal passage operating as a countercurrent heat exchanger (a feature characteristic of many birds and mammals; Schmidt-Nielsen et al., 1970) with alternating flow in opposite directions in a single tube rather than with steady flow in opposite directions in adjacent tubes. Inspired air drawn through the nasal passage causes evaporative cooling of the moist nasal mucosa. This air then becomes warmed and saturated with water in the lungs. During expiration this humid air passes back into the nasal passage and over the cool nasal mucosa, where the air is cooled and loses some of its moisture by condensation on the mucosa. This pattern, repeated with every respiratory cycle, results in cool expired air that is more than 10° C below the body temperature. (In man, by comparison, the temperature of expired air is nearly that of the body, resulting in high rates of pulmonary evaporation.) Although this expired air is saturated with moisture, because it is far cooler than the air in the lungs it contains considerably less water than does saturated air at the high lung temperature. In studies of a kangaroo rat in air of 28° C, the exhaled air was below 24° C; in this case water loss from the respiratory tract was 85 per cent lower than it would have been had exhaled air been the temperature of the body core, about 38° C (Schmidt-Nielsen, 1970). An additional reduction of evaporative water loss occurs when the animal is in its burrow, in which the air is usually quite humid.

MacMillen (1972) has discussed the relationship between evaporative water loss and metabolic water production in desert rodents when these animals are active above ground on cool evenings. Although under conditions of thermal neutrality evaporative water loss surpassed metabolic water production in several species of rodents, when they were subjected to ambient temperatures below their thermoneutral zones (temperatures of 20° to 0° C, a range commonly encountered in deserts at night) these animals produced more metabolic water than they lost by evaporation (Table 20–3). Two factors favor the maintenance of water balance under cool conditions. First, as has been shown by Schmidt-Nielsen et al. (1970), the water-conserving effectiveness of the nasal countercurrent system in rodents varies inversely with temperature; thus, evaporative water loss due to respiration decreases with decreasing temperature. Second, as the ambient temperature drops progressively further below the zone of thermal neutrality, there is a progressive rise in the metabolic rate (this is especially pronounced in small rodents with high surface-volume ratios and poor insulation); with higher metabolic rates, the production of metabolic water increases. In light of these relationships, it is MacMillen's opinion that, while active on the surface at night, most small desert rodents produce more water metabolically than they lose through pulmocutaneous evaporation.

An important reduction in water loss results from the great ability of the kidneys to concentrate urine. The kangaroo rat's urine is roughly five times more concentrated than that of man; in excreting comparable amounts of urea, the kangaroo rat uses one fifth as much water as does man. The concentration of dissolved compounds in the urine of kangaroo rats may be roughly twice that of seawater, and under laboratory conditions these animals have maintained water balance by drinking seawater. The urine of man, on the other hand, has a concentration of dissolved compounds lower than that of sea water. Thus, when man drinks seawater the excretion of the dissolved salts requires the withdrawal of water from the body tissues, resulting in dehydration; this is made more acute by diarrhea caused by the magnesium and sulfate in the seawater.

Fecal water loss is also low in the kangaroo rat. In eliminating wastes from comparable amounts of food, the laboratory white rat (*Rattus*) used five times as much water for the formation of feces as did the kangaroo rat (Table 20–4). This saving in the kangaroo rat is partly the result of more efficient utilization of food and the accompanying reduction in the production of fecal material, but is also due to an exceptional ability to withdraw water from intestinal contents.

Schmidt-Nielsen calculated that, in the kangaroo rat, water intake equalled water losses at atmospheric humidities higher than 20 per cent relative humidity at 25 °C. More advantageous conditions of humidity, in terms of water loss, are faced by kangaroo rats in their burrows and when foraging during the most humid parts of even summer nights. Although the margin for sur-

Table 20–3. Relationships Between Metabolic Water Production (MWP) and Evaporative Water Loss (EWL) in Nocturnal Rodents

T_A °C	P. eremicus R.H.,%	P. eremicus MWP/EWL	L. hermannsburgensis R.H.,%	L. hermannsburgensis MWP/EWL	N. alexis R.H.,%	N. alexis MWP/EWL	N. cervinus R.H.,%	N. cervinus MWP/EWL
5		1.05						
10		1.26	39.9	1.07	25.8	1.14	30.3	1.00
15		1.08						
20		0.88	14.6	0.97	12.4	0.90	13.9	0.89
25		0.69						
28			9.7	0.56	6.9	0.61	6.0	0.84
30		0.43						
33			11.6	0.34	5.1	0.47	5.5	0.39
35		0.40	6.7	0.36	5.2	0.44	6.9	0.32
37		0.19	9.3	0.31	9.4	0.26	8.0	0.27

T_A = ambient temperature. R.H., % = relative humidity. Underlined measurements are of animals in thermal neutrality. *P.*, Peromyscus; *L.*, Leggadina; *N.*, Notomys.
(From MacMillen, 1972. Reprinted with permission of the Zoological Society of London.)

Table 20–4. Amount of Water Lost in Feces of Kangaroo Rat (*Dipodomys merriami*) and White Rat (*Rattus norvegicus*)

	Feces, Grams of Dry Matter/ 100 g Food	Water, mg/g Dry Fecal Matter	Water Lost With Feces Per 100 g Barley Eaten
Kangaroo rat	3.04	834	2.53 g
White rat	6.04	2246	13.6 g

(After Schmidt-Nielsen, 1964.)

vival is slim, it appears that the kangaroo rat's remarkable ability to subsist on dry food and no water can be satisfactorily explained on the basis of the several behavioral and physiological specializations that aid in conserving water.

The ability to be independent of exogenous water is by no means unique to a few rodents. Each major desert region in the world is well populated by rodents, and seemingly each of these regions supports some highly specialized species adapted to dry diets. Such specializations occur in at least some members of the North American heteromyid genera *Perognathus*, *Dipodomys*, and *Microdipodops*. Probably all species of heteromyids that are restricted to deserts can survive on dry diets; not all inhabit deserts, however, and not all are equally well adapted to dry condtions. For example, *Dipodomys agilis*, an inhabitant of coastal southern California and northwestern Baja California, Mexico, occurs largely in coastal sage scrub, where conditions of extreme aridity occur only in midsummer. This kangaroo rat is marginal in its ability to subsist on a dry diet and cannot survive indefinitely without exogenous water (MacMillen, 1964a; Carpenter, 1966).

Among desert rodents of the Old World, two jerboas (*Jaculus jaculus* and *Dipus aegypticus*) and several gerbils of two genera (*Meriones* and *Gerbillus*) can live on dry food, and some surpass kangaroo rats in their ability to concentrate urine (Table 20–5). Adaptations to dry diets have doubtlessly evolved independently in several rodent families (Heteromyidae, Cricetidae,

Dipodidae, Muridae). Striking convergent evolution in these families has led to saltatorial adaptations in some members of each family as well as to similar specializations favoring water conservation. The most concentrated urine yet measured is that of a murid Australian hopping mouse (*Notomys alexis*). This rodent occupies some of the most arid deserts in the world, regions where ten years may pass between rains. This species and two other desert-inhabiting murids of Australia were studied by MacMillen and Lee (1967, 1969), who found that all three species could live on dry seeds with no drinking water. Compared to the kangaroo rat, these murids had higher rates of pulmocutaneous water loss and lost more water in the feces, but in general they had a greater ability to concentrate urine (Table 20–5).

Two species of spiny mice (*Acomys*, Muridae) studied by Shkolnik and Borut (1969) in the desert of Israel are remarkable in their unusual pattern of adaptation to arid conditions. These animals have highly specialized kidneys that can concentrate urine to a greater degree than can the kangaroo rat kidney; but the spiny mice have an evaporative water loss two to three times as great as that in Merriam's kangaroo rat. Probably primarily due to high water loss through the skin, spiny mice are unable to subsist on a diet of dry seeds. Apparently the high cutaneous water loss is important as a means of dissipating heat in a hot climate, and the great ability of the kidney to concentrate urine, coupled with a diet high in land snails (which have a high water content), compensates for the extravagant use of water in thermoregulation.

The kidneys of some bats are specialized to concentrate urine, but these animals are seemingly not independent of drinking water. Carpenter (1969) found that two insectivorous bats that inhabit deserts produced concentrated urine, but that their need for water was increased by high evaporative water losses during flight and when the animals were not torpid. He estimated that the bats lost approximately 3.09 per cent of their body weight through evaporation per

Table 20–5. Concentration of Urea in Urine, Water Loss from Lungs and Skin, and Percentage (By Weight) of Water in Feces of Desert Rodents

Species	Urine MAXIMUM UREA CONCENTRATION (MMOLE/LITER)	Water Loss PULMOCUTANEOUS (MG. H_2O CM.^{3} O_2)	FECES (% H_2O)
Sciuridae			
Ammospermophilus leucurus (North America)	2860	0.53	
Heteromyidae			
Dipodomys merriami (North America)	3840	0.54	45.2
Dipodomys spectabilis (North America)	2710	0.57	
Cricetidae			
Gerbillus gerbillus (Egypt)	3410		
Dipodidae			
Jaculus jaculus (Northern Africa)	4320		
Muridae			
Notomys alexis (Australia)	5430	0.91	48.8
Notomys cervinus (Australia)	3140	0.76	51.8
Leggadina hermannsburgensis (Australia)	3920	1.15	50.4

(Data from MacMillen and Lee, 1967.)

hour of flight. After a careful consideration of the bats' water budgets, Carpenter concluded that the bats were not independent of drinking water, but that their ability to fly long distances to drink water enabled them to maintain water balance in desert areas. A marine fish- and crustacean-eating bat (*Pizonyx vivesi*, Vespertilionidae) that inhabits desert islands and desert coasts of the Gulf of California has the ability to concentrate urine to the extent that it can utilize seawater as a water source (Carpenter, 1968). Because of high evaporative water losses, particularly during flight, the water gained from this bat's food probably is not sufficent to meet its water requirements, and presumably it must drink seawater. The fact that members of the genus *Myotis* have been observed drinking seawater (Dalquest, 1948) suggests that the ability to use seawater as a source of water may be widespread among vespertilionid bats.

SUMMARY

Schmidt-Nielsen (1972) has presented a series of important points bearing on water regulation by desert mammals; these points serve to summarize much of our present knowledge in this area. (1) Most desert mammals have lower metabolic rates than their counterparts living in more moist habitats, and some ungulates respond specifically to heat stress by lowered metabolic rates. With lower metabolic rates go lower rates of water loss. (2) Torpor not only facilitates the conservation of energy but also of water; reduced evaporative water loss during torpor allows an animal to stretch its water reserves. (3) High body temperatures, above ambient temperatures, allow the dissipation of heat to the air, and body temperatures approaching the ambient temperature reduce the thermal gradient and slow the rate of heating of a body. High body

Table 20-6. Temperature and Relative Humidity of Roosting Sites, and Body Weight Loss During Period 8 A.M. to 8 P.M. for Bats (*Myotis*) Caged Singly or in Groups of Four

Species	Ambient Temp. (°C)	Ambient R.H. (Per Cent)	Per Cent Weight Loss Individuals	Per Cent Weight Loss Groups
M. lucifugus (site 1)	26.8	23	10.5(5)	9.9(4)
	15.6–31.1	18–31	7.7–12.8	7.7–11.5
M. lucifugus (site 4)	26.1	32	10.9(7)	11.2(5)
	15.6–30.4	27–40	8.8–13.0	9.0–11.5
M. thysanodes	26.8	23	15.8(8)	10.9(3)
	15.6–31.1	18–31	9.0–21.8	10.1–11.5
M. velifer	22.0	64	8.2(4)	8.4(2)
	20.7–23.3	53–96	5.6–9.8	8.1–8.8

Mean and range are given; sample sizes appear in parentheses.
(From Studier et al., 1970.)

temperatures avoid the water-wasteful device of evaporative cooling by panting or sweating. High body temperatures can be tolerated in some animals because of the system of cooling of blood before it reaches the brain. (4) Lowering the rate of thermal conductance, as, for example, by peripheral vasoconstriction, reduces the rate of heating when air temperature exceeds body temperature and delays panting and the accompanying acceleration of evaporative water loss. (5) The exhalation of air cooler than the core temperature of the body allows for a reduction of pulmonary water loss; water vapor in exhaled air condenses on the cool nasal mucosa.

ECOLOGICAL CONSIDERATIONS

The importance of knowledge of interspecific differences in water metabolism to

Table 20-7. Average Percentage of Body Weight Lost by Four Species of Bats (*Myotis*) Deprived of Water Until Stressed

Species	Sample Size	Per Cent of Weight Loss
M. lucifugus	12	32.3
M. thysanodes	8	31.7
M. yumanensis	3	31.6
M. velifer	4	22.8

(From Studier et al., 1970.)

an understanding of the roles of mammals in the ecosystems of arid regions has only recently been appreciated. The ecological displacement of some species of rodents results in part from their differing means of satisfying water requirements. In a study of a semidesert rodent fauna in California, MacMillen (1964a) found that water metabolism differed significantly between species. Observations in other areas indicate that different preferences with respect to water sources reduce competition between some sympatric and congeneric species. Where *Neotoma stephensi* and *N. albigula* occur together along the Mogollon Rim of Arizona, for example, the former uses juniper as its source of water, whereas the latter uses cactus; competition for food and for nest sites between these woodrats seems slight in this area (Hanson, 1971). In some deserts the patterns of seasonal dormancy in small mammals may be devices that favor water economy by permitting these mammals to avoid periods of intense heat and low humidity. Future research on water metabolism in mammals may contribute importantly to our knowledge of many aspects of mammalian ecology.

Pulmocutaneous water loss is high in some bats exposed to moderately high ambient temperatures, and the choice of roosting sites as well as the geographic distributions of some bats may be limited by the animals' inability to avoid daily dehydration. Clustering during roosting, a common behavior in some bats, reduced pulmocutaneous water loss markedly in one species (*Myotis thysanodes*) studied by Studier et al. (1970), as shown by Table 20–6. It was further found that several species of *Myotis* have an ability to tolerate considerable weight loss caused largely by pulmocutaneous water loss (Table 20–7); but because of high rates of such loss two species were unable to survive two days without access to water.

ACOUSTICAL ORIENTATION AND SPECIALIZATIONS OF THE EAR

21

Because the mammals most familiar to us depend largely on vision for perceiving their environments, it is surprising to note that about 20 per cent of the known species of mammals use acoustical orientation as their primary means, or at least as an important secondary means, of "viewing" their surroundings. Most bats, some members of the orders Insectivora and Carnivora, and probably all odontocete cetaceans use acoustical orientation. Future research may demonstrate that the use of such orientation is even more widespread among mammals.

ECHOLOCATION IN BATS

Foraging insectivorous bats and bats flying within deep caverns face seemingly insurmountable problems: they must perceive tiny prey and obstacles under conditions of nearly complete darkness. The remarkable nocturnal performance of bats indicates that these problems of perception have been effectively solved. Bats are able to capture insects with great efficiency and speed in darkness. The pursuit and capture of an insect take a mustached bat (Pteronotus parnellii; Fig. 21–1) but 1/4 to 1/3 of a second (Novick, 1970). Bats deep in caverns, flying in complete darkness, can not only detect the walls of the cavern but also avoid collisions with hundreds of other bats that are circling and maneuvering abruptly.

Ultrasonic pulses are emitted by the bat, and the echoes of these pulses reflected by objects allow the bat acoustically to "see" in the dark. Even the "nature" of objects can be analyzed to some extent.

When considering echolocation in bats we are dealing almost entirely with members of the suborder Microchiroptera, all of which use echolocation. Most members of the suborder Megachiroptera use visual perception instead of echolocation; but how this serves at low levels of illumination is not known. Among megachiropterans, only members of the genus Rousettus are known to echolocate; their technique is unique, however, in that the pulses are audible and are made by tongue clicking (Mohres and Kulzer, 1956; Kulzer, 1956, 1958; Novick, 1958a).

Although microchiropterans clearly depend on echolocation for perceiving their environments in darkness, the eyes are present in all bats and sight has by no means been totally abandoned. Many phyllostomatids (leaf-nosed bats), for example, have large eyes and obviously make use of them. Olfaction may be of great importance for detecting food by bats that eat fruit, nectar, and pollen or small vertebrates. In species with well-developed visual and olfactory capabilities, echolocation is perhaps primarily used for perceiving nearby obstacles.

An accurate picture of the bat's use of acoustical orientation was long in emerging. As early as 1793, Lazaro Spallanzani performed experiments that suggested that bats use acoustical rather than visual perception when avoiding obstacles and when feeding. Not until the early 1940s,

Figure 21–1. A mustached bat (*Pteronotus parnellii*) in flight (above) and a close-up of the face showing the lips formed into a "megaphone" during echolocation. (Photograph of the flying bat by Timothy Strickler; photograph of the bat face courtesy O. W. Henson.)

however, was the use of echolocation by bats conclusively demonstrated by the careful laboratory experiments of Griffin and Galambos (1940, 1941), and by the observations of Dijkgraaf (1943, 1946). Continued research and the use of refined electronic equipment have contributed to our present detailed, but as yet incomplete, knowledge of echolocation in bats.

EVOLUTION OF ECHOLOCATION. In his excellent studies of vocalization and communication in bats, Gould (1970,

1971) hypothesized that the sonar pulses of bats may have been derived originally from vocalizations that served to establish or maintain spacing or contact between bats. The repetitive communication sounds used by infant bats, and similar pulses perhaps used originally during flight to maintain adequate spacing of foraging individuals, may have secondarily become important in connection with detecting prey and avoiding obstacles. According to Gould (1971:311): "The prominence with which continuous, graded signals pervade the lives of such social and nocturnal mammals as bats suggests that echolocation is an inextricable and integral part of a communication system." This author suggests that some of the vocalizations used by early bats may have been inherited from their insectivore ancestors, in which auditory communication was perhaps as important as it has been shown to be in some living insectivores (Gould, 1969; Eisenberg and Gould, 1970).

ADAPTIVE IMPORTANCE OF ECHO-LOCATION. Probably ever since their appearance in the Paleocene or late Cretaceous, bats have "owned" the aerial adaptive zone during the nocturnal segment of the diel cycle. No birds can match the nocturnal insect-catching ability of bats, nor have birds exploited nocturnal fruit eating or nectar feeding. In many tropical areas the most important flying predators of small vertebrates are seemingly bats, not birds. Bats and birds, the only two flying vertebrates of the Cenozoic, may, early in the era, have divided the diel cycle: birds dominated the aerial zone during daylight, whereas bats dominated this zone during darkness. In tropical areas, which characteristically support a great diversity of both bats and birds, the changing of the guard at sunset is dramatic. Bird activity and bird songs suddenly begin to diminish as the sun disappears; at the same time bats begin to appear and become increasingly more evident as darkness descends. When the twilight glow in the west is gone most birds are inactive and silent, but the cries of bats regularly penetrate the cacophonous blending of frog and insect noises. This chiropteran domination of the nocturnal air is seemingly due largely to their ability to use echolocation. The perfection of highly maneuverable flight and echolocation in bats may have occurred more or less concurrently during their early evolution, and together these abilities were probably responsible for the spectacular nocturnal success of bats.

PRODUCTION OF ULTRASONIC PULSES. The pulses used by microchiropterans for echolocation are produced by the larynx. The cricothyroid muscles (muscles that tense the vocal cords) were shown by Novick (1955) to be essential for normal emission of pulses. These are mostly of frequencies well above those that man can hear. Man can detect frequencies up to roughly 20 kHz (20,000 cycles per second). There are audible components (the so-called "ticklaute") in the vocalizations of many bats, and some sac-winged bats (Emballonuridae) of the Neotropics emit pulses (of 12 kHz) that are audible to man. The pulses, however, are only low intensity components of cries with higher harmonics (harmonics are frequencies that are integral multiples of the fundamental frequency).

The pulses emitted by bats are usually of high intensity (great loudness), and are emitted through either the mouth or the nose. In most vespertilionids, molossids, and noctilionids, the pulses are of the highest intensities recorded for bats, and come from the mouth, which is kept open during flight. The rhinolophids (horseshoe bats) emit high intensity pulses through the nostrils. The Nycteridae, Megadermatidae, and Phyllostomatidae, on the other hand, produce relatively low intensity pulses, which is why these groups are called the whispering bats (Griffin, 1958:232–251). Whereas the bats that emit intense pulses catch flying insects, the whispering bats feed largely on fruit, nectar, small vertebrates, insects on the ground or on vegetation, or combinations of these foods. Some authors (for example, Griffin, 1958:251; Novick, 1970:39,40) have suggested that these soft pulses are well

adapted to close-range perception of surfaces, such as rocks or tree trunks, or
complex tropical environments with interlacing vines and branches and stratified
foliage. Perhaps an advantage is gained by
not receiving echoes from distant objects
and therefore limiting the complexity of
the information from an already complex
perceptual situation.

Just how loud are the pulses emitted by
bats? Inasmuch as we cannot hear them,
their loudness must be expressed in terms
of an energy unit called a dyne (a dyne is
the force necessary to accelerate one gram
of mass one centimeter per second per second). Under ideal circumstances, man's
threshold of hearing is roughly 0.0002
dynes per square centimeter. When recorded 5 cm from a bat's mouth, the least
intense pulses of whispering bats are approximately 1 dyne per square centimeter,
whereas the loudest pulses of other bats
are near 200 dynes per square centimeter.
Such loud pulses are comparable to the
painfully intense noise made by nearby jet
engines.

The strange faces of bats, always a
source of amazement to those unfamiliar
with bats, may have an important function
in connection with echolocation. Mohres
(1953) showed that the horseshoe-like
structure surrounding the nostrils of a
rhinolophid (Fig. 6–48) serves as a horn,
and "beams" the pulses directly forward
from the head. This diminutive megaphone effectively beams the short-
wavelength pulses emitted by these bats;
these 80 to 100 kHz pulses have wavelengths of only 3 or 4 mm. In addition,
because the nostrils are situated almost exactly one-half wavelength apart, the pulses
emitted through the nostrils undergo interference and reinforcement of a sort that
tends to beam the pulses. The bizarre facial patterns of many bats may well function similarly to direct pulses in such a
way that some species can scan their
environment with a concentrated beam of
ultrasonic pulses much as we probe the
darkness with a flashlight beam.

SENSITIVITY OF A BAT TO ITS
OUTGOING PULSES. A number of

structural and physiological specializations reduce the sensitivity of bats to their
own outgoing pulses. Vibrations are transmitted mechanically in mammals by the
ear ossicles (malleus, incus, and stapes)
from the tympanic membrane to the oval
window of the inner ear. Two muscles of
the middle ear of mammals serve to damp-
en the ability of the ossicles to transmit
vibrations when animals are subjected to
unusually loud sounds or when they are
vocalizing. These muscles—the tensor
tympani, which changes the tension on the
tympanic membrane, and the stapedius,
which changes the angle at which the
stapes contacts the oval window—are extremely well developed in bats, and their
contraction reduces the bat's sensitivity to
pulses. Jen and Suga (1976), using highly
sophisticated electronic equipment, found
that action potentials of the cricothyroid
laryngeal muscles were followed 3 msec (a
msec is 1/1000 of a second) later by action
potentials of the middle-ear muscles. This
coordination of laryngeal and middle-ear
muscles ensures the contraction of the latter just prior to vocalization and attenuates
by some 25 per cent the auditory self-
stimulation (Suga and Shimozawa, 1974).
Of additional importance, neural attenuation of the direct reception of sonar sounds
occurs in the brain. Nerve impulses arising
from direct reception and passing from
the cochlea to the inferior colliculus of the
brain are attenuated by the neurons of the
lateral lemniscus of the brain. This, plus
that effected by the middle-ear muscles, attenuates the neural events by some 40 per
cent. Suga and Shimozawa suggest that
similar attenuating mechanisms occur in
man and keep sounds of our own speech
from becoming disturbingly loud. Direct
reception in bats is doubtless reduced also
by the "beaming" of sounds by the lips
and the complex noses and nose leaves
(Fig. 21–2).

An additional structural refinement is
the insulation of the bones that house the
middle and inner ear from the rest of the
skull. This bony otic capsule does not contact other bones of the skull (Fig. 21–3),
and is insulated from the skull by blood-

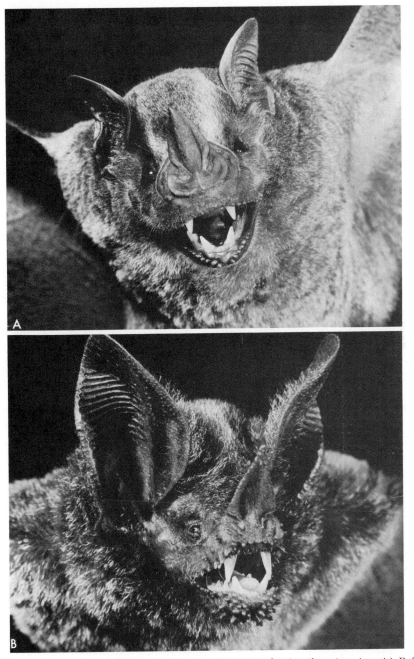

Figure 21-2. Faces of phyllostomatid bats: A, Jamaican fruit-eating bat (*Artibeus jamaicensis*), B, fringe-lipped bat (*Trachops cirrhosis*), a carnivorous-omnivorous species. (Photographs by N. Smythe and F. Bonaccorso.)

filled sinuses or fatty tissue. During the emission of pulses the conduction of vibrations from the larynx and the respiratory passages to the inner ear is thus greatly reduced.

THE ADVANTAGE OF HIGH FREQUENCIES. Because high frequencies are more severely attenuated (weakened in intensity) by air than are low frequencies, one might wonder why echolocation in bats is based almost entirely on high frequencies. Interference from such sounds as insect chirps and frog calls is doubtless partially avoided by the use of high

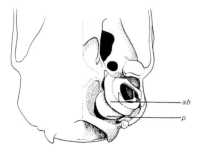

Figure 21–3. Ventral view of the posterior part of the skull of a bat (*Myotis volans*), showing the looseness of attachment of the periotic bone (*p*) and auditory bulla (*ab*) to the skull.

frequencies, but this factor may not be of primary importance. Perhaps of greater importance is the relationship between the size of prey and the wavelength of echolocation pulses. The higher the frequency of a sound, the shorter its wavelength. Frequencies of roughly 30 kHz have wavelengths of approximately 11.5 mm, roughly the size of a small moth; this balance between size of object and wavelength is ideal, for objects with sizes of approximately 1 wavelength reflect sound particularly well. Some species of bats are able to detect wires with as small a diameter as 0.08 mm (some one thirtieth of a wavelength), but in general the wavelengths of the pulses emitted by bats are in the range that is most efficient for the detection of small- to medium-sized insects.

TYPES OF PULSES AND INFORMATION CONTENT OF THEIR ECHOES. Studies of a variety of bats representing a number of families have demonstrated major differences between the characteristics of pulses emitted by different species. Just as styles of foraging and food preferences differ between species, so do the echolocation pulses and the information their echoes convey to the bats. For the discussions that follow I have drawn heavily from the fine treatment of information content of bat sonar echoes by Simmons et al. (1975).

The frequencies of bat-produced ultrasonic pulses span the range from approximately 140 to 30 kHz, and the pulses of microchiropteran bats have frequency-modulated (FM) and constant-frequency (CF) components. The FM signals are typically brief, from 0.5 to 10 msec, but cover a broad bandwidth, usually sweeping through at least an octave. The CF signals may either be short (less than 10 msec) or long (10 to over 50 msec). The echo of the FM signal is best able to convey one type of information to the bat, the CF signal, another. Targets make complex modifications in the "spectrum" of the echoes of broadband FM signals; this type of signal is best suited to detecting size and shape, surface details and range of a target. CF signals, on the other hand, have a narrow bandwidth and, because the amount of information an echo can convey is proportional to the range of frequencies in the signal (bandwidth), are poor for conveying details about the characteristics of the target. But if the target is roughly as long or longer than the wavelength of the CF signal, this signal will indicate the presence of the target. The long CF signal is of importance because it allows the bat to determine whether the target is approaching or going away and at what rate. This discrimination is based on detection of the Doppler shift, discussed later.

Most bats use pulses that combine FM and CF components. There are three common patterns by which these are combined (Fig. 21–4). The first pattern includes a short FM portion that consists of a downward sweep including at least one harmonic, with a terminal brief CF portion. The type can be termed an FM/short-CF orientation sound. The second consists of an initial brief CF signal and a short terminal FM sweep downward. These are short-CF/FM orientation sounds. Harmonics may appear in both parts of this kind of pulse. The third pattern is composed of an initial long CF signal and a terminal short FM sweep downward (long CF/FM orientation sound). Some species, at least part of the time, precede the long CF part of this sound with a very short, upward FM sweep.

A wide variety of bats, including most members of the important family Vespertilionidae, use FM/short-CF echolocation

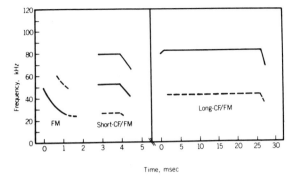

Figure 21–4. Diagrammatic representation of sound spectrograms showing the three major types of echolocation signals used by microchiropteran bats. See text for descriptions. (From Simmons, 1974.)

signals. One vespertilionid, *Eptesicus fuscus*, has a brief FM signal of 1 to 5 msec that sweeps downward from 50 to 25 kHz and may or may not be followed by a CF portion. Those members of the Neotropical Phyllostomatidae that have been studied use similar pulses, but the sounds are complicated by at least two harmonics. Long CF/FM signals are used by members of the Old World Rhinolophidae and by one member of the New World Mormoopidae.

Each of these major patterns of orientation sounds provides a different means of obtaining information, and each exploits in a somewhat different way the information-bearing echoes of FM and CF sounds. In general, each species of bat utilizes a single type of signal, and the auditory receptors of each species are adapted precisely to its own signals. The hearing of FM bats is most acute through the sweep of frequencies of their FM signals. In the mormoopid that uses CF/FM signals, the greatest acuity is at about the level of the CF component. The hearing of rhinolophids is precisely "tuned" to the frequency of the long CF component; the greater the duration of the CF component, the sharper the tuning of the auditory system.

The echolocation performance—the degree of accuracy with which the system allows a bat to assess its immediate environment—has been studied in a number of species. Simmons (1973) tested the range-detecting ability of one species each from the families Vespertilionidae, Phyllostomatidae, Mormoopidae and Rhinolophi-

dae. All of the bats could discriminate range differences of from 1 to 2 cm, and this remarkable performance was apparently achieved by "crosscorrelation" of the transmitted pulse with the returning echo; the essential variable from which the bats estimate distance is the time it takes for a pulse to reach a target and for the echo to return. By comparing the relative target-ranging abilities of the bats in relation to the bandwidths of their echolocation pulses, Simmons determined that the FM components of the pulses were used for target ranging. Laboratory studies on a limited number of species have further demonstrated abilities to discriminate differences in size, shape, and directions of the target. Simmons et al. (1974) used *Eptesicus fuscus* in laboratory experiments to determine this bat's ability to detect details of the shape of a target; the bats could discriminate differences of less than a millimeter in the depths of small holes. The holes modify the spectrum of the echo from the target by absorbing sound energy at certain frequencies within the bandwidth of the bats' FM sweep; with changes in depth of the holes the absorption peaks are shifted to different frequencies. Seemingly, bats can use this ability to associate features of the sonar echo "spectrum" with target shape in selecting appropriate prey from among the variety of nocturnal insects they perceive.

Laboratory studies have demonstrated that some members of the family Rhinolophidae, and the mormoopid *Pteronotus parnellii*, use the long CF component of their pulses for analyzing the velocity of targets relative to that of the bats themselves. This ability is based on the Doppler shifts of the echoes from targets either approaching or going away (Simmons, 1974). For a stationary bat with a fixed target, or for a bat keeping precise pace with a target, the echo returning from the target has the same frequency, and if the prey is going away the echo is Doppler-shifted to a lower frequency. The bats using the long CF/FM pulses bring the frequency of the echo to that to which their ears are "tuned" by altering the frequency of emitted pulses. This remarkable compensatory response occurs within fractions of seconds, often in response to erratic movements of a target (Fig. 21–5). The longer the CF component of the sonar sound, the greater the sensitivity of the bat to target velocities. Computed velocity resolutions listed by Simmons et al. (1975) indicate that the European *Rhinolophus ferrumequinum*, with a very long CF pulse (up to over 60 msec), can perceive relative velocities of targets of less than 0.04 m/sec; the moderately long-pulsed *P. parnellii* (with pulses up to 28 msec) can perceive velocities as low as 0.10 m/sec. The short CF/FM signals of *Pteronotus suapurensis*, however, allow only a poor resolution of velocity, to about 1 m/sec.

Perhaps of greatest interest is the way in which the echolocation pulses and the au-ditory system of a bat are "matched" to its habitat and style of foraging. Because bats exploit a tremendous variety of foods and occupy a broad range of habitats, one would expect echolocation systems to vary broadly. This seems to be the case, and in all probability no two species use precisely the same sonar signals in the same way. Indeed, some bats that are closely similar morphologically can readily be distinguished by their sonar pulses. The East African species *Eptesicus somalicus* and *Pipistrellus nanus* appear virtually identical and their ranges overlap broadly, but Thomas J. O'Shea found (unpublished data) that the pulses of the former sweep from about 70 to 35 kHz, whereas those of the latter sweep from 110 to 70 kHz.

The long CF/FM pulses of bats of the family Rhinolophidae are thought by Simmons et al. (1975) to enable these bats to detect target motion and to track the trajectory of prey with a level of precision not approached by bats with primarily FM pulses. Field observations on a large African rhinolophid—the giant leaf-nosed bat *Hipposideros commersoni*—indicate that this ability is basic to its foraging strategy (Vaughan, 1976). This heavy (up to about 100 g), broad-winged, insectivorous bat is a rapid but unmaneuverable flier; it seems unable to make the abrupt and erratic maneuvers that are the trademark of small

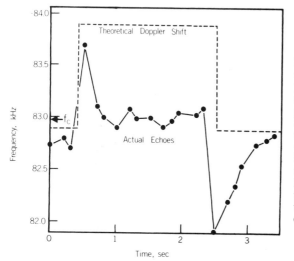

Figure 21–5. An illustration (from Simmons, 1974) of a Doppler-compensating response of *Rhinolophus ferrumequinum* to a 2.2 second Doppler shift in artificial echoes. Simmons explains this illustration as follows: "The magnitude of the Doppler shift, 1.00 kHz upward (based upon a transmitted frequency of 83.00 kHz), corresponds to an approaching target velocity of 2.0 m/second. The dotted line shows the frequencies to be expected in echoes if the bat continued to emit CF signals at a frequency of 82.90 kHz, while the data points and solid line indicate the frequencies actually returned to the bat, showing that the bat decreased its transmitted frequencies to compensate for the simulated Doppler shift."

insectivorous species. *Hipposideros commersoni* is largely a sit-and-wait predator: it hangs from a high branch from which it can "scan" an open area, detects a flying insect at long range (up to some 20 m), and with great precision, and seemingly with no wasted energy or motion, swoops out and intercepts the prey. The bat virtually never pursues insects, but selects ponderous, straight-flying beetles that are vulnerable to a direct interception. The echolocation signal of this bat is of an unusually low frequency for a long CF/FM bat (63 kHz). Because high frequencies are attenuated in air more severely than are low frequencies, the relatively low frequency pulse may be an adaptation to detecting prey at long range. The fairly long CF pulse, up to 20 msec under laboratory conditions, probably allows this bat to perceive the movement and trajectory of prey with the accuracy necessary for precise interceptions.

ECHOLOCATION AND TARGET-DIRECTED FLIGHT. Pulse duration (Table 21–1) and the rate of emission vary widely among different bats, but many follow a basic pattern. This pattern in-

Table 21–1. Pulse Duration Under a Variety of Conditions in Bats

Bat	Pulse Duration	Conditions	References
Megaderma lyra	0.54–1.2 msec (limits) 0.72–1.8 msec (limits)	Approaching goal or obstacle Searching in laboratory	Mohres and Neuweiler, 1966
Rhinolophus ferrumequinum	50–65 msec Decreasing to 10 msec	Take-off and flight in lab Approaching a landing	Schnitzler, 1968
Rhinolophus euryale	30–45 msec Decreasing to 7 msec	Take-off and flight in lab Approaching a landing	Schnitzler, 1968
Noctilio leporinus	7.4 msec (5.9–9.4 msec) 14.3 msec (11.1–16.7 msec) 13.8 msec (10.7–16.0 msec) Decreasing to 1 msec	Searching flight in outdoor cage Searching flight in wild Searching flight in wild End of prey location	Suthers, 1965
Pteronotus parnellii	14–26 msec Rising from 20–21 msec to 28–37 msec Decreasing to 6.8 msec	Searching flight in lab Detecting a fruit fly End of insect pursuit	Novick and Vaisnys, 1964
Pteronotus psilotis	2.9–4.8 msec Decreasing linearly to 0.6–1.0 msec	Searching flight in lab Fruit fly pursuit and capture	Novick, 1965
Vampyrum spectrum	1.5–1.8 msec 0.5–1.5 msec About 0.5 msec	Searching flight Approach Terminal portion of flight	Bradbury, 1970
Myotis lucifugus	2–3 msec Decreasing to 0.3–0.5 msec	Searching flight in lab At end of insect pursuit	Griffin, 1962
Plecotus townsendii	2–5 msec Decreasing to 0.3–0.5 msec	Searching flight in lab At end of insect pursuit	A. D. Grinnell, 1963b
Eptesicus fuscus	2–4 msec 10–15 msec 0.25–0.5 msec	Flying in laboratory Flying in open at 10 m altitude Late terminal phase	Griffin, 1958
Lasiurus borealis	2.4–3.0 msec 0.3–0.5 msec	Pursuit of mealworms in lab Late terminal phase	Webster and Brazier, 1968

(After Novick, 1971.)

volves a pulse rate of approximately 10 per second during "searching" flight. The rate is raised to 25 to 50 per second when prey is located, and to 200 or more per second during the final pursuit of the insect. The pulse duration typical of some bats is remarkably short; in *Megaderma lyra* (Megadermatidae), it is down to 0.72 msec during searching flight. By contrast, *Rhinolophus ferrumequinum* (Rhinolophidae) emits pulses of from 50 to 65 msec in duration under similar conditions.

To a bat approaching a target, several types of information are of critical importance. The distance of the target, its position, the direction of its movement (if it is an insect), its size and shape, and to some extent its nature (whether furry or scaly, for example) are essential to appropriate reactions by the bat. In the case of a bat closing in on an insect, the difficulty of obtaining this information is compounded by the fact that neither the bat nor the insect is stationary, and the means by which the bat is seeking the information is based on pulses that take some time to travel from the bat's larynx to the target and back to the bat's ear. Also, the "processing" of the information — the traveling of the information along sensory pathways to the central nervous system, the bridging of the appropriate synapses, and the resultant initiation of action by pulses relayed along

motor nerves — takes time. Nonetheless, the very survival of insectivorous bats depends on their ability to locate, pursue, and capture insects that they perceive by means of echolocation.

Remarkably long pulses, of 50 to 65 msec duration, are used by the greater horseshoe bat, *Rhinolophus ferrumequinum*, during searching flight. The pulse duration lengthens at from 5.5 to 2.7 m from the target, and then shortens linearly as the target is approached. The briefest pulses, approximately 10 msec, are two or three times as long as the longest pulses of vespertilionids (common, mouse-eared bats). Broad overlap of incoming echoes and outgoing pulses occurs throughout the approach in this bat and in a closely related species, and such overlap may be characteristic of all rhinolophids (horseshoe bats) and mormoopids (leaf-chinned bats) (Fig. 21-6). However, pulse overlap is not characteristic of all patterns of echolocation.

In the little brown bat, *Myotis lucifugus*, and in all other vespertilionids studied, pulse overlap is avoided. In *M. lucifugus* the searching pulse rate is roughly 15/sec (Griffin, 1958:360) and the pulse duration is about 2.6 msec (Webster and Brazier, 1968). A target is detected at some two meters, but the early approach phase begins at approximately 720 mm (Novick,

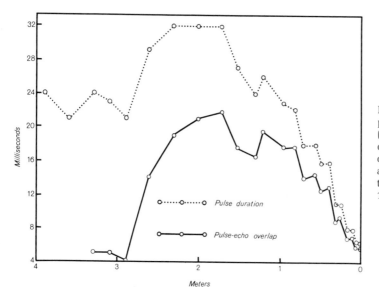

Figure 21-6. An example of the pursuit of a fruit fly by a bat (*Pteronotus parnellii*, Mormoopidae). The pulse duration and calculated pulse overlap are plotted against distance from the insect target. (After Novick and Vaisnys, 1964.)

1971:206). As a bat approaches its target it shortens its pulses by about 1 msec per 260 mm of approach; immediately before contact with the insect, pulse duration may be as short as 0.25 msec and the pulse rate may approach 200/sec. Pulse overlap is avoided by shortening the outgoing pulses before pulse-echo overlap occurs, and then by a linear shortening of the pulses as the bat closes in on the target. Present evidence suggests that avoidance of pulse overlap is probably also typical of bats of the families Megadermatidae, Phyllostomatidae, and Noctilionidae.

COMPLEXITIES AND SPECULATIONS. Novick (1971) suggested that three types of evaluation might be made by a long-CF/FM bat during the approach phase: (1) an evaluation of the relative target movement on the basis of an assessment of the frequency of the CF echo; (2) an evaluation of the distance of the target and the speed with which it is being approached on the basis of the interval between the end of the emission of a pulse and the end of its echo (the shorter this time, the closer the target); (3) an evaluation of the size and direction of travel of the target on the basis of assessment of the FM echo.

Some long-CF/FM bats, adapted to detecting moving prey at long distances and to precise tracking of the prey, enhance their ability to scan their surroundings by coordinated movements of the body, head, and ears. As an example, when the African species *Hipposideros commersoni* hangs from an acacia branch and scans for prey its pendent body revolves continuously, back and forth, through an arc of approximately 180°; the head is in constant motion up and down and the tips of the ears vibrate forward and backward. The body and head movements enable the bat to scan meticulously with its strongly "beamed" pulses the space surrounding its perch. But why the ear movements? Simmons (1975) judged that by moving the directional sensitivity of the ears the long-CF/FM bats scan the vertical plane. The movements of the ears, so rapid as to be perceived by an observer as a vibration or fluttering, are out of phase: the tip of one ear moves forward, toward the target, while the other moves backward, away from the target. Further, the ear movements are in approximate synchrony with the sonar pulses emitted by the bat. These ear movements, alternately toward and away from the target, may heighten the Doppler shift of echoes from moving targets and improve the bats' discrimination of movement. Even without a complete understanding of functional relationships, however, one cannot help but wonder at the elaborate neural coordination of stereotyped body, head, and ear movements with the rate of sonar pulse emission.

Various structural and physiological factors set limits on pulse duration, particularly on the brevity of a pulse. The periods of recovery (refractory periods) of nerve-muscle systems that are necessary between successive muscle contractions, delays associated with reaction time, and time required by the central nervous system for processing information contained in the echoes are all limiting factors. The timing of pulses doubtless partially determines the bat's ability to get appropriate acoustical information. In general, during the approach phase bats wait roughly 25 msec after the termination of the echo of a pulse before they emit the next pulse. This time is presumably necessary for the evaluation of the echo for information about the target. In the terminal phase of the pursuit, when the bat is closing in on its target, many kinds of bats maintain interpulse intervals of some 2.5 to 5.0 msec. The silent interpulse intervals may be important in allowing bats to receive the FM echoes clearly.

The production of ultrasonic pulses of high intensity probably requires considerable energy. The pulse duration and rate of pulse emission, therefore, may in part be determined by the need to conserve energy (Novick, 1971:209). One might expect a bat to use as few pulses or pulses as short as possible within the limits set by the necessity to perceive targets clearly.

AVOIDANCE OF BATS BY MOTHS. Studies by Roeder and his colleagues

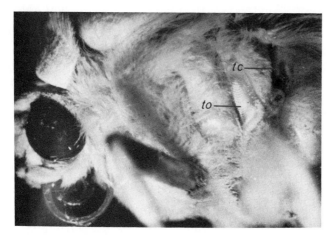

Figure 21-7. Photograph of an arctiid moth, showing the tympanic cavity (*tc*) and timbal organ (*to*). (Photograph by K. D. Roeder.)

(Roeder and Treat, 1961; Dunning and Roeder, 1965; Roeder, 1965; Dunning, 1968) have revealed a remarkable series of adaptations by certain moths in response to predation by bats. Nocturnal moths of the families Noctuidae, Ctenuchidae, Geometridae, and Arctiidae have an ear on each side of the rear part of the thorax (Fig. 21-7). Each ear is a small cavity within which is a transparent membrane; the ears are sensitive to a wide range of frequencies and allow the insects to detect the ultrasonic pulses of foraging bats. Upon detecting the approach of a bat, the moths alter their level flight and adopt various erratic flight patterns. A wide array of loops,

dives, and abrupt turns has been photographed (an example is shown in Figure 21-8). Some members of these families of moths have carried the business of evasion of bats to an even greater extreme. These moths have a noise-making organ on each side of the thorax (Fig. 21-9). When the moths are disturbed these organs produce trains of clicks with prominent ultrasonic components. Under laboratory conditions, flying bats about to capture mealworms tossed into the air regularly turned away from their targets when confronted with recorded trains of moth-produced pulses (Table 21-2). These pulses apparently protect moths from bats; probably the moths

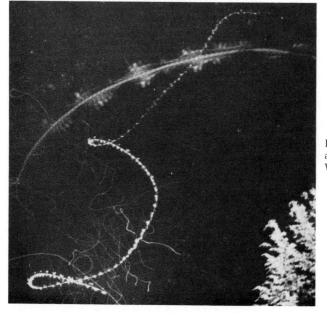

Figure 21-8. The evasive tactics of a moth approached by a bat. (Photograph by F. A. Webster.)

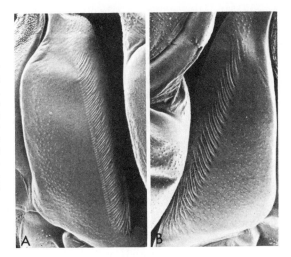

Figure 21-9. Scanning electron micrographs of the microtymbals of two species of moths, *Haploa clymene* (A), and *Pyrrharctia isabella* (B). Contraction of a muscle that inserts on the concave inner surface of the corrugated episternite causes the microtymbals (the corrugations) to buckle in sequence, and the buckling of each microtymbal produces a signal lasting less than one msec and containing ultrasonic frequencies. The number and shape of the microtymbals determine the number and acoustical qualities of the clicks of each sequence. (From Fenton and Roeder, 1974.)

are announcing their identity as bad-tasting prey. Some arctiid moths, and perhaps some noctuids, are unpalatable to bats, and their identification by a bat that had had previous experience with them might be aided by the moth's ultrasonic clicks. Dunning (1968) found that captive *Myotis lucifugus* avoided three species of arctiid

moths when they were sounding off, but bit into them when they were quiet. Two of the moths were rejected when tasted, but the third, presumably a palatable form mimicking the sounds of the unpalatable species, was eaten.

ECHOLOCATION IN CETACEANS

Just as bats must cope with darkness, cetaceans frequently must be able to perceive their underwater environment under conditions that render vision difficult if not useless. In some waters inhabited by cetaceans, suspended material such as soil particles or plankton limits visibility to a few feet or even a few inches. Water transmits light rather poorly, and even under ideal conditions visibility under water is limited. Also, some cetaceans forage at considerable depths where but a trace of light remains. It is not surprising, then, that some cetaceans have developed echolocation.

A growing body of evidence suggests that all odontocetes (toothed whales and dolphins) use echolocation for obstacle avoidance and for the detection of prey. Mysticetes (the baleen whales), however, are not known to echolocate. A tremendous variety of sounds, some having a fascinating musical quality, is made by both mysticetes and odontocetes; the wailing,

Table 21-2. Responses of Bats to Simultaneous Presentation of Mealworms and Sounds of Bats or Moths

Bat Number	Number of Tosses	Number of		
		CONTACTS (%)	DODGES (%)	ATTEMPTS (%)
Bat sounds presented				
3	67	88	7	5
4	150	79	8	12
5	92	65	27	5
Total	309	77	14	8
Moth sounds presented				
3	95	14	86	3
4	249	14	83	3
5	121	11	87	2
Total	465	13	85	2
No sounds presented				
3	141	99	0	1
4	373	98	0	1
5	167	97	1	1
Total	681	98	1	1

(After Dunning, D. C., and Roeder, K. D.: Moth sounds and the insect-catching behavior of bats. *Science,* 147:173–174, 1965. Copyright 1965 by the American Association for the Advancement of Science.)

creaking, and squealing noises of cetaceans have become commonplace to men operating sonar equipment at sea. Payne (1970) has recorded the remarkable "songs" of the humpback whale (Megaptera). The following discussion will primarily consider pulses used by cetaceans in connection with echolocation.

INSULATION OF THE CETACEAN EAR FROM THE SKULL. Each ear of a cetacean functions as a separate hydrophone, allowing the animals to localize the source of sound by discriminating (as we do) between the times at which a sound is received by each ear. The pressure that sound transmitted through water exerts on the bones of the entire skull causes vibrations to be transmitted by the skull. When the bone that houses the middle and inner ear is attached rigidly to the skull, as it is in most mammals, vibrations from water are transmitted through the bones of the skull and reach the ear from various directions. As a consequence, when a mammal with this type of skull is submerged it is unable to localize accurately the source of a sound. Because the localization of sound is of great importance to cetaceans that use echolocation, these animals have evolved several structural features that serve to insulate the bone surrounding the middle and inner ear (tympanic bulla) from the rest of the skull.

First, the tympanic bullae are not fused to the skull in any cetacean, and in the specialized porpoises and dolphins the bullae are separated by an appreciable gap from adjacent bones of the skull. In addition, the bullae are insulated by an extensive system of air sinuses that are unique to cetaceans. These air sinuses surround the bullae and extend forward into the enlarged pterygoid fossae (Fig. 11–5; p. 199), and each sinus is connected by the eustachian tube to the cavity of the middle ear. The sinuses are filled with an oil-mucus emulsion, foamed with air, and are surrounded by fibrous connective tissue and venous networks. These sinuses can apparently retain air even when subjected to pressures of 100 atmospheres (1,470 pounds per square inch), pressures higher than those to which cetaceans are subjected during deep dives. Purves (1966:360) used a foam in which gelatin was substituted for mucus and oil and found that at such pressures the air became dispersed in the mixture as tiny bubbles. The foam in the air sinuses apparently forms a layer around the bullae that retains remarkably constant sound-reflecting and sound-insulating qualities through a wide range of pressures.

Disagreement remains about how sounds reach the cetacean middle ear. Fraser and Purves (1960b) believe the route to be through the external auditory meatus (which may be highly reduced or even covered with skin in odontocetes), the tympanic ligament, and the ear ossicles. But a divergent view is held by Norris (1964, 1968), who regards the extremely thin back part of the lower jaw of delphinids as an acoustical window. Norris holds that sound passes into the skin and blubber overlying the dentary, through the thin part of this bone, which at its thinnest may be only 0.1 mm thick, to the intramandibular fat body. This fat body leads directly to the wall of the auditory bulla, into which the sound presumably passes. Weight is given to the Norris hypothesis by experiments done by Yanagisawa et al. (1966), who found that the jaw is the most acoustically sensitive area of the dolphin's head.

PATTERNS OF ECHOLOCATION IN CETACEANS. Since the account by Schevill and Lawrence (1949) of the underwater noises made by the white whale (Delphinapterus), considerable research has been done on the vocalizations of cetaceans. Much research has dealt with one of the common dolphins, Tursiops truncatus (Kellogg et al., 1953; Wood, 1959; Schevill and Lawrence, 1953; Norris et al., 1961; Lilly, 1962, 1963). Tursiops is able to detect obstacles and to recognize food by means of echolocation, and in the use of short pulses its sonar system resembles that of bats. Tursiops is capable of producing a great variety of sounds, but of primary importance for echolocation are the trains of clicks that it emits. The clicks are audible to man, but cover a wide spectrum of frequencies, some of which are ultrasonic. The rate of emission of clicks varies among different odontocetes: the killer whale (*Orcinus orca*) has a slow repetition rate of from 6 to 18 clicks per second (Schevill and Watkins, 1966), whereas the Amazon dolphin *(Inia geoffrensis)* emits click trains at rates of from 30 to 80 per second (Caldwell et al., 1966). The pulse rate rises as a porpoise approaches a target, and *T. truncatus* can distinguish between a piece of fish and a substitute water-filled capsule with a similar shape (Norris et al., 1961), or even between sheets of the same metal but of different thicknesses (Evans and Powell, 1967).

Despite our rather detailed knowledge of the kinds of sounds made by cetaceans, where and how the sounds are produced is not definitely known (see Norris, 1969:404–407). The larynx of odontocetes lacks vocal cords, but has well-developed muscles and a complicated structure, and is thought by some to be the site of sound production (Purves, 1966). The nasal plugs (muscular valves at the blowhole that close the nares) and the membranes against which they rest have been thought to produce sound, as have the lips of the blowhole (Norris and Prescott, 1961). Because the echolocation sounds produced by delphinids seemingly come from above the margins of the jaws, some of the structures associated with the blowhole may produce sound (Norris et al., 1961).

The sperm whale *(Physeter catodon)* presents an unusually interesting case of echolocation among cetaceans because it uses a unique mode of foraging that is probably made possible by echo ranging. (No actual proof that sperm whales echolocate is available, but this ability has been inferred on the basis of data from other odontocetes.) The click of a sperm whale is known to consist of a series of pulses (Backus and Schevill, 1966). The click lasts roughly 24 msec and is composed of up to 9 separate pulses. These vary in duration from 2 to 0.1 msec and the inter-pulse intervals are some 2 to 4 msec. The clicks are repeated at rates of from less than 1 click per second to 40 per second. Of particular interest is the fact that the sperm whale feeds largely on squid that it takes at depths (down to at least 1000 m) at which prey is scarce and light is virtually absent. It appears, therefore, that the sperm whale is able to utilize deep-sea foraging largely because it is able to use echo scanning to locate food under conditions that require efficient long-range echolocation. Backus and Schevill (1966:525) estimated that sperm whales, by the use of echo scanning, can detect prey up to about 400 m away.

Some small cetaceans that inhabit turbid water have tiny eyes and presumably are dependent on echolocation. One of the most highly specialized of these is the blind river dolphin *(Platanista gangetica)*, an inhabitant of the muddy and murky waters of the Ganges, Indus, and Brahmaputra river systems of India and Pakistan. This unusual dolphin habitually swims on its side with the ventralmost flipper either touching the bottom or moving within 2 or 3 cm of it (Fig. 11–10; p. 208). *Platanista* has greatly reduced eyes that are barely visible externally. The lens is absent, but the retina apparently retains the ability to detect light, although doubtless no image can be formed. The tiny eye opening is surrounded by a sphincter muscle, and another muscle functions to open the sphincter. Blind river dolphins in captivity continuously produced series of pulses at rates of from 20 to 50 per second, primarily in the frequency range between 15 and 60 kHz, and the animals have a remarkable ability to direct their pulses into a narrow beam (Herald et al., 1969).

Studies by Evans and his associates (1964) have demonstrated that the shape of the skull of *Platanista* affects the directional beaming of pulses (these are emitted through the blowhole). The pulses are reflected by the concave front of the skull and are focused by the "melon," a lens-shaped fatty structure that gives a domed profile to the forehead of many odontocetes. The skull of *Platanista* is grotesquely modified by large flanges from the maxillary bones (Fig. 21–10). These promi-

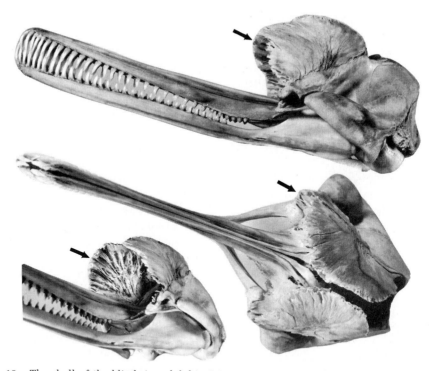

Figure 21–10. The skull of the blind river dolphin (*Platanista gangetica*), showing the highly developed extensions of the maxillary bones. (From Herald, E. S., et al.: Blind river dolphin: first side-swimming cetacean, *Science*, 166:1408–1410, 1969. © 1969 by the American Association for the Advancement of Science.)

nent flanges, rounded on the outside but with an intricate, radial pattern of latticework on the inside, probably serve as acoustical baffles that, with the melon, concentrate the pulses into a narrow beam (Herald et al., 1969). Observations of swimming dolphins, and reception of these pulses with a hydrophone and amplifier system, showed that they were indeed effectively beamed. When a dolphin's snout was directed as little as 10° on either side of the receiver the intensity of the pulses dropped markedly, and a far greater drop occurred when the angle was greater than 40°. As the dolphin swims on its side it moves its head constantly in a sweeping action close to the bottom. One is tempted to speculate that this action serves to scan the area ahead of the porpoise with a beam of pulses, a system similar in some ways to that of the horseshoe bat, which also uses beamed pulses. Such scanning in the dolphin might serve effectively both in determining bottom contours and in finding food.

ECHOLOCATION BY INSECTIVORES, PINNIPEDS AND OTHER MAMMALS

Echolocation is known to occur in four orders of mammals (Insectivora, Chiroptera, Odontoceti, and Carnivora and may have evolved independently in each of these groups. Highly suggestive evidence also points toward the occurrence of echolocation in some other orders.

Many authors have noted high-pitched sounds made by insectivores when they explore unfamiliar surroundings or objects (for example, Swinhoe, 1870; Komarek, 1932; Reed, 1944; Crowcroft, 1957). In a series of carefully controlled laboratory experiments, Gould et al. (1964) demonstrated that three species of *Sorex* could echolocate. These shrews searched around an elevated disc, found a lower platform, and jumped to it, without the use of tactile, visual, or olfactory senses. While the shrews searched their environments they emitted pulses with frequencies between 30 and 60 kHz; the pulse duration was about 5 to 33+ msec. The familiar short-tailed shrew of the eastern United States, *Blarina brevicauda*, produced similar pulses, and all shrews studied produced pulses with the larynx. Laboratory trials revealed that tenrecs (Tenrecidae) from Madagascar could also echolocate (Gould, 1965). These primitive insectivores produced pulses by clicking the tongue, and the pulses were of frequencies audible to man (from 5 to 17 kHz). Further research is necessary to determine how ubiquitous echolocation is within the Insectivora.

Pinnipeds produce a variety of underwater sounds (Ray and Schevill, 1965; Schevill and Watkins, 1965), some of which are of a type that might be used for echolocation. The Weddell seal (*Leptonychotes weddelli*) of Antarctic waters, for example, emits "chirps" with frequencies up to 30 kHz. These chirps are produced consistently as the seals swim between air holes, and Watkins and Schevill (1968) suggested that they may have a navigational value. The California sea lion (*Zalophus californianus*) is probably able to echolocate (Evans and Haugen, 1963; Poulter, 1963; Schusterman and Feinstein, 1965), and makes extensive use of vocal signals for communication both in the water and on land. As in the case of the insectivores, our knowledge of echolocation in pinnipeds is limited.

Diverse lines of evidence indicate that some form of echolocation may occur in still other mammals. Flying lemurs (order Dermoptera) move in a series of jumps interspersed with pauses, and emit series of pulses (Burton, 1949; Tate, 1947); these could be used in echolocation. *Antechinus*, a small, nocturnal, tiny-eyed Australian marsupial that is convergent toward shrews, may possibly echolocate (Gould et al., 1964). Two rodents, the fat dormouse (*Glis*, Gliridae) and the golden hamster (*Mesocricetus*, Cricetidae), are able to locate perches without the use of tactile, visual, or olfactory senses, and can presumably echolocate (Kahmann and Ostermann, 1951). Young mice are known to emit ultrasonic pulses (Zippelius and Schleidt, 1956), but the pulses are important in communication (see page 367) and their use for echolocation has not been demonstrated. Griffin (1958:297–322) discusses the ability of some blind people to echolocate. One is tempted to speculate that a latent ability to echolocate is common to many mammals.

THE SPECIALIZED EAR OF THE KANGAROO RAT

One of the most remarkable cases of specialization of the ear is that of the kangaroo rat (*Dipodomys*). The enormous auditory bullae of these animals, which have a total volume greater than that of the braincase, were a source of wonder to mammalogists for years. Recently, partly due to the studies of Webster (1961; 1963; 1966; Webster et al., 1968; Webster and Stack, 1968), the functional importance of this and other auditory specializations of the kangaroo rat have been clarified.

Before discussing the inflated bullae, let us consider some other important specializations. First, the malleus lacks the anterior and lateral ligaments that in most mammals brace this bone, and therefore rotates unusually freely. In addition, the manubrium of the malleus, which rests against the tympanic membrane, is greatly lengthened; it thus serves as a lever arm that transforms the relatively weak vibrations of the tympanic membrane into more powerful movements transmitted to the incus and stapes. Further, the tympanic membrane is exceptionally large, and the footplate of the stapes, which rests against the oval window of the inner ear, is small. This forms a piston system of sorts. Because the force per unit area is increased in proportion to the difference in surface areas of the two structures, relatively weak pressure on the large tympanic membrane is transformed into relatively great pressure on the fluid within the inner ear *via* the small footplate of the stapes. There is, then, great amplification of force by the combined means of the piston system and the long lever arm of the malleus. The degree of this amplification is expressed by the *transformer ratio*, which in Merriam's kangaroo rat is extremely high, about 97:1. (This ratio in man is roughly 18:1.)

The inflation of the auditory bullae of kangaroo rats (Fig. 10–21; p. 187) results in a great increase in the volume of the air-filled chambers surrounding the middle ear. This increase in volume reduces the resistance the enclosed air offers to the compression caused by inward movement of the tympanic membrane. Consequently, the damping effect on vibrations of the tympanic membrane is diminished. The advantage gained during transmission of low frequency sounds is especially great, for these sounds cause relatively great movements of the membrane. Laboratory experiments, involving filling the bullae of kangaroo rats with plasticine, demonstrated that the transmission ability of the tympano-ossicular system in individuals lacking use of the enlarged middle ear chambers is seriously lowered. Both experimental and control animals, however, were most sensitive to low intensity sounds with frequencies between 1 and 3 kHz.

In an attempt to simulate natural predator-prey confrontations, Webster tested the reactions of kangaroo rats to two predators adapted to hunting in darkness. He used the sidewinder rattlesnake *(Crotalus cerastes)*, which detects its prey by using heat-sensitive facial "pits," and the barn owl *(Tyto alba)*, which can use hearing to locate the position of prey. By using delicate recording equipment, Webster found that the wings of owls produce low frequency (below 3 kHz) whirring sounds

as the birds swoop toward prey, and a rattlesnake produces a short burst of low frequency sound (perhaps when scales rub against the substrate) just before it strikes. Under experimental conditions, when a kangaroo rat heard these low intensity and low frequency sounds produced by attacking predators, it made a sudden vertical leap and avoided capture. In contrast, individuals with artificially reduced middle ear volumes, and presumably reduced ability to hear faint sounds, could not evade capture. Clearly, the beautifully coordinated specializations of the tympano-ossicular system and the auditory bullae are highly adaptive: in darkness kangaroo rats can detect the faint sounds of attacking predators, make evasive leaps, and go on to reproduce their sharp-eared kind.

Kangaroo rats are not alone in having enlarged bullae. Many other rodents, including other heteromyids, some South American hystricomorphs, gerbils (Cricetidae), jerboas (Dipodidae), the springhaas (Pedetidae), and Australian hopping mice *(Notomys,* Muridae), have this specialization, as do elephant shrews (Macroscelididae). Whether or not the adaptive importance of enlarged bullae is the same among all these mammals is not known. Many kinds of mammals that live in deserts have auditory bullae that are relatively larger than those of relatives occupying cooler and more humid areas. Sound is known to be absorbed to different degrees under different conditions of temperature and humidity (Knudsen, 1931, 1935), and the enlargement of the bullae may compensate for the poor sound-carrying qualities of warm, dry desert air.

COMMENTS ON MAMMALS AND MAN

Man has long been interested in his fellow mammals and has long exploited them. More than a million years ago *Australopithecus* was killing and eating baboons and antelope, and the use of mammals for food remains characteristic of most cultures of men today. Many kinds of mammals have been domesticated by man, and some are taught to work for him: the trained Indian elephant lifts and drags teak logs in the tropical forests of Ceylon, where the periodically saturated soil limits the usefulness of vehicles; dogs help some African natives capture antelope and other game; and trained rhesus monkeys pick coconuts from tall trees nearly inaccessible to man and drop them to their masters. Even the vicious and rebellious camel has been trained to carry man and his burdens. The raising of various kinds of mammals—from homozygous strains of mice to be used for medical research to stocky beef cattle—are important enterprises today. The very distribution of early man was probably influenced by his ability to kill his fellow mammals, for the skins and furs of mammals may have enabled primitive man, probably endowed with a hopelessly inadequate insulation of his own, to penetrate cool or cold regions.

Wild mammals are still under pressure from man in nearly all parts of the world. In many primitive areas men either are basically hunters and partly depend on mammals for food, or at least hunt to supplement limited food supplies. Even in the United States, most boys who live on the farm or away from towns regularly hunt small game, and in Mexico the "veinte-dos" (.22 rifle) is used year round in many areas for hunting anything from squirrels to deer. In the United States, trapping for furs has long been important. Not only was the exploration of the western United States largely accomplished by trappers seeking new territory, but in more recent years trapping has been an important source of seasonal income to some. In the early 1920s Henry Duval, trapping marten in the vicinity of Mineral King in the Sierra Nevada of California, took as many as 96 marten in a winter (Grinnell et al., 1937:205). This probably brought Duval over $1,500 for his winter's work (the average price of a marten fur in this period was $15.71), more than the average worker made in a full year. In 1929, a top year with respect to demand for furs, the return to North American trappers was estimated at $60 million (Hamilton, 1939:393). Fur farming has provided stiff competition to the trapper in the United States, but trapping, particularly for muskrat, continues in some areas. To many boys the excitement of running a trapline is as attractive as is the monetary return.

In some states big game hunting is a major source of income. In Colorado in 1968, for example, expenditures of deer and elk hunters were estimated at over $60 million. Many states realize considerable returns from the sale of deer licenses alone (Table 22–1). Perhaps equally important is the recreational value of hunting to the individual. Only during the hunting season do some men get away from their work in the city to experience the sights, smells, and sounds of primitive country. Many spend months anticipating and planning the yearly deer hunt.

Table 22-1. Revenue Received by Several States from the Sale of Hunting Licenses for Deer

State	Year	Number of Deer Licenses Sold (Resident and Nonresident)	Revenue From Sales
Arizona	1969	101,337	$ 790,855.00
Colorado	1969	161,919	3,172,527.00
Idaho	1970	168,624*	420,759.00
Michigan	1970		3,767,702.00
Nebraska	1970	36,139	389,510.00
New York	1970	670,239*	2,803,792.65
Utah	1970	92,021	1,320,830.75
Wyoming	1968	100,961*	1,786,830.00

*Includes deer-bear permits.
(Data from material supplied by these states.)

But wild animals may also be costly. Pocket gophers, rabbits, meadow voles, ground squirrels, and even deer and elk may cause local damage to crops or rangeland, and efforts to combat these losses are frequently expensive. In addition, the Federal Government supports considerable research on economically important mammals, and finances the local control of some mammals. In 1971, Federal animal control programs cost U.S. taxpayers over $8 million.

MAN'S IMPACT

Unhappily, the long-term exploitation of mammals by man has had a tremendous impact. In the last 400 years, some 36 species of mammals have become extinct, and today over 100 species are threatened by extinction. Some species were disposed of remarkably summarily. The Steller's sea cow was pushed to extinction some 27 years after its first discovery by white man. Sea otters, which were hunted at least as early as 1786, were ruthlessly killed for their valuable fur; probably more than 200,000 were killed between 1786 and 1868 (Evermann, 1923:524). By 1900 the animals were rare over much of their range, and they were seemingly lucky to have survived until protected by legislation in the early 1900s. Not so lucky was the grizzly bear in California. In the 1890s grizzlies still persisted in the San Gabriel Mountains near Los Angeles, but the last known southern California grizzly was killed in 1916, and the last verified occurrence in California was in 1922, in the foothills of the Sierra Nevadas in central California (Grinnell et al., 1937:93). Only about 60 years were required to bring the grizzly in California from fair abundance to total extirpation. In Mexico, a population of grizzlies that survived in a small mountain range in central Chihuahua in 1957 was probably wiped out by about 1963, the very year when funds were raised by the World Wildlife Fund to set aside a refuge for the animals.

The quagga, a zebra that inhabited southern Africa, was extirpated in the wild about 1860, and another type of zebra was exterminated by roughly 1910. The Arabian oryx, well on its way toward extinction, has been hunted in recent years with machine guns mounted on jeeps—surely a manly style of hunting. The impact of man has nowhere been more strongly felt than by some marine mammals. Through persistent hunting by man, the blue whale, the largest animal of all times, has been reduced to a total population of probably no more than several hundred individuals. Other activities of man have left their mark: DDT has seemingly been responsible for decimating bat populations in some areas, and a freeway in California has transected the unusually small range of the handsome Morro Bay kangaroo rat. The list of mammals extirpated or endangered by man is depressingly long.

Since I wrote the first edition of this text I had the privilege of spending a 14-month sabbatical leave in Kenya, East Africa. My family and I lived at Bushwhackers Safari Camp, a group of thatch-roofed cabins situated well away from any village in southern Kenya, and visited by amateur naturalists, scientists, and others who enjoyed unspoiled bush country. When we arrived in May of 1973, Bushwhackers was surrounded by almost unbroken tropical deciduous woodland, which supported not only a rich assemblage of smaller mammals, but also rhinos, giraffes, hippos, and buffalo. At night we occasionally heard lions, and we became accustomed to seeing waterbucks, bushbucks, dik-diks, and even lesser kudus about and frequently within the camp.

During our stay the land surrounding the camp became progressively more heavily settled by Kamba families, whose "shambas" were scattered through the woodland. The shambas are small, several-acre plots, cleared by cutting and burning vegetation, where families live and plant patches of corn. Although the practice is illegal, many men set snares for game in openings in the thorn fences surrounding the shambas or in game trails in the woodland. Some families have goats or cattle. When we left in August of 1974, the rhinos, giraffes and buffalo were largely gone, and during even a brief walk through the bush one could find the remains of animals snared and butchered for food. Less than three years after we left Kenya, the area around Bushwhackers is a mosaic of shambas and cleared patches of woodland; the large ungulates are gone, lions are not heard at night, and lesser kudus no longer appear in the evening in the arroyo just north of camp.

I place no blame on the Kamba families. These are good people who are trying to improve their lot, or in some cases simply to survive, by planting corn in new land and exploiting the wildlife in an effort to supplement their monotonous diet. Nor can the poor people be blamed who are having the same effect on the land in such places as Mexico, other parts of Africa, or

Southeast Asia. But in these areas, the clearing of plots of land and the use of subsistence agriculture, while providing only a marginal living for the people, is rapidly restricting many types of wildlife habitat to islands surrounded by land under pressure from the heavy hand of man.

There are some rays of light. The national parks systems in Kenya, Tanzania, Zambia, and a number of other African countries deserve high praise for their efforts to maintain areas in which wildlife is protected. The African game rangers in many of the national parks and game reserves are doing a dedicated job in controlling poaching and the intrusion of domestic livestock. Through tourism, these natural areas bring large amounts of revenue to many countries, a situation that helps reinforce the countries' growing regard for their wildlife. Regrettably, not all countries are able to maintain large areas where wildlife can thrive, and in many places the situation is desperate.

It is obvious that the fate of the wildlife depends on the persistence of appropriate habitat, which is being destroyed over broad areas at an ever-increasing rate by increasing populations of man. In many nonindustrial parts of the world where the focal point of man's existence is the day-to-day search for the resources to assure survival, the pressure on the land and its wildlife is intense. This pressure becomes vastly more acute when revolution, struggles for independence, or strife between political factions is associated with reigns of lawlessness.

The situation in Angola, Africa, eloquently described by Huntley (1976), provides a tragic case in point. Here, in Quissama National Park, for example, hundreds of red buffalo (*Syncerus caffer nanus*), members of a primitive race of African buffalo, have been lawlessly slaughtered with machine guns in the hands of soldiers supplying meat to themselves and their supporters; and in Luando District Nature Reserve the future of the giant sable antelope has been uncertain since the Conservation Department personnel fled after repeated threats on their lives by undisciplined bands of liberation soldiers. During an aerial survey of Quissama National Park in November of 1974, only bleaching skeletons and a network of jeep tracks were seen on the plains where red buffalo, eland, roan antelope, and reedbuck had once been abundant. As seems typically to be the case, white men were major exploiters. Their attitude is embodied in the statement of an incorrigible white poacher charged with killings four elephants: "If the blacks are to be given Angola, I'm not leaving them any ivory." The only professional conservationist still in Angola (as of early 1976) is Dr. Serodio Almeida, who leads a small and inexperienced group of African rangers.

The Angolan wildlife tragedy, unhappily not a unique and isolated case, is summed up by Huntley: "The dark cloud that has moved over Angola will surely pass, but in its wake little of a once-rich fauna and flora will remain. We cannot expect that many of the already threatened species—giant sable, gorilla, chimpanzee, manatees and so on—will survive, but we can only hope that the small nuclei that escape the holocaust will suffice to repopulate the national parks. As hard as the realities of the situation are to bear, those who are devoted to the conservation of the country's magnificent wildlife heritage must face the truth and speak it, that the world may record, if not learn from its mistakes."

We have clearly reached the eleventh hour—if we do not learn from such mistakes the wildlife of the world will pay a devastating price.

But it is impossible to lament the decimation of wildlife without recognizing the terrible human tragedies that accompany such upheavals as that occurring in Angola. Man's inhumanity to man must be of prime concern. For many centuries, human history has been dominated by the overriding inability of man to develop social systems that allow people to live together peacefully and countries to coexist peacefully. Some statesmen have suggested (with little justification, it

would seem) that fear of our modern instruments of war would hasten development of peaceful ways of solving our social problems.

The need for man to set his own house in order is basic to both our own survival and to the perpetuation of the biological richness of the world. Clearly, man's ability to solve his social problems and the future of wildlife are tightly linked. People not under the pressures and stresses occasioned by high populations and limited resources, and at peace with one another, can work toward the saving of biotas of the world—fearful people with empty stomachs make poor conservationists. On the success of the statesmen, scientists, and teachers of the world in halting the rise in human populations and stopping strife between peoples hinges the survival of our biotic heritage.

A NEW CONSERVATION ETHIC

Ecologists and conservationists have argued that each element of a biota plays an essential part in the ecosystem that it occupies, and that the loss of even a seemingly insignificant species might tip the delicate biotic balance. We do not begin to know the degree of pressure that most ecosystems can tolerate before collapse. Biologists have stressed practical problems: the pollution of water should be avoided not only because of the potential for serious public health problems, but also because the ecosystem of a stream might be drastically altered and species of fishes from which we derive pleasure or some monetary return might disappear. Range managers have emphasized economic problems: unwise grazing practices alter a grassland ecosystem to the point at which its economic importance is reduced; in concrete terms, the numbers of pounds that each head of cattle gains per day may be reduced to a point at which a rancher can no longer realize a profit from his operation. But can we, with due respect for honesty, justify conservation with only these kinds of arguments?

In his excellent discussion entitled *The Conservation of Non-resources*, Ehrenfeld (1976) points out that attempts to justify the conservation of many species on the basis of their economic importance, such as those that are small, insignificant, and totally unknown to the general public, are unjustifiable scientifically. The conservation doctrines lose force if a species is destroyed without the disruption of its ecosystem; but if the ecosystem is destroyed following the loss of the species it is not only too late to save the day but the cause and effect relationship involved can never be proved and may not even be hypothesized.

Perhaps we must look to a new and honest conservation ethic. Each species cannot be proved to be essential to the survival of its ecosystem nor to have some economic value; nonetheless, each species forms a part of a biological richness developed over millions of years, and each is worthy of perpetuation at least in part because of what Ehrenfeld terms its "natural art value." Two moving presentations of this view are quoted by Ehrenfeld. In his book *Ulendo: Travels of a Naturalist in and out of Africa*, Carr (1964) states, "It would be cause for world fury if the Egyptians should quarry the pyramids, or the French should loose urchins to throw stones in the Louvre. It would be the same if the Americans dammed the Valley of the Colorado. A reverence for original landscape is one of the humanities. It was the first humanity. Reckoned in terms of human nerves and juices, there is no difference in the value of a work of art and a work of nature. There is this difference though...Any art might somehow, some day be replaced—the full symphony of the savanna landscape never."

Regarding specific nonresource species, in this case small primates called lion tamarins (three species of the genus *Leontideus*), Coimbra-Filho et al. (1975) write: "In purely economic terms, it really doesn't matter if three Brazilian monkeys vanish into extinction. Although they can be (and previously were) used as laboratory animals in biomedical research, other far more abundant species from other parts of South America serve equally well or better in laboratories. Lion tamarins can be effectively exhibited in zoos, but it is doubtful that the majority of zoo-goers would miss them. No, it seems that the main reason for trying to save them and other animals like them is that the disappearance of any species represents a great esthetic loss for the entire world. It can perhaps be compared to the destruction of a great work of art by a famous painter or sculptor, except that, unlike a man-made work of art, the evolution of a single species is a process that takes many millions of years and can never again be duplicated."

Although effort expended on the perpetuation of some species can be justified on the basis of their economic importance, I would regard this natural art value of many species as their greatest importance to man. A more basic argument would insist that a species should be preserved because of a reverence for its vast evolutionary history, and that each species has a right to play out its evolutionary role; man has no moral right to set himself up as the instrument of their destruction. Certainly, present attitudes must change if our biotas are to survive, and a yet unformulated rationale or conservation ethic may serve to unite enough of mankind behind the conservation cause to turn the tide.

If the present decimation of the earth's wildlife is to be curtailed, at least the following steps must be taken:

1. Man's population growth must be halted and his need for space and resources stabilized.
2. Large tracts of land undisturbed by man must be maintained for wildlife.
3. Man's exploitation of many species (whales are notable examples) must be drastically reduced.
4. A broad understanding of the meaning of such biotic interactions as predation must underlie an interest in preserving balanced faunas.
5. Control of animals threatening crops and livestock must be local, with no attempt to exterminate a species over

wide areas with little respect to the damage it is doing.

6. The use of biocides must be carefully controlled.

7. Man must accept some types of economic losses and inconvenience caused by wildlife, and must feel that these are more than compensated for by his enjoyment of a rich and balanced biota.

Because the survival of the wildlife of the world, and indeed the survival of mankind itself, is in our hands, one cannot help but fervently hope that the word *sapiens* (meaning wise) becomes a justly earned part of the name *Homo sapiens*.

BIBLIOGRAPHY

Note: This list includes, in addition to publications cited in the text, a few additional important references by cited authors.

Abbott, J. H., and A. E. von Doenhoff. 1949. Theory of wing sections. McGraw-Hill Aeronautical Science Series, New York.

Abbott, K. D. 1971. Water economy of the canyon mouse, *Peromyscus crinitus stephensi*. Comp. Biochem. Physiol., *38A*:37–52.

Ables, E. D. 1969. Home range studies of red foxes (*Vulpes vulpes*). J. Mamm., *50*:108–120.

Adcock, E. W., F. Teasdale, C. S. August, S. Cox, G. Meschia, F. C. Battaglia, and M. A. Naughton. 1973. Human chorionic gonadotropin: its possible role in maternal lymphocyte suppression. Science, *181*:845–847.

Alcorn, S. M., S. E. McGregor, G. D. Butler, Jr., and E. B. Kurtz, Jr. 1959. Pollination requirements of the saguaro (*Carnegiea gigantea*). Cactus Succ. J. Amer., *31*:39–41.

Alexander, R. D., 1974. The evolution of social behavior. Ann. Rev. Ecol. System., 5.325–383.

Allee, W. C., A. E. Emerson, O. Park, T. Park, and K. P. Schmidt, 1949. Principles of animal ecology. W. B. Saunders Co., Philadelphia. 837 pp.

Allen, G. M. 1940. The mammals of China and Mongolia, Part 2. Amer. Mus. Nat. Hist., New York. pp. 621–1350.

Allen, J. A. 1924. Carnivora collected by the American Museum Congo Expedition. Bull. Amer. Mus. Nat. Hist., *47*:73–281.

Altenbach, J. S. 1977. Functional morphology of two bats: *Leptomycteris* and *Eptesicus*. Amer. Soc. Mamm., Spec. Publ. No. 5. (In press.)

Altman, P. L., and D. S. Dittmer, 1964. Biology data book. Fed. Amer. Soc. Exp. Biol., Washington, D.C. 633 pp.

Amoroso, E. C., and J. H. Matthews. 1951. The growth of the grey seal (*Halichoerus grypus*) from birth to weaning. J. Anat., *85*:426–428.

Andersen, H. T. (ed.). 1969. The biology of marine mammals. Academic Press, New York, 511 pp.

Anderson, J. M. 1972. Nature's transplant. The transplantation immunology of viviparity. Appleton-Century-Crofts, New York. 145 pp.

Anderson, J. W. 1954. The production of ultrasonic sounds by laboratory rats and other mammals. Science, *119*:808–809.

Anderson, S. 1967a. Primates, pp. 151–177. *In* S. Anderson and J. K. Jones, Jr. (eds.), Recent mammals of the world. Ronald Press, New York.

_____, 1967b. Introduction to the rodents, pp. 206–209. *In* S. Anderson and J. K. Jones, Jr. (eds.), Recent mammals of the world. Ronald Press, New York.

_____, and J. K. Jones, Jr. (eds.). 1967. Recent mammals of the world—a synopsis of families. Ronald Press, New York. 453 pp.

Andersson, A. 1969. Communication in the lesser bushbaby (*Galago senegalensis moholi*). Unpublished M. S. thesis, University of Witwatersrand.

Ansell, W. F. H. 1971. Order Artiodactyla, Part 15, pp. 1–84. *In* J. Meester and H. W. Setzer (eds.), The mammals of Africa, an identification manual. Smithsonian Institution Press, Washington, D.C.

Arey, L. B. 1974. Developmental anatomy, Rev. 7th ed. W. B. Saunders Co., Philadelphia. 695 pp.

Armitage, K. B. 1962. Social behaviour of a colony of the yellow-bellied marmot (*Marmota flaviventris*). Anim. Behav., *10*:319–331.

Asdell, S. A. 1964. Patterns of mammalian reproduction, 2nd ed. Cornell Univ. Press, Ithaca, New York, 670 pp.

Augee, M. L., and E. J. M. Ealey, 1968. Torpor in the echidna, *Tachyglossus aculeatus*. J. Mamm., *49*:446–454.

Aumann, G. D. 1965. Microtine abundance and soil sodium levels. J. Mamm., *46*:594–612.

Axelrod, D. I., and H. P. Bailey. 1968. Cretaceous dinosaur extinction. Evolution, *22*:595–611.

Backhouse, K. M. 1954. The grey seal. Univ. Durham Coll. Med. Gaz., *48(2)*:9–16.

Backus, R. H., and W. E. Schevill. 1966. *Physeter* clicks, pp. 510–528. *In* K. S. Norris (ed.), Whales, dolphins and porpoises. Univ. California Press, Berkeley, California.

Bailey, T. N. 1974. Social organization in a bobcat population. J. Wildl. Manage., *38*:435–446.

Baker, H. G. 1961. The adaptation of flowering plants to nocturnal and crepuscular pollinators. Quart. Rev. Biol., *36*:64–73.

_____. 1973. Evolutionary relationships between flowering plants and animals in American and African tropical forests, pp. 145–159. In B. J. Meggers, E. S. Ayensu, and W. D. Duckworth (eds.), Tropical forest ecosystems in Africa and South America, a comprehensive review. Smithsonian Institution Press, Washington, D.C.

———, and B. J. Harris, 1957. The pollination of Parkia by bats and its attendant evolutionary problems. Evolution, 11:449–460.

Baker, J. R., and R. M. Ranson. 1933. Factors affecting the breeding of the field mouse (Microtus agrestis). Proc. Roy. Soc. London, 113B:486–495.

Baker, R. H. 1971. Nutritional strategies of myomorph rodents in North American grasslands. J. Mamm., 52:800–805.

Baker, R. J. 1967. Karyotypes of bats of the family Phyllostomatidae and their taxonomic implications. Southwestern Nat., 12:407–428.

Baker, W. W., S. G. Marshall, and V. B. Baker. 1968. Autumn fat deposition in the evening bat (Nycticieus humeralis). J. Mamm., 49:314–317.

Balinsky, B. I. 1975. An introduction to embryology. 4th ed. W. B. Saunders Co., Philadelphia. 648 pp.

Balph, D. E., and A. W. Stokes. 1963. On the ethology of a population of Uinta ground squirrels. Amer. Midl. Nat., 69:106–126.

Banfield, A. W. F. 1954. Preliminary investigation of the barren ground caribou. Part II. Life history, ecology and utilization. Can. Wildl. Serv., Wildl. Manage. Bull. ser. 1., no. 10B. 112 pp.

Barash, D. P. 1973. The social biology of the Olympic marmot. Anim. Behav. Monogr., 6:172–245.

———. 1974. The evolution of marmot societies: a general theory. Science, 185:415–420.

———, 1976. Social behavior and individual differences in free-living Alpine marmots (Marmota marmota). Anim. Behav., 24:27–35.

Barfield, R. J., and L. A. Geyer. 1972. Sexual behavior: ultrasonic postejaculatory song of the male rat. Science, 176:1349–1350.

Barghusen, H. R., and J. A. Hopson. 1970. Dentary-squamosal joint and the origin of mammals. Science, 168:573–575.

Bartholomew, G. A. 1968. Body temperature and energy metabolism, pp. 290–354. In M. S. Gordon, G. A. Bartholomew, A. D. Grinnell, C. B. Jorgensen, and F. N. White (eds.), Animal function: principles and adaptations, Macmillan Company, New York.

———, and T. J. Cade. 1957. Temperature regulation, hibernation, and estivation in the little pocket mouse, Preognathus longimembris. J. Mamm., 38:60–72.

———, and N. E. Collias. 1962. The role of vocalization in the social behavior of the northern elephant seal. Anim. Behav., 10:7–14.

———, W. R. Dawson, and R. C. Lasiewski. 1970. Thermoregulation and heterothermy in some of the smaller flying foxes (Megachiroptera) of New Guinea. Z. Vergl. Physiol., 70:196–209.

———, and J. W. Hudson. 1960. Aestivation in the Mohave ground squirrel, Citellus mohavensis. Bull. Mus. Comp. Zool. Harvard, 124:353–372.

———, ———. 1961. Desert ground squirrels. Sci. Amer., 205(5) 107–116.

———, ———. 1962. Hibernation, estivation, temperature regulation, evaporative water loss, and heart rate of the pigmy opossum, Cercartetus nanus. Physiol. Zool., 29:26–40.

———, P. Leitner, and J. E. Nelson. 1964. Body temperature, oxygen consumption, and heart rate in three species of Australian flying foxes. Physiol. Zool., 37:179–198.

———, and M. Rainy. 1971. Regulation of body temperature in the rock hyrax, Heterohyrax brucei. J. Mamm., 52:81–95.

Bateman, G. C. and T. A. Vaughan. 1974. Nightly activities of mormoopid bats. J. Mamm., 55:45–65.

Bateman, J. A. 1959. Laboratory studies of the golden mole and the mole rat. Afr. Wildlife, 13.

Bateson, G. 1966. Problems in cetacean and other mammalian communication, pp. 569–579. In K. S. Norris (ed.), Whales, dolphins and porpoises. Univ. California Press, Berkeley.

———, and B. Gilbert. 1966. Whaler's Cove dolphin community: an interim report, p. 5. The Oceanic Institute, Makpuu Point, Waimanalo, Oahu, Hawaii.

Batzli, G. A., and F. A. Pitelka. 1970. Influence of meadow mouse populations on California grassland. Ecology, 51:1027–1039.

———, and F. H. Pitelka. 1971. Condition and diet of cycling populations of the California vole, Microtus californicus. J. Mamm., 52:141–163.

Beatley, J. C. 1969. Dependence of desert rodents on winter annuals and precipitation. Ecology, 50:721–724.

Bee, J. W., and E. R. Hall. 1956. Mammals of Northern Alaska. Univ. Kansas Publ., Mus. Nat. Hist., Misc. Publ. No. 8. 309 pp.

Beer, J. R. 1961. Seasonal reproduction in the meadow vole. J. Mamm., 42:483–489.

———, R. Lukens, and D. Olson. 1954. Small mammal populations on the islands of Basswood Lake, Minn. Ecology, 35:437–445.

———, and R. K. Meyer. 1951. Seasonal changes in the endocrine organs and behavior patterns of the muskrat. J. Mamm., 32:173–191.

Bell, R. H. V. 1970. The use of the herb layer by grazing ungulates in the Serengeti, pp. 111–125. In Adam Watson (ed.), Animal populations in relation to their food resources. Blackwell Scientific Publications, Oxford. 477 pp.

———. 1971. A grazing ecosystem in the Serengeti. Sci. Amer., 225(1):86–93.

Benson, S. B. 1933. Concealing coloration among some desert rodents of the southwestern United States. Univ. Calif. Publ. Zool., 40:1–70.

_____, and A. E. Borell. 1931. Notes on the life history of the red tree mouse, Phenacomys longicaudus. J. Mamm., 12:226–233.

Berg, L. S. 1950. Natural regions of the U.S.S.R. Macmillan Company, New York.

Bertram, B. C. R. 1973. Lion population regulation. E. Afr. Wildl. J., 11:215–225.

_____. 1975. The social system of lions. Sci. Amer., 232:54–65.

Bertram, G. C. L. 1940. The biology of the Weddell and crabeater seals, with a study of the comparative behavior of the Pinnipedia. Brit. Mus. (Nat. Hist.) Sci. Repts. Brit. Graham Land Exped. 1934–1937, 1:1–139.

Black, H. L. 1974. A North Temperate bat community: structure and prey populations. J. Mamm., 55 138–157.

Blair, W. F. 1939. Some observed effects of stream-valley flooding on mammalian populations in eastern Oklahoma. J. Mamm., 20:304–306.

_____. 1951. Evolutionary significance of geographic variation in population density. Texas J. Sci., 1:53–57.

Blake, E. W., Jr. 1887. The coast fox. West. Amer. Sci., 3:49–52.

Blatt, C. M., C. R. Taylor, and M. B. Habal. 1972. Thermal panting in dogs: the lateral nasal gland, a source of water for evaporative cooling. Science, 177:804–805.

Bloedel, P. 1955. Hunting methods of fish-eating bats, particularly Noctilio leporinus. J. Mann., 36:390–399.

Bodemer, C. W. 1968. Modern embryology. Holt, Rinehart and Winston, Inc., New York. 475 pp.

Bodenheimer, F. S. 1949. Problems of vole populations in the Middle East. Report on the population dynamics of the Levant vole (Microtus quentheri D.). Azriel Print. Works, Jerusalem, 77 pp.

Bond, R. M. 1945. Range rodents and plant succession. Trans. N. Amer. Wildl. Conf., 10:229–234.

Bourliere, F. 1956. The natural history of mammals. Alfred A. Knopf, Inc., New York. 364 pp.

_____. 1973. Comparative ecology of rain forest mammals, pp. 279–292. In B. J. Meggers, E. S. Ayensu, and W. D. Duckworth (eds.), Tropical forest ecosystems in Africa and South America: a comparative review. Smithsonian Institution Press, Washington, D.C.

Boving, B. G. 1959. Implantation. Ann. N.Y. Acad. Sci., 75:700–725.

Bowker, M. 1977. Behavior of Kirk's dik-dik, Rhynchotragus kirki. Unpublished Ph.D Thesis, Northern Arizona University, Flagstaff.

Bradbury, J. W. 1970. Target discrimination by echolocating bat Vampyrum spectrum. J. Exp. Zool., 173:23–46.

_____, 1977. Social organization and communication. In W. Wimsatt (ed.), Biology of bats. Vol. 3. Academic Press, New York.

_____, and L. H. Emmons. 1974. Social organization of some Trinidad bats, I. Emballonuridae. Z. Tierpsychol., 36:137–183.

Bradshaw, G. V. R. 1962. Reproductive cycle of the California leaf-nosed bat, Macrotus californicus. Science, 136:645–646.

Breadon, G. 1932. The flying fox in the Punjab. J. Bombay Nat. Hist. Soc., 35:670.

Broadbooks, H. E. 1970. Home ranges and territorial behavior of the yellow-pine chipmunk, Eutamias amoenus. J. Mamm., 51:310–326.

Brockie, R. 1976. Self-anointing by wild hedgehogs, Erinaceus europaeus. Anim. Behav., 24:68–71.

Brodie, P. F. 1975. Cetacean energetics, an overview of intraspecific size variation. Ecology, 56:152–161.

Bromley, P. T. 1969. Territoriality in pronghorn bucks on the National Bison Range, Moiese, Montana. J. Mamm., 50:81–89.

Brosset, A. 1968. Permutation du cycle chez Hipposideros caffer au voisinage de l'equateur. Biologica Gabonica, 4.

Brown, J. H. 1968. Adaptation to environmental temperature in two species of woodrats, Neotoma cinerea and N. albigula. Misc. Publ. Mus. Zool., Univ. Mich., 135:1–48.

_____, and G. A. Bartholomew. 1969. Periodicity and energetics of torpor in the kangaroo mouse, Microdipodops pallidus. Ecology, 50:705–709.

_____, and R. C. Lasiewski. 1972. Metabolism of weasels: the cost of being long and thin. Ecology, 53:939–943.

Brown, L. N. 1967. Ecological distribution of six species of shrews and comparison of sampling methods in the central Rocky Mountains. J. Mamm., 48:617–623.

Brown, P. 1976. Vocal communication in the pallid bat, Antrozous pallidus. Z. Tierpsychol., 41:34–54.

Bruce, V. G. 1960. Environmental entrainment of circadian rhythms. Cold Spr. Harb. Symp. Quant. Biol., 25:29–48.

Bruce, W. C. 1915. Measurements and weights of antarctic seals (part II, pp. 159–174,2 pls.) In Report on the scientific results of the voyage of S. Y. "Scotia" during the years 1902, 1903, and 1904 . . . Edinburgh, Scottish Oceanogr. Lab.

Buckner, C. H. 1969. Some aspects of the population ecology of the common shrew, Sorex araneus, near Oxford, England, J. Mamm., 50:326–332.

Buechner, H. K. 1961. Territorial behavior in Uganda kob. Science, 133:698–699.

———. 1963. Territoriality as a behavioral adaptation to environment in the Uganda kob. Proc. XVI Int. Congr. Zool., 3:59–65.

———. 1974. Implications of social behavior in the management of the Uganda kob, pp. 853–870. In V. Geist and F. R. Walther (eds.), The behavior of ungulates and its relation to management. International Union for Conservation of Nature and Natural Resources, Morges, Switzerland.

———, and H. D. Roth. 1974. The lek system of the Uganda kob. Amer. Zool., 14:145–162.

Bugge, J. 1974. The cephalic arteries of hystricomorph rodents, pp. 61–78. In I. W. Rowlands and B. Weir (eds.), The biology of hystricomorph rodents. Symp. Zool. Soc. London. Academic Press, New York.

Bullard, E. 1969. The origin of the oceans. Sci. Amer., 221:66–75.

Burckhardt, D. 1958. Kindliches Verhalten als Ausdrucksvewegung im Fortpflanzungszeremoniell einiger Wiederkauer. Rev. suisse Zool., 65:311–316.

Burrell, H. 1927. The platypus. Angus and Robertson Ltd., Sydney.

Burt, W. H. 1940. Territorial behavior and populations of some small mammals in southern Michigan. Misc. Publ. Mus. Zool., Univ. Michigan, 45:1–58.

———. 1943. Territoriality and home range concepts as applied to mammals. J. Mamm., 24:346–352.

Burton, M. 1949. Wildlife of the world. Long Acre, London. 384 pp.

Busnel, R. G. (ed.). 1966. Animal sonar systems, biology and bionics, Tomes I and II. Laboratoire de Physiologie Acoustique, Paris.

Butler, P. M. 1969. In L. S. B. Leakey (ed.), Fossil vertebrates of Africa, Vol. 1. Academic Press, New York.

———. 1972. Some functional aspects of molar evolution. Evolution, 26:474–483.

———, and Z. Kielan-Jaworowska. 1973. Is Deltatheridium a marsupial? Nature, 245:105–106.

Cadenat, J. 1959. Bull. Inst. Franc. Afrique Noire, 21:1137–1143.

Calaby, J. H. 1971. The current status of Australian Macropodidae. Australian Zool., 16:17–29.

Calder, N. 1972. The restless earth: a report on the new geology. Viking Press, New York. 152 pp.

Calder, W. A. 1969. Temperature relations and underwater endurance of the smallest homeothermic diver, the water shrew. Comp. Biochem. Physiol., 30:1075–1082.

Caldwell, D. K., J. H. Prescott, and M. C. Caldwell, 1966. Bull. So. California Acad. Sci., 65:245–248.

Caldwell, M. C., and D. K. Caldwell. 1970. Further studies on audible vocalizations of the Amazon freshwater dolphin, Inia geoffrensis. Contrib. Sci., 187:1–5.

———. 1972. Behavior of marine mammals: sense and communication, pp. 419–502. In S. H. Ridgway (ed.), Mammals of the sea: biology and medicine. Charles C Thomas, Springfield, Ill.

Camp, C. L., and N. S. Smith. 1942. Phylogeny and functions of the digital ligaments of the horse. Mam. Univ. California, 13:69–124.

Campbell, C. B. G. 1966. Taxonomic status of tree shrews. Science, 153:436.

Carpenter, R. E. 1966. A comparison of thermoregulation and water metabolism in the kangaroo rats Dipodomys agilis and Dipodomys marriami. Univ. California Publ. Zool., 78:1–36.

———. 1968. Salt and water metabolism in the marine fish-eating bat, Pizonyx vivesi. Comp. Biochem. Physiol., 24:951–964.

———. 1969. Structure and function of the kidney and the water balance of desert bats. Physiol. Zool., 42:288–302.

———, and J. B. Graham. 1967. Physiological responses to temperature in the long-nosed bat, Leptonycteris sanborni. Comp. Biochem. Physiol., 22:709–722.

Carr, A. 1974. Ulendo: travels of a naturalist in and out of Africa. Knopf, New York.

Chaffee, R. R. J., and J. C. Roberts. 1971. Temperature acclimation in birds and mammals. Ann. Rev. Physiol., 33:155–202.

Chapskiy, K. K. 1936. The walrus of the Kara Sea. Trans. Arct. Inst., Leningrad, Tom. 67. 111 pp. (In Russian; resumé in English, pp. 112–124.)

Chitty, D. 1954. Tuberculosis among wild voles with a discussion of other pathological conditions among certain mammals and birds. Ecology, 35:227–237.

———. 1958. Self-regulation of numbers through changes in viability. Cold Spr. Harb. Symp. Quant. Biol., 22:277–280.

———. 1960. Population processes in the vole and their relevance to general theory. Can. J. Zool., 38:99–113.

———, and H. Chitty. 1962. Population trends among the voles at Lake Vyrnwy, 1932–1960. J. Anim. Ecol., 35:313–331.

Christian, J. J. 1950. The adreno-pituitary system and population cycles in mammals. J. Mamm., 31:247–259.

———. 1954. The relation of the adrenal cortex to population size in rodents. Doctoral dissertation, Johns Hopkins School of Hygiene and Public Health, Baltimore.

———. 1959a. Control of population growth in rodents by interplay between population density and endocrine physiology. Wildl. Dis., 1:1–38.

———. 1959b. The roles of endocrine and behavioral factors in the growth of mammalian populations, pp. 71–97. In A. Gorbman (ed.), Comparative endocrinology. Columbia Univ. Symposium, New York.

———. 1963. Endocrine adaptive mechanisms and the physiologic regulation of population growth, pp. 189–353. In W. V. Mayer and R. G. Van Gelder (eds.), Physiological mammalogy. Academic Press, New York.

_____, and D. E. Davis. 1956. The relationship between adrenal weight and population status in Norway rats. J. Mamm., 37:475–486.

_____, _____. 1966. Adrenal glands in female voles (Microtus pennsylvanicus) as related to reproduction and population size. J. Mamm., 47:1–18.

_____, V. Flyger, and D. E. Davis. 1960. Factors in mass mortality of a herd of sika deer. Chesapeake Sci., 1:79–95.

Clark, W. E. LeGros. 1959. The antecedents of man. Edinburgh Univ. Press, Edinburgh. 374 pp.

Clemens, W. A. 1968. Origin and early evolution of marsupials. Evolution, 22:1–18.

_____. 1970. Mesozoic mammalian evolution, pp. 357–390. In R. F. Johnston (ed.), Annual Review of Ecology and Systematics, Vol. I. Annual Reviews, Inc., Palo Alto.

_____, and L. G. Marshall. 1976. Fossilium Catalogus: American and European Marsupialia. W. Junk (The Hague), Pars. 123:1–114.

_____, and M. Plane. 1974. Mid-Tertiary Thylacoleonidae (Marsupialia, Mammalia). J. Paleont., 48(4):652–660.

Clough, G. C. 1965. Lemmings and population problems. Amer. Sci., 53:199–212.

Coblentz, B. E. 1976. Functions of scent-urination in ungulates with special reference to feral goats (Capra hircus L.). Amer. Nat., 110:549–557.

Coe, M. J. 1967. Preliminary notes on the spring hare Pedetes surdaster larvalis in East Africa. E. Afr. Wildl. J., 5.

Coimbra-Filho, A. F., A. Magnanini, and R. A. Mittermeier. 1975. Vanishing gold: last chance for Brazil's lion tamarins. Animal Kingdom, Dec., pp. 20–26.

Colbert, E. H. 1948. The mammal-like reptile Lycaenops. Bull. Amer. Mus. Nat. Hist., 89:353–404.

_____. 1949. The ancestors of mammals. Sci. Amer., 180:40–43.

_____. 1969. Evolution of the vertebrates, 2nd ed. John Wiley & Sons. New York. 535 pp.

_____. 1973. Wandering lands and animals. E. P. Dutton and Company, New York. 323 pp.

Cole, R. W. 1970. Pharyngeal and lingual adaptations in the beaver. J. Mamm., 51:424–425.

Conaway, C. H. 1958. Maintenance, reproduction and growth of the least shrew in captivity. J. Mamm., 39:507–512.

Conley, W. H. 1971. Behavior, demography and competition in Microtus longicaudus and M. mexicanus. Ph.D. Thesis, Texas Tech. Univ.

Cott, H. B. 1966. Adaptive coloration in animals. Methuen and Company, Ltd., London. 508 pp.

Coulombe, H. N. 1970. The role of succulent halophytes in water balances of salt marsh rodents. Oecologia (Berl.) 4:223–247.

Cowan, I. M. 1947. The timber wolf in the Rocky Mountain national parks of Canada. Can. J. Res., 25:139–174.

_____. 1956. Life and times of the coast black-tailed deer, 523–617. In W. P. Taylor (ed.), The deer of North America. Stackpole Company, Harrisburg, Pennsylvania.

Crawford, E. C., Jr. 1962. Mechanical aspects of panting in dogs. J. Appl. Physiol., 17:249–251.

Crisler, L. 1956. Observations of wolves hunting caribou. J. Mamm., 37:337–346.

_____. 1958. Arctic wild. Harper. New York. 301 pp.

Crompton, A. W. 1971. The origin of the tribosphenic molar, pp. 65–87. In D. M. Kermack and K. A. Kermack (eds.), Early mammals. J. Linn. Soc. (London) Zool., 50:(1):165–180.

_____. 1972. The evolution of the jaw articulation in cynodonts, pp. 231–251. In K. A. Joysey, and T. S. Kemp (eds.), Studies in vertebrate evolution. Oliver and Boyd, Edinburgh. 284 pp.

_____. 1974. The dentitions and relationships of the southern African mammals, Erythrotherium parringtoni and Megazostrodon rudnerae. Bull. Brit. Mus. Nat. Hist. (Geol.), 24:399–437.

_____, and K. Hiiemae. 1969. How mammalian molar teeth work. Discovery, 5(1):23–34.

_____, _____. 1970. Molar occlusion and mandibular movements during occlusion in the American opossum, Didelphus marsupialis L. J. Linn. Soc. (London) Zool., 49:21–47.

_____, and F. A. Jenkins, Jr. 1968. Molar occlusion in Late Triassic mammals. Biol. Rev., 43:427–458.

_____, _____. 1973. Mammals from reptiles: a review of mammalian origins. Ann. Rev. Earth Planetary Sci., Vol. I. Annual Reviews, Inc., Palo Alto.

Crowcroft, P. 1953. The daily cycle of activity in British shrews. Proc. Zool. Soc. London, 123:715–729.

_____. 1957. The life of the shrew. Max Reinhart, London. 166 pp.

_____. 1966. Mice all over. G. T. Foulis & Company, London.

Curry-Lindahl, K. 1962. The irruption of the Norway lemmings in Sweden during 1960. J. Mamm., 43:171–184.

Dagg, A. I. 1962. The role of the neck in the movements of the giraffe. J. Mamm., 43:88–97.

Dalquest, W. W. 1948. The mammals of Washington. Univ. Kansas Publ., Mus. Nat. Hist., 2:1–444.

Daniel, J. C., Jr. 1970. Dormant embryos of mammals. Bio. Sci., 20(7):411–415.

Darling, F. F. 1937. A herd of red deer. Oxford Univ. Press, London. 215 pp.

Darlington, P. J. 1957. Zoogeography: the geographical distribution of animals. John Wiley & Sons, New York, 675 pp.

Dasmann, R. F., and A. S. Mossman. 1962. Population studies of impala in Southern Rhodesia. J. Mamm., 43:375–395.

David, A. 1968. Can young bats communicate with their parents at a distance? J. Bombay Nat. Hist. Soc., 65:210.

Davis, D. E. 1951. The relation between level of population and pregnancy of Norway rats. Ecology, 32:459, 461.

———. 1967. The annual rhythm of fat deposition in woodchucks (Marmota monax). Physiol. Zool., 40:391–402.

Davis, R. B., C. F. Herreid, Jr., and H. L. Short. 1962. Mexican free-tailed bats in Texas. Ecol. Monogr., 32:311–346.

Davis, W. H. 1970. Hibernation: ecology and physiological ecology, pp. 265–300. In W. A. Wimsatt (ed.), Biology of bats. Academic Press, New York.

———, and H. B. Hitchcock. 1965. Biology and migration of the bat. Myotis lucifugus, in New England. J. Mamm., 46:296–313.

———, and W. Z. Lidicker, Jr. 1956. Winter range of the red bat. J. Mamm., 37:280–281.

Dawe, A. R., and W. A. Spurrier. 1969. Hibernation induced in ground squirrels by blood transfusion. Science, 163:298–299.

Dawson, M. R. 1958. Later Tertiary Leporidae of North America. Univ. Kansas Paleont. Contributions, Vertebrata, Art. 6, pp. 1–75.

———. 1967a. Fossil history of the families of Recent mammals, pp. 12–53. In S. Anderson and J. K. Jones, Jr. (eds.), Recent mammals of the world. Ronald Press, New York.

———. 1967b. Lagomorph history and stratigraphic record. Essays in Paleontology and Stratigraphy, Raymond C. Moore Commemorative Volume, Univ. Kansas, Department of Geology Spec. Publ. 2, pp. 287–316.

Dawson, T. J. 1973a. Primitive mammals. In C. G. Whittow (ed.), Comparative physiology of thermoregulation, Vol. III. Academic Press, New York.

———. 1973b. Thermoregulatory responses of the aris zone kangaroos Megaleia rufa and Macropus robustus. Comp. Biochem. Physiol., 46:153–169.

———, and A. J. Hulbert. 1970. Standard metabolism, body temperature, and surface areas of Australian marsupials. Amer. J. Physiol., 218:1233–1238.

———, D. Robertshaw, and C. R. Taylor. 1974. Sweating in the red kangaroo: an evaporative cooling mechanism during exercise but not heat. Amer. J. Physiol., 227:494–498.

DeCoursey, P. 1960. Phase control of activity in a rodent. Cold Spr. Harb. Symp. Quant. Biol., 25:49–55.

———. 1961. Effect of light on the circadian activity rhythm of the flying squirrel, Glaucomys volans. Z. Vergl. Physiol., 44:331–354.

Deevey, E. S., Jr. 1947. Life tables for natural populations of animals. Quart. Rev. Biol., 22:283–314.

Degerbøl, M., and P. Freuchen. 1935. Mammals, vol. 2, p. 278. In Report of the fifth Thule expedition, 1921–1924. Copenhagen, Nordisk Forlag. Part I. Systematic notes, by Degerbøl, Part II. Field notes and biological observations, by Freuchen.

Delany, M. J. 1971. The biology of small rodents in Mayanja Forest, Uganda. J. Zool., London, 165:85–129.

DeVore, I (ed.). 1965. Primate behavior. Holt, Rinehart and Winston, New York. 654 pp.

———, and K. R. L. Hall. 1965. Baboon ecology, pp. 20–52. In I. DeVore (ed.), Primate behavior. Holt, Rinehart and Winston, New York.

Dietz, R. S., and J. C. Holden, 1970. The breakup of Pangaea. Sci. Amer., 223(4):30–41.

Dijkgraaf, S. 1943. Over een merkwaardige functie wan den gehoorzin bij vleermuizen. Verslagen Nederlandsche Akademie Wetenschappen Afd. Naturkunde, 52:622–627.

———. 1946. Die Sinneswelt der Fledermäuse. Experientia, 2:438–448.

———. 1960. Spallanzani's unpublished experiments on the sensory basis of object perception in bats. Isis, 51:9–20.

Dobzhansky, T. 1950. Mendelian populations and their evolution. Amer. Nat., 84:401–418.

Douglas-Hamilton, I. 1972. On the ecology and behavior of the African elephant: the elephants of Lake Manyara. Ph.D. thesis, Oxford Univ., Oxford. 268 pp.

———. 1973. On the ecology and behavior of the Lake Manyara elephants. E. Afr. Wildl. J., 11:401–403.

Dreher, J. J. 1966. Cetacean communication; small-group experiments, pp. 529–543. In K. S. Norris (ed.), Whales, dolphins and porpoises. Univ. California Press, Berkeley.

———, and W. E. Evans. 1964. Cetacean communication, pp. 373–393. In W. N. Tavolga (ed.), Marine bio-acoustics. Pergamon Press, Oxford.

Dryden, G. L. 1975. Establishment and maintenance of shrew colonies. Zool. Yearbook, 15:12–18.

———. 1976. Personal communication.

Dubost, G. 1968. Aperçu sur le rythme annuel de reproduction des muridés du nord-est du Gabon. Biol. Gabonica, 4:227–239.

Dücker, G. 1957. Fard- und Helligkeitssehen und Instinkte bei Viverriden und Feliden. Zool Beitr. (Berl.), 3:25–99.

Dunning, D. C. 1968. Warning sounds of moths. Z. Tierpsychol., 25:129–138.

———, and K. D. Roeder. 1965. Moth sounds and the insect-catching behavior of bats. Science, 147:173–174.

Durrell, G. M. 1954. The Bafut beagles. Rupert Hart-Davies, London.

DuToit, A. L. 1957. Our wandering continents. Oliver and Boyd, Edinburgh. 366 pp.

Eadie, W. R. 1952. Shrew predation and vole populations on a localized area. J. Mamm., 33:185–189.

Edwards, R. L. 1946. Some notes on the life history of the Mexican ground squirrel in Texas. J. Mamm., 27:105–115.

Egorov, Y. E. 1966. Relationships between American mink and otter in Bashkiria. Acclimatization of animals in the U.S.S.R., pp. 57–58. Jerusalem.

Ehrenfeld, D. W. 1976. The conservation of non-resources. Amer. Sci., 64:648–656.

Eibl-Eibesfeldt, I. 1958. Das Verhalten der Nagetiere. Handb. Zool., 8:1–88.

Einarsen, A. S. 1948. The pronghorn antelope. Wildl. Manage. Inst., Washington, D.C. 238 pp.

Eisenberg, J. F. 1966. The social organization of mammals. Handb. Zool. (Berl.), 10:1–92.

———, and E. Gould. 1970. The tenrecs: a study in mammalian behavior and evolution. Smithson. Contrib. Zool., 27:1–138.

———, and D. G. Kleiman. 1972. Olfactory communication in mammals. Ann. Rev. Ecol. System., 3:1–32.

Eisentraut, M. 1934. Der Winterschlaf der Fledermäuse mit besonderer Berücksichtigung der Warmeregulation. Z. Morphol. Oekol Tiere, 29:231–267.

———. 1957. Aus dem Leben der Fledermäuse and Flughunde. Jena, Veb. Gustav Fischer Verlag. 175 pp.

———. 1960. Heat regulation in primitive mammals and in tropical species. Bull. Mus. Comp. Zool. Harvard, 124:31–43.

Eisner, T., and J. A. Davis. 1967. Mongoose throwing and smashing millipedes. Science, 155:577–579.

Ellerman, J. R. 1940. The families and genera of living rodents. Vol. I. British Mus. Nat. Hist. 689 pp.

———. 1941. Ibid. Vol. II. 690 pp.

———. 1949. Ibid. Vol. III. 210 pp.

Elliot, P. W. 1969. Dynamics and regulation of a Clethrionomys population in central Alberta. Ph.D. Thesis, Univ. Alberta.

Eloff, F. C. 1967. Personal communication.

Eloff, G. 1958. The functional and structural degeneration of the eye in the African rodent moles Cryptomys bigalkei and Bathyergus maritimus. S. Afr. J. Sci., 54.

Elsner, R. 1965. Hvalradets Skrifter Norske Videnskaps-Akad. Oslo, 48:24.

———. 1969. Cardiovascular adjustments to diving, pp. 117–143. In H. T. Andersen (ed.), The biology of marine mammals. Academic Press, New York.

Elton, C. 1942. Voles, mice and lemmings. Clarendon Press, Oxford.

Ende, P. Van den. 1973. Predator-prey interactions in continuous culture. Science, 181:562–564.

Erickson, A. B. 1944. Helminth infections in relation to population fluctuations in snowshoe hares. J. Wildl. Manage., 8:134–153.

Erlinge, S. 1972. Interspecific relations between otter Lutra lutra and mink Mustela vision in Sweden. Oikes, 23:327–335.

———. 1974. Distribution, territoriality and numbers of the weasel Mustela navalis in relation to prey abundance. Oikos, 25:308–314.

———. 1975. Feeding habits of the weasel Mustela navalis in relation to prey abundance. Oikos, 26:378–384.

Errington, P. L. 1937. What is the meaning of predation? Smithsonian Rep. for 1936, pp. 243–252.

———. 1943. An analysis of mink predation upon muskrats in north-central United States. Agric. Exp. Sta. Iowa State Coll. Res. Bull., 320:797–924.

———. 1946. Predation and vertebrate populations. Quart. Rev. Biol., 21:144–177, 221–245.

———. 1957. Of populations, cycles and unknowns. Cold. Spr. Harb. Symp. Quant. Biol., 22:287–300.

———. 1963. Muskrat populations. Iowa State Univ. Press, Ames, Iowa.

———. 1967. Of predation and life. Iowa State Univ. Press, Ames, Iowa. 277 pp.

Espenshade, E. B. (ed.). 1971. Goode's world atlas. 13th ed. Rand-McNally, Chicago.

Estes, J. A., and J. F. Palmisano. 1974. Sea otters: their role in structuring nearshore communities. Science, 185:1058–1060.

Estes, R. D. 1966. Behavior and life history of the wildebeest (Connochaetes taurinus Burchell). Nature, Lond., 212:999–1000.

———. 1967. The comparative behavior of Grant's and Thompson's gazelles. J. Mamm., 48:189–209.

———. 1974. Social organization of the African Bovidae, pp. 166–205. In V. Geist and F. R. Walther (eds.), The behavior of ungulates and its relation to management. International Union for Conservation of Nature and Natural Resources, Morges, Switzerland.

———. and J. Goddard. 1970. Prey selection and hunting behavior of the African wild dog. J. Wildl. Manage., 31(1):52–70.

Evans, F. G. 1942. The osteology and relationships of elephant shrews (Macroscelididae). Bull. Amer. Mus. Nat. Hist., 80(4):85–125.

Evans, W. E., and J. Bastian. 1969. Marine mammal communication: social and ecological factors, pp. 425–475. In H. T. Andersen (ed.), The biology of marine mammals. Academic Press, New York.

———, and R. M. Haugen. 1963. An experimental study of the echolocation ability of a California sea lion, Zalophus californianus (Lesson). Bull. So. Calif. Acad. Sci., 62:165–175.

———, and B. A. Powell. 1967. Proc. symp. bionic models of animal sonar systems, Frascati, Italy, 1966, pp. 363–398. Labor. d'Acoustique Animal, Jouy-en-Josas, France.

———, W. W. Sutherland, and R. G. Beil. 1964. pp. 353–372, vol. 1. In W. N. Tavolga (ed.), Marine bioacoustics. Pergamon Press, Oxford.

Evermann, B. W. 1923. The conservation of marine life of the Pacific. Sci. Mon., 16:521–538.

Evernden, J. F., D. E. Savage, G. H. Curtis, and G. T. Jones. 1964. Potassium-argon dates and the Cenozoic mammalian chronology of North America. Amer. J. Sci., *262*:145–198.

Ewer, R. F., 1968. Ethology of mammals. Plenum Press, New York. 418 pp.

———. 1973. The carnivores. Cornell Univ. Press, Ithaca, N.Y. 494 pp.

Ewing, W. G., E. H. Studier, and M. J. O'Farrell. 1970. Autumn fat deposition and gross body composition in three species of *Myotis*. Comp. Biochem. Physiol., *36*:119–129.

Faegri, K., and L. Van Der Pijl. 1966. The principles of pollination ecology. Pergamon Press, New York.

Fenton, M. B., and K. D. Roeder. 1974. The microtymbals of some Arctiidae. J. Lepidopt. Soc., *28*:205–211.

Fields, R. W. 1957. Histricomorph rodents from late Miocene of Colombia, South America. Univ. California Publ. Geol. Sci., *32*:273–404.

Findley, J. S. 1967. Insectivores and dermopterans, pp. 87–108. *In* S. Anderson and J. K. Jones (eds.), Recent mammals of the world. Ronald Press, New York.

———. 1969. Biogeography of Southwestern boreal and desert mammals, pp. 113–128. *In* J. K. Jones, Jr. (ed.), Contributions in mammalogy. Univ. Kansas, Mus. Nat. Hist., Misc. Publ. No. 51.

———, and D. E. Wilson. 1974. Observations on the Neotropical disk-winged bat, *Thyroptera tricolor* Spix. J. Mamm., *55*:562–571.

Fink, B. D. 1959. Observation of porpoise predation on a school of Pacific sardines. California Fish and Game, *45(3)*:216–217.

Finley, R. B., Jr. 1969. Cone caches and middens of *Tamasciurus* in the Rocky Mountain region, pp. 233–273. *In* J. K. Jones, Jr. (ed.), Contributions in mammalogy. Univ. Kansas, Mus. Nat. Hist., Misc. Publ. No. 51.

Fischer, G. A., J. C. Davis, F. Iverson, and F. P. Cronmiller. 1944. The winter range of the Interstate Deer Herd. U.S. Dept. Agric., Forest Serv., Region 5:1–20 (mimeo).

Fisher, E. M. 1939. Habits of the southern sea otter. J. Mamm., *20*:21–36.

Fitch, H. S. 1948. Ecology of the California ground squirrel on grazing lands. Amer. Midl. Nat., *39*:513–596.

———, R. Goodrum, and C. Newman. 1952. The armadillo in the southeastern United States. J. Mamm., *33*:21–37.

Fitzgerald, B. M. 1972. The role of weasel predation in cyclic population changes in the montane vole *(Microtus montanus)*. Ph.D. Thesis, Univ. California, Berkeley.

Fleming, T. H. 1970. Notes on the rodent faunas of two Panamanian forests. J. Mamm., *51*:473–490.

———. 1971. *Artibeus jamaicensis*: delayed embryonic development in a Neotropical bat. Science, *171*:402–404.

———, E. T. Hooper and D. E. Wilson. 1972. Three Central American bat communities: structure, reproductive cycles, and movement patterns. Ecology, *53*:555–569.

Flessa, K. W. 1975. Area, continental drift and mammalian diversity. Paleobiology, *1*:189–194.

Flexner, L. B., D. B. Crowie, L. M. Hellman, W. S. Wilde, and G. J. Vosburgh, 1948. The permeability of the human placenta to sodium in normal and abnormal pregnancies and the supply of sodium to the human fetus as determined with radioactive sodium. Amer. J. Obst. Gyn., *55*:469–480.

Fogden, A. 1974. A preliminary field study of the Western tarsier, *Tarsius bancanus* Horsefield, pp. 151–156. *In* R. D. Martin, G. A. Doyle, and A. C. Walker (eds.), Prosimian biology. Univ. Pittsburgh Press, Pittsburgh.

Fons, P. R. 1974. Méthodes de capture et d'élevage de la Pachyure étrusque *Suncus etruscus* (Savi, 1822) (Insectivora, Soricidae). Z. Saugetierk., *39*:204–210.

Forman, G. L. 1971. Comparative morphological and histochemical studies of gastrointestinal tracts of selected North American bats. Univ. Kansas Sci. Bull., *49*:591–729.

———. 1973. Studies of gastric morphology in North American Chiroptera (Emballonuridae, Noctilionidae, and Phyllostomatidae). J. Mamm., *54*:909–923.

Formozov, A. N. 1946. The covering of snow as an integral factor of the environment and its importance in the ecology of mammals and birds. Material for Fauna and Flora of the USSR, New Series Zool., *5*:1–141.

———. 1966. Adaptive modifications of behavior in mammals of the Eurasian steppes. J. Mamm., *47(2)*:208–222.

Fossey, D. 1972. Living with mountain gorillas, pp. 209–229. *In* P. R. Marler (ed.), The marvels of animal behavior. National Geographic Soc., Washington, D.C.

Foster, J. B. 1964. Evolution of mammals on islands. Nature, *202*:234–235.

Fox, R. C. 1964. The adductor muscles of the jaw in some primitive reptiles. Univ. Kansas. Publ., Mus. Nat. Hist., *12*:657–680.

Fox, R. F. 1971. Marsupial mammals from the early Campanian Milk River Formation, Alberta, Canada, pp. 145–164. *In* D. M. Kermack and K. A. Kermack (eds.), Early mammals. Academic Press, New York.

Fraser, F. C., and P. E. Purves. 1955. The blow of whales. Nature, *176*:1221–1222.

———,———. 1960a. Anatomy and function of the cetacean ear. Proc. Roy. Soc. (London), B, *152*:62–77.

———,———. 1960b. Hearing in cetaceans. Bull. Brit. Mus. Nat. Hist. Zool., *7*:1–140.

Freeland, W. J., and D. H. Janzen. 1974. Strategies in herbivory by mammals: the role of plant secondary compounds. Amer. Nat., *108*:269–289.

French, N. R., W. E. Grant, W. Grodzinski, and D. M. Swift. 1976. Small mammal energetics in grassland ecosystems. Ecol. Monogr., 46:201–220.

Freudenthal, M. 1972. *Deinogalerix koenigswaldi* nov. gen., Nov. Spec., a giant insectivore from the Neogene of Italy. Scripta Geologica, Leiden, 14:1–10.

Fries, S. 1879. Uber die Fortpflanzung der einheimischen Chiropteren. Zool. Anzeiger, Bd. 2:355–357.

Frith, H. J., and S. H. Calaby. 1969. Kangaroos. F. W. Cheshire, Melbourne. 209 pp.

Fuller, W. A. 1967. Écologie hivernale des lemmings et fluctuations de leurs populations. Terre Vie, 114:97–115.

_____. 1969. Changes in numbers of three species of small rodents near Great Slave Lake, N.W.T., Canada, 1964–1967, and their significance for general population theory. Ann. Zool. Fennici, 6:113–144.

Gaines, M. S., and C. J. Krebs. 1971. Genetic changes in fluctuating vole populations. Evolution, 25:702–723.

Galambos, R., and D. R. Griffin. 1942. Obstacle avoidance by flying bats; the cries of bats. J. Exp. Zool., 89:475–490.

Gawn, R. L. W. 1948. Aspects of the locomotion of whales. Nature, 161:44.

Geist, V. 1971. Mountain-sheep: a study in behavior and evolution. Univ. Chicago Press, Chicago. 383 pp.

_____. 1974. On the relationship of ecology and behavior in the evolution of ungulates: theoretical considerations, pp. 235–246. *In* V. Geist and F. R. Walther (eds.), The behavior of ungulates and its relation to management. International Union for Conservation of Nature and Natural Resources, Morges, Switzerland.

Genelly, R. E. 1965. Ecology of the common mole-rat (*Cryptomys hottentotus*) in Rhodesia. J. Mamm., 46:647–665.

Gill, E. D. 1955a. The problem of extinction, with special reference to the Australian marsupials. Evolution, 9:87–92.

_____. 1955b. The Australian "arid period." Australian J. Sci., 17:204–206.

_____. 1957. The stratigraphical occurrence and paleontology of some Australian Tertiary marsupials. Mem. Nat. Mus., Victoria, 21:135–203.

Gilmore, R. M. 1961. Whales, porpoises, and the U.S. Navy. Norsk Hvalfangst-tid., 3:1–9.

Glander, K. E. 1977. Poison in a monkey's Garden of Eden. Nat. Hist., 86:35–41.

Goodall, J. 1965. Chimpanzees of the Gombe stream reserve, pp. 425–473. *In* I. DeVore (ed.), Primate behavior. Holt, Rinehart and Winston, New York.

Goodwin, G. G. 1954. Mammals of the air, land, and waters of the world, Book 1, pp. 1–680 (vol. 1) and pp. 681–874 (vol. 2). *In* F. Drimmer (ed.), The animal kingdom. Doubleday & Company, Garden City, New York.

Gorman, M. L. 1976. A mechanism for individual recognition by odor in *Herpestes auropunctatus* (Carnivora, Viverridae). Anim. Behav., 24:141–145.

_____, D. B. Nedwell, and R. M. Smith. 1974. An analysis of the anal scent pockets of *Herpestes auropunctatus* (Carnivora: Viverridae). J. Zool., London, 172:389–399.

Gould, E. 1955. The feeding efficiency of insectivorous bats. J. Mamm., 36:399–407.

_____. 1965. Evidence for echolocation in the Tenrecidae of Madagascar. Proc. Amer. Phil. Soc., 109(6):352–360.

_____. 1969. Communication in three genera of shrews (Soricidae): *Suncus*, *Blarina*, and *Cryptotis*. Comm. Behav. Biol., Part A, 3(1):11–31.

_____. 1970. Echolocation and communication in bats, pp. 144–161. *In* B. H. Slaughter and D. W. Walton (eds.), About bats. Southern Methodist Univ. Press, Dallas.

_____. 1971. Studies of maternal-infant communication and development of vocalization in the bats *Myotis* and *Eptesicus*. Comm. Behav. Biol., Part A, 5(5):263–313.

_____, and J. F. Eisenberg. 1966. Notes on the biology of the Tenrecidae. J. Mamm., 47:660–686.

_____, N. C. Negus, and A. Novick. 1964. Evidence for echolocation in shrews. J. Exp. Zool., 156:19–38.

Gould, L. M. 1973. Foreword, pp. xv–xvii. *In* E. H. Colbert, Wandering lands and animals. E. P. Dutton and Company, New York. 323 pp.

Graf, W. 1955. The Roosevelt elk. Port Angeles, Wash.: Port Angeles Evening News. 105 pp.

Grand, T. I., and R. Lorenz. 1968. Functional analysis of the hip joint in *Tarsius bancanus* (Horsefield, 1821) and *Tarsius syrichta* (Linnaeus, 1758). Folia Primat., 9:161–181.

Grant-Taylor, T. L. and Rafter, T. A. 1963. New Zealand natural radiocarbon measurements I-V. Radiocarbon, 5:118–162.

Green, R. G., and C. L. Larson. 1938. A description of shock disease in the snowshoe hare. Amer. J. Hyg., 28:190–212.

_____, _____, and J. F. Bell. 1939. Shock disease as the cause of the periodic decimation of the snowshoe hare. Amer. J. Hyg., 30B:83–102.

Greenwald, G. S. 1956. The reproductive cycle of the field mouse, *Microtus californicus*. J. Mamm., 37:213–222.

_____. 1957. Reproduction in a coastal California population of the field mouse *Microtus californicus*. Univ. California Publ. Zool., 54:421–446.

Griffin, D. R. 1951. Audible and ultrasonic sounds of bats. Experientia, 7:448–453.

————. 1953. Bat sounds under natural conditions with evidence for echolocation of insect prey. J. Exp. Zool., *36*:399–407.

————. 1958. Listening in the dark. Yale Univ. Press, New Haven, Connecticut. 413 pp.

————. 1962. Comparative studies of the orientation sounds of bats. Symp. Zool. Soc. London, *7*:61–72.

————. 1970. Migrations and homing of bats, pp. 233–264. *In* W. A. Wimsatt (ed.), Biology of bats. Academic Press, New York.

————, D. Dunning, D. A. Cahlander, and F. A. Webster, 1962. Correlated orientation sounds and ear movements of horseshoe bats (Part I). Nature, *196*:1185–1186.

————, and R. Galambos. 1940. Obstacle avoidance by flying bats. Anat. Rec., *78*:95.

————, ————. 1941. The sensory basis of obstacle avoidance by flying bats. J. Exp. Zool., *86*:481–506.

————, and H. B. Hitchcock. 1965. Probable 24-year longevity records for *Myotis lucifugus*. J. Mamm., *46*:332.

————, and A. Novick. 1955. Acoustic orientation of Neotropical bats. J. Exp. Zool., *130*:251–300.

————, F. A. Webster, and C. R. Michael. 1960. The echolocation of flying insects by bats. Anim. Behav., *8*:141–154.

Grinnell, A. D. 1963a. The neurophysiology of audition in bats: intensity and frequency parameters. J. Physiol., *167*:38–66.

————. 1963b. The neurophysiology of audition in bats: temporal parameters. J. Physiol., *167*:67–96.

Grinnell, J. 1914a. An account of the mammals and birds of the lower Colorado Valley with especial reference to the distributional problems presented. Univ. California Publ. Zool., *12*:51–294.

————. 1914b. Barriers to distribution as regards to birds and mammals. Amer. Nat., *48*:248–254.

————. 1914c. The Colorado River as a hindrance to the dispersal of species. Univ. California Publ. Zool., *12*:100–107.

————. 1922. A geographical study of the kangaroo rats of California. Univ. California Publ. Zool., *24*:1–124.

————. 1926. Geography and evolution in the pocket gopher. Univ. California Chron., *30*:429–450.

————. 1933. Review of the Recent mammal fauna of California. Univ. California Publ. Zool., *40*:71–234.

————, J. S. Dixon, and J. M. Linsdale. 1937. Fur-bearing mammals of California. 2 vols. Univ. California Press, Berkeley. 777 pp.

————, and T. I. Storer. 1924. Animal life in the Yosemite. Univ. California Press, Berkeley. 752 pp.

Grummon, R. A., and A. Novick. 1963. Obstacle avoidance in the bat *Macrotus mexicanus*. Physiol. Zool., *36*:361–369.

Guilday, J. E. 1958. The prehistoric distribution of the opossum. J. Mamm., *39*:39–43.

Guthrie, M. J. 1933. The reproductive cycles of some cave bats. J. Mamm., *14*:199–216.

Guthrie, R. D., and R. G. Petocz. 1970. Weapon automimicry among mammals. Amer. Nat., *104*:585–588.

Guyton, A. C. 1976. Textbook of medical physiology, 5th ed. W. B. Saunders Company, Philadelphia. 1194 pp.

Hall, E. R. 1946. Mammals of Nevada. Univ. California Press, Berkeley. 710 pp.

————. 1951. American weasels. Univ. Kansas Publ., Mus. Nat. Hist., Vol 4:1–466.

————. 1958. Introduction, Part. II, pp 371–373. *In* C. L. Hubbs (ed.), Zoogeography. Amer. Assoc. Adv. Sci. Publ., 51.

————, and W. W. Dalquest. 1963. The mammals of Veracruz. Univ. Kansas Publ., Mus. Nat. Hist., 14:165–362.

————, and K. R. Kelson. 1959. The mammals of North America. Ronald Press, New York. 2 vols.

Hall, K. R. L. 1965. Behaviour and ecology of the Wild Patas Monkeys, *Erythrocebus patas*, in Uganda. J. Zool. Soc. London, *148*:15–87.

————. 1968. Behaviour and ecology of the Wild Patas monkey, pp. 32–119. *In* P. C. Jay (ed.), Primates, studies in adaptation and variability. Holt, Rinehart and Winston, New York.

————, and I. DeVore. 1965. Baboon social behavior, pp. 53–110. *In* I. Devore (ed.), Primate behavior, Holt, Rinehart and Winston, New York.

————, and G. B. Schaller. 1964. Tool-using behavior of the California sea otter. J. Mamm., *45*:287–298.

Hamilton, W. J., Jr. 1937. The biology of microtine cycles. J. Agric. Res., *54*:779–790.

————1939. American mammals. McGraw-Hill Book Company, Inc. 434 pp.

Hansen, R. M. 1962. Movements and survival of *Thomomys talpoides* in a mima-mound habitat. Ecology, *43*:151–154.

————, and A. L. Ward. 1966. Some relations of pocket gophers to rangelands on Grand Mesa, Colorado, Colo. Agric. Exp. Sta., Tech. bull., 88. 20 pp.

Hanson, D. D. 1971. The food habits and energy dynamics of *Neotoma stephensi*. M. S. Thesis. Northern Arizona Univ., Flagstaff, Arizona.

Hardy, R. 1945. The influence of types of soil upon the local distribution of some mammals in southwestern Utah. Ecol. Monogr., *15*:71–108.

Harrington, J. E. 1976. Discrimination between individuals by scent in *Lemur fulvus*. Anim. Behav., *24*:207–212.

Hart, F. M., and J. A. King. 1966. Distress vocalizations of young in two subspecies of *Peromyscus*. J. Mamm., *47*:287–293.

Hart, J. S. 1956. Seasonal changes in insulation of the fur. Can. J. Zool., 34:53–57.

Hartman, C. G. 1933. On the survival of spermatozoa in the female genital tract of the bat. Quart. Rev. Biol., 8:185–193.

Harvey, M. J., and R. W. Barbour. 1965. Home ranges of *Microtus ochrogaster* as determined by a modified minimum area method. J. Mamm., 46:398–402.

Hart, R. T. 1932. The vertebral columns of ricochetal rodents. Bull. Amer. Mus. Nat. Hist., 58:599–738.

———. 1934. The pangolins and aard-varks collected by the American Museum Congo Expedition. Bull. Amer. Mus. Nat. Hist., 66:643–672.

———. 1936. Hyraxes collected by the American Museum Congo Expedition. Bull. Amer. Mus. Nat. Hist., 72:117–141.

Hayward, J. S., and C. P. Lyman. 1967. Non-shivering heat production during arousal from hibernation and evidence for the contribution of brown fat, pp. 346–355. In K. C. Fisher et al. (eds.), Mammalian hibernation, III. Oliver & Boyd, Edinburgh and London.

———, ———, and C. R. Taylor. 1965. The possible role of brown fat as a source of heat during arousal from hibernation. Ann. N.Y. Acad. Sci., 131:441–446.

Hecht, M. K. 1975. The morphology and relationships of the largest known terrestrial lizard, *Megalania prisca* Owen, from the Pleistocene of Australia. Proc. Roy. Soc. Victoria, 87:239–251.

Hediger, H. 1950. Gefangenschaftsgeburt ein afrikanischen Springhasen. Zool. Gart. Leipzig, 17(5).

Heezen, B. C. 1957. Whales entangled in deep-sea cables. Deep Sea Res., 4:105–115.

Heim de Balsac, H. 1954. Un genre inédit et inattendu de mammifera (Insectivore Tenrecidae) d'Afrique Occidentale. Compt. Rend. Acad. Sci., Paris, 239.

Heinroth-Berger, K. 1959. Beobachtungen an handaufgezogenen Mantelpavianen (*Papio hamadryas* L.), Z. Tierpsychol., 16:706–732.

Heller, H. C., and T. L. Poulson, 1970. Circannian rhythms. II. Endogenous and exogenous factors controlling reproduction and hibernation in chipmunks (*Eutamias*) and ground squirrels (*Spermophilus*). Comp. Biochem. Physiol., 33:357–383.

Hellwing, S. 1971. Maintenance and reproduction in the white-toothed shrew, *Crocidura russula monacha* Thomas. in captivity. Z. Saügetierk, 36:103–113.

Hendrichs, H., and U. Hendrichs. 1971. Dikdik und Elephanten. R. Piper, Munich. 173 pp.

Henshaw, R. E. 1970. Thermoregulation in bats, pp. 188–232. In B. H. Slaughter, and D. W. Walton (eds.), About bats. Southern Methodist Univ. Press, Dallas.

———, and G. E. Folk, Jr. 1966. Relation of thermoregulation to seasonally changing microclimate in two species of bats (*Myotis lucifugus* and *M. sodalis*). Physiol. Zool., 39:223–236.

Henson, O. W., Jr. 1961. Some morphological and functional aspects of certain structures of the middle ear in bats and insectivores. Univ. Kansas Sci. Bull., 42:151–225.

———. 1965. The activity and function of the middle-ear muscles in echo-locating bats. J. Physiol., 180:871–887.

Heppes, J. B. 1958. The white rhinoceros in Uganda. Afr. Wildlife, 12:273–280.

Herald, E. S., R. L. Brownell, Jr., F. L. Frye, E. J. Morris, W. E. Evans, and A. B. Scott. 1969. Blind river dolphin: first side-swimming cetacean. Science, 166:1408–1410.

Herman, L. M., M. F. Peacock, M. P. Yunker, and C. J. Madsen. 1975. Bottlenosed dolphin: double-slit pupil yields equivalent aerial and underwater diurnal acuity. Science, 189:650–652.

Herreid, C. F., II. 1963. Temperature regulation and metabolism in Mexican free-tail bats. Science, 142:1573–1574.

———. 1967. Temperature regulation, temperature preference and tolerance, and metabolism of young and adult free-tail bats. Physiol. Zool., 40:1–22.

Hershkovitz, P. 1969. The Recent mammals of the Neotropical Region: a zoogeographic and ecological review. Quart. Rev. Biol., 44:1–70.

———. 1972. The Recent mammals of the Neotropical Region: a zoogeographic and ecological review, pp. 311–431. In A. Keast, F. C. Erk, and B. Glass (eds.), Evolution, mammals, and southern continents. State Univ. New York Press, Albany.

Hertel, A. 1969. Hydrodynamics of swimming and wave-riding dolphins, pp. 31–63. In H. T. Andersen (ed.), The biology of marine mammals. Academic Press, New York.

Hesse, R., W. C. Allee, and K. P. Schmidt, 1951. Ecological animal geography, 2nd ed. John Wiley & Sons, New York.

Hibbard, C. W., D. E. Ray, D. E. Savage, D. W. Taylor, and J. E. Guilday. 1965. Quaternary mammals of North America, pp. 509–525. In H. E. Wright, Jr. and D. G. Frey (eds.), The Quaternary of the United States. Princeton Univ. Press, Princeton, New Jersey.

———, and G. C. Rinker. 1942. A new bog-lemming (*Synaptomys*) from Meade County, Kansas. Univ. Kansas Sci. Bull., 28:25–35.

Hiiemae, K. 1967. Masticatory function in mammals. J. Dent. Res., 46(2):883–893.

Hilborn, R. 1975. Similarities in dispersal tendency among siblings in four species of voles. Ecology, 56:1221–1225.

Hildebrand, M. 1959. Motions of the running cheetah and horse. J. Mamm., 40:481–496.

———. 1960. How animals run. Sci. Amer., 202(5):148–156.

———. 1962. Walking, running, and jumping. Amer. Zool., 2:151–155.

———. 1965. Symmetrical gaits of horses. Science, 150:701–708.

———. 1974. Analysis of vertebrate structure. John Wiley & Sons, New York. 710 pp.

Hill, J. E. 1974. A new family, genus and species of bat (Mammalia: Chiroptera) from Thailand. Bull. Brit. Mus. Nat. Hist., 27:301–336.

———, and T. D. Carter, 1941. The mammals of Angola, Africa. Bull. Amer. Mus. Nat. Hist., 78:1–211.

Hill, R. W., and J. H. Veghte. 1976. Jackrabbit ears: surface temperatures and vascular responses. Science, *194*:436–438.

Hill, W. C. O., and J. Meester. 1971. Suborder Prosimii, Infraorder Lorisiformes, Part 3.2. In J. Meester and H. W. Setzer (eds.), The mammals of Africa, an identification manual. Smithsonian Institution Press, Washington, D.C.

Hinde, R. A. 1970. Animal behavior: a synthesis of ethology and comparative psychology. 2nd ed. McGraw-Hill Book Company, New York. 876 pp.

Hisaw, F. L. 1924. The absorption of the public symphysis of the pocket gopher, *Geomys bursarius* (Shaw). Amer. Nat., *58*:93–96.

Hock, R. J. 1951. The metabolic rates and body temperatures of bats. Biol. Bull., *101*:289–299.

Hoese, H. D. 1971. Dolphin feeding out of water in a salt marsh. J. Mamm., *52*:222–223.

Hoffmann, R. S. 1958. The role of reproduction and mortality in population fluctuations of voles (*Microtus*). Ecol. Monogr., *28*:79–109.

Holling, C. S. 1959. The components of predation as revealed by a study of small mammal predation of the European pine sawfly. Can. Entomol., *91*:293–320.

———. 1961. Principles of insect predation. Ann. Rev. Entomol., *6*:163–182.

Hooper, E. T. 1952. A systematic review of the harvest mice (genus *Reithrodontomys*) of Latin America. Misc. Publ. Mus. Zool., Univ. Michigan, *77*:1–255.

———. 1968. Anatomy of middle-ear walls and cavities in nine species of microtine rodents. Univ. Michigan Occ. Papers, *657*:1–28.

———, and J. H. Brown. 1968. Foraging and breeding in two sympatric species of Neotropical bats, genus *Noctilio*. J. Mamm., *49*:310–312.

Hopkins, D. M. 1959. Cenozoic history of the Bering land bridge. Science, *129*:1519–1528.

Hopson, J. A. 1966. The origin of the mammalian middle ear. Amer. Zool., *6*:437–450.

———. 1970. The classification of non-therian mammals. J. Mamm., *51*:1–9.

———, and A. W. Crompton. 1969. Origin of mammals, pp. 15–72. In T. Dobzhansky (ed.), Evolutionary biology. Appleton-Century-Crofts, New York.

Hornocker, M. G. 1970a. The American lion. Nat. Hist., *79*:40–49, 68–71.

———. 1970b. An analysis of mountain lion predation upon mule deer and elk in the Idaho Primitive Area. Wildl. Monogr. No. 21, 39 pp.

Horst, R. 1969. Observations on the structure and function of the kidney of the vampire bat (*Desmodus rotundus murinus*), pp. 73–83. In C. C. Hoff and M. L. Riedesel (eds.), Physiological systems in semiarid environments. Univ. New Mexico Press, Albuquerque.

Houlihan, R. T. 1963. The relationship of population density to endocrine and metabolic changes in the California vole, *Microtus californicus*. Univ. Calif. Publ. Zool., *65*:327–362.

Howell, A. B. 1930. Aquatic mammals. Charles C Thomas, Springfield, Illinois.

———. 1944. Speed in animals. Univ. Chicago Press, Chicago. 270 pp.

Howell, D. J. 1974a. Acoustic behavior and feeding in glossophagine bats. J. Mamm., *55*:293–308.

———. 1974b. Bats and pollen: physiological aspects of the syndrome of chiropterophily. Comp. Biochem. Physiol., *48A*:263–276.

———, and D. Burch. 1974. Food habits of some Costa Rican bats. Rev. Biol. Trop., *21*:281–294.

———, and J. Pylka. 1976. Why bats hang upside down: a biomechanical hypothesis. (In manuscript.)

Hudson, J. W. 1962. The role of water in the biology of the antelope ground squirrel. Univ. Calif. Publ. Zool., *64*:1–56.

———. 1965. Temperature regulation and torpidity in the pigmy mouse, *Baiomys taylori*. Physiol. Zool., *38*:243–254.

———. 1973. Torpidity in mammals, pp. 97–165. In Comparative physiology of thermoregulation, Vol. III. Academic Press, New York.

———. 1974. The estrous cycle, reproduction, growth, and development of temperature regulation in the pigmy mouse, *Baiomys taylori*. J. Mamm., *55*:572–588.

———, and T. J. Dawson. 1975. Role of sweating from the tail in the thermal balance of the rat-kangaroo *Potorous tridactylus*. Aust. J. Zool., *23*:453–461.

——— D. R. Deavers, and S. R. Bradley. 1972. A comparative study of temperature regulation in ground squirrels with special reference to the desert species. Symp. Zool. Soc. London, *31*:191–213.

Huey, R. B. 1969. Winter diet of the Peruvian desert fox. Ecology, *50*:1089–1091.

Hughes, R. L. 1974. Morphological studies on implantation in marsupials. J. Reprod. Fertil., *39*:173–186.

Humboldt, A. von, and A. Bonpland. 1852–53. Personal narrative of travels to the equinoctial regions of America during the years 1799–1804. 3 vols. Henry G. Bohn, London.

Humphrey, S. R. 1974. Zoogeography of the nine-banded armadillo (*Dasypus novemcinctus*) in the United States. BioScience, *24*:457–462.

———. 1975. Nursery roosts and community diversity in Nearctic bats. J. Mamm., *56*:321–346.

Huntley, B. 1976. Angola, a situation report. Afr. Wildlife, *30(1)*:10–14.

Hurley, P. M. 1968. The confirmation of continental drift. Sci. Amer., *218(4)*:52–64.

Hutchinson, G. E. 1957. Concluding remarks. Cold Spr. Harb. Symp. Quant. Biol., *22*:415–427.

Ingles, L. G. 1949. Ground water and snow as factors affecting the seasonal distribution of pocket gophers, *Thomomys monticola*. J. Mamm., *30*:343–350.

Irving, L. 1966. Adaptations to cold. Sci. Amer., *214(1)*:94–101.

———. 1969. Temperature regulation in marine mammals, pp. 147–173. *In* H. T. Andersen (ed.), The biology of marine mammals. Academic Press, New York.

———, and J. S. Hart. 1957. The metabolism and insulation of seals as bare-skinned mammals in cold water. Can. J. Zool., *35*:497–511.

———, H. Krog, and M. Monson. 1955. The metabolism of some Alaskan animals in winter and summer. Physiol. Zool., *28*:173–185.

Izawa, K. 1970. Unit groups of chimpanzees and their nomadism in the savannah woodland. Primates, *11*:1–46.

Izawa, K., and J. Itani. 1966. Chimpanzees in Kasakati Basin, Tanganyika: I, ecological study in the rainy season 1963–1964. Kyoto Univ. Afr. Stud., *1*:73–156.

Jaeger, E. C. 1950. The coyote as a seed distributor. J. Mamm., *31*:452–453.

———, and R. A. Mead. 1964. Seasonal changes in body fat, water and basic weight in *Citellus lateralis*, *Eutamios speciosus* and *E. amoenus*. J. Mamm., *45*:359–365.

Jameson, E. W., Jr. 1952. Food of deer mice, *Peromyscus maniculatus* and *P. boylei*, in the northern Sierra Nevada, California. J. Mamm., *33*:50–60.

———. 1953. Reproduction of deer mice (*Peromyscus maniculatus* and *P. boylei*) in the Sierra Nevada, California. J. Mamm., *34*:44–58.

Jarman, M. V. 1970. Attachment to home area in impala. E. Afr. Wildl. J., *8*:198–200.

Jarman, R. J. 1974. The social organization of antelope in relation to their ecology. Behavior, *58*:215–267.

———, and M. V. Jarman. 1974. Impala behavior and its relevance to management, pp. 871–881. *In* V. Geist and F. R. Walther (eds.), The behavior of ungulates and its relation to management. International Union for Conservation of Nature and Natural Resources, Morges, Switzerland.

Jarvis, J. U. M. 1973. The structure of a population of mole-rats, *Tachyoryctes splendens* (Rodentia: Rhizomyidae). J. Zool., London, *171*:1–14.

———, and J. B. Sale. 1971. Burrowing and burrow patterns of East African mole-rats *Tachyoryctes*, *Heliophobius* and *Heterocephalus*. J. Zool., London, *163*:451–479.

Jay, P. C. (ed.). 1968. Primates, studies in adaptation and variability. Holt, Rhinehart and Winston, New York. 529 pp.

Jen, P. H.-S., and N. Suga. 1976. Coordinated activities of middle-ear and laryngeal muscles in echolocating bats. Science, *191*:950–952.

Jenkins, F. A., Jr. 1970. Limb movement in a monotreme (*Tachyglossus aculeatus*): a cineradiographic analysis. Science, *168*:1473–1475.

———, 1971. Limb posture and locomotion in the Virginia opossum (*Didelphis marsupialis*) and in other non-cursorial mammals. J. Zool., London, *165*:303–315.

———, and F. R. Parrington. 1976. The postcranial skeletons of the Triassic mammals *Eozostrodon*, *Megazostrodon*, and *Erythrotherium*. Phil. Trans. Roy. Soc. London, in press.

Jenkins, H. O. 1948. A population study of the meadow mice (*Microtus*) in three Sierra Nevada meadows. Proc. Calif. Acad. Sci., ser. 4, *26*:43–67.

Jepsen, G. L. 1966. Early Eocene bat from Wyoming. Science, *154*:1333–1339.

———. 1970. Bat origins and evolution, pp. 1–64. *In* W. A. Wimsatt (ed.), Biology of bats. Academic Press, New York.

———, and M. O. Woodburne. 1969. Paleocene hyracothere from Polecat Bench Formation, Wyoming. Science, *164*:543–547.

Johannessen, C. L., and J. A. Harder. 1960. Sustained swimming speeds of dolphins. Science, *132*:1550–1551.

Johansen, K., and J. Krog. 1959. Diurnal body temperature variations in the birch mouse, *Sicista betulina*. Amer. J. Physiol., *196*:1200.

Johnson, D. R. 1961. The food habits of rodents on rangelands of southern Idaho. Ecology, *42*:407–410.

———. 1964. Effects of range treatment with 2,4-D on food habits of rodents. Ecology, *45*:241–249.

Jolly, A. 1966. Lemur behavior: a Madagascar field study. Univ. Chicago Press, Chicago.

———. 1972. Troop continuity and troop spacing in *Propithecus verreauxi* and *Lemur catta* at Berenty (Madagascar). Folia Primat., *17*:335–362.

Jones, C. 1967. Growth, development, and wing loading in the evening bat, *Nycticeius humeralis* (Rafinesque). J. Mamm., *48*:1–19.

Jones, F. W. 1923. The mammals of South Australia. Part I: the monotremes and carnivorous marsupials, pp. 1–131. Government Printer, Adelaide.

———. 1924. The mammals of South Australia. Part II: the bandicoots and the herbivorous marsupials, pp. 133–270. Government Printer, Adelaide.

———. 1925. The mammals of South Australia. Part III: the monodelphia pp. 271–458. Government Printer, Adelaide.

Jones, J. K., Jr. 1964. Distribution and taxonomy of mammals of Nebraska. Univ. Kansas Publ., Mus. Nat. Hist., *16*:1–356.

———, and R. R. Johnson. 1967. Sirenians, pp. 367–373. *In* S. Anderson and J. K. Jones, Jr. (eds.), Recent mammals of the world. Ronald Press, New York.

Jones, R. 1968. The biographical background to the arrival of man in Australia and Tasmania. Arch. Phys. Anthrop. Oceania, III*(3)*:186–215.

Kahmann, H., and K. Ostermann. 1951. Wahrnehmen und Hervorbringen hoher Tone bei kleiner Saugetieren. Experientia, *7*:268–269.

Kalela, O. 1957. Regulation of reproduction rate in subarctic populations of the vole *Clethrionomys rufocanus* (Sund.). Ann. Acad. Sci. Fennicae, Ser. A, *4(34)*:1–60.

———. 1961. Seasonal change of habitat in the Norwegian lemming *Lemmus lemmus* L. Ann. Acad. Sci. Fennicae, Ser. A, *4(55)*:1–72.

———. 1962. On the fluctuations in the numbers of arctic and boreal small rodents as a problem of production biology. Ann. Acad. Sci. Fennicae, Ser. A, *4(66)*:1–38.

Kangas, E. 1949. On the damage to the forests caused by the moose and its significance in the economy of the forest. Eripainos: Suomen Riista, vol. 4, pp. 62–90 (English summary, pp. 88–90.)

Kanwisher, J., and H. Leivestad. 1957. Thermal regulation in whales. Norsk Hvalfangst-tid., *1*:1–5.

———, and G. Sundnes. 1966. Thermal regulation in cetaceans, pp. 397–409. *In* K. S. Norris (ed.), Whales, dolphins and porpoises. Univ. California Press, Berkeley.

Kaufman, J. H. 1974a. Habitat use and social organization of nine sympatric species of macropod marsupials. J. Mamm., *55*:66–80.

———. 1974b. Social ethology of the whiptail wallaby, *Macropus parryi*, in northeastern New South Wales. Anim. Behav., *22*:281–369.

Keast, A. 1972. Australian mammals: zoogeography and evolution, pp. 195–246. *In* A. Keast, F. C. Erk, and B. Glass (eds.), Evolution, mammals, and southern continents. State Univ. New York Press, Albany.

Kellogg, W. N. 1961. Porpoises and sonar. Univ. Chicago Press, Chicago. 177 pp.

———, R. Kohler, and H. N. Morris. 1953. Porpoise sounds as sonar signals. Science, *117*:239–243.

Kelsall, J. P. 1970. Migration of the barren-ground caribou. Nat. Hist., *79*:98–106.

Kenagy, G. J. 1972. Saltbush leaves: excision of hypersaline tissues by a kangaroo rat. Science, *178*:1094–1096.

———. 1973. Daily and seasonal patterns of activity and energetics in a heteromyid rodent community. Ecology, *54*:1201–1219.

Kendeigh, S. C. 1961. Animal ecology. Prentice-Hall, Inc., Englewood Cliffs, New Jersey. 468 pp.

Kennerly, T. E., Jr. 1964. Microenvironmental conditions of the pocket gopher burrow. Texas J. Sci., *14(4)*:397–441.

Kennerly, T. R. 1971. Personal communication.

Kermack, K. A. 1963. The cranial structure of triconodonts. Phil. Trans. Roy. Soc. London, Ser. B, *246*:83–103.

———. and Z. Kielan-Jaworowska. 1971. Therian and non-therian mammals. pp. 103–115. *In* D.M. Kermack and K. A. Kermack (eds.), Early mammals. Academic Press, London.

———, and F. Musset. 1958. The jaw articulation of the docodonta and the classification of Mesozoic mammals. Proc. Roy. Soc. London, B, *149*:204–215.

Kevan, P. G. 1975. Sun-tracking solar furnaces in high arctic flowers: significance for pollination and insects. Science, *189*:723–726.

Kielan-Jaworowska, Z. 1969. Preliminary data on the Upper Cretaceous eutherian mammals from Bayn Dzak Gobi Desert. Palaeontol. Polon., *19*:171–191.

———. 1975. Late Cretaceous mammals and dinosaurs from the Gobi Desert. Amer. Sci., *63*:150–159.

King, J. A. 1955. Social behavior, social organization and population dynamics in a black-tail prairie dog town in the Black Hills of South Dakota. Contrib. Lab. Vert. Biol. Univ. Michigan, No. 67:1–123.

———. 1959. The social behavior of prairie dogs. Sci. Amer., *201(4)*:128–140.

———. (ed.). 1968. Biology of *Peromyscus* (Rodentia). Spec. Publ. No. 2, Amer. Soc. Mammal. 593 pp.

Kingdon, J. 1971. East African mammals. Vol. I. Academic Press, New York. 446 pp.

———. 1974a. East African mammals. Vol. IIA. Academic Press, New York, 341 pp.

———. 1974b. East African mammals. Vol. IIB. Academic Press, New York. pp. 342–704.

Kinnear, J. E., K. G. Purohit, and A. R. Main. 1968. The ability of the tammar wallaby *(Macropus eugenii*, Marsupialia) to drink seawater. Comp. Biochem. Physiol., *25*:761–782.

Kitchen, D. W. 1974. Social behavior and ecology of the pronghorn. Wildl. Monogr., *38*:1–96.

Klingel, H. 1967. Soziale Organisation und Verhalten freilebender Steppenzebras. Z. Tierpsychol., *24*:580–624.

Klingener, D. 1964. The comparative myology of four dipodoid rodents (Genera *Zapus, Napaeozapus, Sicista* and *Jaculus*). Misc. Publ. Mus. Zool., Univ. Michigan, *124*:1–100.

Knappe, H. 1964. Zur Funktion des Jacobsonschen Organs *(Organon vomeronasale Jacobsoni)*. Zool. Gart. (Leipzig), *28*:188–194.

Knudsen, V. O. 1931. The effect of humidity upon the absorption of sound in a room, and determination of the coefficients of absorption of sound in air. J. Acoustical Soc. Am., *3*:126–138.

———. 1935. Atmospheric acoustics and the weather. Sci. Mon., *40*:485–486.

Komarek, E. V. 1932. Notes on mammals of Menominee Indian Reservation, Wisconsin, J. Mamm., 13:203–209.

Kooyman, G. L., and H. T. Andersen, 1969. Deep diving, pp. 65–94. In H. T. Andersen (ed.), The biology of marine mammals. Academic Press, New York.

Koshkina, T. V. 1965. Population density and its importance in regulating the abundance of the red vole. (Russian translated by W. A. Fuller.) Bull. Moscow Soc. Nat. Biol., 70:5–19.

———, and A. S. Kholansky. 1962. Reproduction of the Norwegian lemming (Lemmus lemmus L.) on the Kola Penninsula. (Russian translated by W. A. Fuller.) Zool. Zh., 41:604–615.

Kramer, M. O. 1960. J. Am. Soc. Naval Engrs. pp. 25–33.

Krassilov, V. 1973. Mesozoic plants and the problem of angiosperm ancestry. Lethaia, 6:163–178.

Krebs, C. J. 1963. Lemming cycle at Baker Lake, Canada during 1959–62. Science, 146:1559–1560.

———. 1964a. Cyclic variation in skull-body regressions of lemmings. Can. J. Zool., 42:631–643.

———. 1964b. The lemming cycle at Baker Lake, Northwest Territories, during 1959–62. Arctic Inst. N. Amer. Tech. Paper No. 15.

———. 1966. Demographic changes in fluctuating populations of Microtus californicus. Ecol. Monogr., 36:239–273.

———. 1970. Microtus population biology: behavioral changes associated with the population cycle in M. ochrogaster and M. pennsylvanicus. Ecology, 51:34–52.

———, and K. T. DeLong. 1965. A Microtus population with supplemental food. J. Mamm., 46:566–573.

———, M. S. Gains, B. L. Keller, J. H. Myers, and R. H. Tamarin. 1973. Population cycles in small rodents. Science, 179:35–41.

———, B. L. Keller, and R. Tamarin. 1969. Microtus population biology: demographic changes in fluctuating populations of M. ochrogaster and M. pennsylvanicus in southern Indiana. Ecology, 50:587–607.

———, and J. H. Myers. 1974. Population cycles in small mammals, pp. 267–399. In MacFadyen (ed.), Advances in ecological research. Academic Press, New York.

Krebs, H. A. 1950. Body size and tissue metabolism. Biochem. Biophys. Acta, 4:249–269.

Krishnan, R. S., and J. C. Daniel, 1967. "Blastokinin"–an inducer and regulator of blastocyst development in the rabbit uterus. Science, 158:490–492.

Kruger, L. 1966. Specialized features of the cetacean brain, pp. 232–254. In K. S. Norris (ed.), Whales, dolphins and porpoises. Univ. California Press, Berkeley.

Krumrey, W. A., and I. O. Buss. 1968. Age estimation, growth, and relationships between body dimensions of the female African elephant. J. Mamm., 49:22–31.

Kruuk, H. 1966. Clan-system and feeding habits of spotted hyaenas (Crocuta crocuta Erxleben). Nature, Lond., 209:1257–1258.

———. 1970. Interactions between populations of spotted hyaenas (Crocuta crocuta) and their prey species, pp. 359–374. In A. Watson (ed.), Animal populations in relation to their food resources. Blackwell Scientific Publications, Oxford.

———. 1972. The spotted hyena: a study of predation and social behavior. Univ. Chicago Press, Chicago. 335 pp.

———, and H. Van Lawick. 1968. Hyaenas, the hunters nobody knows. National Geographic, 134(1):44–57.

———, and W. A. Sands. 1972. The aardwolf (Proteles cristatus Sparrman, 1783) as predator of termites. E. Afr. Wildl. J., 10:211–227.

Krzanowski, A. 1960. Investigations of flights of Polish bats, mainly Myotis myotis. Acta Theriol., 4:175–184.

———. 1961. Weight dynamics of bats wintering in the cave at Pulway (Poland). Acta Theriol., 4:249–264.

———. 1964. Three long flights by bats. J. Mamm., 45:152.

Kühme, W. 1965. Freilandstudien zur Soziologie des Hyäenenhundes (Lycaon pictus lupinus Thomas 1902). Z. Tierpsychol., 225:495–541.

———. 1966. Beobachtungen zur Soziologie des Löwens in der Serengeti-Steppe Ostafrikas. Z. Saugetierk., 31:205–213.

Kulzer, E. 1956. Flughunde erzeugen Oreintierung durch Zungenschlag. Naturwiss., 43:117–118.

———. 1958. Untersuchungen über die Biologie von Flughunden der Gattung Rousettus Gray. Z. Morph. Ökol Biere., 47:374–402.

———. 1960. Physiologische und morphologische Untersuchungen über die Erzeugung der Orientierungslaute von Flughunden der Gattung Rousettus. Z. Vergl. Physiol., 43:231–268.

———. 1961. Über die Biologie der Nil-Flughunde (Rousettus aegyptiacus). Natur Volk., 91:219–228.

———. 1963. Temperaturregulation bei Flughunden der Gattung Rousettus Gray. Z. Vergl. Physiol., 46:595–618.

———. 1965. Temperaturregulation bei Fledermäusen (Chiroptera) aus berschiedenen Klimazonen. Z. Vergl. Physiol., 50:1–34.

Kummer, H. 1968a. Two variations in the social organization of baboons, pp. 293–312. In P. C. Jay (ed.), Primates, studies in adaptation and variability. Holt, Rinehart and Winston, New York.

———. 1968b. Social organization of hamadryas baboons. Univ. Chicago Press, Chicago. 189 pp.

Kunz, T. H. 1971. Ecology of the cave bat, Myotis velifer, in southcentral Kansas and northwestern Oklahoma. Unpublished Ph.D. dissertation, Univ. Kansas, Lawrence. 148 pp.

_____. 1973a. Population studies of the cave bat *(Myotis velifer)*: reproduction, growth, and development. Occas. Papers Mus. Nat. Hist., Univ. Kansas, *15*:1–43.

_____. 1973b. Resource utilization: temporal and spatial components of bat activity in central Iowa. J. Mamm., *54*:14–32.

_____. 1974. Feeding ecology of a temperate insectivorous bat *(Myotis velifer)*. Ecology, *55*:693–711.

Kurten, B. 1969. Continental drift and evolution. Sci. Amer., *220(3)*:54–64.

Lack, D. 1948. The significance of litter size. J. Anim. Ecol., *17*:45–50.

_____. 1954a. The natural regulation of animal numbers. Oxford Univ. Press, London. 343 pp.

_____. 1954b. Cyclic mortality. J. Wildl. Manage., *18*:25–37.

_____. 1966. Population studies of birds. Clarendon Press, Oxford. 341 pp.

Lackey, J. A. 1967. Growth and development of *Dipodomys stephensi*. J. Mamm., *48*:624–632.

Lamprey, H. F., G. Halevy, and S. Makacha. 1974. Interactions between Acacia, bruchid seed beetles and large herbivores. E. Afr. Wildl. J., *12*:81–85.

Landry, S. O. 1957. The interrelationships of the New World and Old World histricomorph rodents. Univ. California Publ. Zool., *56*:1–118.

Lang, H., and J. P. Chapin. 1917. The American Museum Congo Expedition Collection of Bats III: Field notes. Bull. Amer. Mus. Nat. Hist., *37*.

Lang, T. G. 1966. Hydrodynamic analysis of cetacean performance, pp. 410–432. *In* K. S. Norris (ed.), Wales, dolphins and porpoises. Univ. California Press, Berkeley.

Langworthy, M., and R. Horst. 1971. Reproductive behavior in a captive colony of *Molossus ater*. In manuscript.

Lavocat, R. 1962. Réflexions sur l'origine et la structure du groupe des rongeurs, pp. 287–299. *In* Problèmes actuels de paleontologie. Colloq. Int. Cent. Nat. Rech. Sci., Paris.

_____. 1973. Les rongeurs du Miocène d'Afrique Orientale. I, Miocene inférieur. Trav. Mém. Inst. E.P.A.E. Monpellier, *1*:1–284.

_____. 1974. What is an hystricomorph, pp. 7–20. *In* I. W. Rowlands, and B. Weir (eds.), The biology of hystricomorph rodents. Symp. Zool. Soc. London. Academic Press, New York.

Lawlor, T. E. 1973. Aerodynamic characteristics of some Neotropical bats. J. Mamm., *54*:71–78.

Lawrence, B., and A. Novick. 1963. Behavior as a taxonomic clue: relationships of *Lissonycteris* (Chiroptera). Mus. Comp. Zool., *184*:1–16.

Laws, R. M. 1953. The elephant seal *(Mirounga leonina, Linn.)* I. Growth and age. Falkland Is. Depend. Surv. Sci. Repts., *8*:1–62.

_____. 1970. Elephants as agents of habitat and landscape change in East Africa. Oikos, *21*:1–15.

_____, and I. S. C. Parker. 1968. Recent studies on elephant populations in East Africa. Symp. Zool. Soc. London, *21*:319–359.

Layne, J. N. 1958. Observations on freshwater dolphins in the upper Amazon. J. Mamm., *39*:1–22.

_____. 1965. Observations on marine mammals in Florida waters. Bull. Florida State Mus., *9*:131–181.

_____. 1968. Ontogeny, pp. 148–253. *In* J. A. King (ed.), Biology of *Peromyscus* (Rodentia). Spec. Publ. No. 2, Amer. Soc. Mamm.

Lear, J. 1970. The bones on Coalsack Bluff: a story of drifting continents. Sat. Rev., *53(6)*:46–51.

Le Boeuf, B. J., and R. S. Peterson. 1969. Social status and mating activity in elephant seals. Science, *163*:91–93.

Lechleitner, R. R. 1958a. Certain aspects of behavior of the black-tailed jackrabbit. Amer. Midl. Nat., *60*:145–155.

_____. 1958b. Movements, density and mortality in a black-tailed jackrabbit population. J. Wildl. Manage., *22*:371–384.

_____. 1959. Sex ratio, age classes and reproduction of the black-tailed jackrabbit. J. Mamm., *40*:63–81.

_____, J. V. Tileston, and L. Kartman. 1962. Die-off of a Gunnison's prairie dog colony in central Colorado. I. Ecological observations and description of the epizootic. Zoonoses Res., *1*:185–199.

Lee, A. K. 1963. The adaptations to arid environments in wood rats of the genus *Neotoma*. Univ. Calif. Publ. Zool., *64*:57–96.

Leitner, P. 1966. Body temperature, oxygen consumption, heart rate and shivering in the California mastiff bat, *Eumops perotis*. Comp. Biochem. Physiol., *19*:431–443.

_____, and J. E. Nelson, 1967. Body temperature, oxygen consumption and heart rate in the Australian false vampire bat, *Macroderma gigas*. Comp. Biochem. Physiol., *21*:65–74.

Lenfant, C. 1969. Physiological propterties of blood marine mammals, pp. 95–115. *In* H. T. Andersen (ed.), The biology of marine mammals. Academic Press, New York.

Leopold, A. S., T. Riney, R. McCain, and L. Tevis, Jr. 1951. The jawbone deer herd. California Div. Fish and Game, Game Bull., *4*:1–139.

_____, L. K. Sowls, and D. L. Spencer. 1947. A survey of overpopulated deer ranged in the United States. J. Wildl. Manage., *11*:162–177.

Leyhausen, P. 1956. Verhaltensstudien an Katzen. Z. Tierpsychol., *2*:1–120.

_____. 1964. The communal organization of solitary animals. Symp. Zool. Soc. London, *14*:249–263.

Lidicker, W. Z., Jr. 1968. A phylogeny of New Guinea rodent genera based on phallic morphology. J. Mamm., *49*:609–643.

_____. 1973. Regulation of numbers in an island population of the California vole; a problem in community dynamics. Ecol. Monogr., 43:271–302.

_____, and P. K. Anderson. 1962. Colonization of an island by *Microtus californicus*, analyzed on the basis of runway transects. J. Anim. Ecol., 31:503–517.

Lillegraven, J. A. 1969. Latest Cretaceous mammals of upper part of Edmonton Formation of Alberta, Canada, and review of marsupial-placental dichotomy in mammalian evolution. Univ. Kansas Paleontol. Contrib., 50:1–122.

_____. 1974. Biogeographical considerations of the marsupial–placental dichotomy. In R. F. Johnston (ed.), Annual review of ecology and systematics, Vol. 5, Annual Reviews, Inc., Palo Alto.

_____. 1976. Biological considerations of the marsupial-placental dichotomy. Evolution, 29:707–722.

Lilly, J. C. 1961. Man and dolphin. Doubleday, New York. 191 pp.

_____. 1962. Vocal behavior of the bottle-nosed dolphin. Proc. Amer. Phil. Soc., 106:520–529.

_____. 1963. Distress call of the bottle-nosed dolphin: stimuli and evoked behavioral responses. Science, 139:116–118.

_____. 1967. Mind of the dolphin: a nonhuman intelligence. Doubleday, New York. 286 pp.

Lindlöf, B., E. Lindström, and A. Pehrson. 1974. Nutrient content in relation to food preferred by mountain hare. J. Wildl. Manage., 38:875–879.

Linsdale, J. M. 1946. The California ground squirrel. Univ. California Press, Berkeley. 475 pp.

Linzey, D. W., and A. V. Linzey. 1967. Maturational and seasonal molts in the golden mouse, *Ochrotomys nuttalli*. J. Mamm., 48:236–241.

Litchfield, C., A. J. Greenberg, D. K. Caldwell, M. C. Caldwell, J. C. Sipos, and R. G. Ackman. 1975. Comparative lipid patterns in acoustical and nonacoustical fatty tissues of dolphins, porpoises and toothed whales. Comp. Biochem. Physiol., 508:591–597.

Lockie, J. D. 1959. Estimation of the food of foxes. J. Wildl. Manage., 23:224–227.

_____. 1966. Territory in small carnivores. Symp. Zool. Soc. London, 18:143–165.

Long, A., R. M. Hansen, and P. S. Martin. 1974. Extinction of the Shasta ground sloth. Geol. Soc. Amer. Bull., 85:1843–1848.

Lorenz, K. 1950. The comparative method of studying innate behavior patterns. Symp. Soc. Exp. Biol., 4:229–269.

_____. 1963. Das sogenannte Böse. G. Borotha-Schoeler, Vienna. (English version, 1966, On aggression. Methuen, London.)

Louch, C. D. 1958. Adrenocortical activity in two meadow vole populations. J. Mamm., 39:109–116.

Luckens, M. M., and W. H. Davis. 1964. Bats: sensitivity to DDT. Science, 146:948.

Lund, R. D., and J. S. Lund. 1965. The visual system of the mole, *Talpa europaea*. Exp. Neurol., 13:302–316.

Lyman, C. P. 1954. Activity, food consumption and hoarding in hibernators. J. Mamm., 35:545–552.

_____. 1970. Thermoregulation and metabolism in bats, pp. 301–330. In W. A. Wimsatt (ed.), Biology of bats, vol. 1. Academic Press, New York.

_____, and W. A. Wimsatt. 1966. Temperature regulation in the vampire bat, *Desmodus rotundus*. Physiol. Zool., 39:101–109.

MacArthur, R. H. 1955. Fluctuations of animal populations and a measure of community stability. Ecology, 36:533–536.

_____, and E. R. Pianka. 1966. On optimal use of a patchy environment. Amer. Nat., 100:603–609.

MacKay, M. R. 1970. Lepidoptera in Cretaceous amber. Science, 167:379–380.

MacLulich, D. A. 1937. Fluctuations in the numbers of the varying hare (*Lepus americanus*). Univ. Toronto Studies, Biol. Ser., No. 43.

MacMillen, R. E. 1964a. Population ecology, water relations, and social behavior of a southern California semidesert rodent fauna. Univ. Calif. Publ. Zool., 71:1–66.

_____. 1964b. Water economy and salt balance in the western harvest mouse, *Reithrodontomys megalotis*. Physiol. Zool., 37(1):45–56.

_____. 1965. Aestivation in the cactus mouse, *Peromyscus eremicus*. Comp. Biochem. Physiol., 16:227–248.

_____. 1972. Water economy of nocturnal desert rodents. In G. M. O. Maloiy (ed.), Comparative physiology of desert animals. Symp. Zool. Soc. London. Academic Press, New York.

_____, and A. K. Lee. 1967. Australian desert mice: independence of exogenous water. Science, 158(3799):383–385.

_____, _____. 1969. Water metabolism of Australian hopping mice. Comp. Biochem. Physiol., 28:493–514.

_____, _____. 1970. Energy metabolism and pulmocutaneous water loss of Australian hopping mice. Comp. Biochem. Physiol., 35:355–369.

_____, and J. E. Nelson. 1969. Bioenergetics and body size in dasyurid marsupials. Amer. J. Physiol., 217:1246–1251.

Maglio, V. J. 1973. Origin and evolution of the Elephantidae. Trans. Amer. Phil. Soc., 63:1–149.

Maher, W. J. 1967. Predation by weasels on a winter population of lemmings, Banks Island, Northwest Territories. Can. Field Nat., 81:248–250.

_____. 1970. The pomarine jaeger as a brown lemming predator in northern Alaska. Wilson Bull., 82:130–157.

Maloiy, G. M. O. 1973. The water metabolism of a small East African antelope: the dik-dik. Proc. Roy. Soc. London, B., *184*:167–178.

Marler, P. R. 1965. Communication in monkeys and apes, pp. 544–584. *In* I. DeVore (ed.), Primate behavior. Holt, Rinehart and Winston, New York.

———, and W. J. Hamilton III. 1966. Mechanisms of animal behavior. John Wiley & Sons, Inc., New York. 771 pp.

Marshall L. G. 1972. A study of the peramelid tarsus. Aust. Mammal., *1*:67.

———. 1974. Why kangaroos hop. Nature, Lond., *248*:174–176.

———. 1976. Evolution of the Thylacosmilidae, extinct saber-tooth marsupials of South America. PaleoBios., *23*:1–30.

———. 1977a. Evolution of the Borhyaenidae, extinct South American predaceous marsupials. Univ. Calif. Publ. Geol. Sci. In press.

———. 1977b. A new species of *Lycopsis* (Borhyaenidae: Marsupialia) from the La Venta fauna (Miocene) of Colombia, South America. J. Paleont., *51(3)*. In press.

———, and G. J. Weisenberger. 1971. A new dwarf shrew locality for Arizona. Plateau, *43*:132–137.

Martin, E. P. 1956. A population study of the prairie vole *(Microtus ochrogaster)* in northeastern Kansas. Univ. Kansas Publ. Mus. Nat. Hist., *8*:361–416.

Martin, P. 1971. Movements and activities of the mountain beaver *(Aplodontia rufa)*. J. Mamm., *52*:717–723.

Martin, R. D. 1973. A review of the behavior and ecology of the lesser mouse lemur *(Microcebus murinus*, J. F. Miller 1777), pp. 1–68. *In* R. P. Michael, and J. H. Crook (eds.), Comparative ecology and behavior of primates.

Martinsen, D. L. 1968. Temporal patterns in the home ranges of chipmunks. J. Mamm., *49*:83–91.

———. 1969. Energetics and activity patterns of short-tailed shrews *(Blarina)* on restricted diets. Ecology, *50*:505–510.

Matschie, P. 1899. Beitrage zur Kenmtnis von *Hypsignathus monstrosus* Allen. Sitz. Ber. Ges. Naturf. Freunde, Berlin.

Maynard Smith, J. 1976. Evolution and theory of games. Amer. Sci., *64*:41–45.

Mayr, E. 1942. Systematics and the origin of species. Columbia Univ. Press, New York. 334 pp.

———. 1963. Animal species and evolution. Harvard Univ. Press, Cambridge, Massachusetts. 797 pp.

McCabe, T. T., and B. D. Blanchard. 1950. Three species of *Peromyscus*. Rood Associates, Santa Barbara, California. 136 pp.

McCarley, H. 1959. The effect of flooding on a marked population of *Peromyscus*. J. Mamm., *40*:57–63.

McCullough, D. R. 1969. The tule elk, its history, behavior, and ecology. Univ. Calif. Publ. Zool., *88*:1–209.

McKenna, M. C. 1972. Possible biological consequences of plate tectonics. BioScience, *22*:519–525.

———. 1975. Fossil mammals and Early Eocene North Atlantic land continuity. Ann. Missouri Bot. Gard., *62*:335–353.

McLaren, A. 1970. The fate of the zona pellucida in mice. J. Embryol. Exp. Morph., *23*:1–19.

McLean, D. C. 1944. The prong-horned antelope in California. Bureau Game Cons., Calif. Div. Fish Game, San Francisco, *30(4)*:221–241.

McNab, B. K. 1966. The metabolism of fossorial rodents; a study of convergence. Ecology, *47*:712–733.

McNaughton, S. J. 1976. Serengeti migratory wildebeest: facilitation of energy flow by grazine. Science, *191*:92–94.

Mead, R. A. 1968. Reproduction in western forms of the spotted skunk (genus *Spilogale)*. J. Mamm., *49*:373–390.

Mech, L. D. 1966. The wolves of Isle Royale. U.S. Nat. Park Serv., Fauna ser. 7.210 pp.

Menaker, M. 1961. The free-running period of the bat clock; seasonal variations at low body temperature. J. Cell Comp. Physiol., *57*:81–86.

Menard, H. W. 1969. The deep-ocean floor. Sci. Amer., *221*:126–142.

Merriam, C. H. 1894. Laws of temperature control of the geographic distribution of terrestrial animals and plants. Nat. Geogr., *6*:229–238.

———. 1899. Life zones and crop zones of the United States. Bull. U.S. Biol. Surv., *10*:1–79.

Merrilees, D. 1968. Man the destroyer: Late Quaternary changes in the Australian marsupial fauna. J. Roy. Soc. West. Aust., *51*:1–24.

Michael, R. P., E. B. Keverne, and R. W. Bonsall, 1971. Pheromones: isolation of male sex attractants from a female primate. Science, *172*:964–966.

Miller, G. S., Jr. 1907. The families and genera of bats. Bull. U.S. Nat. Mus., *57*. 282 pp.

Miller, R. S. 1964. Ecology and distribution of pocket gophers (Geomyidae) in Colorado. Ecology, *45*:256–272.

———. 1967. Pattern and process in competition. Adv. Ecol. Res., *4*:1–74.

———. 1969. Competition and species diversity. Brookhaven Symp. Biol., *22*:63–70.

Misonne, X. 1959. Analyse zoogéographique des mammifères de l'Iran. Bruxelles, Inst. Royal Sci. Nat. Belgique, Mémoires, 2me sér. 59. 157 pp.

Mizuhara, H. 1957. The Japanese monkey, its social structure. Kyota: San-ichi-syobo (in Japanese).

Mobius, K. 1877. Die Auster und die Austernwirtschaft. Berlin. pp. 22, 35, 436, 508. (Transl., 1880, The oyster and oyster culture, Rept. U.S. Fish. Comm., *1880*:683–751.

Mohr, E. 1941. Schwanzverlust und Schwanzregeneration bei Nagetieren. Zool. Anzeiger, *135*:49–65.

Mohres, F. P. 1953. Über die Ultraschallorientierung der Hufeisennasen (Chiroptera—Rhinolophidae). Z. Vergl. Physiol., *34*:547–588.

———. 1966. Communicative characters of sonar signals in bats, pp. 939–945. In R. G. Busnel (ed.), Animal sonar systems, biology and bionics, Tome II. Laboratoire de Physiologie Acoustique, Paris.

———, and E. Kulzer, 1956. Über die Orientierung der Flughunde (Chiroptera—Pteropodidae). Z. Vergl. Physiol., 38:1–29.

———, and G. Neuweiler. 1966. Ultrasonic orientation in megadermid bats, pp. 115–128. In R. G. Busnel (ed.), Animal sonar systems, biology and bionics, Tome I. Laboratoire de Physiologie Acoustique, Paris.

Mooser, O., and W. W. Dalquest. 1975. Pleistocene mammals from Aguascalientes, Central Mexico. J. Mamm., 56:781–820.

Morhardt, J. E. 1970. Body temperatures of white-footed mice (Peromyscus sp.) during daily torpor. Comp. Biochem. Physiol., 33:423–439.

———, and D. M. Gates. 1974. Energy-exchange analysis of the Belding ground squirrel and its habitat. Ecol. Monogr., 44:17–44.

Morrison, P. 1959. Body temperatures in some Australian mammals. I. Chiroptera. Biol. Bull., 116:484–497.

———. 1962. Body temperatures in some Australian mammals. III. Cetacea (Megaptera). Biol. Bull., 123:154–169.

———, and B. K. McNab. 1962. Daily torpor in a Brazilian murine opossum (Marmosa). Comp. Biochem. Physiol., 6:57–68.

———, ———. 1967. Temperature regulation in some Brazilian phyllostomid bats. Comp. Biochem. Physiol., 21:207–221.

———, and F. A. Ryser, 1952. Weight and body temperature in mammals. Science, 116:231–232.

———, ———, and A. R. Dawe. 1959. Studies on the physiology of the masked shrew Sorex cinereus. Physiol. Zool., 32:256–271.

Moynihan, M. H. 1966. Communication in the titi monkey, Callicebus. J. Zool., London, 150:77–127.

Muller, J. 1970. Palynological evidence on early differentiation of angiosperms. Biol. Rev., 45:417–450.

Müller-Schwarze, D. 1971. Pheromones in the black-tailed deer (Odocoileus hemionus columbianus). Anim. Behav., 19:141–152.

Murdoch, H. W., and A. Oaten. 1975. Predation and population stability, pp. 1–125. In A. MacFadyen (ed.), Advances in ecological research. Academic Press, New York.

Murie, A. 1940. Ecology of the coyote in the Yellowstone, U.S. Dept. Int. Nat. Park Serv., Fauna ser. 4. 206 pp.

———. 1944. The wolves of Mount McKinley, U.S. Dept. Int., Nat. Park Serv., Fauna ser. 5. 238 pp.

Myers, G. T., and T. A. Vaughan. 1964. Food habits of the plains pocket gopher in eastern Colorado. J. Mamm., 45:588–598.

Myers, J. H., and C. J. Krebs. 1971. Genetic, behavioral, and reproductive attributes of dispersing field voles, Microtus pennsylvanicus and Microtus ochrogaster. Ecol. Monogr., 41:53–78.

Mykytowycz, R. 1968. Territorial marking by rabbits. Sci. Amer., 218:116–126.

Nadler, R. D. 1975. Sexual cyclicity in captive lowland gorillas. Science, 189:813–814.

Nagy, J. G., and R. P. Tengerdy. 1967. Antibacterial action of essential oils of Artemisia as an ecological factor. II. Antibacterial action of the volatile oils of Artemisia tridentata (big sagebrush) on bacteria of the rumen of the mule deer. Appl. Microbiol., 16:441–444.

Needham, A. D., T. J. Dawson, and J. R. S. Hales. 1974. Forelimb blood flow and saliva spreading in the thermoregulation of the red kangaroo, Megaleia rufa. Comp. Biochem. Physiol., 49:555–565.

Negus, N. C., and A. J. Pinter, 1965. Litter sizes of Microtus montanus in the laboratory. J. Mamm., 46(3):434–437.

———, ———. 1966. Reproductive responses of Microtus montanus to plants and plant extracts in the diet. J. Mamm., 47(4):596–601.

Neumann, C. A. 1965/1966. Geo-marine Technol., 2:1; as cited by Ridgway (1966).

Nicholson, P. J. 1963. Wombats. Timbertop Magazine. Geelong Grammar School, No. 8. pp. 28–32.

Nishida, T., and K. Kawanaka. 1972. Interunit-group relationships among wild chimpanzees of the Mahali Mountains. Kyoto Univ. Afr. Stud., 7:131–169.

Noirot, E. 1969. Sound analysis of ultrasonic distress calls of mouse pups as a function of their age. Anim. Behav., 17:340–349.

Norberg, U. M. 1969. An arrangement giving a stiff leading edge to the hand wing in bats. J. Mamm., 50:766–770.

———. 1972. Bat wing structures important for aerodynamics and rigidity. Z. Morph. Tiere, 73:45–61.

Norris, K. S. 1964. Some problems in echolocation in cetaceans, pp. 317–336. In W. N. Tavolga (ed.), Marine bio-acoustics. Pergamon Press, Oxford.

———. (ed.). 1966. Whales, dolphins and porpoises. Univ. California Press, Berkeley. 789 pp.

———. 1968. The evolution of acoustic mechanisms in odontocete cetaceans. Peabody Museum Centenary Celebration Volume, Yale Univ.

———. 1969. The echolocation of marine mammals, pp. 391–423. In H. T. Andersen (ed.), The biology of marine mammals. Academic Press, New York.

———, H. A. Baldwin, and D. J. Samson. 1965. Deep Sea Res., 12:505–509.

———, and J. H. Prescott, 1961. Observations on Pacific cetaceans of California and Mexican waters. Univ. Calif. Publ. Zool., 63:291–402.

————, A. Prescott, D. V. Asa-Doran, and P. Perkins. 1961. An experimental demonstration of echolocation behavior in the porpoise, Tursiops truncatus (Montagu). Biol. Bull., 120:163–176.

Novick, A. 1955. Laryngeal muscles of the bat and production of ultrasonic sounds. Amer. J. Physiol., 183:648.

————. 1958a. Orientation in paleotropical bats. II Megachiroptera. J. Exp. Zool., 137:443–462.

————. 1958b. Orientation in paleotropical bats. I Microchiroptera. J. Exp. Zool., 138:81–254.

————. 1962. Orientation in neotropical bats. I Natalidae and Emballonuridae. J. Mamm., 43:449–455.

————. 1963a. Orientation in neotropical bats. II Phyllostomatidae and Desmodontidae. J. Mamm., 44:44–56.

————. 1963b. Pulse duration in the echolocation of insects by the bat, Pteronotus. Ergeb. Biol., 261–26.

————. 1965. Echolation of flying insects by the bat. Chilonycteris psilotis. Biol. Bull., 128:297–314.

————. 1970. Echolocation in bats. Nat. Hist., 79(3):32–41.

————. 1971. Echolocation in bats: some aspects of pulse design. Amer. Sci., 59(2):198–209.

————, and D. R. Griffin. 1961. Laryngeal mechanisms in bats for production of orientation sounds. J. Exp. Zool., 148:125–146.

————, and J. R. Vaisnys. 1964. Echolocation of flying insects by the bat Chilonycteris parnellii. Biol. Bull., 127:478–488.

Odum, E. P. 1971. Fundamentals of ecology, 3rd ed. W. B. Saunders Company, Philadelphia. 574 pp.

O'Farrell, M. J., and E. H. Studier. 1970. Fall metabolism in relation to ambient temperatures in three species of Myotis. Comp. Biochem. Physiol., 35:697–703.

O'Farrell, T. P. 1965. Home range and ecology of snowshoe hares in interior Alaska. J. Mamm., 46:406–418.

O'Gara, B. W., and G. Matson. 1975. Growth and casting of horns by pronghorns and exfoliation of horns by bovids. J. Mamm., 56:829–846.

————, R. F. Moy, and G. D. Bear. 1971. The annual testicular cycle and horn casting in the pronghorn (Antilocapra americana). J. Mamm., 52:537–544.

Orr, R. T. 1940. The rabbits of California. Occ. Papers California Acad. Sci., 19:1–207.

————. 1976. Vertebrate biology, 4th ed. W. B. Saunders Company, Philadelphia. 472 pp.

Osborn, H. F. 1936–1942. Proboscidea: A monograph of the discovery, evolution, migration, and extinction of the mastodonts and elephants of the world. Vol. 1. Moeritheroidea, Deinotheroidea, Mastodontoidea. Vol. 2. Stegodontoidea, Elephantoidea. Amer. Mus. Nat. Hist., New York. 1675 pp. (Although the taxonomic schemes and the phylogenetic patterns presented in this paper have been seriously questioned, the figures and discussions of structure are useful.)

Otte, D. 1974. Effects and functions in the evolution of signaling systems. Ann. Rev. Ecol. System., 5:385–418.

Paine, R. T. 1969. A note on trophic complexity and community stability. Amer. Nat., 103:91–93.

Parrington, F. R. 1971. On the Upper Triassic mammals. Phil. Trans. Roy. Soc. London, B, 261:231–272.

Parry, D. A. 1949. The structure of whale blubber and its thermal properties. Quart. J. Microbiol. Sci., 90:13–26.

Patterson, B., and R. Pascual. 1968. The fossil mammal fauna of South America. Quart. Rev. Biol., 43:409–451.

————, ————. 1972. The fossil mammal fauna of South America, pp. 247–309. In A. Keast, F. C. Erk, and B. Glass (eds.), Evolution, mammals, and southern continents. State Univ. New York Press, Albany.

Payne, R. S. 1961. The acoustical location of prey by the barn owl (Tyto alba). Amer. Zool., 1:379.

————. 1970. Songs of the humpback whale. An LP Record by CRM Records, Del Mar, California.

————, and S. McVay. 1971. Songs of the humpback whales. Science, 173:585–597.

Pearson, O. P. 1942. On the cause and nature of a poisonous action produced by the bite of a shrew (Blarina brevicauda). J. Mamm., 23:159–166.

————. 1947. The rate of metabolism of some small mammals. Ecology, 28:127–145.

————. 1948. Metabolism of small mammals with remarks on the lower limit of mammalian size. Science, 108:44.

————. "1959" 1960. Biology of the subterranean rodents, Ctenomys, in Peru. Mem. del Museo de Hist. Nat. "Javier Prado," 9:1–56.

————. 1963. History of two local outbreaks of feral house mice. Ecology, 44:540–549.

————. 1964. Carnivore-mouse predation: an example of its intensity and bioenergetics. J. Mamm., 45:177–188.

————. 1966. The prey of carnivores during one cycle of mouse abundance. J. Anim. Ecol., 35:217–233.

————. 1971. Additional measurements of the impact of carnivores on California voles (Microtus californicus). J. Mamm., 52:41–49.

————, M. R. Koford, and A. K. Pearson. 1952. Reproduction of the lump-nosed bat (Corynorhinus rafinesquei) in California. J. Mamm., 33:273–320.

Pengelley, E. T. 1967. The relation of external conditions to the onset and termination of hibernation and estivation, pp. 1–25. In K. C. Fisher, et al. (eds.), Mammalian hibernation, Vol. III. Oliver and Boyd, London.

_____. 1968. Interrelationships of circannian rhythms in the ground squirrel Citellus lateralis. Comp. Biochem. Physiol., 24:915–919.

_____, and K. C. Fisher. 1963. The effect of temperature and photoperiod on the yearly hibernating behavior of captive golden-mantled ground squirrels (Citellus lateralis tescorum). Can. J. Zool., 41:1103–1120.

Pequegnat, W. E. 1951. The biota of the Santa Ana Mountains. J. Entomol. Zool., 42:1–84.

Perkins, J. 1945. Biology at Little America III, the west base of the United States Antarctic service expedition 1939–1941. Proc. Amer. Phil. Soc., 89:270–284.

Perry, J. S. 1954. Some observations on growth and tusk weight in male and female African elephants. Proc. Zool. Soc. London, 124:97–104.

Peters, R. P., and L. D. Mech. 1975. Scent-marking in wolves. Sci. Amer., 63:628–637.

Peterson, Randolph S. 1955. North American moose. Univ. Toronto Press, Toronto. 280 pp.

Peterson, Richard S. 1965. Behavior of the northern fur seal. D.Sc. Thesis. Johns Hopkins Univ., Baltimore. 214 pp.

_____, and G. A. Bartholomew. 1967. The natural history and behavior of the California sea lion. Amer. Soc. Mamm., Spec. Publ. No. 1. 79 pp.

Petter, J. J. 1962a. Ecological and behavioral studies of Madagascar lemurs in the field. Ann. N.Y. Acad. Sci., 102:267–281.

_____. 1962b. Recherches sur l'écologie et l'éthologie des lémuriens malgaches. Mém. Mus. Nat. Hist. Paris, ser. A. (Zool.), 27:1–146.

_____. 1965. The lemurs of Madagascar, pp. 292–319. In I. DeVore (ed.), Primate behavior. Holt, Rinehart and Winston, New York.

Pianka, E. R. 1966. Convexity, desert lizards and spatial heterogeneity. Ecology, 47:1055–1059.

Pieper, R. D. 1964. Production and chemical composition of arctic tundra vegetation and their relation to the lemming cycle. Ph.D. Thesis, Univ. California, Berkeley.

Pitelka, F. A. 1957a. Some characteristics of microtine cycles in the arctic. Eighteenth Ann. Biol. Coll., Oregon State College. pp. 73–88.

_____. 1957b. Some aspects of population structure in the short-term cycle of the brown lemming in northern Alaska. Cold Spr. Harb. Symp. Quant. Biol., 22:237–251.

_____. 1958. Some aspects of population structure in the short-term cycle of the brown lemming in northern Alaska. Cold Spr. Harb. Symp. Quant. Biol., 22:237–251.

_____. 1964. The nutrient-recovery hypothesis for Arctic microtine cycles. I. Introduction, pp. 55–56. In P. J. Crisp (ed.), Grazing in terrestrial and marine environments. Brit. Ecol. Soc. Symp. No. 4. Blackwell, Oxford.

_____, P. Q. Tomich, and G. W. Treichel. 1955. Ecological relations of jaegers and owls as lemming predators near Barrow, Alaska. Ecol. Monogr., 25:85–117.

Poole, E. L. 1936. Relative wing ratios of bats and birds. J. Mamm., 17:412–413.

Porter, F. L. 1975. Aspects of social behavior in Carollia. Paper presented at Sixth No. Amer. Symp. on Bat Research, Las Vegas, Nevada.

Poulter, T. C. 1963. Sonar signals of the sea lion. Science, 139:753–755.

Pournelle, G. H. 1968. Classification, biology, and description of the venom apparatus of insectivores of the genera Solenodon, Neomys, and Blarina. In W. Bucherl, E. A. Buckley and V. Deulofeu (eds.), Venomous animals and their venoms. Academic Press, New York.

Powers, J. B., and S. S. Winans. 1975. Vomeronasal organ: critical role in mediating sexual behavior of the male hamster. Science, 187:961–963.

Prakash, I. 1959. Foods of the Indian false vampire. J. Mamm., 40:545–547.

Pucek, M. 1968. Chemistry and pharmacology of insectivore venoms, pp. 43–50. In W. Bucherl, E. A. Buckley and V. Deulofeu (eds.), Venomous animals and their venoms. Academic Press, New York.

Pulliam, H. R. 1974. On the theory of optimal diets. Amer. Nat., 108:59–74.

Purves, P. E. 1966. Anatomy and physiology of the outer and middle-ear in cetaceans, pp. 320–380. In K. S. Norris (ed.), Whales, dolphins and porpoises. Univ. California Press, Berkeley.

Quilliam, T. A. 1966. The problem of vision in ecology of Talpa europaea. Exp. Eye Res., 5:63–78.

Rahm, U. 1970. Note sur la reproduction des sciuridés et muridés dans forêt équatoriale au Congo. Rev. Suisse Zool., 77:635–646.

Ralls, K. 1971. Mammalian scent marking. Science, 171:443–449.

Ratcliff, H. M. 1941. Winter range conditions in Rocky Mountain National Park. Trans. Sixth Amer. Wildl. Conf., pp. 132–139.

Ratcliffe, F. N. 1932. Notes on the fruit bat of Australia. J. Anim. Ecol., 1:32–37.

Rathbun, G. 1973. The golden-rumped elephant shrew. A.W.L.F. News, 8, No. 3.

_____. 1976. The ecology and social structure of the elephant shrews Rhynchocyon chrysopygus and Elephantulus rufescens. Ph.D. Thesis, Univ. Nairobi, Kenya. 263 pp.

Rausch, R. 1950. Observations on a cyclic decline of lemmings (Lemmus) on the Arctic coast of Alaska during the spring of 1949. Arctic, 3:166–177.

Ray, C., and W. E. Schevill. 1965. The noisy underwater world of the Weddell seal. Animal Kingdom, New York Zool. Soc., 68:34–39.

Redman, J. P., and J. A. Sealander, 1958. Home ranges of deer mice in southern Arkansas. J. Mamm., 39:390–395.

Reed, C. A. 1944. Behavior of a shrew mole in captivity. J. Mamm., 25:196–198.

———. 1951. Locomotion and appendicular anatomy in three soricoid insectivores. Amer. Midl. Nat., 45:513–671.

Reeder, W. G., and R. B. Cowles. 1951. Aspects of thermoregulation in bats. J. Mamm., 32:389–403.

Reichman, O. J. 1975. Relation of desert rodent diets to available resources. J. Mamm., 56:731–751.

———, and K. M. Van De Graaff. 1973. Seasonal activity and reproductive patterns of five species of Sonoran Desert rodents. Amer. Midl. Nat., 90:118–126.

———, ———. 1975. Association between ingestion of green vegetation and desert rodent reproduction. J. Mamm., 56:503–506.

Reig, O. A. 1970. Ecological note on the fossorial octodont rodent *Spalacopus cyanus* (Molina). J. Mamm., 51:592–601.

Reysenbach de Haan, F. W. 1966. Listening underwater: thoughts on sound and cetacean hearing, pp. 583–596. *In* K. S. Norris (ed.), Whales, dolphins and porpoises. Univ. California Press, Berkeley.

Rice, D. W. 1967. Cetaceans, pp. 291–324. *In* S. Anderson and J. K. Jones, Jr. (eds.), Recent mammals of the world. Ronald Press, New York.

———, and A. A. Wolman. 1971. The life history and ecology of the gray whale *(Eschrichtius robustus)*. Spec. Publ. No. 3, Amer. Soc. Mammal.

Ricklefs, R. E. 1973. Ecology. Chiron Press, Newton, Massachusetts. 861 pp.

Ride, W. D. L. 1970. A guide to the mammals of Australia. Oxford Univ. Press, New York and London. 249 pp.

Ridgway, S. H. 1966. Proc., Third Ann. Conf. Biol. Sonar Diving Mammals, pp. 151–158. Stanford Res. Inst., Menlo Park, California.

———, B. L. Scronce, and J. Kanwisher. 1969. Respiration and deep diving in the bottle-nosed dolphin. Science, 166:1651–1653.

Rinker, G. C. 1954. The comparative myology of the mammalian genera *Sigmodon, Oryzomys, Neotoma,* and *Peromyscus* (Cricetinae), with remarks on their intergeneric relationships. Misc. Publ. Mus. Zool., Univ. Michigan, 83:1–124.

Robertshaw, D., and C. R. Taylor. 1969. A comparison of sweat gland activity in eight species of East African bovids. J. Physiol., London, 203:135–143.

Robinson, K., and D. H. K. Lee. 1941. Reactions of the cat to hot atmospheres. Proc. Roy Soc. Queensland, 53:159–170. Reactions of the dog to hot atmospheres. 53:171–188.

Robinson, P., C. Black, and M. Dawson. 1964. Late Eocene multituberculates and other mammals from Wyoming. Science, 145:809–811.

Rodin, L. E., N. I. Bazilevich, and N. N. Rozov. 1975. Productivity of the world's ecosystems, pp. 13–26. *In* Productivity of world ecosystems. NAS, Washington, D.C.

Roeder, K. D. 1965. Moths and ultrasound. Sci. Amer., 212(4):94–102.

———, and A. E. Treat. 1961. The detection and evasion of bats by moths. Amer. Sci., 49(2):135–148.

Romer, A. S. 1966. Vertebrate paleontology, 3rd ed. Univ. Chicago Press, Chicago. 468 pp.

———. 1968. Notes and comments on vertebrate paleontology. Univ. Chicago Press, Chicago. 304 pp.

———. 1969. Cynodont reptile with incipient mammalian jaw articulation. Science, 166:881–882.

———, and Parsons, T. S. 1977. The vertebrate body, W. B. Saunders Company, Philadelphia.

Rood, J. P. 1958. Habits of the short-tailed shrew in captivity. J. Mamm., 39:499–507.

———. 1970. Notes on the behavior of the pygmy armadillo. J. Mamm., 51:179.

Rosenzweig, M. R., D. A. Riley, and K. Krech. 1955. Evidence for echolocation in the rat. Science, 121:600.

Rosevear, D. R. 1969. The rodents of West Africa. British Museum Natural History, London. 604 pp.

Rowan, W. 1950. Winter habits and numbers of timber wolves. J. Mamm., 31:167–169.

Rowell, T. E. 1962. Agonistic noises of the rhesus monkey *(Macaca mulatta)*. Symp. Zool Soc. London, 8:91–96.

Ruff, F. J. 1938. Trapping deer on the Pisgah National Game Preserve, North Carolina. J. Wildl. Manage., 2:151–161.

Rutherford, W. H. 1953. Effects of a summer flash flood upon a beaver population. J. Mamm., 34:261–262.

Sale, J. B. 1970. The behavior of the resting rock hyrax in relation to its environment. Zool. Afr. 5:87–99.

Sargent, A. B., and D. W. Warner. 1972. Movements and denning habits of a badger. J. Mamm., 53:207–210.

Saunders, J. K., Jr., 1963. Movements and activities of the lynx in Newfoundland. J. Wildl. Manage., 27:390–400.

Savage, J. M. 1974. The isthmian link and the evolution of Neotropical mammals. Contrib. Sci., Nat. Hist. Mus., Los Angeles County, 260:1–51.

Schaeffer, B. 1947. Notes on the origin and function of the artiodactyl tarsus. Amer. Mus. Novitates, 1356:1–24.

Schaller, G. B. 1963. The mountain gorilla: ecology and behavior. Univ. Chicago Press, Chicago. 431 pp.

_____. 1964. The year of the gorilla. Univ. Chicago Press, Chicago. 260 pp.

_____. 1965a. The behavior of the mountain gorilla, pp. 324–367. *In* I. De Vore (ed.), Primate behavior: field studies of monkeys and apes. Holt, Rinehart and Winston, New York.

_____. 1965b. The year of the gorilla. Ballantine Books, New York. 285 pp.

_____. 1967. The deer and the tiger: a study of wildlife in India. Univ. Chicago Press, Chicago. 370 pp.

_____. 1972. The Serengeti lion: a study of predator-prey relations. Univ. Chicago Press, Chicago. 480 pp.

Scheffer, V. B. 1958. Seals, sea lions and walruses. Stanford Univ. Press, Stanford, California. 179 pp.

_____, and J. W. Slipp. 1944. The harbor seal in Washington state. Amer. Midl. Nat., *32*:373–416.

Schenkel, R. 1966a. Play, exploration and territoriality in the wild lion. Symp. Zool. Soc. London, 18:11–22.

_____. 1966b. On sociology and behaviour in impala (*Aepyceros melampus suara* Matschie). Z. Säugetierk., *31*:177–205.

_____. 1967. Submission: its features and functions in the wolf and dog. Amer. Zool., *7*:319–329.

Schevill, W. E., and B. Lawrence. 1949. Underwater listening to the white porpoise, *Delphinapterus leucas*. Science, *109*:143–144.

_____, _____. 1953. Auditory response of a bottle nosed porpoise, *Tursiops truncatus*, to frequencies above 100 kc. J. Exp. Zool., *124*:147–165.

_____, and C. Ray. 1965. The Weddell seal at home. Animal Kingdom, N.Y. Zool. Soc., *68*:151–154.

_____, and W. A. Watkins. 1965. Underwater calls of *Leptonychotes* (Weddell seal). Zoologica, N.Y. Zool. Soc., *50*:45–47.

_____, _____. 1966. Sound structure and directionality in *Orcinus* (killer whale). Zoologica, N.Y. Zool. Soc., *51(2)*:71–76.

_____, _____, and C. Ray. 1963. Underwater sounds of pinnipeds. Science, *141*:50–53.

_____, _____, _____. 1966. Analysis of underwater *Odobenus* calls with remarks on the development and function of the pharyngeal pouches. Zoologica, N.Y. Zool. Soc., *51(3)*:103–106.

Schmidt-Nielsen, B., and K. Schmidt-Nielsen. 1950a. Do kangaroo rats thrive when drinking sea water? Amer. J. Physiol., *160*:291–294.

_____, _____. 1950b. Evaporative water loss in desert rodents in their natural habitat. Ecology, *31*:75–85.

_____, _____. 1951. A complete account of water metabolism in kangaroo rats and experimental verification. J. Cell Comp. Physiol., *38*:165–182.

_____, _____, T. R. Houpt, and S. A. Jarnum. 1956. Water balance of the camel. Amer. J. Physiol., *185*:185–194.

_____, _____, _____. 1957. Urea excretion in the camel. Amer. J. Physiol., *188*:477–484.

Schmidt–Nielsen, K. 1959. The physiology of the camel. Sci. Amer., *201*:140–151.

_____. 1964. Desert animals, physiological problems of heat and water. Oxford Univ. Press, New York and Oxford. 277 pp.

_____. 1972. Recent advances in the comparative physiology of desert animals, pp. 371–382. *In* G. M. O. Maloiy (ed.), Comparative physiology of desert animals. Academic Press, New York.

_____, W. L. Bretz, and C. R. Taylor. 1970. Panting in dogs: unidirectional air flow over evaporative surfaces. Science, *169*:1102–1104.

_____, F. R. Hainsworth, and D. E. Murrish. 1970. Counter-current heat exchange in the respiratory passages: effects on water and heat balance. Resp. Physiol., *9*:263–276.

_____, and A. E. Newsome. 1962. Water balance in the mulgara (*Dasycercus cristicauda*), a carnivorous desert marsupial. Aust. J. Biol. Sci., *15*:683–689.

_____, and B. Schmidt-Nielsen. 1952. Water metabolism of desert mammals. Physiol. Rev., *32*:135–166.

_____, _____. 1953. The desert rat. Sci. Amer., *189(1)*:73–78.

_____, _____. 1954. Heat regulation in small and large desert animals, pp. 182–187. *In* J. L. Cloudsley-Thompson (ed.), Biology of deserts. Inst. Biol., London. 182 pp.

_____, _____, T. A. Houpt, and S. A. Jarnum. 1956. The question of water storage in the stomach of the camel. Mammalia, *20*:1–15.

_____, _____, _____. 1957. Body temperature of the camel and its relation to water economy. Amer. J. Physiol., *188*:103–112.

Schneider, K. M. 1930. Das Flehmen. Zool. Gart. (Leipzig), *3*:183–198; *4*:349–364; *5*:200–226, 287–297.

Schneider, R., H. Jurg Kugn and G. Kelemen. 1967. De Larynx der *Hypsignathus monstrosus* Allen 1861. Ein Unifum in der Morphologie des Kehlkopfes. Z. Wiss. Zool.

Schnitzler, H. U. 1968. Echoortung bei der Ortungslaute der Hufeisen-Fledermause (Chiroptera-Rhinolophidae) in verschiedenen Orientierungssituationen. Z. Vergl. Physiol., *57*:376–408.

Schoener, T. W. 1971. Theory of feeding strategies. Ann. Rev. Ecol. System., *2*:369–404.

Scholander, P. F. 1940. Hvalradets Skrifter Norske Videnskaps-Akad. Oslo, *22*:1–131.

_____. 1955. Evolution of climatic adaptation in homeotherms. Evolution, *9*:15–26.

_____, R. Hock, V. Walters, and L. Irving. 1950. Adaptation to cold in arctic and tropical mammals and birds in relation to body temperature, insulation, and basal metabolic rate. Biol. Bull., *99*:259–271.

———, ———, ———, F. Johnson, and L. Irving. 1950. Heat regulation in some arctic and tropical mammals and birds. Biol. Bull., 99:237–258.

———, and J. Krog. 1957. Countercurrent and vascular heat exchange; sloths. J. Appl. Physiol., 10:404–411.

———, and W. E. Schevill. 1955. Countercurrent and vascular heat exchange; whales. J. Appl. Physiol., 8:279–282.

———, V. Walters, R. Hock, and L. Irving. 1950. Body insulation of some arctic and tropical mammals and birds. Biol. Bull., 99:225–236.

Schubert, G. H. 1953. Ponderosa pine cone cutting by squirrels. J. Forestry, 51:202.

Schultz, A. M. 1964. The nutrient-recovery hypothesis for Arctic microtine cycles. II Ecosystem variables in relation to Arctic microtine cycles, pp. 57–58. In P. J. Crisp (ed.), Grazing in terrestrial and marine environments. Brit. Ecol. Soc. Symp. No. 4. Blackwell, Oxford.

———. 1965. The tundra as a homeostatic system. Presented at A.A.A.S. Meeting, Dec. 1965. (Mimeo.)

———. 1969. A study of an ecosystem: the Arctic tundra, pp. 77–93. In G. M. Van Dyne (ed.), The ecosystem concept in natural resource management. Academic Press, New York.

Schultz, C. B., M. R. Schultz, and L. D. Martin. 1970. A new tribe of saber-toothed cats (Barbourofelini) from the Pliocene of North America. Bull. Univ. Nebraska State Mus., 9:1–31.

Schultze-Westrum, T. 1965. Nochweis differenzierter Duftstoffe beim Gleitbeulter Petaurus breviceps papuanus Thomas (Marsupialia, Phalangeridae). Naturwiss., 51(9):226–227.

Schusterman, T. J., and S. N. Feinstein. 1965. Shaping and discriminative control of underwater click vocalization in a California sea lion. Science, 150:1743–1744.

Sclater, W. L., and P. L. Sclater. 1899. The geography of mammals. Kegan, Paul, Trench, Trubner, London.

Selander, R. K. 1966. Sexual dimorphism and differential niche utilization in birds. Condor, 68:113–151.

Selye, H. 1955. Stress and disease. Science, 122:625–631.

———, and J. B. Collip. 1936. Fundamental factors in the interpretation of stimuli influencing the endocrine glands. Endocrin., 20:667–672.

Semeonoff, R., and F. W. Robertson. 1968. A biochemical and ecological study of plasma esterase polymorphism in natural populations of the field vole, Microtus agrestis. Biochem. Genet., 1:205–227.

Sewell, G. D. 1968. Ultrasound in rodents. Nature, 217:682–683.

Sharman, G. B. 1970. Reproductive physiology of marsupials. Science, 167:1221–1228.

Shaw, W. T. 1934. The ability of the giant kangaroo rat as a harvester and storer of seeds. J. Mamm., 15:275–286.

———. 1936. Moisture and its relation to the cone-storing habit of the western pine squirrel. J. Mamm., 17:337–349.

Sheldrick, D. 1972. Death of the Tsavo elephant. Sat. Rev., Sept. 30, pp. 29–35.

———. 1973. The Tsavo story. Collins and Harvill Press, London. 288 pp.

Shirer, H. W., and H. S. Fitch. 1970. Comparison from radiotracking of movements and denning habits of the raccoon, striped skunk, and opossum in northeastern Kansas. J. Mamm., 51:491–503.

Shkolnik, A., and Borut, A. 1969. Temperature and water relations in two species of spiny mice (Acomys). J. Mamm., 50:245–255.

Short, H. L. 1966. Effects of cellulose levels on the apparent digestibility of seeds eaten by mule deer. J. Wildl. Manage., 30:163–167.

———, D. R. Dietz, and E. E. Remmenga. 1966. Selected nutrients in mule deer plants. Ecology, 47:222–229.

Shortridge, G. C. 1934. The mammals of Southwest Africa. 2 vols. Heinemann, London.

Sige, B. 1971. Anatomie du membre anterieur chez un chiroptere molosside (Tadarida sp.) du Stampien de Cereste (Alpes-de-Haute-Provence). Palaeovert., 4:1–38.

Siivonen, L. 1954. Features of short-term fluctuations. J. Wildl. Manage., 18:38–45.

Simmons, J. A. 1971. Echolocation in bats: signal processing of echoes for target range. Science, 171:925–927.

———. 1973. The resolution of target range by echolocating bats. J. Acoust. Soc. Amer., 54:157–173.

———. 1974. Response of the Doppler echolocation system in the bat, Rhinolophus ferrumequinum. J. Acoust. Soc. Amer., 56:672–682.

———, D. J. Howell, and N. Suga. 1975. Information content of bat sonar echoes. Amer. Sci., 63:204–215.

———, W. A. Lavender, B. A. Lavender, C. A. Doroshow, S. W. Kiefer, R. Livingston, A. C. Scallet, and D. E. Crowley. 1974. Target structure and echo spectral discrimination by echolocating bats. Science, 186:1130–1132.

Simpson, G. G. 1937. Skull structure of the multituberculata. Bull. Amer. Mus. Nat. Hist., 73:727–763.

———. 1940. Mammals and land bridges. J. Wash. Acad. Sci., 30:137–163.

———. 1943. Mammals and the nature of continents. Amer. J. Sci., 24:1–31.

———. 1944. Tempo and mode in evolution. Columbia Univ. Press, New York. 237 pp.

———. 1945. The principles of classification and a classification of mammals. Bull. Amer. Mus. Nat. Hist., 85:1–350.

———. 1947. Evolution, interchange, and resemblance of North American and Eurasian Cenozoic mammalian faunas. Evolution, 1:218–220.

_____. 1950. History of the fauna of Latin America. Amer. Sci., *38*:361–389.

_____. 1951. Horses. Oxford Univ. Press, New York. 247 pp.

_____. 1952. Probabilities of dispersal in geologic time. Bull. Amer. Mus. Nat. Hist., *99*:163–176.

_____. 1953. The major features of evolution. Columbia Univ. Press, New York, 434 pp.

_____. 1959. Mesozoic mammals and the polyphyletic origin of mammals. Evolution, *13*:405–414.

_____. 1961. Historic zoogeography of Australian mammals. Evolution, *15*:431–446.

_____. 1965a. Attending marvels. A Patagonian journal. Time Inc., New York. 289 pp.

_____. 1965b. The geography of evolution. Chilton Books, Philadelphia. 249 pp.

_____. 1969. South American mammals, pp. 879–909. *In* E. J. Fihkan, et al. (eds.), Biogeography and ecology in South America. Mono. Biol., 19. W. Junk, The Hague.

_____. 1970a. Additions to knowledge of the Argyrolagidae (Mammalia, Marsupialia) from the late Cenozoic of Argentina. Breviora, Mus. Comp. Zool., *361*:1–9.

_____. 1970b. The Argyrolagidae, extinct South American Marsupials. Bull. Mus. Comp. Zool., *139*:1–86.

Sinclair, A. R. E. 1970. Studies of the ecology of the East African buffalo. Ph.D. thesis, Oxford Univ., Oxford.

_____. 1974. The social organization of the East African buffalo, pp. 676–689. *In* V. Geist and F. R. Walther (eds.), The behavior of ungulates and its relation to management. International Union for Conservation of Nature and Natural Resources, Morges, Switzerland.

Slaughter, B. H. 1968. Earliest known marsupials. Science, *162(3850)*:254–255.

_____, and D. W. Walton (eds.). 1970. About bats. Southern Methodist Univ. Press, Dallas. 339 pp.

Slijper, E. J. 1936. Die Cetacean, vergleichend-anatomisch und systematisch. Capita Zool., *7*:1–590.

_____. 1962. Whales. Basic Books, Inc. New York. 475 pp.

Slobodchikoff, C. 1977. Experimental studies of predation on tenebrionid beetles by skunks. (In manuscript.)

Smith, H. M. 1960. Evolution of chordate structure. Holt, Rinehart, and Winston, New York. 529 pp.

Smith, J. D. 1972. Systematics of the chiropteran family Mormoopidae. Univ. Kansas Publ., Mus. Nat. Hist., Misc. Publ. No. 56, pp. 1–132.

Smith, P. W. 1965. Recent adjustments in animals' ranges, pp. 633–642. *In* H. E. Wright, Jr. and D. G. Frey (eds.), The Quaternary of the United States. Princeton Univ. Press, Princeton, New Jersey.

Smith, R. B. 1971. Seasonal activities and ecology of terrestrial vertebrates in a Neotropical monsoon environment. M. S. Thesis, Northern Arizona Univ., Flagstaff, Arizona. 127 pp.

Soholt, L. S. 1973. Consumption of primary production by a population of kangaroo rats (*Dipodomys merriami*) in the Mojave Desert. Ecol. Monogr., *43*:358–376.

Sondaar, P. Y. 1977. Insularity and its effects on mammal evolution. NATO Advanced Study Institute, Plenum Press, New York. (In press.)

Sorenson, M. W., and C. H. Conaway. 1968. The social and reproductive behavior of *Tupaia montana* in captivity. J. Mamm., *49*:502–512.

Southern, H. N. 1964. Handbook of British mammals. Blackwell, Oxford.

Sowls, L. K. 1974. Social behavior of the collared peccary *Dicotyles tajacu*, pp. 144–165. *In* V. Geist and F. R. Walther (eds.), The behavior of ungulates and its relation to management. International Union for Conservation of Nature and Natural Resources, Morges, Switzerland.

Spencer, A. W., and H. W. Steinhoff. 1968. An explanation of geographic variation in litter size. J. Mamm., *49*:281–286.

Spencer, D. A. 1958a. Preliminary investigations on the northwestern *Microtus* irruption. U. S. Fish and Wildl. Serv., Denver Wildl. Res. Lab. Spec. Report.

_____. 1958b. Biological and control aspects, pp. 15–25. *In* The Oregon meadow mouse irruption of 1957–1958. Federal Cooperative Extension Service, Oregon State College, Corvallis, Oregon.

Spitz, F. 1963. Estude des densities de population de *Microtus arvalis*. Pall. A Saint-Michal-en L'Hern (Vendu). Mammalia, *27*:497–531.

Sprankel, H. 1965. Untersuchungen an *Tarsus*. I Morphologie des Schwanzes nebst ethologischen Bemerkungen. Folia Primat., *3*:153–188.

Steiniger, F. 1950. Beiträge zur Soziologie und sonstigen Biologie der Wanerratte. Z. Tierpsychol., *7*:356–379.

Stenlund, M. H. 1955. A field study of the timber wolf (*Canis lupus*) on the Superior National Forest, Minnesota. Minn. Dept. Cons. Tech. Bull., 4. 55 pp.

Stephenson, A. B. 1969. Temperatures within a beaver lodge in winter. J. Mamm., *50*:134–136.

Sterling, I. 1969. Ecology of the Weddell seal in McMurdo Sound, Antarctica. Ecology, *50*:573–586.

Stevenson-Hamilton, J. 1947. Wild life in South Africa. Cassell, London. 364 pp.

Stirton, R. A., R. H. Tedford, and M. O. Woodburne. 1967. A new Tertiary formation and fauna from the Tirari Desert, South Australia. Rec. South Aust. Mus., *15*:427–462.

Stock, C. 1949. Rancho La Brea: a record of Pleistocene life in California. L. A. County Mus., Sci. Ser., No. 13:1–81.

Stones, R. C., and J. E. Wiebers. 1965. A review of temperature regulation in bats (Chiroptera). Amer. Midl. Nat., *74*:155–167.

_____, _____. 1967. Temperature regulation in the little brown bat, *Myotis lucifugus*, pp. 97–109. *In* K. C. Fisher, A. R. Dawe, C. P. Lyman, E. Schonbaum, and F. E. South, Jr. (eds.), Mammalian hibernation, Vol. III. Oliver & Boyd and Amer. Elsevier, New York.

Storer, R. W. 1955. Weight, wing area, and skeletal proportions in three accipiters. Acta 11th Cong. Int. Ornithol., pp. 278–290.

Storer, T. I., and R. L. Usinger. 1965. General zoology. McGraw-Hill Book Co., New York. 741 pp.

Struhsaker, T. T. 1967. Behavior of elk *(Cervus canadensis)* during the rut. Z. Tierpsychol., *24(1)*:80–114.

———. 1976. The red colobus monkey. Univ. Chicago Press, Chicago. 312 pp.

Studier, E. H., and D. J. Howell. 1969. Heart rate of female big brown bats in flight. J. Mamm., 50:842–845.

———, J. W. Proctor, and D. J. Howell. 1970. Diurnal body weight loss and tolerance of weight loss in five species of *Myotis*. J. Mamm., 51:302–309.

Suga, N., and T. Shimozawa. 1974. Site of neural attenuation of responses to self-vocalized sounds in echolocating bats. Science, *177*:1211–1213.

Sugiyama, Y. 1973. Social organization of wild chimpanzees, pp. 68–80. *In* C. R. Carpenter (ed.), Behavioral regulators of behavior in primates. Bucknell Univ. Press, Lewisburg, Pennsylvania.

Suthers, R. A. 1965. Acoustic orientation by fish-catching bats. J. Exp. Zool., *158*:319–348.

———. 1967. Comparative echolocation by fishing bats. J. Mamm., *48*:79–87.

Swank, W. G. 1958. The mule deer in Arizona chaparral. Arizona Game and Fish Dept., Wildl. Bull. No. 3. 109 pp.

Sweeney, R. C. H. 1956. Notes on *Manis temmincki*. Ann. Mag. Nat. Hist., London.

Swinhoe, R. 1870. On the mammals of Hainan. Proc. Zool. Soc. London, *1870*:224–239.

Szalay, F. S. 1968. The beginnings of primates. Evolution, *22*:19–36.

———. 1969. Origin and evolution of function of the mesonychid condylarth feeding mechanism. Evolution, *23(4)*:703–720.

Taber, R. D., and R. F. Dasmann. 1957. The dynamics of three natural populations of deer *Odocoileus hemionus columbianus*. Ecology, *38*:233–246.

Talbot, L. M., and M. H. Talbot. 1963. The wildebeest in western Masailand. Wildl. Monogr., Chestertown, No. *12*:1–88.

Talmage, R. V., and G. D. Buchanan. 1954. The armadillo *(Dasypus novemcinctus)*: a review of its natural history, ecology, anatomy and reproductive physiology. The Rice Institute Pamphlet, Monograph in Biology, Vol. 41, 135 pp.

Tappen, N. C. 1960. Problems of distributions and adaptations of the African monkeys. Cur. Anthrop., 1:91–120.

Tate, G. H. H. 1933. A systematic revision of the marsupial genus *Marmosa*. Bull. Amer. Mus. Nat. Hist., *66*:1–250.

———, 1947. Mammals of eastern Asia. Macmillan Company, New York. 366 pp.

———, and R. Archbold. 1937. Results of the Archbold Expeditions. 16, Some marsupials of New Guinea and Celebes. Bull. Amer. Mus. Nat. Hist., *73*:331–476.

Tavolga, M. C. 1966. Behavior of the bottlenose dolphin *(Tursiops truncatus)*: social interaction in a captive colony, pp. 718–730. *In* K. S. Norris (ed.), Whales, dolphins and porpoises. Univ. California Press, Berkeley.

Taylor, C. R. 1968a. Hygroscopic food: a source of water for desert antelopes? Nature, *219*:181–182.

———. 1968b. The minimum water requirements of some East African bovids. Symp. Zool. Soc. London. *21*:195–206.

———.1969a. The eland and the oryx. Sci. Amer., *220*:89–95.

———. 1969b. Metabolism, respiratory changes, and water balance of an antelope, the eland. Amer. J. Physiol., *217(1)*:317–320.

———. 1970. Dehydration and heat: effect on temperature regulation of East African ungulates. Amer. J. Physiol., *219*:1136–1139.

———. 1972. The desert gazelle: a paradox resolved, pp. 215–227. *In* G. M. O. Maloiy (ed.), Comparative physiology of desert animals. Symp. Zool. Soc. London, *31*. Academic Press, New York.

———, and C. P. Lyman. 1967. A comparative study of the environmental physiology of an East African antelope, the eland, and the Hereford steer. Physiol. Zool., *40(3)*:280–295.

———,———. 1972. Heat storage in running antelopes: independence of brain and body temperatures. Amer. J. Physiol., *222*:114–117.

———, K. Schmidt-Nielsen, R. Dmi'el, and M. Fedak. 1971. Effect of hyperthermia on heat balance during running in the African hunting dog. Amer. J. Physiol., *220*:823–827.

———, C. A. Spinage, and C. P. Lyman. 1969. Water relations of the waterbuck, an East African antelope. Amer. J. Physiol., *217(2)*:630–634.

Taylor, W. T., and R. J. Weber. 1951. Functional mammalian anatomy. D. Van Nostrand Company, New York. 575 pp.

Tedford, R. H. 1967. The fossil Macropodidae from Lake Menindee, New South Wales. Univ. Calif. Publ. Geol. Sci., *64*:1–165.

———. 1974. Marsupials and the new paleogeography, pp. 109–126. *In* C. A. Ross (ed.), Paleogeographic provinces and provinciality. Soc. Econ. Paleontol. Mineral., Publ. 21.

———, M. R. Banks, N. R. Kemp, I. McDougall, and F. L. Sutherland. 1975. Recognition of the oldest known fossil marsupials from Australia. Nature, Lond., *255*:141–142.

Tevis, L. 1950. Summer behavior of a family of beavers in New York State. J. Mamm., *31*:40–65.

Thiessen, D. D., K. Owen, and G. Lindzey. 1971. Mechanisms of territorial marking in the male and female Mongolian gerbils *(Meriones unguiculatus)*. J. Comp. Physiol. Psychol., *77*:38–47.

———. F. E. Regnier, M. Rice, M. Goodwin, N. Isaacks, and N. Lawson. 1974. Identification of a ventral scent marking pheromone in the male Mongolian gerbil (*Meriones unguiculatus*). Science, 184:83–84.

Thomas, S. P., and R. A. Suthers. 1972. The physiology and energetics of bat flight. J. Exp. Physiol., 57:317–335.

Thompson, D. Q. 1955. The role of feed and cover in population fluctuations of the brown lemming at Point Barrow, Alaska. Trans. North Amer. Wildl. Conf., 20:166–176.

Tinbergen, N. 1963. On aims and methods of ethology. Z. Tierpsychol., 20:410–433.

Tinkle, D. W., and I. G. Patterson. 1965. A study of hibernating populations of *Myotis velifer* in Northwest Texas. J. Mamm., 46:612–633.

Tomich, P. Q. 1962. The annual cycle of the California ground squirrel, *Citellus beecheyi*. Univ. Calif. Publ. Zool., 65:213–282.

Torrey, T. W. 1971. Morphogenesis of the vertebrates. John Wiley & Sons, New York. 529 pp.

Townsend, M. T. 1935. Studies on some small mammals of central New York. Roosevelt Wildl. Ann., 4:1–120.

Troughton, E. 1947. Furred animals of Australia. Charles Scribner's Sons, New York, 374 pp.

Trumler, E. 1959. Das "Rossighkeitsgesicht" und ähnliches Ausdrucksverhalten bei Einhufern. Z. Tierpsychol., 16:478–488.

Tucker, V. A. 1962. Diurnal torpidity in the California pocket mouse. Science, 136:380–381.

———. 1965. The relation between the torpor cycle and heat exchange in the California pocket mouse *Perognathus californicus*. J. Cell Comp. Physiol., 65:405–414.

Turner, B. N. 1971. The annual cycle of aggression in male *Microtus pennsylvanicus*, and its relation to population parameters. M.S. Thesis, Univ. North Dakota.

Udvardy, M. D. F. 1969. Dynamic zoogeography. Van Nostrand Reinhold Company, New York. 445 pp.

Vachrameev, V. A., and M. A. Akhmet'yev. 1972. The development of floras on the boundary of the Late Cretaceous and the Paleocene (on data from the study of leaf remains), pp. 21–26. In V. N. Shimansiy, A. N. Solov'yev (eds.), The development and replacement of the organic world on the boundary of the Mesozoic and Cenozoic. (Conf. April, 1972.) Abstr. pap. methodol. mater., Moscow: Akad. Nauk SSSR, Moscow Soc. Natur. (In Russian.)

Valentine, J. W., and E. M. Moores. 1970. Plate-tectonic regulation of faunal diversity and sea level: a model. Nature, 228:657–659.

Van Couvering, J. A., and J. A. H. Van Couvering. 1975. African isolation and the Tethys seaway. Sixth Congress Regional Committee on Mediterranean Neogene Stratigraphy, Bratislava, pp. 363–367.

Van De Graaff, K., and R. P. Balda. 1973. Importance of green vegetation for reproduction in the kangaroo rat, *Dipodomys merriami merriami*. J. Mamm., 54:509–512.

Van den Ende, P. 1973. Predator-prey interactions in continuous culture. Science, 181:562–564.

Van Deusen, H. M. 1967. Personal communication.

———. 1969. Results of the Archbold Expeditions. No. 90. Notes on the echidnas (Mammalia, Tachyglossidae) of New Guinea. Amer. Mus. Novitates, 2383:1–23.

———. 1971. *Zaglossus*, New Guinea's egg-laying anteater. Fauna, Vol. 1:12–19.

———, and J. K. Jones, Jr. 1967. Marsupials, pp. 61–86. In S. Anderson and J. K. Jones, Jr. (eds.), Recent mammals of the world. Ronald Press, New York.

Van Gelder, R. G. 1953. The egg-opening technique of a spotted skunk. J. Mamm., 34:255–256.

Van Lawick-Goodall, J. 1968. A preliminary report on expressive movements and communication in the Gombe Stream Chimpanzees, pp. 313–374. In P. C. Jay (ed.), Primates, studies in adaptation and variability. Holt, Rinehart and Winston, New York.

Van Valen, L. 1965a. The earliest primates. Science, 150:743–745.

———. 1965b. Treeshrews, primates and fossils. Evolution, 19:137–151.

———. 1966. Deltatheridia, a new order of mammals. Bull. Amer. Mus. Nat. Hist., 132:1–126.

———, and R. E. Sloan. 1966. The extinction of the multituberculates. Syst. Zool., 15:261–278.

Vaughan, T. A. 1954. Mammals of the San Gabriel mountains of California. Univ. Kansas Publ., Mus. Nat. Hist., 7:513–582.

———. 1959. Functional morphology of three bats: *Eumops, Myotis, Macrotus*. Univ. Kansas Publ., Mus. Nat. Hist., 12:1–153.

———. 1961. Vertebrates inhabiting pocket gopher burrows in Colorado. J. Mamm., 42:171–174.

———. 1966a. Morphology and flight characteristics of molossid bats. J. Mamm., 47:249–260.

———. 1966b. Food-handling and grooming behaviors in the plains pocket gopher. J. Mamm., 47:132–133.

———. 1967. Food habits of the northern pocket gopher on shortgrass prairie. Amer. Midl. Nat., 77:176–189.

———. 1969. Reproduction and population densities in a montane small mammal fauna, pp. 51–74. In J. K. Jones, Jr. (ed.), Contributions in mammalogy. Misc. Publ., Mus. Nat. Hist., Univ. Kansas, No. 51.

———. 1970a. Adaptations for flight in bats, pp. 127–143. In B. H. Slaughter and D. W. Walton (eds.), About bats. Southern Methodist Univ. Press, Dallas.

————. 1970b. The skeletal system. The muscular system. Flight patterns and aerodynamics, pp. 97–216. *In* W. A. Wimsatt (ed.), Biology of bats. Academic Press, New York.

————. 1974. Resource allocation in some sympatric, subalpine rodents. J. Mamm., 55:764–795.

————. 1976. Nocturnal behavior of the African false vampire bat *(Cardioderma cor)*. J. Mamm., 57:227–248.

————. 1977. Foraging behavior of the giant leaf-nosed bat. E. Afr. Wildl. J., 15. (In press.)

————, and G. C. Bateman. 1970. Functional morphology of the forelimbs of mormoopid bats. J. Mamm., 51:217–235.

————, and T. J. O'Shea. 1976. Roosting ecology of the pallid bat, *Antrozous pallidus*. J. Mamm., 57:19–42.

Vessey-FitzGerald, D. F. 1960. Grazing succession among East African game animals. J. Mamm., 41:161–172.

Villa-R., B. 1966. Los Murciélagos de Mexico. Inst. Biol., UNAM. 491 pp.

————, E. L. Cockrum. 1962. Migration in the guano bat, *Tadarida brasiliensis mexicana*. J. Mamm., 43:43–64.

Vogel, V. B., and geb. El-Kareh. 1969. Vergleichende Untersuchungen über den Wasserhaushalt von Fledermäusen *(Rhinopoma, Rhinolophus* und *Myotis)*. Z. Vergl. Physiol., 64:324–345.

Vogl, R. J. 1973. Ecology of the knobcone pine in the Santa Ana Mountains, California. Ecol. Monogr., 43:125–143.

Walker, E. P. 1968. Mammals of the world, 2nd ed. 2 vols. Johns Hopkins Press, Baltimore.

Wallace, A. R. 1876. The geographical distribution of animals. 2 vols. Harper, New York. 503 pp. and 553 pp. Reprinted by Hafner Publ. Company, New York.

Walther, F. 1958. Zum Kampf- und Paarungsverhalten einiger Antilopen. Z. Tierpsychol., 15:340–382.

————. 1965. Verhaltensstudien an der Grantgazell *(Gazella granti* Brooke, 1872) im Ngorongoro-Krater. Z. Tierpsychol., 22:167–208.

————. 1966. Zum Liegeverhalten des Weisschwanzgnus *(Connochaetes gnou* Zimmerman, 1780). Z. Säugetierk., 31:1–16.

Ward, A. L., and J. O. Keith. 1962. Feeding habits of pocket gophers on mountain grasslands. Ecology, 43:744–749.

Watkins, W. A., and W. E. Schevill. 1968. Underwater playback of their own sounds to *Leptonychotes* (Weddell seals). J. Mamm., 49:287–296.

Watson, A. 1958. The behavior, breeding and food-ecology of the snowy owl *Nyctea scandiaca*. Ibis, 99:419–462.

Watson, R. M. 1968. Report on aerial photographic studies of vegetation carried out in the Tsavo area of Kenya. 15 pp., typescript. (Cited by Laws, 1970.)

Weber, N. S., and J. S. Findley. 1970. Warm-season changes in fat content of *Eptesicus fuscus*. J. Mamm., 51:160–162.

Webster, D. B. 1961. The ear apparatus of the kangaroo rat, *Dipodomys*. Amer. J. Anat., 108:23–148.

————. 1962. A function of the enlarged middle-ear cavities of the kangaroo rat, *Dipodomys*. Physiol. Zool., 35:248–255.

————. 1963. A case of parallel evolution of the ear apparatus. Anat. Rec., 145:297.

————. 1966. Ear structure and function in modern mammals. Amer. Zool., 6:451–466.

————, R. F. Ackermann, and G. C. Longa. 1968. Central auditory system of the kangaroo rat, *Dipodomys merriami*. J. Comp. Neurol., 133(4):477–494.

————, and C. R. Stack. 1968. Comparative histochemical investigation of the organ of Corti in the kangaroo rat, gerbil and guinea pig. J. Morph., 129(4):413–434.

Webster, F. A., and O. G. Brazier, 1968. Experimental studies on echolocation mechanisms in bats. Aerospace Medical Research Laboratories, Wright-Patterson Air Force Base, Ohio. 156 pp.

————, and D. R. Griffin. 1962. The role of flight membranes in insect capture by bats. Anim. Behav., 10:332–340.

Wegener, A. 1915. Die Entstehung der Kontinente und Ozeane. Sammlung Vieweg. No. 23, 144 pp. Brunswick. (Translation, 1924. The origin of continents and oceans. Methuen, London. 212 pp.)

————. 1966. The origin of continents and oceans. (Translated from the fourth edition by John Birum.) Dover Publications, New York. 246 pp.

Wegge, P. 1975. Reproduction and early calf mortality in Norwegian red deer. J. Wildl. Manage., 39:92–100.

Weichert, C. K. 1965. Anatomy of the chordates. McGraw-Hill Book Company, New York, 758 pp.

Weil, W. L. P. 1968. Food habits of the western jumping mouse in north-central Colorado. M.S. Thesis, Colorado State Univ., Fort Collins. 23 pp.

Weir, B. J. 1974. The tuco-tuco and plains viscacha, pp. 113–127. *In* I. W. Rowlands and B. J. Weir (eds.), The biology of the histricomorph rodents. Academic Press, New York.

Wells, P. V., and C. D. Jorgensen. 1964. Pleistocene wood rat middens and climatic change in the Mojave Desert: a record of juniper woodlands. Science, 143:1171–1174.

Wetzel, R. M., R. E. Dubos, R. L. Martin, and P. Myers. 1975. *Catagonus*, an "extinct" peccary alive in Paraguay. Science, 189:379–381.

Wharton, C. H. 1950. Notes on the life history of the flying lemur, *Cynocephalus volans*. J. Mamm., 31:269–273.

Whitaker, J. O., Jr. 1963. Food, habitat and parasites of the woodland jumping mouse in central New York. J. Mamm., 44:316–321.

White, A. C. 1948. The call of the Bushveld. A. C. White P. & P. Company, Bloemfontein, S. Africa. 269 pp.

Whittaker, R. H. 1970. Communities and ecosystems. Macmillan Company, New York. 162 pp.

Whittow, G. C. 1974. Sun, sand, and sea lions. Nat. Hist., 83:56–63.

Wickler, W. Von, and D. Uhrig. 1969. Verhalten und okologische Nische der Gelbfügelfledermaus, *Lavia frons* (Geoffroy) (Chiroptera, Megadermatidae). Z. Tierpsychol., 26:726–736.

Williams, G. C. 1967. Natural selection, the costs of reproduction, and a refinement of Lack's principle. Amer. Nat., 100:687–690.

Williams, T. C., L. C. Ireland, and J. M. Williams. 1973. High altitude flights of the free-tailed bat, *Tadarida brasiliensis*, observed with radar. J. Mamm., 54:807–821.

Wilson, D. E., and J. S. Findley. 1970. Reproductive cycle of a Neotropical insectivorous bat, *Myotis nigricans*. Nature, 225:1155.

Wilson, E. O. 1975. Sociobiology: the new synthesis. Univ. Chicago press, Chicago. 697 pp.

Wilson, R. W. 1972. Evolution and extinction in Early Tertiary rodents. Proc. Int. Geol. Congr., 24(Sec. 7):217–224.

———. 1960 Early Miocene rodents and insectivores from northeastern Colorado. Univ. Kansas Paleont. Cont., Vertebrata, Art. 7, pp. 1–92.

Wimsatt, W. A. 1944. Further studies on the survival of spermatozoa in the female reproductive tract of the bat. Anat. Rec., 88:193–204.

———. 1945. Notes on breeding behavior, pregnancy, and parturition in some vespertilionid bats of eastern United States. J. Mamm., 26:23–33.

———. 1969a. Transient behavior, nocturnal activity patterns, and feeding efficiency of vampire bats *(Desmodus rotundus)* under natural conditions. J. Mamm., 50:223–244.

———. 1969b. In Society of Experimental Biology, Symposium No. 23. H. W. Woolhouse (ed.), Academic Press, New York.

———, (ed.). 1970. Biology of bats. Academic Press, New York, 406 pp.

———, and B. Villa-R. 1970. Locomotor adaptations in the disc-winged bat. Amer. J. Anat., 129:89–120.

Winge, H. 1941. The interrelationships of the mammalian genera. Vol. 1. C. A. Reitzels Forlag, Copenhagen. 418 pp. (Danish translated by E. Deichmann and G. M. Allen.)

Wirtz, W. O., II. 1968. Reproduction, growth and development, and juvenile mortality in the Hawaiian monk seal. J. Mamm., 49:229–238.

———. 1971. Personal communication.

Wislocki, G. B. 1942. Studies on the growth of deer antlers. I. On the structure and histogenesis of the antlers of the Virginia deer *(Odocoileus virginianus borealis)*. Amer. J. Anat., 71:371–415.

———. 1943. Studies on growth of deer antlers, pp. 629–653. In Essays in Biology. Univ. California Press, Berkeley.

———, J. C. Aub, and C. M. Waldo. 1947. The effects of gonadectomy and the administration of testosterone proprionate on the growth of antlers in male and female deer. Endocrin., 40:202–224.

Wood, A. E. 1937. Parallel radiation among the geomyoid rodents. J. Mamm., 18:171–176.

———. 1955. A revised classification of the rodents. J. Mamm., 36:165–187.

———. 1959. Are there rodent suborders? Syst. Zool., 7:169–173.

———. 1965. Grades and clades among rodents. Evolution, 19:115–130.

———. 1974. The evolution of the Old World and New World hystricomorphs, pp. 21–60. In I. W. Rowlands and B. Weir (eds.), The biology of hystricomorph rodents. Symp. Zool. Soc. London. Academic Press, New York.

———. 1975. The problem of the hystricognathous rodents. Univ. Mich. Papers Paleontol., 12:75–80.

Wood, F. G., Jr. 1959. Underwater sound production and concurrent behavior of captive porpoises, *Tursiops truncatus* and *Stenella plagiodon*. Bull. Mar. Sci. Gulf Caribbean, 3:120–133.

Woodard, T. N., R. J. Gutierrez, and W. H. Rutherford. 1974. Bighorn lamb production, survival, and mortality in south-central Colorado. J. Wildl. Manage., 38:771–774.

Woodburne, M. E., and R. H. Tedford. 1975. The first Tertiary monotreme from Australia. Amer. Mus. Novitates, No. 2588:1–11.

Wunder, B. A. 1970. Temperature regulation and the effects of water restriction on Merriam's chipmunk. Comp. Biochem. Physiol., 33:385–403.

Wynn, R. M. 1971. Immunological implications of comparative placental ultrastructure, pp. 495–514. In R. J. Blandau (ed.), The biology of the blastocyst. Univ. Chicago Press, Chicago.

Wynne-Edwards, V. C. 1959. The control of population density through social behavior: a hypothesis. Ibis, 101:436–441.

———. 1960. The overfishing principle applied to natural populations and their food resources: and a theory of natural conservation. Proc. Int. Orn. Congr., 12:790–794.

———. 1962. Animal dispersion in relation to social behavior. Hafner Publ. Company, New York. 653 pp.

Yanagisawa, K., G. Sata, M. Nomoto, Y. Katsuki, E. Ibezono, A. D. Grinnell, and T. H. Bullock, 1966. Fed. Proc., Physiol., p. 1539 (abstr.).

Yoakum, J. 1958. Seasonal food habits of the Oregon pronghorn antelope (*Antilocapra americana oregona* Bailey). Inter. Antelope Conf. Trans., 9:47–59.

Young, J. Z. 1957. The life of mammals. Clarendon Press, Oxford. 820 pp.

Young, S. P., and H. H. T. Jackson. 1951. The clever coyote. Wildl. Manage. Inst., Washington, D.C. 411 pp.

Zimmerman, E. G. 1965. A comparison of habitat and food of two species of *Microtus*. J. Mamm., 46:605–612.

Zippelius, H., and W. M. Schleidt. 1956. Ultraschalllaute bei jungen Mausen. Naturwisse., 21:1–2.

INDEX

Page numbers in *italics* indicate illustrations. The symbol *t* indicates a table.